Third
Edition

THE BASICS OF OCCUPATIONAL SAFETY

David L. Goetsch

*Vice President Emeritus and Professor,
Northwest Florida State College*

 Pearson

330 Hudson Street, NY, NY 10013

Vice President, Portfolio Management: Andrew Gilfillan

Editorial Assistant: Lara Dimmick

Senior Vice President, Marketing: David Gesell

Marketing Coordinator: Elizabeth MacKenzie-Lamb

Director, Digital Studio and Content Production: Brian Hyland

Managing Producer: Jennifer Sargunar

Content Producer (Team Lead): Faraz Sharique Ali

Manager, Rights Management: Johanna Burke

Operations Specialist: Deidra Smith

Cover Design: Cenveo Publisher Services

Cover Photo: Lucidio Studio, Inc./ Moment/ Getty Images

Full-Service Project Management: Integra Software Services Pvt. Ltd.

Composition: Integra Software Services Pvt. Ltd.

Printer/Binder: LSC Communications, Inc.

Cover Printer: Phoenix Color/Hagerstown

Text Font: Times Ten LT Std

Library of Congress Cataloging-in-Publication Data

Names: Goetsch, David L., author.
Title: The basics of occupational safety / David L. Goetsch, vice president emeritus and professor, Northwest Florida State College.
Description: Third edition. | Boston : Pearson, [2019] | Includes index.
Identifiers: LCCN 2017018773 | ISBN 9780134678719 | ISBN 0134678710
Subjects: LCSH: Industrial safety. | Industrial hygiene.
Classification: LCC T55 .G5848 2019 | DDC 363.11—dc23
   LC record available at https://lccn.loc.gov/   2017018773

8 2019

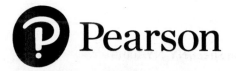

ISBN-10:   0-13-467871-0
ISBN-13: 978-0-13-467871-9

Preface

BACKGROUND

The field of occupational safety and health has undergone significant change over the past three decades. There are many reasons for this. Some of the more prominent reasons include technological changes that have introduced new hazards in the workplace; proliferation of health and safety legislation and corresponding regulations; increased pressure from regulatory agencies; realization by executives that workers in a safe and healthy workplace are typically more productive; healthcare and workers' compensation cost increases; increased pressure from environmental groups and the public; a growing interest in ethics and corporate social responsibility; professionalization of health and safety occupations; increased pressure from labor organizations and employees in general; rapidly mounting costs associated with product safety and other types of litigation; and increasing incidents of workplace violence.

All of these factors, when combined, have made the job of the modern safety and health professional more challenging and more important than it has ever been. These factors have also created a need for an up-to-date book on the basics of workplace safety and health that contains the latest information needed by people who will have responsible positions in today's fast-paced, competitive workplace.

WHY IS THIS BOOK WRITTEN AND FOR WHOM?

This book is written to fulfill the need for an up-to-date, practical teaching resource that focuses on the basic safety-related needs of people in the workplace. It is intended for use in universities, colleges, community colleges, technical schools, and corporate training settings that offer programs, courses, workshops, and seminars in occupational safety and health. Educators and students in such disciplines as safety engineering, engineering, industrial technology, manufacturing technology, industrial engineering, safety engineering, engineering technology, occupational safety, management, and supervision will find this book both valuable and easy to use. The direct, straightforward presentation of material focuses on making the theories and principles of occupational safety and health practical and useful in a real-world setting. Up-to-date research has been integrated throughout in a down-to-earth manner.

ORGANIZATION OF THE BOOK

The text contains 20 chapters, each focusing on a major area of concern in workplace safety and health. The chapters are presented in an order that is compatible with the typical organization of a college-level safety and health course. A standard chapter format is used throughout the book. There is a list of learning objectives at the beginning of each chapter. All chapters include review questions, key terms and concepts, and endnotes. These materials are provided to encourage review, stimulate additional thought, and provide opportunities for applying what has been learned.

INSTRUCTOR RESOURCES

To access supplementary materials online, instructors need to request an instructor access code. Go to **www.pearsonhighered.com/irc** to register for an instructor access code. Within 48 hours of registering, you will receive a confirming e-mail including an instructor access code. Once you have received your code, locate your text in the online catalog and click on the Instructor Resources button on the left side of the catalog product page. Select a supplement and a login page will appear. Once you have logged in, you can access instructor material for all Pearson textbooks. If you have any difficulties accessing the site or downloading a supplement, please contact Customer Service at **http://support.pearson.com/getsupport**.

HOW THIS BOOK DIFFERS FROM OTHERS

This book is written because, in the age of global competition, safety and health in the workplace have changed drastically. Many issues, concerns, and factors relating specifically to modern workplace environments have been given more attention, greater depth of coverage, and more illumination here than in other textbooks. Some of the areas receiving more attention and specific occupational examples include the following:

- The Occupational Safety and Health Act (OSH Act) and Occupational Safety and Health Administration (OSHA)
- Standards and codes
- Laws and liability
- Stress-related problems
- Fire hazards and life safety
- The evolving roles of health and safety professionals

- Health and safety training
- Human factors in safety
- Bloodborne pathogens in the workplace
- Ergonomics and safety
- Workers' compensation
- Repetitive strain injuries (RSIs)

NEW TO THIS EDITION

The third edition of *The Basics of Occupational Safety* is a major revision encompassing new regulations, revised regulations, and other new and updated material of importance for the students of occupational safety and health. Specifically, the following revisions are made in the third edition:

CHAPTER 1:

1. Added a section about safety applying to all work sectors: manufacturing, retail, hospitality, healthcare, etc.
2. Added information on the Chemical Safety Board.
3. Added information on ergonomics as a trend in safety in the 1990s.
4. Added information on the West Fertilizer Company tragedy.
5. Added information on whether the accident rate has decreased because America has lost so many manufacturing jobs.

CHAPTER 2:

1. Moved "Heat Burns and Chemical Burns" sections to Chapter 15.
2. Moved "Repetitive Strain/Soft Tissue Injuries" section to Chapter 8.
3. Added a reference year to Figures 2–1, 2–2, and 2–3 so students know when the data was compiled.
4. Added a brief section on OSHA reports and logs here (and reference where they appear in detail in Chapter 5).
5. Converted Figure 2–4 into a chart.
6. Added a link to Bureau of Labor Statistics for more detailed information.

CHAPTER 3:

1. Bolded the sentence about Heinrich's Theory being discounted by contemporary research for emphasis.
2. Added information on James Reason's Swiss Cheese Model of accident causation.
3. Changed the section on "Drugs and Accident Causation" to "Individual Factors and Accident Causation" (includes drugs, depression, obesity, fatigue, personality, etc.).

CHAPTER 4:

1. Added information on the importance of the employee on safety teams and the employee's role in safety.
2. Replaced Figures 4–9 and 4–10 with photographs.

CHAPTER 5:

1. Added information on indirect costs of OSHA fines (bad PR, loss of goodwill, corporate image, legal fees, paperwork, etc.).
2. Replaced Figure 5–4 with instructions for finding consultation services in your state.
3. Revised Figure 5–5 to include website addresses for each agency instead of street addresses.
4. Updated the OSHA standard subparts listed to ensure they are up to date.
5. Added information on where the fines go when OSHA collects them and discussion plus examples about the size of fines.

CHAPTER 6:

1. Replaced Figure 6–1 with a figure showing how students can access workers' compensation information for their states.
2. Added information on workers' compensation abuse (employees trying to take advantage of the system).
3. Replaced Figure 6–2 with a photograph.

CHAPTER 7:

1. Added information about why safety professionals who do accident reports need to be familiar with common causes of accidents (added to the "Common Causes of Accidents" section).
2. Added a section on "Who is responsible for reporting."
3. Added an accident investigation exercise to the end of the chapter.

CHAPTER 8:

1. Added information on ergonomic assessment tools such as NIOSH guidelines, RULA, and REBA.
2. Moved section on "lifting" from Chapter 11 to this chapter and expanded the content of the section.
3. Expanded the section on "Human Factors and Safety."
4. Added information on choosing which workstations/operations to evaluate for ergonomics to the section on "Worksite Analysis Program."
5. In the section titled "Training and Education," referred students to using recommended training materials from OSHA.
6. In the section on "Identifying Specific Ergonomic Problems" added material on "anthropometry" (body size).
7. In the section on "Helpful Assessment Tools: NIOSH, RULA, REBA, and HAL" (added information on analysis tools including RULA, REBA, HAL, and Strain Index).

CHAPTER 9:

1. Explained how the list of "common causes" can be used and how it relates to the rest of the chapter.
2. Added explanations for the strategies for dealing with stress.
3. Added physiological measures of stress (heart rate, pupil dilation, perspiration, etc.).
4. Added information about 24/7 use of technology and multitasking as causes of stress.
5. Added information about Employee Assistance Programs (EAPs) and company wellness programs for dealing with stress.

CHAPTER 10:

1. Added information about how a given type of machine guard is chosen.
2. Added information on advanced sensors and Bluetooth technology.

CHAPTER 11:

1. Dropped "Lifting" from the title to this chapter, and moved the section on lifting to Chapter 8.
2. Changed title to "Falling, Impact, Acceleration, and Vision Hazards with Appropriate PPE."

CHAPTER 12:

1. Added information on Clo as a unit for measure for PPE temperature protection.
2. Added a section on OSHA recommendations and guidelines for temperature hazards.
3. Added a note to "Chemical Burns" explaining why it is in this chapter and not in the chapter on fire safety.

CHAPTER 13:

1. Added a note that refers students to Chapter 16 for coverage of "Confined Spaces."
2. Added information on 29 CFR 1910 Subpart H (OSHA's standards on pressure hazards).

CHAPTER 14:

1. Added information about power strips and daisy chains.
2. Made minor updates corrections to the text.

CHAPTER 15:

1. Added information about the number of fire extinguishers needed, how to choose the type, and where they should be located in a facility.
2. Moved sections on "Chemical Burns" from Chapter 2 to this chapter.
3. Added information to the "egress" section about lighting and signage.
4. Strengthened the material on "egress."

CHAPTER 16:

1. Added information to describe TWA in more detail.
2. Removed the underline from the TWA side of the equation.

CHAPTER 17:

1. Expanded the explanations of risk reduction strategies.

CHAPTER 18:

1. Added a section about "earbuds" from handheld devices and potential hazards.
2. In the section on "Hazard Levels and Risks" added information from Chapter 16 on calculating TWA.
3. In the section on "Vibration Hazards" added information about tools insulation, tool mounting, and job rotation.

4. In the section on "Noise Control Strategies" added information on specific engineering controls (e.g., mufflers, insulation, wall panels, and sound absorption).
5. Added information about calculating Noise Reduction Rating (NRR) and how to evaluate PPE based on the NRR.
6. Moved the section on "Fit Testing" to earlier in the chapter.

CHAPTER 19:

1. Added a section on "active shooter" response.
2. Added a section on "reporting suspicious activities."

ABOUT THE AUTHOR

David L. Goetsch is vice president Emeritus of Northwest Florida State College and professor of safety, quality, and management. In addition, Dr. Goetsch is president and CEO of the Institute for Organizational Excellence (IOE), a private consulting firm dedicated to the continual improvement of organizational competitiveness, safety, and quality. Dr. Goetsch is cofounder of The Quality Institute, a partnership of the University of West Florida, Northwest Florida State College, and the Okaloosa Economic Development Council and founder of the Leadership Institute of Northwest Florida State College and founder of the Leadership Institute of Northwest Florida State College.

ACKNOWLEDGMENTS

The author acknowledges the invaluable assistance of the following people in developing this book: Dr. Lissa Galbraith, Florida A&M/Florida State University, for the material she contributed on electrical and fire hazards in the first edition.

Introduction

SAFETY VERSUS HEALTH

Throughout the text, the titles "safety and health professional" and "safety and health manager" are used. This, too, is done by design. This approach underscores the point that the field of occupational safety has been broadened to encompass both safety and health. Consequently, managers, technical personnel, and engineers in this field must be knowledgeable about safety and health and be prepared to oversee a program that encompasses both areas of responsibility.

Safety and health, although closely related, are not the same. One view is that safety is concerned with injury-causing situations, whereas health is concerned with disease-causing conditions. Another view is that safety is concerned with hazards to humans that result from sudden severe conditions; health deals with adverse reactions to prolonged exposure to dangerous, but less intense, hazards. Both of these views are generally accurate in portraying the difference between safety and health. However, the line between these two concepts is not always clearly marked.

For example, stress is a hazard that can cause both psychological and physiological problems over a prolonged period. In this case, it is a health concern. On the other hand, an overly stressed worker may be more prone to unintentionally overlook safety precautions and thus may cause an accident. In this case, stress is a safety concern.

Because personnel in this evolving profession are likely to be responsible for safety *and* health, it is important that they have a broad academic background covering both. This book attempts to provide that background.

Brief Contents

Contents

CHAPTER

1

SAFETY AND HEALTH MOVEMENT, THEN AND NOW

Learning Objectives

■ Summarize key developments relating to workplace safety and health prior to the Industrial Revolution

■ List the most important milestones in the safety movement

■ Explain how workplace tragedies have changed the safety movement

■ Describe the role of organized labor in the safety movement

■ Describe the roles specific health problems have played in the safety movement

■ Explain how safety and health standards apply to more than just manufacturing

■ Describe the development of accident prevention programs as part of the safety movement

■ Describe the safety and health movement today

■ Explain the integrated approach to safety and health

■ Describe how new materials, new processes, and new problems are affecting the safety and health movement today

■ Summarize the rapid growth that has occurred in the safety and health profession

■ Explain how the movement of manufacturing jobs overseas has affected the accident rate in the United States

The **safety movement** in the United States has developed steadily since the early 1900s. In that time period, industrial accidents were commonplace in this country; for example, in 1907, more than 3,200 people were killed in mining accidents. Legislation, precedent, and public opinion all favored management. There were few protections for workers' safety.

Working conditions for industrial employees today have improved significantly. The chance of a worker being killed in an industrial accident is less than half of what it was 60 years ago.[1] According to the National Safety Council (NSC), the current death rate from work-related injuries is approximately 4 per 100,000, or less than a third of the rate 50 years ago.[2]

Improvements in safety until now have been the result of pressure for legislation to promote safety and health, the steadily increasing costs associated with accidents and injuries, and the professionalization of safety as an occupation. Improvements in the future are likely to come as a result of greater awareness of the cost-effectiveness and resultant competitiveness gained from a safe and healthy workforce.

This chapter examines the history of the safety movement in the United States and how it has developed over the years. Such a perspective will help practicing and prospective safety professionals form a better understanding of both their roots and their future.

DEVELOPMENTS BEFORE THE INDUSTRIAL REVOLUTION

It is important for students of occupational health and safety to first study the past. Understanding the past can help safety and health professionals examine the present and future with a sense of perspective and continuity. Modern developments in health and safety are neither isolated nor independent. Rather, they are part of the long continuum of developments in the safety and health movement.

The continuum begins with the days of the ancient Babylonians. During that time, circa 2000 BC, their ruler, Hammurabi, developed his **Code of Hammurabi**. The code encompassed all the laws of the land at that time, showed Hammurabi to be a just ruler, and set a precedent followed by other Mesopotamian kings. The significance of the code from the perspective of safety and health is that it contained clauses dealing with injuries, allowable fees for physicians, and monetary damages assessed against those who injured others.[3] This clause from the code illustrates Hammurabi's concern for the proper handling of injuries: "If a man has caused the loss of a gentleman's eye, his own eye shall be caused to be lost."[4]

This movement continued and emerged in later Egyptian civilization. As evidenced from the temples and pyramids that still remain, the Egyptians were an industrious people. Much of the labor was provided by slaves, and there is ample evidence that slaves were not treated well—that is, unless it suited the needs of the Egyptian taskmasters.

One such case occurred during the reign of Rameses II (circa 1500 BC), who undertook a major construction project, the Ramesseum. To ensure the maintenance of a workforce sufficient to build this huge temple bearing his name, Rameses created an industrial medical service to care for the workers. They were required to bathe daily in the Nile and were given regular medical examinations. Sick workers were isolated.[5]

The Romans were vitally concerned with safety and health, as can be seen from the remains of their construction projects. The Romans built aqueducts, sewerage systems, public baths, latrines, and well-ventilated houses.[6]

As civilization progressed, so did safety and health developments. In 1567, Philippus Aurcolus produced a treatise on the pulmonary diseases of miners. Titled

On the Miners' Sickness and Other Miners' Diseases, the treatise covered diseases of smelter workers and metallurgists and diseases associated with the handling of and exposure to mercury. Around the same time, Georgius Agricola published his treatise *De Re Metallica*, emphasizing the need for ventilation in mines and illustrating various devices that could be used to introduce fresh air into mines.[7]

The eighteenth century saw the contributions of Bernardino Ramazzini, who wrote *Discourse on the Diseases of Workers*. Ramazzini drew conclusive parallels between diseases suffered by workers and their occupations. He related occupational diseases to the handling of harmful materials and to irregular or unnatural movements of the body. Much of what Ramazzini wrote is still relevant today.[8]

The Industrial Revolution changed forever the methods of producing goods. According to J. LaDou, the changes in production brought about by the Industrial Revolution can be summarized as follows:

- Introduction of **inanimate power** (i.e., steam power) to replace people and animal power
- Substitution of machines for people
- Introduction of new methods for converting raw materials
- Organization and specialization of work, resulting in a division of labor[9]

These changes necessitated a greater focusing of attention on the safety and health of workers. Steam power increased markedly the potential for life-threatening injuries, as did machines. The new methods used for converting raw materials also introduced new risks of injuries and diseases. Specialization, by increasing the likelihood of boredom and inattentiveness, also made the workplace a more dangerous environment.

MILESTONES IN THE SAFETY MOVEMENT

Just as the United States traces its roots to Great Britain, the safety movement in this country traces its roots to England. During the Industrial Revolution, child labor in factories was common. The hours were long, the work hard, and the conditions often unhealthy and unsafe. Following an outbreak of fever among the children working in their cotton mills, the people of Manchester, England, began demanding better working conditions in the factories. Public pressure eventually forced a government response, and in 1802 the Health and Morals of Apprentices Act was passed. This was a milestone piece of legislation: It marked the beginning of governmental involvement in workplace safety.

When the industrial sector began to grow in the United States, hazardous working conditions were commonplace. Following the Civil War, the seeds of the safety movement were sown in this country. Factory inspection was introduced in Massachusetts in 1867. In 1868, the first barrier safeguard was patented. In 1869, the Pennsylvania legislature passed a mine safety law requiring two exits from all mines. The Bureau of Labor Statistics (BLS) was established in 1869 to study industrial accidents and report pertinent information about those accidents.

The following decade saw little new progress in the safety movement until 1877, when the Massachusetts legislature passed a law requiring safeguards for hazardous machinery. This year also saw passage of the Employer's Liability Law, establishing the potential for **employer liability** in workplace accidents. In 1892, the first recorded safety program was established in a Joliet, Illinois, steel plant in response to a scare caused when a flywheel exploded. Following the explosion, a committee of managers was formed to investigate and make recommendations. The committee's recommendations were used as the basis for the development of a safety program that is considered to be the first safety program in American industry.

Around 1900, Frederick Taylor began studying efficiency in manufacturing. His purpose was to identify the impact of various factors on efficiency, productivity, and profitability. Although safety was not a major focus of his work, Taylor did draw a connection between lost personnel time and management policies and procedures. This connection between safety and management represented a major step toward broad-based safety consciousness.

In 1907, the U.S. Department of the Interior created the Bureau of Mines to investigate accidents, examine health hazards, and make recommendations for improvements. Mining workers definitely welcomed this development, since more than 3,200 of their fellow workers were killed in mining accidents in 1907 alone.[10]

One of the most important developments in the history of the safety movement occurred in 1908 when an early form of **workers' compensation** was introduced in the United States. Workers' compensation actually had its beginnings in Germany. The practice soon spread throughout the rest of Europe. Workers' compensation as a concept made great strides in the United States when Wisconsin passed the first effective workers' compensation law in 1911. In the same year, New Jersey passed a workers' compensation law that withstood a court challenge.

The common thread among the various early approaches to workers' compensation was that they all provided some amount of compensation for on-the-job injuries regardless of who was at fault. When the workers' compensation concept was first introduced in the United States, it covered a very limited portion of the workforce and provided only minimal benefits. Today, all 50 states have some form of workers' compensation that requires the payment of a wide range of benefits to a broad base of workers. Workers' compensation is examined in more depth in Chapter 6.

The Association of Iron and Steel Electrical Engineers (AISEE), formed in the early 1900s, pressed for a national conference on safety. As a result of the AISEE's efforts, the first meeting of the **Cooperative Safety Congress (CSC)** took place in Milwaukee in 1912. What is particularly significant about this meeting is that it planted the seeds for the eventual establishment of the NSC. A year after the initial meeting of the CSC, the **National Council of Industrial Safety (NCIS)** was established in Chicago. In 1915, this organization changed its name to the National Safety Council. It is now the premier safety organization in the United States.

From the end of World War I (1918) through the 1950s, safety awareness grew steadily. During this period, the federal government encouraged contractors to implement and maintain a safe work environment. Also during this period, industry in the United States arrived at two critical conclusions: (1) there is a definite connection between quality and safety, and (2) off-the-job accidents have a negative impact on productivity. The second conclusion became painfully clear to manufacturers during World War II when the call-up and deployment of troops had employers struggling to meet their labor needs. For these employers, the loss of a skilled worker due to an injury or for any other reason created an excessive hardship.[11]

The 1960s saw the passage of a flurry of legislation promoting workplace safety. The Service Contract Act of 1965, the Federal Metal and Nonmetallic Mine Safety Act, the Federal Coal Mine and Safety Act, and the Contract Workers and Safety Standards Act all were passed during the 1960s. As their names indicate, these laws applied to a limited audience of workers.

These were the primary reasons behind passage of the **Occupational Safety and Health Act (OSH Act)** of 1970 and the Federal Mine Safety Act of 1977. These federal laws, particularly the OSH Act, represent the most significant legislation to date in the history of the safety movement. During the 1990s, the concept of Total Safety Management (TSM) was introduced and adopted by firms that were already practicing Total Quality Management (TQM). TSM encourages organizations to take a holistic approach to safety management in which the safety of employees, processes, and products is considered when establishing safe and healthy work practices.

At the turn of the century, workplace violence, including terrorism, began to concern safety and health professionals. In addition, the twenty-first century saw a trend in which older people were returning to work to supplement their retirement income. This trend led to a special emphasis on the safety and health of older workers. A more recent trend is greater concern of U.S. companies for the safety and health of employees in foreign countries that manufacture goods that are sold in the United States.

The Superfund Amendments and Reauthorization Act was passed by Congress in 1986, followed by the Amended Clean Air Act in 1990; both were major pieces of environmental legislation. Another milestone that occurred in the decade of the 1990s was the trend toward safety professionals making ergonomics part of their overall approach for preventing accidents and injuries. Ergonomics involves fitting the work to the worker rather than the worker to the work. It is concerned with, among other things, the prevention of musculoskeletal disorders (MSDs) and injuries.

Figure 1–1 summarizes some significant milestones in the development of the safety movement in the United States.

TRAGEDIES THAT HAVE CHANGED THE SAFETY MOVEMENT

The pace of the safety and health movement in the United States has been accelerated by the occurrence of workplace tragedies. These tragedies could have been prevented had appropriate safety and health measures been followed. Unfortunately, they were not. This section summarizes some of the more significant of these workplace tragedies; tragedies that have had a lasting effect on the safety and health movement.

Hawk's Nest Tragedy

In the 1930s, the public began to take notice of the health problems suffered by employees who worked in dusty environments. The Great Depression was indirectly responsible for the attention given to an occupational disease that came to be known as *silicosis*. As the economic crash spread, business after business shut down and laid off its workers. Unemployed miners and foundry workers began to experience problems finding new jobs when physical examinations revealed that they had lung damage from breathing silica. Cautious insurance companies recommended preemployment physicals as a way

FIGURE 1–1   Milestones in the safety movement.

1867	Massachusetts introduces factory inspection.
1868	Patent is awarded for first barrier safeguard.
1869	Pennsylvania passes law requiring two exits from all mines, and the Bureau of Labor Statistics is formed.
1877	Massachusetts passes law requiring safeguards on hazardous machines, and the Employer's Liability Law is passed.
1892	First recorded safety program is established.
1900	Frederick Taylor conducts first systematic studies of efficiency in manufacturing. Followed by the motion studies of Frank Gilbreth.
1907	Bureau of Mines is created by U.S. Department of the Interior.
1908	Concept of workers' compensation is introduced in the United States.
1911	Wisconsin passes the first effective workers' compensation law in the United States, and New Jersey becomes the first state to uphold a workers' compensation law.
1912	First Cooperative Safety Congress meets in Milwaukee.
1913	National Council of Industrial Safety is formed.
1915	National Council of Industrial Safety changes its name to National Safety Council.
1916	Concept of negligent manufacture is established (product liability).
1924	Hawthorne Light Experiments.
1936	National Silicosis Conference convened by the U.S. Secretary of Labor.
1970	Occupational Safety and Health Act passes.
1977	Federal Mine Safety Act passes.
1978	OSHA offers education and training grants.
1980	OSHA coverage extended to federal employees.
1986	Superfund Amendments and Reauthorization Act pass.
1990	Amended Clean Air Act of 1970 passes. Also, safety professionals begin to apply the principles of ergonomics.
1996	Total safety management (TSM) concept is introduced.
2000	U.S. firms begin to pursue ISO 14000 registration for environmental safety management.
2003	Workplace violence and terrorism are an ongoing concern of safety and health professionals.
2007	Safety of older people reentering the workplace becomes an issue.
2009	Global Harmonization System for chemicals established.
2010	Off-the-job safety becomes an issue.
2017	Pressure on foreign companies that produce goods sold in the United States to improve their safety standards.

to prevent future claims based on preexisting conditions. Applicants with silica-damaged lungs were refused employment. Many of them sued. This marked the beginning of industry-wide interest in what would eventually be called the "king" of occupational diseases.

Lawsuits and insurance claims generated public interest in silicosis, but it was the Hawk's Nest tragedy that solidified public opinion in favor of protecting workers from this debilitating disease.[12] A company was given a contract to drill a passageway through a mountain located in the Hawk's Nest region of West Virginia (near the city of Gauley Bridge). Workers spent as many as 10 hours per day breathing the dust created by drilling and blasting. It turned out that this particular mountain had an unusually high silica content. Silicosis is a disease that normally takes 10–30 years to show up in exposed workers. At Hawk's Nest, workers began dying in as little time as a year. By the time the project was completed, hundreds had died. To make matters even worse, the company often buried an employee who died from exposure to silica in a nearby field without notifying the family. Those who inquired were told that their loved one left without saying where he was going.

A fictitious account of the Gauley Bridge disaster titled *Hawk's Nest*, by Hubert Skidmore, whipped the public outcry into a frenzy, forcing Congress to respond.

This tragedy and the public outcry that resulted from it led a group of companies to form the Air Hygiene Foundation to conduct research and develop standards for working in dusty environments. Soon thereafter, the U.S. Department of Labor provided the leadership necessary to make silicosis a compensable disease under workers' compensation in most states. Today, dust-producing industries use a wide variety of administrative controls, engineering controls, and personal protective equipment to protect workers in dusty environments. However, silicosis is still a problem. Approximately 1 million workers in the United States are still exposed to silica every year, and 250 people die annually from silicosis.

Asbestos Menace

Asbestos was once considered a "miracle" fiber, but in 1964, Dr. Irving J. Selikoff told 400 scientists at a conference on the biological effects of asbestos that this widely used material was killing workers. This conference changed how Americans viewed not just asbestos but also workplace hazards in general. Selikoff was the first to link asbestos to lung cancer and respiratory diseases.[13]

At the time of Selikoff's findings, asbestos was one of the most widely used materials in the United States. It was found in homes, schools, offices, factories, ships, and even in the filters of cigarettes. Selikoff continued to study the effects of asbestos exposure from 1967 to 1986. During this time, he studied the mortality rate of 17,800 workers who had been exposed to asbestos. He found asbestos-related cancer in the lungs, gastrointestinal tract, larynx, pharynx, kidneys, pancreas, gall bladder, and bile ducts of workers.

Finally, in the 1970s and 1980s, asbestos became a controlled material. Regulations governing the use of asbestos were developed, and standards for exposure were established. Asbestos-related lawsuits eventually changed how industry dealt with this tragic material. In the 1960s, industry covered up or denied the truth about asbestos. Now, there is an industry-wide effort to protect workers who must remove asbestos from old buildings and ships during remodeling, renovation, or demolition projects.

Bhopal Tragedy

On the morning of December 3, 1984, over 40 tons of methyl isocyanate (MIC) and other lethal gases, including hydrogen cyanide, leaked into the northern end of Bhopal, killing more than 3,000 people in its aftermath.[14] After the accident, it was discovered that the protective equipment that could have halted the impending disaster was not in full working order. The refrigeration system that should have cooled the storage tank was shut down, the scrubbing system that should have absorbed the vapor was not immediately available, and the flare system that would have burned any vapor that got past the scrubbing system was out of order.[15]

The International Medical Commission visited Bhopal to assess the situation and found that as many as 50,000 other people had been exposed to the poisonous gas and may still suffer disability as a result. This disaster shocked the world. Union Carbide Corporation, the owner of the chemical plant in Bhopal, India, where the incident occurred, was accused of many things, including the following:

- **Criminal negligence.**
- **Corporate prejudice.** Choosing poverty-stricken Bhopal, India, as the location for a hazardous chemical plant on the assumption that few would care if something went wrong.
- **Avoidance.** Putting its chemical plant in Bhopal, India, to avoid the stricter safety and health standards of the United States and the Occupational Safety and Health Administration (OSHA) in particular.

In February 1989, India's Supreme Court ordered Union Carbide India Ltd. to pay $470 million in compensatory damages. The funds were paid to the Indian government to be used to compensate the victims. This disaster provided the impetus for the passage of stricter safety legislation worldwide. In the United States, it led to the passage of the Emergency Planning and Community Right-to-Know Act (EPCRA) of 1986.

Factory Fire in Bangladesh

In November 2012, a garment-factory fire in Bangladesh killed 112 employees. The magnitude of the tragedy was enhanced when it was discovered that the factory produced garments to sell in several major retail outlets in the United States. Fire inspectors suspect that an electrical short circuit caused the blaze, which spread quickly because of the flammable nature of material used to produce T-shirts in the factory. There were complaints that well-known retailers in the United States, and elsewhere in the Western world, were partially culpable in the tragedy because there was evidence that they knew of the unsafe conditions beforehand.

The garment factory in question had a functioning fire alarm and the alarm did go off properly. Unfortunately, supervisors demanded that workers go back to their sewing machines and even blocked an exit door workers could have used to escape the conflagration. It was learned in the subsequent investigation that the factory's fire extinguishers did not work and were displayed only to fool inspectors. A follow-up investigation revealed that 100 workers had been burned to death inside the factory while another 12 jumped to their deaths to escape the flames. This tragedy added to the mounting pressure for U.S. companies that contract with offshore manufacturers to pressure those manufacturers to implement safe and healthy work practices.

ROLE OF ORGANIZED LABOR

Organized labor has played a crucial role in the development of the safety movement in the United States. From the outset of the Industrial Revolution in this country, organized labor has fought for safer working conditions and appropriate compensation for workers injured on the job. Many of the earliest developments in the safety movement were the result of long and hard-fought battles by organized labor.

Although the role of unions in promoting safety is generally acknowledged, one school of thought takes the opposite view. Proponents of this dissenting view hold that union involvement actually slowed the development of the safety movement. Their theory is that unions allowed their demands for safer working conditions to become entangled with their demands for better wages; as a result, they met with resistance from management. Regardless of the point of view, there is no question that working conditions in the earliest years of the safety movement were often reflective of an insensitivity to safety concerns on the part of management.

Among the most important contributions of organized labor to the safety movement was their work to overturn antilabor laws relating to safety in the workplace. These laws were the fellow servant rule, the statutes defining contributory negligence, and the concept of assumption of risk.[16] The **fellow servant rule** held that employers were not liable for workplace injuries that resulted from the negligence of other employees. For example, if Worker X slipped and fell, breaking his back in the process, because Worker Y spilled oil on the floor and left it there, the employer's liability was removed. In addition, if the actions of employees contributed to their own injuries, the employer was absolved of any liability. This was the doctrine of **contributory negligence**. The concept of **assumption of risk** was based on the theory that people who accept a job assume the risks that go with it. It says employees who work voluntarily should accept the consequences of their actions on the job rather than blame the employer.

Because the overwhelming majority of industrial accidents involve negligence on the part of one or more workers, employers had little to worry about. Therefore, they had little incentive to promote a safe work environment. Organized labor played a crucial role in bringing deplorable working conditions to the attention of the general public. Public awareness and, in some cases, outrage eventually led to these **employer-biased laws** being overturned in all states except one. In New Hampshire, the fellow servant rule still applies.

West Fertilizer Company Explosion

On April 17, 2013, an explosion occurred at the storage and distribution center of the West Fertilizer Company in West, Texas. The explosion occurred when fire set off ammonium nitrate while emergency personnel were battling the blaze. The explosion killed 15 people and injured 160 others. More than 150 surrounding buildings were damaged or destroyed. In the ensuing investigation, it was determined that the fire had been deliberately set. Prior to the fire and explosion, OSHA had fined the company for improper storage of anhydrous ammonia and cited the company for violating its respiratory protection standards. The company had also been fined by the Environmental Protection Agency (EPA) in 2006 for failing to file a risk management program plan in a timely manner. Further, the company had been fined in 2012 by the U.S. Department of Transportation for violations pertaining to the improper storage of anhydrous ammonia.

ROLE OF SPECIFIC HEALTH PROBLEMS

Specific health problems that have been tied to workplace hazards have played significant roles in the development of the modern safety and health movement. These health problems contributed to public awareness of dangerous and unhealthy working conditions that, in turn, led to legislation, regulations, better work procedures, and better working conditions.

Lung disease in coal miners was a major problem in the 1800s, particularly in Great Britain, where much of the Western world's coal was mined at the time. Frequent contact with coal dust led to a widespread outbreak of anthracosis among Great Britain's coal miners. Also known as the *black spit*, this disease persisted from the early 1800s, when it was first identified, until around 1875, when it was finally eliminated by such safety and health measures as ventilation and decreased work hours.

In the 1930s, Great Britain saw a resurgence of lung problems among coal miners. By the early 1940s, British scientists were using the term *coal-miner's pneumoconiosis*, or CWP, to describe a disease from which many miners suffered. Great Britain designated CWP a separate and compensable disease in 1943. However, the United States did not immediately follow suit, even though numerous outbreaks of the disease had occurred among miners in this country.

The issue was debated in the United States until Congress finally passed the Coal Mine Health and Safety Act in 1969. The events that led up to the passage of this act were tragic. An explosion in a coal mine in West Virginia in 1968 killed 78 miners. This tragedy focused attention on mining health and safety, and Congress responded by passing the Coal Mine Health and Safety Act. The act was amended in 1977 and again in 1978 to broaden the scope of its coverage.

Over the years, the diseases suffered by miners were typically lung diseases caused by the inhalation of coal dust particulates. However, health problems were not limited to coal miners. Other types of miners developed a variety of diseases, the most common of which was silicosis. Once again, it took a tragic event—the Gauley Bridge disaster, discussed earlier—to focus attention on a serious workplace problem.

Congress held a series of hearings on the matter in 1936. That same year, representatives from business, industry, and government attended the National Silicosis Conference, convened by the U.S. secretary of labor. Among other outcomes of this conference was a finding that silica dust particulates did, in fact, cause silicosis.

Mercury poisoning is another health problem that has contributed to the evolution of the safety and health movement by focusing public attention on unsafe conditions in the workplace. The disease was first noticed among the citizens of a Japanese fishing village in the early 1930s. A disease with severe symptoms was common in Minamata, but extremely rare throughout the rest of Japan. After much investigation into the situation, it was determined that a nearby chemical plant periodically dumped methyl mercury into the bay that was the village's primary source of food. Consequently, the citizens of this small village ingested hazardous dosages of mercury every time they ate fish from the bay.

Mercury poisoning became an issue in the United States after a study was conducted in the early 1940s that focused on New York City's hat-making industry. During that time, many workers in this industry displayed the same types of symptoms as the citizens of Minamata, Japan. Because mercury nitrate was used in the production of hats, enough suspicion was aroused to warrant a study. The study linked the symptoms of workers with the use of mercury nitrate. As a result, the use of this hazardous chemical in the hat-making industry was stopped, and a suitable substitute—hydrogen peroxide—was found.

As discussed earlier, asbestos was another important substance in the evolution of the modern safety and health movement. By the time it was determined that asbestos is a hazardous material, the fibers of which can cause asbestosis or lung cancer (mesothelioma), thousands of buildings contained the substance. As these buildings began to age, the asbestos—particularly that used to insulate pipes—began to break down. As asbestos breaks down, it releases dangerous microscopic fibers into the air. These fibers are so hazardous that removing asbestos from old buildings has become a highly specialized task requiring special equipment and training.

More recently, concern over the potential effects of bloodborne pathogens in the workplace has had a significant impact on the safety and health movement. Diseases such as acquired immunodeficiency syndrome (AIDS) and pathogens such as human immunodeficiency virus (HIV) and Hepatitis B (HBV) have caused changes to how safety and health professionals respond to medical emergencies and injuries in which blood and other bodily fluids may be present. Concern over the potential effects of bloodborne pathogens has introduced a whole new set of precautions as well as fears—some rational and some irrational—into the realm of workplace safety. Chapter 20 is devoted to the concept of bloodborne pathogens as it relates to workplace safety.

SAFETY AND HEALTH STANDARDS APPLY TO MORE THAN JUST MANUFACTURING

There is a misconception in some quarters that safety and health procedures such as those presented in this book apply only to manufacturing. While it is important for employers in the manufacturing sector to provide a safe and healthy work environment for their personnel, it is equally important for employers in all work sectors to do the same. The safe and healthy work practices advocated in this book apply to all work sectors, including natural resources and mining, manufacturing, construction, services, trade, transportation, utilities, information, financial, professional and business, education, government, healthcare, maritime, leisure, retail, and hospitality. As you will learn in Chapter 5, certain industry sectors such as construction and maritime have their own specific safety and health standards. But other sectors are covered by the rules set forth in 29 CFR 1910, OSHA's General Industry Standard.

DEVELOPMENT OF ACCIDENT PREVENTION PROGRAMS

In the modern workplace, there are many different types of **accident prevention** programs ranging from the simple to the complex. Widely used accident prevention techniques include failure minimization, fail-safe designs, isolation, lockouts, screening, personal protective equipment, redundancy, timed replacements, and many others. These techniques are individual components of broader safety programs. Such programs have evolved since the late 1800s.

In the early 1800s, employers had little concern for the safety of workers and little incentive to be concerned. Consequently, organized safety programs were nonexistent, a situation that continued for many years. However, between World War I and World War II, industry discovered the connection between quality and safety. Then, during World War II, troop call-ups and deployments created severe labor shortages. Faced with these shortages, employers could not afford to lose workers to accidents or for any other reason. This realization created a greater openness toward giving safety the serious consideration that it deserved. For example, according to the Society of Manufacturing Engineers (SME), around this time industry began to realize the following:

- Improved engineering could prevent accidents.
- Employees were willing to learn and accept safety rules.
- Safety rules could be established and enforced.
- Financial savings from safety improvement could be reaped by savings in compensation and medical bills.[17]

With these realizations came the long-needed incentive for employers to begin playing an active role in creating and maintaining a safe workplace. This, in turn, led to the development of organized safety programs sponsored by management. Early safety programs were based on the **three E's of safety**: Engineering, Education, and Enforcement (see Figure 1–2). The engineering aspects of a safety program involve making design improvements to both product and process. By altering the design of a product, the processes used to manufacture it can be simplified and, as a result, made less dangerous. In addition, the manufacturing processes for products can be engineered in ways that decrease potential hazards associated with the processes.

The education aspect of a safety program ensures that employees know how to work safely, why it is important to do so, and that safety is expected by management. Safety education typically covers the what, when, where, why, and how of safety.

The enforcement aspect of a safety program involves making sure that employees abide by safety policies, rules, regulations, practices, and procedures. Supervisors and fellow employees play a key role in the enforcement aspects of modern safety programs.

FIGURE 1–2   Three E's of safety.

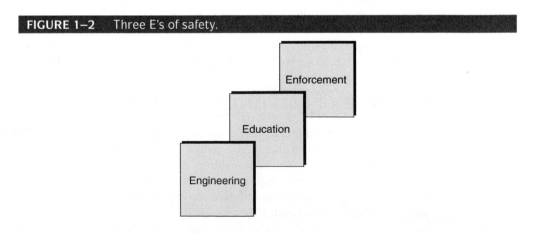

DEVELOPMENT OF SAFETY ORGANIZATIONS

Today, numerous organizations are devoted in full, or at least in part, to the promotion of safety and health in the workplace. Figure 1–3 lists organizations with workplace safety as part of their missions. Figure 1–4 lists several governmental agencies and two related organizations concerned with safety and health. These lists are extensive now, but this has not always been the case. Safety organizations in this country had humble beginnings.

The grandfather of them all is the NSC. The SME traces the genesis of this organization as follows:

The Association of Iron and Steel Electrical Engineers was organized in the first decade of the 20th century and devoted much attention to safety problems in its industry. In 1911, a request came from this association to call a national industrial safety conference. The first Cooperative Safety Congress met in Milwaukee in 1912. A year later, at a meeting in New York City, the National Council of Industrial Safety was formed. It began operation in a small office in Chicago. At its meeting in 1915, the organization's name was changed to the National Safety Council (NSC).[18]

Today, the NSC is the largest organization in the United States devoted solely to safety and health practices and procedures. Its purpose is to prevent the losses, both

FIGURE 1–3 Organizations concerned with workplace safety.

Alliance for American Insurers
American Board of Industrial Hygiene
American Conference of Government Industrial Hygienists
American Industrial Hygiene Association
American Insurance Association
American National Standards Institute
American Occupational Medical Association
American Society for Testing and Materials
American Society of Mechanical Engineers
American Society of Safety Engineers
Chemical Transportation Emergency Center
Human Factors Society
National Fire Protection Association
National Safety Council
National Safety Management Society
Society of Automotive Engineers
System Safety Society
Underwriters Laboratories, Inc.

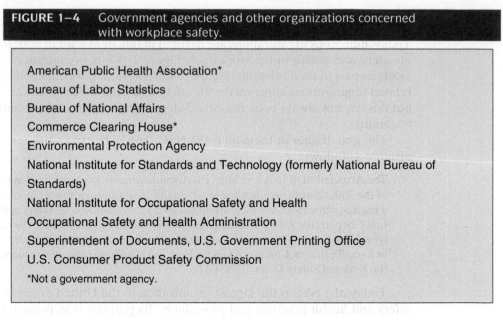

FIGURE 1–4 Government agencies and other organizations concerned with workplace safety.

American Public Health Association*

Bureau of Labor Statistics

Bureau of National Affairs

Commerce Clearing House*

Environmental Protection Agency

National Institute for Standards and Technology (formerly National Bureau of Standards)

National Institute for Occupational Safety and Health

Occupational Safety and Health Administration

Superintendent of Documents, U.S. Government Printing Office

U.S. Consumer Product Safety Commission

*Not a government agency.

direct and indirect, arising out of accidents or from exposure to unhealthy environments. Although it is chartered by an act of Congress, the NSC is a nongovernmental, not-for-profit, public service organization.

The **Occupational Safety and Health Administration (OSHA)** is the government's administrative arm for the Occupational Safety and Health Act (OSH Act). Formed in 1970, OSHA sets and revokes safety and health standards, conducts inspections, investigates problems, issues citations, assesses penalties, petitions the courts to take appropriate action against unsafe employers, provides safety training, provides injury prevention consultation, and maintains a database of health and safety statistics.

Another governmental organization is the **National Institute for Occupational Safety and Health (NIOSH)**. This organization is part of the Centers for Disease Control and Prevention (CDC) of the Department of Health and Human Services. NIOSH is required to publish annually a comprehensive list of all known toxic substances. NIOSH will also provide on-site tests of potentially toxic substances so that companies know what they are handling and what precautions to take.

An organization of note is the **Chemical Safety Board (CSB)**. The CSB is a federal agency charged with the responsibility to investigate industrial chemical accidents. The CSB conducts root cause analyses of chemical accidents that occur at fixed-site facilities (as opposed to accidents involving chemicals that are being transported). The agency does not issue fines or citations, but does make recommendations to OSHA and the EPA. CSB is independent of these and other regulatory agencies, so its investigations can be used to measure the effectiveness of regulations relating to chemical safety.

SAFETY AND HEALTH MOVEMENT TODAY

The safety and health movement has come a long way since the Industrial Revolution. Today, there is widespread understanding of the importance of providing a safe and healthy workplace. The tone was set during and after World War II when all the various practitioners of occupational health and safety began to see the need for cooperative efforts. These practitioners included safety engineers, safety managers, industrial hygienists, occupational health nurses, and physicians. Integration has led to better sharing of knowledge among these practitioners concerning safety and health problems in the workplace, brought a greater level of expertise to bear on evaluating the causes of safety and health problems, generated a large and growing database of helpful information about safety and health problems, enhanced the focus on accident prevention, and encouraged employers to make safety and health a high priority.

INTEGRATED APPROACH TO SAFETY AND HEALTH

The integrated approach has become the norm that typifies the safety and health movement of today. By working together and drawing on their own respective areas of expertise, safety and health professionals are better able to identify, predict, control, and correct safety and health problems.

OSHA reinforces the integrated approach by requiring companies to have a plan for doing at least the following: (1) providing appropriate medical treatment for injured or ill workers, (2) regularly examining workers who are exposed to toxic substances, and (3) having a qualified first-aid person available during all working hours.

Smaller companies may contract out the fulfillment of these requirements. Larger companies often maintain a staff of safety and health professionals. The safety and health staff of a modern company can now include such positions as safety engineer, safety manager, ergonomist, industrial hygiene professionals, **radiation control specialists**, occupational nurses, physicians, psychologists, counselors, educators, and dieticians.

NEW MATERIALS, NEW PROCESSES, AND NEW PROBLEMS

The job of the safety and health professional is more complex than it has ever been. The materials out of which products are made have become increasingly complex and exotic. Engineering metals now include carbon steels, alloy steels, high-strength low-alloy steels, stainless steels, managing steels, cast steels, cast irons, tungsten, molybdenum, titanium, aluminum, copper, magnesium, lead, tin, zinc, and powdered metals. Each of these metals requires its own specialized processes.

Nonmetals are more numerous and have also become more complex. Plastics, plastic alloys and blends, advanced composites, fibrous materials, elastomers, and ceramics also bring their own potential hazards to the workplace.

In addition to the more complex materials being used in modern industry and the new safety and health concerns associated with them, modern industrial processes are also becoming more complex. As these processes become automated, the potential hazards associated with them often increase. Computers; lasers; industrial robots; non-traditional processes such as explosive welding, photochemical machining, laser beam machining, ultrasonic machining, and chemical milling; automated material handling; water-jet cutting expert systems; flexible manufacturing cells; and computer-integrated manufacturing have all introduced new safety and health problems in the workplace and new challenges for the safety and health professional.

RAPID GROWTH IN THE PROFESSION

The complexities of the modern workplace have made safety and health a growing profession. Associate and baccalaureate degree programs in industrial technology typically include industrial safety courses. Some engineering degree programs have safety and health tracks. Several colleges and universities offer full degrees in occupational safety and health.

The inevitable result of the increased attention given to safety and health is that more large companies are employing safety and health professionals and more small companies are assigning these duties to existing employees. This is a trend that is likely to continue as employers see their responsibilities for safety and health spread beyond the workplace to the environment, the community, the users of their products, and the recipients of their by-products and waste.

DOES MOVING MANUFACTURING JOBS OVERSEAS REDUCE THE ACCIDENT RATE?

If you would like to start a debate among occupational safety and health professionals, ask this question: Is moving manufacturing jobs overseas reducing the accident rate in America? Some say "yes" and others say "no." Here are some facts:

- Over the past decade, thousands of jobs in the manufacturing sector have migrated to China, India, Malaysia, Pakistan, and other countries that maintain low labor rates.
- Since 2003, the recordable injury and illness rate has declined in the United States every year.
- The largest year-to-year declines have been in the manufacturing sector. However, even the construction industry has seen declines during the same time period and those jobs are not being exported.

Some claim that the workplace accident rate is declining because America has *exported* its accidents along with its jobs, particularly in the manufacturing sector. Others claim that the declines can be attributed to more and better safety and health regulations, oversight, education, and enforcement. The more likely scenario is that both the loss of manufacturing jobs to foreign countries and better safety and health practices have contributed to the decline.

Key Terms and Concepts

- Accident prevention
- Asbestos menace
- Assumption of risk
- Bhopal tragedy
- Chemical Safety Board (CSB)
- Code of Hammurabi
- Contributory negligence
- Cooperative Safety Congress (CSC)
- Employer-biased laws
- Employer liability
- Fellow servant rule
- Hawk's Nest tragedy
- Inanimate power
- National Council of Industrial Safety (NCIS)
- National Institute for Occupational Safety and Health (NIOSH)
- Occupational Safety and Health Act (OSH Act)
- Occupational Safety and Health Administration (OSHA)
- Organized labor
- Radiation control specialists
- Safety movement
- Three E's of safety
- West Fertilizer Company Explosion
- Workers' compensation

Review Questions

1. To what cause(s) can the improvements in workplace safety made to date be attributed?
2. Explain the significance of the Code of Hammurabi in terms of the safety movement.
3. Describe the circumstances that led to the development of the first organized safety program.
4. What is Frederick Taylor's connection to the safety movement?
5. Explain the development of the National Safety Council.
6. What impact did labor shortages in World War II have on the safety movement?
7. Explain how workplace tragedies have affected the safety movement. Give examples.
8. Explain the primary reasons behind the passage of the OSH Act.
9. Summarize briefly the role that organized labor has played in the advancement of the safety movement.
10. Define the following terms: fellow servant rule, contributory negligence, and assumption of risk.
11. Explain the three E's of safety.
12. Explain the term *integration* as it relates to modern safety and health.

Endnotes

1. S. Minter, "The Birth of OSHA," *Occupational Hazards*, July 1998, 59.
2. National Safety Council, *Accident Facts* (Chicago: National Safety Council, 2016).
3. J. LaDou, ed., *Introduction to Occupational Health and Safety* (Chicago: National Safety Council, 1997), 28.
4. Ibid., 28.
5. A. Soubiran, "Medical Services under the Pharaohs," *Abbottempo* 1: 19–23. 2011.
6. LaDou, *Occupational Health and Safety*, 31.
7. Ibid., 34.
8. Ibid., 35.
9. Ibid., 37.

10. S. Minter and V. Sutcliff, "Fighting Two Wars," *Occupational Hazards*, July 1998, 41–42.
11. Ibid., 41.
12. Ibid., 41–42.
13. *OSHA Fact Sheet: Asbestos*. Retrieved from osha.gov on May 6, 2017.
14. E. Broughton, *The Bhopal Disaster and Its Aftermath: A Review*. Retrieved from ncbi.nlm .nih.gov/pmc/articles/PMC1142333/ on August 19, 2017.
15. Ibid., 67.
16. Minter and Sutcliff, "Fighting Two Wars," 41.
17. Ibid.
18. Ibid., 42.

2

ACCIDENTS AND THEIR EFFECTS

Learning Objectives

- Summarize the cost of accidents in the United States
- Summarize the figures for accidents and deaths in the United States
- Compare workplace deaths with other causes of deaths in the United States
- Describe the costs of workplace accidents
- Explain the amount of time that is lost in the workplace because of accidents
- Summarize the figures for deaths that occur in workplace accidents
- Categorize workplace injuries by type of accident
- Categorize death rates according to industry
- Summarize the figures for injuries to different body parts in the workplace
- Explain how to estimate the cost of accidents
- Describe the global impact of accidents and injuries in the workplace
- Describe the types of accident/injury reports and logs required by OSHA

There is a long history of debate in this country concerning the effect of **accidents** on industry (the workers and the companies) and the cost of preventing accidents. Historically, the prevailing view was that **accident prevention** programs were too costly. The more contemporary view is that accidents are too costly and that accident prevention makes sense economically. As a result, accident prevention, which had been advocated on a moral basis, is now justified in economic terms.

Accidents are the fourth leading cause of death in this country after heart disease, cancer, and strokes. This ranking is based on all types of accidents, including motor vehicle accidents, drownings, fires, falls, natural disasters, and work-related accidents.

Although deaths from **natural disasters** tend to be more newsworthy than workplace deaths, their actual impact is substantially less. For example, natural disasters in the

FIGURE 2–1 Approximate accident costs (in dollars) by accident type (in billions, 2017).

Motor vehicle accidents	$72.5
Workplace accidents	50
Home accidents	20
Public accidents	13

United States cause fewer than 100 deaths per year on average. **Workplace accidents**, on the other hand, cause more than 10,000 deaths every year in the United States.[1]

This chapter provides prospective and practicing **safety and health professionals** with the information they need to have a full understanding of workplace accidents and their effect on industry in the United States. Such an understanding will help professionals play a more effective role in keeping both management and labor focused appropriately on safety and health in the workplace.

COSTS OF ACCIDENTS

To gain a proper perspective on the economics of workplace accidents, we must view them in the overall context of all accidents. The overall cost of accidents in the United States is approximately $155 billion. These costs include such factors as **lost wages**, **medical expenses**, **insurance administration**, **fire-related losses**, motor vehicle **property damage**, and **indirect costs**.

Figure 2–1 breaks down this overall amount by categories of accidents. Figure 2–2 breaks them down by cost categories. Notice in Figure 2–1 that workplace accidents rank second behind motor vehicle accidents in cost. Figure 2–2 shows that the highest cost category is wages lost by workers who are either injured or killed. The category of indirect losses from work accidents consists of costs associated with responding to accidents (i.e., giving first aid, filling out accident reports, and handling production slowdowns).

Clearly, accidents on and off the job cost the U.S. industry dearly. Every dollar that is spent responding to accidents is a dollar that could have been reinvested in modernization, research and development, facility upgrades, and other competitiveness-enhancing activities.

FIGURE 2–2 Approximate accident costs by categories, 2017.

Wages lost	$40
Medical expenses	25
Insurance administration	30
Property damage (motor vehicle)	30
Fire losses	10
Indirect losses for work accidents	40

ACCIDENTAL DEATHS IN THE UNITED STATES

Accidental deaths in the United States result from a variety of causes, including motor vehicle accidents, falls, poisoning, drowning, fire-related injuries, **suffocation** (ingested object), firearms, medical complications, air transport accidents, interaction with machinery, mechanical suffocation, and the impact of falling objects. The National Safety Council (NSC) periodically computes death totals and death rates in each of these categories. The statistics for a typical year are as follows:

- **Motor vehicle accidents.** Motor vehicle accidents are the leading cause of accidental deaths in the United States each year. They include deaths resulting from accidents involving mechanically or electrically powered vehicles (excluding rail vehicles) that occur on or off the road. In a typical year, there are approximately 47,000 such deaths in the United States.
- **Falls.** This category includes all deaths from **falls** except those associated with transport vehicles. For example, a person who is killed as the result of falling while boarding a bus or train would not be included in this category. In a typical year, there are approximately 13,000 deaths in the United States from falls.
- **Poisoning.** The **poisoning** category is divided into two subcategories: (1) poisoning by solids and liquids and (2) poisoning by gases and vapors. The first category includes deaths that result from the ingestion of drugs, medicine, widely recognized solid and liquid poisons, mushrooms, and shellfish. It does not include poisoning from spoiled food or salmonella. The second category includes deaths caused by incomplete combustion (e.g., gas vapors from an oven or unlit pilot light) or from carbon monoxide (e.g., exhaust fumes from an automobile). In a typical year, there are approximately 6,000 deaths in the first category and 1,000 in the second.
- **Drowning.** This category includes work-related and non-work-related **drownings** but excludes those associated with floods or other natural disasters. In a typical year, there are approximately 5,000 deaths from drowning in the United States.
- **Fire-related injuries.** This category includes deaths from burns, asphyxiation, falls, and those that result from falling objects in a fire. In a typical year, there are more than 4,000 fire-related deaths in the United States.
- **Suffocation (ingested object).** This category includes deaths from the ingestion of an object that blocks the air passages. In many such deaths, the ingested object is food. In a typical year, there are approximately 4,000 suffocation deaths in the United States.
- **Firearms.** This category includes deaths that result when recreational activities involving firearms or household accidents involving firearms result in death. For example, a person killed in the home while cleaning a firearm would be included in this category. However, a person killed in combat would not be included. In a typical year, there are approximately 2,000 such deaths in the United States.
- **Others.** This category includes deaths resulting from **medical complications** arising out of mistakes made by healthcare professionals, air transport injuries, interaction with machinery, **mechanical suffocation**, and the impact of falling objects. In a typical year, there are more than 14,000 deaths in these subcategories.[2]

FIGURE 2–3   Causes of accidents (ages 25–44 years, 2017).

Accidents	30,000
Cancer	21,000
Motor vehicle	17,000
Heart disease	16,000
Poison (solid, liquid)	3,000
Drowning	1,500
Falls	1,200
Fire related	1,000

ACCIDENTS VERSUS OTHER CAUSES OF DEATH

Although there are more deaths every year from **heart disease**, **cancer**, and **strokes** than from accidents, these causes tend to be concentrated among people at or near retirement age. Among people 37 years of age or younger—prime working years— accidents are the number one cause of death. Figure 2–3 summarizes the causes of death for persons from 25 to 44 years of age. Notice that the leading cause is accidents.

Figure 2–3 shows that accidents represent a serious detriment to productivity, quality, and competitiveness in today's workplace. Yet accidents are the one cause of death and injury that companies can most easily control. Although it is true that companies may have some success in decreasing the incidence of heart disease and stroke among their employees through such activities as corporate wellness programs, their impact in this regard is limited. However, employers can have a significant impact on preventing accidents.

WORK ACCIDENT COSTS AND RATES

Workplace accidents cost employers millions every year. Consider the following examples from the recent past. ARCO Chemical Company was ordered to pay $3.48 million in fines as a result of failing to protect workers from an explosion at its petrochemical plant in Channelview, Texas. The steel-making division of USX paid a $3.25 million fine to settle numerous health and safety violation citations. BASF Corporation agreed to pay a fine of $1.06 million to settle Occupational Safety and Health Administration (OSHA) citations associated with an explosion at a Cincinnati chemical plant that caused two deaths and 17 injuries.

These examples show the costs of fines only. In addition to fines, these employers incurred costs for safety corrections, medical treatment, survivor benefits, death and burial costs, and a variety of indirect costs. Clearly, work accidents are expensive. However, the news is not all bad. The trend in the rate of accidents is downward.

Work **accident rates** in this century are evidence of the success of the safety movement in the United States. As the amount of attention given to workplace safety and health has increased, the accident rate has decreased.

As Figure 2–1 shows, the cost of these 10,000 work deaths and work injuries was $50 billion. This translates into a cost of $500 per worker in the United States, computed as the value-add required per worker to offset the cost of work injuries.

Although statistics are not available to document the supposition, many safety and health professionals believe that the major cost of accidents and injuries on the job results from damage to morale. Employee morale is a less tangible factor than documentable factors such as lost time and medical costs. However, it is widely accepted among management professionals that few factors affect productivity more than employee morale. Employees with low morale do not produce up to their maximum potential. This is why so much time and money are spent every year to help supervisors and managers learn different ways to help improve employee morale.

Because few things are as detrimental to employee morale as seeing a fellow worker injured, accidents can have a devastating effect on morale. Whenever an employee is injured, his or her colleagues silently think, "That could have been me," in addition to worrying about the employee. Morale is damaged even more if the injured employee is well-liked and other employees know his or her family.

TIME LOST BECAUSE OF WORK INJURIES

An important consideration when assessing the effect of accidents on industry is the amount of **lost time** due to **work injuries**.[3] According to the NSC, approximately 35 million hours are lost in a typical year as a result of accidents. This is actual time lost from disabling injuries and does not include additional time lost for medical checkups after the injured employee returns to work. Accidents that occurred in previous years often continue to cause lost time in the current year.

DEATHS IN WORK ACCIDENTS

Deaths on the job have decreased markedly over the years. However, they still occur. For example, in a typical year, there are 10,400 work deaths in the United States. The causes of death in the workplace vary. They include those related to motor vehicles, falls, electric current, drowning, fires, air transport, poisoning, water transport, machinery, falling objects, rail transport, and mechanical suffocation.[4] Figure 2–4 gives a complete breakdown of the percentages for the various categories of causes.

WORK INJURIES BY TYPE OF ACCIDENT

Work injuries can be classified by the type of accident from which they resulted. The most common causes of work injuries are as follows:

- Overexertion
- Impact accidents
- Falls
- Bodily reaction (to chemicals)
- Compression

FIGURE 2–4 Work deaths by cause for a typical year.

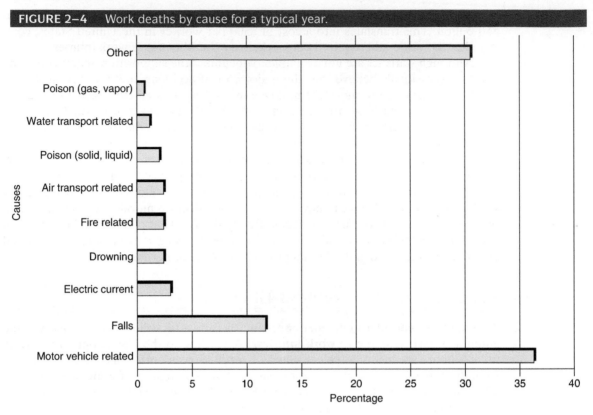

- Motor vehicle accidents
- Exposure to radiation or caustics
- Rubbing or abrasions
- Exposure to extreme temperatures

Overexertion, the result of employees working beyond their physical limits, is the leading cause of work injuries. According to the NSC, almost 31 percent of all work injuries are caused by overexertion. **Impact accidents** involve a worker being struck by or against an object. The next most prominent cause of work injuries is falls.[5] The remaining accidents are distributed fairly equally among the other causes listed above.

DEATH RATES BY INDUSTRY

A variety of agencies and organizations, including the Bureau of Labor Statistics, the National Center for Health Statistics, and the NSC, collect data on **death rates** within industrial categories.[6] Such information can be used in a variety of ways, not the least of which is in assigning workers' compensation rates. The most widely used industrial categories are agriculture, including farming, forestry, and fishing; mining/ quarrying, including oil and gas drilling and extraction; construction; manufacturing;

transportation/public utilities; trade, both wholesale and retail; services, including finance, insurance, and real estate; and federal, state, and local governments.

The rate of fatalities across all occupations fluctuates somewhat from year to year and from industry to industry. However, the rate of 3.5 fatalities per year per 100,000 workers holds fairly constant over time. The greater fluctuation is in the industry sector that experiences the most fatalities in a given year. The following order of ranking will change from year to year, but the industry sectors that comprise the top sectors by annual death rate remain fairly constant. Occupations that typically have the highest rate of fatalities on the job are as follows (not necessarily in order since the order changes every year):

- Fishing
- Logging
- Mining/quarrying
- Agriculture
- Construction
- Transportation and warehousing
- Manufacturing
- Aircraft pilots, flight engineers, and maintenance
- Refuse and recycling collectors
- Law enforcement
- Professional and business services
- Retail
- Public administration/government

Figures for costs, work-related deaths, accidents, and so on change over time. For the most up-to-date figures, go to the website of the Bureau of Labor Statistics: www.bls.gov

---

PARTS OF THE BODY INJURED ON THE JOB

To develop and maintain an effective safety and health program, it is necessary to know not only the most common causes of death and injury but also the parts of the body most frequently injured. The NSC stated the following:

> Disabling work injuries in the entire nation totaled approximately 1.75 million in 1998. Of these, about 10,400 were fatal and 60,000 resulted in some permanent impairment. Injuries to the back occurred most frequently, followed by thumb and finger injuries and leg injuries.[7]

Typically, the most frequent injuries to specific parts of the body are as follows (from most frequent to least):

1. Back
2. Legs and fingers
3. Arms and multiple parts of the body
4. Trunk
5. Hands
6. Eyes, head, and feet
7. Neck, toes, and body systems

The back is the most frequently injured part of the body. Legs and fingers are injured with approximately the same frequency, as are arms and multiple parts of the body; the hands are next in frequency, followed by the eyes, the head, and feet; and neck, toes, and body systems. This ranking shows that one of the most fundamental components of a safety and health program should be instruction on how to lift without hurting the back (see Chapter 11).

ESTIMATING THE COST OF ACCIDENTS

Even decision makers who support accident prevention must consider the relative costs of such efforts. Clearly, accidents are expensive. However, to be successful, safety and health professionals must be able to show that accidents are more expensive than prevention. To do this, they must be able to estimate the cost of accidents. The procedure for estimating costs set forth in this section was developed by Professor Rollin H. Simonds of Michigan State College working in conjunction with the Statistics Division of the NSC.

Cost-Estimation Method

Professor Simonds states that in order to have value, a cost estimate must relate directly to the specific company in question. Applying broad industry cost factors will not suffice. To arrive at company-specific figures, Simonds recommends that costs associated with an accident be divided into *insured* and *uninsured* costs.[8]

Determining the insured costs of accidents is a simple matter of examining accounting records. The next step involves calculating the uninsured costs. Simonds recommends that accidents be divided into the following four classes:

- **Class 1 accidents.** Lost workdays, permanent partial disabilities, and temporary total disabilities.
- **Class 2 accidents.** Treatment by a physician outside the company's facility.
- **Class 3 accidents.** Locally provided first aid, property damage of less than $100, or the loss of less than eight hours of work time.
- **Class 4 accidents.** Injuries that are so minor that they do not require the attention of a physician, result in property damage of $100 or more, or cause eight or more work hours to be lost.[9]

Average uninsured costs for each class of accident can be determined by pulling the records of all accidents that occurred during a specified period and by sorting the records according to class. For each accident in each class, record every cost that was not covered by insurance. Compute the total of these costs by class of accident and divide by the total number of accidents in that class to determine an average uninsured cost for each class, specific to the particular company.

Figure 2–5 is an example of how the average cost of a selected sample of Class 1 accidents can be determined. In this example, there were four Class 1 accidents in the pilot test. These four accidents cost the company a total of $554.23 in uninsured costs, or an average of $138.56 per accident. Using this information, accurate cost estimates of an accident can be figured, and accurate predictions can be made.

FIGURE 2–5 Uninsured costs worksheet.

Class of Accident	Accident Number							
Class 1	1	2	3	4	5	6	7	8
Cost A	$ 16.00	$ 6.95	$ 15.17	$ 3.26				
Cost B	72.00	103.15	97.06	51.52				
Cost C	26.73	12.62	—	36.94				
Cost D	—	51.36	—	38.76				
Cost E	—	11.17	—	24.95				
Cost F	—	—	—	–13.41				
Cost G	—	—	—	—				
Total	114.73	185.25	112.23	142.02				

Grand Total: $554.23

Average Cost per Accident: $138.56 (Grand Total ÷ Number of Accidents)

Signature: _____   Date: _____

Other Cost-Estimation Methods

The costs associated with workplace accidents, injuries, and incidents fall into broad categories such as the following:

- Lost work hours
- Medical costs
- Insurance premiums and administration
- Property damage
- Fire losses
- Indirect costs

Calculating the direct costs associated with lost work hours involves compiling the total number of lost hours for the period in question and multiplying the hours times the applicable loaded labor rate. The loaded labor rate is the employee's hourly rate plus benefits. Benefits vary from company to company but typically inflate the hourly wage by 20–35 percent. A sample cost-of-lost-hours computation follows:

$$\text{Employee Hours Lost (4th quarter)} \times \text{Average Loaded Labor Rate} = 386 \times 13.48 = \$5,203.28$$

In this example, the company lost 386 hours due to accidents on the job in the fourth quarter of its fiscal year. The employees who actually missed time at work formed a pool of people with an average loaded labor rate of $13.48 per hour ($10.78 average hourly wage plus 20 percent for benefits). The average loaded labor rate multiplied by the 386 lost hours reveals an unproductive cost of \$5,203.28 to this company.

By studying records that are readily available in the company, a safety professional can also determine medical costs, insurance premiums, property damage, and fire losses for the time period in question. All these costs taken together result in a subtotal cost. This figure is then increased by a standard percentage to cover indirect costs to determine the total cost of accidents for a specific time period. The percentage used to calculate indirect costs can vary from company to company, but 20 percent is a widely used figure.

Estimating Hidden Costs

Safety professionals often use the *iceberg analogy* when talking about the real costs of accidents. Accident costs are like an iceberg in that their greatest portion is hidden from view.[10] In the case of icebergs, the larger part is hidden beneath the surface of the water. In the case of an accident, the larger part of the actual cost is also hidden beneath the surface.

There are many different models that can be used for estimating both the direct and indirect costs of accidents. Some of these models are so complex that their usefulness is questionable. The checklist in Figure 2–6 is a simple and straightforward tool that can be used to estimate the hidden costs of accidents.

FIGURE 2–6   Some accident costs that might be overlooked.

Checklist for Estimating the Hidden Costs of Accidents
√   Paid time to the injured employee on the day of the accident.
√   Paid time of any emergency-responder personnel involved (including ambulance driver).
√   Paid time of all employees who were interviewed as part of the accident investigation.
√   Paid time of the safety personnel who conducted the accident investigation.
√   Paid time of the human resources personnel who handled the workers' compensation and medical aspects of the accident.
√   Paid time of the supervisor involved in the accident investigation and accident response.
√   Paid time to employees near the accident working (or slowed down) temporarily as a result of the accident.
√   Paid time to employees who spent time talking about the accident as news of it spread through the company's grapevine.

GLOBAL IMPACT OF ACCIDENTS AND INJURIES

According to the International Labour Organization (ILO) of the United Nations, approximately 2.2 million people die every year of work-related injuries and occupational diseases.[11] Actually, due to poor record keeping and reporting in underdeveloped countries, it is estimated that this figure is low. Rapid development and the pressure of global competition are resulting in increased workplace fatalities in China and the Pacific Rim countries.

What is missing in many of the developing countries that are becoming industrialized is a safety and health infrastructure. Such an infrastructure would include government-enforced safety and health regulations, company-sponsored safety and health training, record-keeping and reporting systems, and management practices that make occupational safety and health a fully integrated component of work processes and the competitive philosophy of organizations. Occupational safety and health must come to be seen as a strategy for sustaining economic growth and social development in emerging countries.

There is much to be done if developing countries are going to provide safe and healthy working conditions for their citizens. The ILO reports:

- Record-keeping and reporting systems in developing countries are deteriorating instead of improving. Consequently, only a fraction of the real toll of workplace accidents and injuries is being reported.
- Men in developing countries tend to die as the result of accidents, lung diseases, and work-related cancers such as those caused by asbestos. Women in developing countries suffer more from musculoskeletal disorders, communicable diseases, and psychosocial problems.
- Occupational injuries in developing countries are more prevalent in such high-risk industries as mining, construction, and agriculture.
- Younger workers in developing countries are more likely to suffer nonfatal injuries, while older workers are more likely to suffer fatal injuries.
- In developing countries, more than half of retirements are taken early to collect pensions based on work-related disabilities rather than normal retirement.

OSHA REPORTS AND LOGS

Accident investigation and reporting are important aspects of the safety professional's responsibilities. OSHA has specific requirements concerning accident reporting and specific logs for making reports. Safety professionals must know how to complete OSHA logs 300, 300A, and 301, when to use each log, how to submit these logs, and when to submit them. OSHA Form 300 is a log of work-related injuries and illnesses. OSHA Form 300A is a summary report of work-related injuries and illnesses. OSHA Form 301 is used to report individual incidents of injuries and illnesses. The concept of using OSHA reports and logs is introduced here only for informational purposes. Readers will learn all of OSHA's requirements for accident reporting as which forms to use, how to complete them, and when to submit them in Chapter 5.

Accident rates are especially high in developing countries because these countries are responding to the pressures of global competition without first putting a safety and health infrastructure in place (e.g., regulations, training, and record keeping).

Key Terms and Concepts

- Accident prevention
- Accident rates
- Accidents
- Cancer
- Death rates
- Drownings
- Falls
- Fire-related losses
- Heart disease

- Impact accidents
- Indirect costs
- Insurance administration
- Lost time
- Lost wages
- Mechanical suffocation
- Medical complications
- Medical expenses
- Natural disasters

- Overexertion
- Poisoning
- Property damage
- Safety and health professionals
- Strokes
- Suffocation
- Work injuries
- Workplace accidents

Review Questions

1. What are the leading causes of death in the United States?
2. When the overall cost of an accident is calculated, what elements make up the cost?
3. What are the five leading causes of accidental deaths in the United States?
4. What are the leading causes of death in the United States of people between the ages of 25 and 44?
5. Explain how today's rate of accidental work deaths compares with the rate in the early 1900s.
6. What are the five leading causes of work deaths?
7. What are the five leading causes of work injuries by type of accident?
8. When death rates are classified by industry type, what are the three leading industry types?
9. Rank the following body parts according to frequency of injury from highest to lowest: neck, fingers, trunk, back, and eyes.
10. Explain the reasons for high accident rates in developing countries.

Endnotes

1. National Safety Council, *Accident Facts* (Chicago: National Safety Council, 2017), 37.
2. Ibid., 4–5.
3. Ibid., 35.
4. Ibid., 36.
5. Ibid.
6. Ibid., 37.
7. Ibid., 38.
8. National Safety Council, *Accident Prevention Manual for Business and Industry*, 13th ed. (Chicago: National Safety Council, 2009), 158.
9. Ibid.
10. Daniel Corcoran, "The Hidden Value of Safety," *Occupational Health & Safety* 71, no. 6: 20–22. 2009.
11. Occupational Health & Safety Online, *Study: 2.2 Million People Die Worldwide of Work-related Accidents, Occupational Diseases*. Retrieved from ohsonline.com on January 6, 2017, 1–3.

THEORIES OF ACCIDENT CAUSATION

Learning Objectives

- Explain the Domino Theory of accident causation
- Explain the Human Factors Theory of accident causation
- Explain the Accident/Incident Theory of accident causation
- Explain the Epidemiological Theory of accident causation
- Explain the Systems Theory of accident causation
- Explain the Combination Theory of accident causation
- Explain the Behavioral Theory of accident causation
- Describe the individual factors that can be part of accident causation
- Describe how depression can play a role in accident causation
- Summarize how management failures can play a role in accident causation
- Describe how obesity can be a factor in accident causation
- Explain the Swiss Cheese Model and how it relates to accident causation

Each year, work-related accidents cost the United States almost $50 billion.[1] This figure includes costs associated with lost wages, medical expenses, insurance costs, and indirect costs. The number of persons injured in **industrial place accidents** in a typical year is 7,128,000, or 3 per 100 persons per year.[2] In the workplace, there is one accidental death approximately every 51 minutes and one injury every 19 seconds.[3]

Why do accidents happen? This question has concerned safety and health decision makers for decades, because in order to prevent accidents we must know why they happen. Over the years, several theories of accident causation have evolved that attempt to explain why accidents occur. Models based on these theories are used to predict and prevent accidents.

The most widely known theories of accident causation are the domino theory, the human factors theory, the accident/incident theory, the epidemiological theory, the systems theory, the combination theory, and the behavioral theory. This chapter provides practicing and prospective safety professionals with the information they need to understand fully and apply these theories.

DOMINO THEORY OF ACCIDENT CAUSATION

An early pioneer of accident prevention and industrial safety was Herbert W. Heinrich, an official with the Travelers Insurance Company. In the late 1920s, after studying the reports of 75,000 industrial accidents, Heinrich concluded that

- 88 percent of industrial accidents are caused by unsafe acts committed by fellow workers.
- 10 percent of industrial accidents are caused by unsafe conditions.
- 2 percent of industrial accidents are unavoidable.[4]

Heinrich's study laid the foundation for his *Axioms of Industrial Safety* and his theory of accident causation, which came to be known as the **domino theory**. **So much of Heinrich's theory has been discounted by more contemporary research that it is now considered outdated.** However, because some of today's more widely accepted theories can be traced back to Heinrich's theory, students of industrial safety should be familiar with his work.

Heinrich's Axioms of Industrial Safety

Heinrich summarized what he thought health and safety decision makers should know about industrial accidents in 10 statements he called **Axioms of Industrial Safety**. These axioms can be paraphrased as follows:

1. Injuries result from a completed series of factors, one of which is the accident itself.
2. An accident can occur only as the result of an unsafe act by a person and/or a physical or mechanical hazard.
3. Most accidents are the result of unsafe behavior by people.
4. An unsafe act by a person or an unsafe condition does not always immediately result in an accident/injury.
5. The reasons why people commit unsafe acts can serve as helpful guides in selecting corrective actions.
6. The severity of an accident is largely fortuitous, and the accident that caused it is largely preventable.
7. The best accident prevention techniques are analogous with the best quality and productivity techniques.
8. Management should assume responsibility for safety because it is in the best position to get results.
9. The supervisor is the key person in the prevention of industrial accidents.
10. In addition to the direct costs of an accident (e.g., compensation, liability claims, medical costs, and hospital expenses), there are also hidden or indirect costs.[5]

According to Heinrich, these axioms encompass the fundamental body of knowledge that must be understood by decision makers interested in preventing accidents. Any accident prevention program that takes all 10 axioms into account is more likely to be effective than a program that leaves out one or more axioms.

Heinrich's Domino Theory

Perhaps you have stood up a row of dominoes, tipped the first one over, and watched as each successive domino topples the one next to it. This is how Heinrich's theory of accident causation works. According to Heinrich, there are five factors in the sequence of events leading up to an accident. These factors can be summarized as follows:

1. **Ancestry and social environment.** Negative character traits that may lead people to behave in an unsafe manner can be inherited (**ancestry**) or acquired as a result of the **social environment** (e.g., alcoholism).
2. **Fault of person.** Negative character traits, whether inherited or acquired, are why people behave in an unsafe manner and why hazardous conditions exist.
3. **Unsafe act/mechanical or physical hazard.** **Unsafe acts** committed by people and **mechanical** or **physical hazards** are the direct causes of accidents.
4. **Accident.** Typically, accidents that result in injury are caused by falling or being hit by moving objects.
5. **Injury.** Typical injuries resulting from accidents include lacerations and fractures.[6]

Heinrich's theory has two central points: (1) injuries are caused by the action of **preceding factors** and (2) removal of the **central factor** (unsafe act/**hazardous condition**) negates the action of the preceding factors and, in so doing, prevents accidents and injuries.

Heinrich's Theory and Corrective Action

The three "E's" of safety—Engineering, Education, and Enforcement—(Chapter 1) work well in taking corrective action in conjunction with Heinrich's theory. The engineering component recommends that hazards be controlled through better product design or process improvements. The education component recommends training workers on all aspects of safe work practices and management on their role in providing a safe working environment. The enforcement component recommends that safety professionals, managers, supervisors, and fellow workers act immediately when someone is observed ignoring safe work practices. Safety rules and procedures have no value unless they are followed, and they will not be followed unless they are enforced.

Domino Theory in Practice

Construction Products Company (CPC) is a distributor of lumber, pipe, and concrete products. Its customers are typically small building contractors. CPC's facility consists of an office in which orders are placed and several large warehouses. Contractors place their orders in the office. They then drive their trucks through the appropriate warehouses to be loaded by CPC personnel.

Because the contractors are small operators, most of their orders are also relatively small and can be loaded by hand. Warehouse personnel go to the appropriate bins, pull out the material needed to fill their orders, and load the materials on customers' trucks. Even though most orders are small enough to be loaded by hand, many of the materials purchased are bulky and cumbersome to handle. Because of this, CPC's loaders are required to wear such personal protection gear as hard hats, padded gloves, steel-toed boots, and lower-back-support belts.

For years, CPC's management team had noticed an increase in minor injuries to warehouse personnel during the summer months. Typically, these injuries consisted of nothing worse than minor cuts, scrapes, and bruises. However, this past summer had been different. Two warehouse workers had sustained serious back injuries. These injuries have been costly to CPC both financially and in terms of employee morale.

An investigation of these accidents quickly identified a series of events and a central causal behavior that set up a *domino effect* that, in turn, resulted in the injuries. The investigation revealed that CPC's warehouses became so hot during the summer months that personal protection gear was uncomfortable. As a result, warehouse personnel simply discarded it. Failure to use appropriate personal protection gear in the summer months had always led to an increase in injuries. However, because the injuries were minor in nature, management had never paid much attention to the situation. It was probably inevitable that more serious injuries would occur eventually.

To prevent a recurrence of the summer-injury epidemic, CPC's management team decided to remove the causal factor—failure of warehouse personnel to use their personal protection gear during the summer months. To facilitate the removal of this factor, CPC's management team formed a committee consisting of one executive manager, one warehouse supervisor, and three warehouse employees.

The committee made the following recommendations: (1) provide all warehouse personnel with training on the importance and proper use of personal protection gear; (2) require warehouse supervisors to monitor the use of personal protection gear more closely; (3) establish a company policy that contains specific and progressive disciplinary measures for failure to use required personal protection gear; and (4) implement several heat reduction measures to make warehouses cooler during the summer months.

CPC's management team adopted all the committee's recommendations. In doing so, it removed the central causal factor that had historically led to an increase in injuries during the summer months.

HUMAN FACTORS THEORY OF ACCIDENT CAUSATION

The **human factors theory** of accident causation attributes accidents to a chain of events ultimately caused by **human error**. It consists of the following three broad factors that lead to human error: overload, inappropriate response, and inappropriate activities (see Figure 3–1). These factors are explained in the following paragraphs.

FIGURE 3–1   Factors that cause human errors.

Human Error Factors
• Overload
• Inappropriate activities
• Inappropriate responses

Overload

Overload amounts to an imbalance between a person's capacity at any given time and the load that person is carrying in a given state. A person's capacity is the product of such factors as his or her natural ability, training, state of mind, fatigue, stress, and physical condition. The load that a person is carrying consists of tasks for which he or she is responsible and added burdens resulting from **environmental factors** (noise, distractions, and so on), **internal factors** (personal problems, emotional stress, and worry), and **situational factors** (level of risk, unclear instructions, and so on). The state in which a person is acting is the product of his or her motivational and arousal levels.

Inappropriate Response and Incompatibility

How a person responds in a given situation can cause or prevent an accident. If a person detects a hazardous condition but does nothing to correct it, he or she has responded inappropriately. If a person removes a safeguard from a machine in an effort to increase output, he or she has responded inappropriately. If a person disregards an established safety procedure, he or she has responded inappropriately. Such responses can lead to accidents. In addition to **inappropriate responses**, this component includes workstation incompatibility. The incompatibility of a person's workstation with regard to size, force, reach, feel, and similar factors can lead to accidents and injuries.

Inappropriate Activities

Human error can be the result of **inappropriate activities**. An example of an inappropriate activity is a person who undertakes a task that he or she doesn't know how to do. Another example is a person who misjudges the degree of risk involved in a given task and proceeds based on that misjudgment. Such inappropriate activities can lead to accidents and injuries. Figure 3–2 summarizes the various components of the human factors theory.[7]

FIGURE 3–2 Human factors theory.

Human Factors Theory in Practice

Kitchenware Manufacturing Incorporated (KMI) produces aluminum kitchenware for commercial settings. After 10 years of steady, respectable growth in the U.S. market, KMI suddenly saw its sales triple in less than six months. This rapid growth was the result of KMI's successful entry into European and Asian markets.

The growth in sales, although welcomed by both management and employees, quickly overloaded and, before long, overwhelmed the company's production facility. KMI responded by adding a second shift of production personnel and approving unlimited overtime for highly skilled personnel. Shortly after the upturn in production, KMI began to experience a disturbing increase in accidents and injuries. During his accident investigations, KMI's safety manager noticed that human error figured prominently in the accidents. He grouped all the human errors identified into three categories: (1) overload, (2) inappropriate response, and (3) inappropriate activities.

In the category of *overload*, he found that the rush to fill orders was pushing production personnel beyond their personal limits in some cases, and beyond their capabilities in others. Stress, insufficient training of new employees, and fatigue all contributed to the overload. In the category of *inappropriate response*, the safety manager determined that many of KMI's production personnel had removed safeguards from their machines in an attempt to speed up production. All the machines involved in accidents had had safeguards removed.

In the category of *inappropriate activities*, the safety manager found that new employees were being assigned duties for which they weren't yet fully trained. As a result, they often misjudged the amount of risk associated with their work tasks.

With enough accident investigations completed to identify a pattern of human error, the safety manager prepared a presentation containing a set of recommendations for corrective measures for KMI's executive management team. His recommendations were designed to prevent human-error-oriented accidents without slowing production.

ACCIDENT/INCIDENT THEORY OF ACCIDENT CAUSATION

The **accident/incident theory** is an extension of the human factors theory. It was developed by Dan Petersen and is sometimes referred to as the Petersen accident/incident theory.[8] Petersen introduced such new elements as **ergonomic traps**, the decision to err, and systems failures, while retaining much of the human factors theory. A model based on his theory is shown in Figure 3–3.

In this model, overload, ergonomic traps, or a decision to err leads to human error. The decision to err may be conscious and based on logic, or it may be unconscious. A variety of pressures such as deadlines, peer pressure, and budget factors can lead to **unsafe behavior**. Another factor that can influence such a decision is the "it won't happen to me" syndrome.

The systems failure component is an important contribution of Petersen's theory. First, it shows the potential for a **causal relationship** between management decisions or management behavior and safety. Second, it establishes management's role in accident prevention as well as the broader concepts of safety and health in the workplace.

FIGURE 3–3   Accident/incident theory.

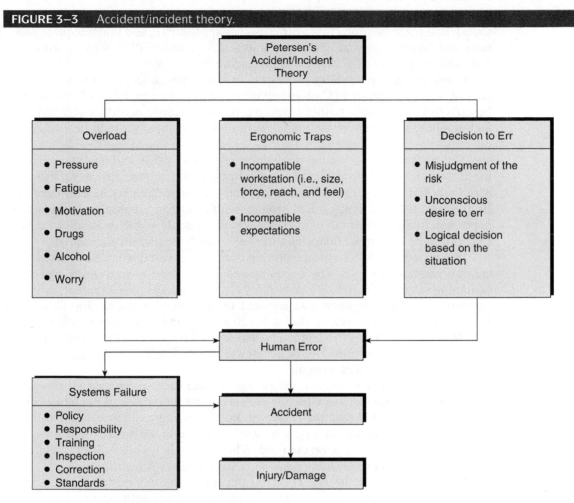

Following are just some of the different ways that systems can fail, according to Petersen's theory:

- Management does not establish a comprehensive safety policy.
- Responsibility and authority with regard to safety are not clearly defined.
- Safety procedures such as measurement, inspection, correction, and investigation are ignored or given insufficient attention.
- Employees do not receive proper orientation.
- Employees are not given sufficient safety training.

Accident/Incident Theory in Practice

Poultry Processing Corporation (PPC) processes chickens and turkeys for grocery chains. Poultry processing is a labor-intensive enterprise involving a great deal of handwork. A variety of different knives, shears, and cleavers are used. Much of the work is monotonous and repetitive. Selected parts of the overall process must be done in cold conditions.

PPC has gone to great lengths to ensure that workstations are ergonomically sound, that personal protection gear is used as appropriate, and that adequate precautions are taken to prevent illness and injuries. As a result, PPC is an award-winning company in the area of workplace safety and health.

Consequently, the poultry-processing industry was shocked when a class action lawsuit was filed against PPC on behalf of over 50 employees, all of whom claimed to be suffering from carpal tunnel syndrome (CTS). Because of PPC's excellent safety and health record, most observers felt sure that the company would be vindicated in the end.

The company's policies and procedures relating to safety and health were investigated thoroughly by consultants brought in by both PPC and the attorney for the plaintiffs. Over 100 witnesses gave depositions, and several preliminary hearings were held. By the time the trial finally rolled around, both sides had accumulated mountains of paper and filing cabinets full of evidence. Then, suddenly and without advance notice, PPC offered a substantial financial settlement, which the plaintiffs accepted.

It was one of PPC's outside consultants who discovered what had caused the increased incidence of CTS. The company had always used a centralized approach to managing safety and health. Responsibility for such tasks as measurement, inspection, correction, and investigation was assigned to the safety manager, Joe Don Huttle. Huttle had an excellent record during his 20 years in the poultry-processing industry, with the last 5 spent at PPC. In fact, he was so well respected in the industry that his peers had elected him president of a statewide safety organization. This, as it turned out, is where PPC's troubles began.

When Huttle took it over, the safety organization had experienced a three-year decline in membership and was struggling to stay afloat financially. He had been elected as "the man who could save the organization." Intending to do just that, Huttle went right to work. For months at a time he worked seven days a week, often spending as much as two weeks at a time on the road. When he was in his office at PPC, Huttle was either on the telephone or doing paperwork for the safety organization.

Within six months, he had reversed the organization's downhill slide, but not without paying a price at home. During the same six-month period, his duties at PPC were badly neglected. Measurement of individual and group safety performance had come to a standstill. The same was true of inspection, correction, investigation, and reporting.

It was during this time of neglect that the increased incidence of CTS occurred. Safety precautions that Huttle had instituted to guard against this particular problem were no longer observed properly once the workers realized that he had stopped observing and correcting them. Measurement and inspection may also have prevented the injuries had Huttle maintained his normal schedule of these activities.

PPC's consultant, in a confidential report to executive managers, cited the *accident/incident theory* in explaining his view of why the injuries occurred. In this report, the consultant said that Huttle was guilty of applying "it won't happen here" logic when he made a conscious decision to neglect his duties at PPC in favor of his duties with the professional safety organization. Of course, the employees themselves were guilty of not following clearly established procedures. However, because Huttle's neglect was also a major contributing factor, PPC decided to settle out of court.

EPIDEMIOLOGICAL THEORY OF ACCIDENT CAUSATION

Traditionally, safety theories and programs have focused on accidents and the resulting injuries. However, the current trend is toward a broader perspective that also encompasses the issue of industrial hygiene. **Industrial hygiene** concerns environmental factors that can lead to sickness, disease, or other forms of impaired health.

This trend has, in turn, led to the development of an epidemiological theory of accident causation. Epidemiology is the study of causal relationships between environmental factors and disease. The **epidemiological theory** holds that the models used for studying and determining these relationships can also be used to study causal relationships between environmental factors and accidents or diseases.[9]

Figure 3–4 illustrates the epidemiological theory of accident causation. The key components are **predispositional characteristics** and **situational characteristics**. These characteristics, taken together, can either result in or prevent conditions that may result in an accident. For example, if an employee who is particularly susceptible to peer pressure (predispositional characteristic) is pressured by his coworkers (situational characteristic) to speed up his operation, the result will be an increased probability of an accident.

Epidemiological Theory in Practice

Jane Andrews was the newest member of the loading unit for Parcel Delivery Service (PDS). She and the other members of her unit were responsible for loading 50 trucks every morning. It was physically demanding work, and she was the first woman ever

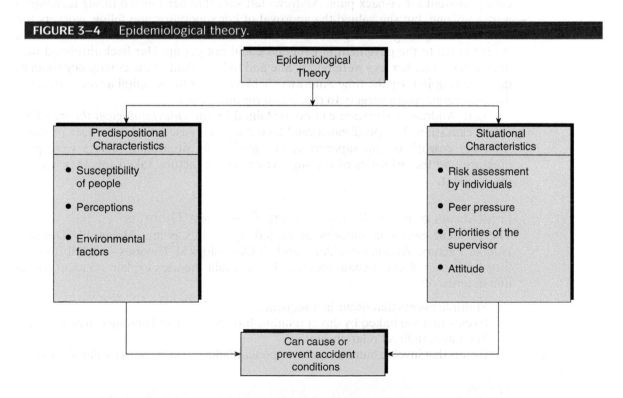

FIGURE 3–4 Epidemiological theory.

selected by PDS to work in the loading unit. She had gotten the job as part of the company's upward mobility program. She was excited about her new position because within PDS the loading unit was considered a springboard to advancement. Consequently, she was eager to do well. The responsibility she felt toward other female employees at PDS only intensified her anxiety. Andrews felt that if she failed, other women might not get a chance to try in the future.

Before beginning work in the loading unit, employees must complete two days of training on proper lifting techniques. The use of back-support belts is mandatory for all loading dock personnel. Consequently, Andrews became concerned when the supervisor called her aside on her first day in the unit and told her to forget what she had learned in training. He said, "Jane, nobody wants a back injury, so be careful. But the key to success in this unit is speed. The lifting techniques they teach in that workshop will just slow you down. You've got the job, and I'm glad you're here. But you won't last long if you can't keep up."

Andrews was torn between following safety procedures and making a good impression on her new supervisor. At first, she made an effort to use proper lifting techniques. However, when several of her coworkers complained that she wasn't keeping up, the supervisor told Andrews to "keep up or get out of the way." Feeling the pressure, she started taking the same shortcuts she had seen her coworkers use. Positive results were immediate, and Andrews received several nods of approval from fellow workers and a "good job" from the supervisor. Before long, Andrews had won the approval and respect of her colleagues.

However, after two months of working in the loading unit, she began to experience persistent lower-back pain. Andrews felt sure that her hurried lifting techniques were to blame, but she valued the approval of her supervisor and fellow workers too much to do anything that might slow her down. Finally, one day while loading a truck, Andrews fell to the pavement in pain and could not get up. Her back throbbed with intense pain, and her legs were numb. She had to be rushed to the emergency room of the local hospital. By the time Andrews checked out of the hospital a week later, she had undergone major surgery to repair two ruptured discs.

Jane Andrews's situation can be explained by the *epidemiological theory* of accident causation. The predispositional factor was her susceptibility to peer pressure from her coworkers and supervisor. The applicable situational factors were peer pressure and the priorities of the supervisor. These factors, taken together, caused the accident.

Limitations of Event-Chain Accident Causation Theories

The accident causation theories presented up to this point—Heinrich's Domino, Human Factors, Accident/Incident, and Epidemiological Theories—all fall into the broad category of event-chain theories. Event-chain theories explain accident causation in terms of:

- Multiple events that occur in a sequence.
- Events that are linked by direct relationships between and among causal factors but ignore indirect relationships.
- Events that involve human error, component failure, and/or energy-related factors.

All of these event-chain theories have value, but they also have shortcomings—shortcomings that safety and health professionals should be aware of. These shortcomings include the following:

- Neglect broader social and organizational factors.
- Do not adequately account for physical context, social context, personal values, or the dynamics of work processes when assessing human error.
- Neglect human adaptation (the tendency of more experienced workers to behave in ways that involve higher levels of risk).

The shortcomings of event-chain theories of accident causation led to the development of additional theories that attempt to take a more holistic approach to the analysis of accident causation. These additional theories are explained in the remainder of this chapter.

SYSTEMS THEORY OF ACCIDENT CAUSATION

A *system* is a group of regularly interacting and interrelated components that together form a unified whole. This definition is the basis for the **systems theory** of accident causation. This theory views a situation in which an accident may occur as a system composed of the following components: person (host), machine (agency), and **environment**.[10] The likelihood of an accident occurring is determined by how these components interact. Changes in the patterns of interaction can increase or reduce the probability of an accident.

For example, an experienced employee who operates a numerically controlled five-axis machining center in a shop environment may take a two-week vacation. Her temporary replacement may be less experienced. This change in one component of the system (person/host) increases the probability of an accident. Such a simple example is easily understood. However, not all changes in patterns of interaction are this simple. Some are so subtle that their analysis may require a team of people, each with a different type of expertise.

The primary components of the systems model are the person/machine/environment, information, decisions, risks, and the task to be performed.[11] Each of the components has a bearing on the probability that an accident will occur. The systems model is illustrated in Figure 3–5.

As this model shows, even as a person interacts with a machine within an environment, three activities take place between the system and the task to be performed. Every time a task must be performed, there is the risk that an accident may occur. Sometimes the risks are great; at other times, they are small. This is where information collection and decision making come in.

Based on the information that has been collected by observing and mentally noting the current circumstances, the person weighs the risks and decides whether to perform the task under existing circumstances. For example, say a machine operator is working on a rush order that is behind schedule. An important safety device has malfunctioned on his machine. Simply taking it off will interrupt work for only five minutes, but it will also increase the probability of an accident. However, replacing it could take up to an hour. Should the operator remove the safety guard and proceed with the task or take the time to replace it? The operator and his supervisor may assess the situation (collect information), weigh the risks, and make a decision to proceed. If their information was

FIGURE 3–5   Systems theory model.

right and their assessment of the risks accurate, the task will probably be accomplished without an accident.

However, the environment in which the machine operator is working is unusually hectic, and the pressure to complete an order that is already behind schedule is intense. These factors are **stressors** that can cloud the judgment of those collecting information, weighing risks, and making the decision. When stressors are introduced between points 1 and 3 in Figure 3–5, the likelihood of an accident increases.

For this reason, the following five factors should be considered before beginning the process of collecting information, weighing risks, and making a decision:

- Job requirements
- The workers' abilities and limitations
- The gain if the task is successfully accomplished
- The loss if the task is attempted but fails
- The loss if the task is not attempted[12]

These factors can help a person achieve the proper perspective before performing the above-mentioned tasks. It is particularly important to consider these factors when stressors such as noise, time constraints, or pressure from a supervisor may tend to cloud one's judgment.

Systems Theory in Practice

Precision Tooling Company (PTC) specializes in difficult orders that are produced in small lots, and in making corrections to parts that otherwise would wind up as expensive rejects in the scrap bin. In short, PTC specializes in doing the types of work that other companies cannot, or will not, do. Most of PTC's work comes in the form of subcontracts from larger manufacturing companies. Consequently, living up to its reputation as a high-performance, on-time company is important to PTC.

Because much of its work consists of small batches of parts to be reworked, PTC still uses several manually operated machines. The least experienced machinists operate these machines. This causes two problems. The first problem is that it is difficult for even a master machinist to hold to modern tolerance levels on these old machines. Consequently, apprentice machinists find holding to precise tolerances quite a challenge. The second problem is that the machines are so old that they frequently break down.

Complaints from apprentice machinists about the old machines are frequent. However, their supervisors consider time on the old "ulcer makers" to be one of the rites of passage that upstart machinists must endure. Their attitude is, "We had to do it, so why shouldn't you?" This was where things stood at PTC when the company won the Johnson contract.

PTC had been trying for years to become a preferred supplier for H. R. Johnson Company. PTC's big chance finally came when Johnson's manufacturing division incorrectly produced 10,000 copies of a critical part before noticing the problem. Simply scrapping the part and starting over was an expensive solution. Johnson's vice president for manufacturing decided to give PTC a chance.

PTC's management team was ecstatic! Finally, they had an opportunity to partner with H. R. Johnson Company. If PTC could perform well on this one, even more lucrative contracts were sure to follow. The top managers called a company-wide meeting of all employees. Attendance was mandatory. The CEO explained the situation as follows:

> Ladies and gentlemen, we are faced with a great opportunity. I've just signed a contract with H. R. Johnson Company to rework 10,000 parts that their manufacturing folks produced improperly. The rework tasks are not that complicated, but every part has got to go through several manual operations at the front end of the rework process. This means our manual machining unit is going to have to supply the heroes on this job. I've promised the manufacturing VP at Johnson that we would have his parts ready in 90 days. I know that's a lot to do in so short a period of time, but Johnson is in a real bind here. If we can produce on this one, they won't forget us in the future.

This put PTC's apprentice machinists on the spot. If PTC didn't perform on this contract, it would be their fault. They cursed their old machines and got to work. The CEO had said the rework tasks would not be "that complicated," but, as it turned out, the processes weren't that simple either. The problem was tolerances. Holding to the tolerances specified in the Johnson contract took extra time and a special effort on every single part. Before long, the manual machining unit was behind schedule, and management was getting nervous. The situation was made even worse by the continual breakdowns and equipment failures experienced. The harder the unit supervisor pushed, the more stressed the employees and machines became.

Predictably, it wasn't long before safety procedures were forgotten, and unreasonable risks were being taken. The pressure from management, the inexperience of the apprentice machinists, and the constant equipment failures finally took their toll. In a hurry to get back on schedule, and fearing that his machine would break down again, one machinist got careless and ran his hand into the cutter on his milling machine. By the time the machine had been shut down, his hand was badly mutilated. In the aftershock of this accident, PTC was unable to meet the agreed-upon completion schedule. Unfortunately, PTC did not make the kind of impression on H. R. Johnson's management team that it had hoped.

This accident can be explained by the systems theory. The *person–machine–environment* chain has a direct application in this case. The person involved was relatively inexperienced. The machine involved was old and prone to breakdowns. The environment was especially stressful and pressure packed. These three factors, taken together, resulted in this serious and tragic accident.

COMBINATION THEORY OF ACCIDENT CAUSATION

There is often a degree of difference between any theory of accident causation and reality. The various models presented with their corresponding theories in this chapter attempt to explain why accidents occur. For some accidents, a given model may be very accurate. For others, it may be less so. Often the cause of an accident cannot be adequately explained by just one model or theory. Thus, according to the **combination theory**, the actual cause may combine parts of several different models. Safety personnel should use these theories as appropriate for both accident prevention and accident investigation. However, they should avoid the tendency to try to apply one model to all accidents.

Combination Theory in Practice

Crestview Grain Corporation (CGC) maintains 10 large silos for storing corn, rice, wheat, barley, and various other grains. Because stored grain generates fine dust and gases, ventilation of the silos is important. Consequently, all of CGC's silos have several large vents. Each of these vents uses a filter similar to the type used in home air conditioners that must be changed periodically.

There is an element of risk involved in changing the vent filters because of two potential hazards. The first hazard comes from unvented dust and gases that can make breathing difficult, or even dangerous. The second hazard is the grain itself. Each silo has a catwalk that runs around its inside circumference near the top. These catwalks give employees access to the vents that are also near the top of each silo. The catwalks are almost 100 feet above ground level, they are narrow, and the guardrails on them are only knee high. A fall from a catwalk into the grain below would probably be fatal.

Consequently, CGC has well-defined rules that employees are to follow when changing filters. Because these rules are strictly enforced, there had never been an accident in any of CGC's silos—that is, not until the Juan Perez tragedy occurred. Perez was not new to the company. At the time of his accident, he had worked at CGC for over five years. However, he was new to the job of silo maintenance. His inexperience, as it turned out, would prove fatal.

It was time to change the vent filters in silo number 4. Perez had never changed vent filters himself. He hadn't been in the job long enough. However, he had served as the required "second man" when his supervisor, Bao Chu Lai, had changed the filters in silos 1, 2, and 3. Because Chu Lai was at home recuperating from heart surgery and would be out for another four weeks, Perez decided to change the filters himself. Changing the filters was a simple enough task, and Perez had always thought the second-man concept was overdoing it a little. He believed in taking reasonable precautions as much as the next person, but in his opinion, CGC was paranoid about safety.

Perez collected his safety harness, respirator, and four new vent filters. Then he climbed the external ladder to the entrance–exit platform near the top of silo number 4. Before going in, Perez donned his respirator and strapped on his safety harness. Opening the hatch cover, he stepped inside the silo onto the catwalk. Following procedure, Perez attached a lifeline to his safety harness, picked up the new vent filters, and headed for the first vent. He changed the first two filters without incident. It was while he was changing the third filter that tragedy struck.

The filter in the third vent was wedged in tightly. After several attempts to pull it out, Perez became frustrated and gave the filter a good jerk. When the filter suddenly broke loose, the momentum propelled Perez backward and he toppled off the catwalk. At first, it appeared that his lifeline would hold, but without a second person to pull him up or call for help, Perez was suspended by only the lifeline for over 20 minutes. He finally panicked, and in his struggle to pull himself up, knocked open the buckle of his safety harness. The buckle gave way, and Perez fell over 50 feet into the grain below. The impact knocked off his respirator, the grain quickly enveloped him, and Perez was asphyxiated.

The accident investigation that followed revealed that several factors combined to cause the fatal accident—the combination theory. The most critical of these factors were as follows:

- Absence of the supervisor
- Inexperience of Perez
- A conscious decision by Perez to disregard CGC's safety procedures
- A faulty buckling mechanism on the safety harness
- An unsafe design (only a knee-high guardrail on the catwalk)

BEHAVIORAL THEORY OF ACCIDENT CAUSATION

The behavioral theory of accident causation and prevention is often referred to as **behavior-based safety (BBS)**. BBS has both proponents and critics. One of the most prominent proponents of BBS is E. Scott Geller, a senior partner of Safety Performance Solutions, Inc., and a professor of psychology. It is appropriate that Geller is a professional psychologist because BBS is the application of behavioral theories from the field of psychology to the field of occupational safety.

According to Geller, there are seven basic principles of BBS: (1) intervention that is focused on employee behavior; (2) identification of external factors that will help understand and improve employee behavior (from the perspective of safety in the workplace); (3) direct behavior with activators or events antecedent to the desired behavior, and motivation of the employee to behave as desired with incentives and rewards that will follow the desired behavior; (4) focus on the positive consequences that will result from the desired behavior as a way to motivate employees; (5) application of the scientific method to improve attempts at behavioral interventions; (6) use of theory to integrate information rather than to limit possibilities; and (7) planned interventions with the feelings and attitudes of the individual employee in mind.[13]

Those who have studied psychology will recognize BBS as an innovative and practical application of standard behavioral theory to the field of occupational safety. These theories are relevant in any situation in which certain types of human behaviors are desired while others are to be avoided. Positive reinforcement in the form of incentives and rewards is used to promote the desired (safe) behaviors and to discourage undesirable (unsafe) behaviors.

Behavioral Theory in Action

Mark Potter is the safety manager for Excello Corporation. Several months ago, he became concerned because employees seemed to have developed a lax attitude toward wearing hard hats. What really troubled Potter was that there was more than the usual potential for head injuries because of the type of work done in Excello's plant, and he had personally witnessed two near misses in less than a week. An advocate of BBS, he decided to apply the ABC model in turning around this unsafe behavior pattern.

His first step was to remove all the old "Hard Hat Area" signs from the plant and replace them with newer, more noticeable signs. Then he scheduled a brief seminar on head injuries and cycled all employees through it over a two-week period. The seminar took an unusual approach. It told a story of two employees. One was in a hospital bed surrounded by family members he did not even recognize. The other was shown enjoying a family outing with happy family members. The clear message of the video was "the difference between these two employees is a hard hat." These two activities were the antecedents to the behavior he hoped to produce (all employees wearing hard hats when in a hard hat area).

The video contained a powerful message and it had the desired effect. Within days, employees were once again disciplining themselves to wear their hard hats (the desired behavior). The consequence was that near misses stopped, and no head injuries have occurred at Excello in months. The outcome of this is that Excello's employees have been able to continue enjoying the fruits of their labor and the company of loved ones.

INDIVIDUAL FACTORS AND ACCIDENT CAUSATION

One of the most pernicious causes of accidents on the job is chemicals—but not the kind industrial hygienists generally concern themselves with.[14] The chemicals alluded to here are the illicit drugs and alcohol used by employees. Drugs and alcohol are the root cause or contributing cause of many accidents on the job every year. Consequently, safety professionals need to be on guard for employees who are drug and alcohol abusers.

This is why so many companies implement drug-free workplace programs. In fact, since 1989 federal contractors have been required to do so. Such programs typically include the following components: drug-free workplace policy, supervisory training, employee education, employee assistance programs (EAPs), and alcohol and drug testing.

Establishing drug-free workplace programs is typically the responsibility of the human resources department. However, safety and health professionals should be aware of the workplace problems that can be caused by alcohol and drug abuse. Further, if a cross-functional team of representatives from various departments is convened by the human resources department for the purpose of developing a drug-free workplace program, the chief safety and health professional for the organization should be a member of that team. An invisible problem in today's workplace is **clinical depression**. People who suffer from clinical depression are seriously impaired and, as a result, they pose a clear and present safety risk to themselves, fellow workers, and their employer.[15] Mental health professionals estimate that up to 10 percent of the adult population in the United States suffers from clinical depression. This translates to 1 in every 20 people on the job.

The causes of clinical depression are many and varied, but the most common causes are biological (too few or too many of the brain chemicals known as **neurotransmitters**), cognitive (negative thought processes), genetic (family history of depression), and concurring illnesses (strokes, cancer, heart disease, Alzheimer's, and other diseases can increase the incidence of depression).

Warning Signs
Safety and health professionals are not mental health professionals and should not attempt to play that role. However, they should be alert to the warning signs of clinical depression in employees. These signs are as follows:

- Persistent dreary moods (sadness, anxiety, and nervousness)
- Signs of too little sleep

- Sleeping on the job or persistent drowsiness
- Sudden weight loss or gain
- General loss of interest, especially in areas of previous interest
- Restlessness, inability to concentrate, or irritability
- Chronic physical problems (headaches, digestive disorders, etc.)
- Forgetfulness or an inability to make simple decisions
- Persistent feelings of guilt
- Feelings of low self-worth
- Focus on death or talk of suicide

Safety and health professionals who recognize any or all of these symptoms in an employee should avoid the natural human tendency to *help the employee deal with the problems*. Rather, the appropriate action is to get the employee into the hands of competent mental health professionals right away. The best way to do this is to approach the employee's supervisor and recommend that he or she refer the employee to the organization's EAP or to the human resources department. Other individual factors that can be the cause or a contributing cause of accidents in the workplace include obesity, fatigue, memory, and personality. Obesity not only restricts the movement and potential reaction time of individuals which can, in turn, cause accidents, but it can be a major contributing factor in overexertion related accidents. Obesity has become such an all-pervasive problem in the United States that this potential cause of accidents is treated in greater detail in a separate section later in this chapter.

Fatigue can cause or be a contributing factor in accidents because it impairs judgment, attention, and reasoning power. Memory can be a causal factor in accidents when a worker simply forgets to employ the safe work learned. Personality can contribute causally to accidents in several ways. Some people have personalities that lend themselves to following safety rules and employing safe work practices, while others have personalities that lend themselves to laxity in these areas. Some people are hot-tempered and prone to act rashly. Others are curious and, as a result, prone to disregard safe work practices and experiment. Some people are attentive, while others are prone to let their minds wander. The list of personal factors that can affect workplace safety, although not infinite, is long.

Sources of Help

Because clinical depression in employees has become such an all-pervasive problem that increases the risk of accidents and injuries on the job, safety and health professionals need to learn all they can about this problem and keep up-to-date with the latest information concerning it. The following sources may help:

Employee Assistance Professionals Association, 703-522-6272, www.eapa.org.
National Institute of Mental Health, 800-421-4211, www.nimh.nih.gov.
National Mental Health Association, 800-969-NMHA, www.nmha.org.

MANAGEMENT FAILURES AND ACCIDENT CAUSATION

One of the leading causes of accidents in the workplace is the failure of management to do its part to ensure a safe and healthy work environment.[16] Different levels of management have different levels of responsibility. The level of management with the most direct, hands-on, day-to-day responsibility for workplace safety and health is the supervisory level. Supervisors play a critical role in making sure that employees work in a safe and healthy environment.

Role of the Supervisor in Workplace Safety and Health

Safety and health professionals cannot do their jobs effectively without the full cooperation and day-to-day assistance of first-line supervisors. Supervisors and safety professionals must be partners when it comes to providing a safe and healthy workplace for employees. Supervisors should be assigned responsibility for the work environment and for the safety of employees in their units. Safety and health professionals should be readily available to help supervisors fulfill this responsibility.

Key responsibilities of supervisors relating to workplace safety and health include the following:

- Orienting new employees to the safe way to do their jobs
- Ensuring that new and experienced employees receive the safety and health training they need on a continual basis
- Monitoring employee performance and enforcing safety rules and regulations
- Assisting safety and health professionals in conducting accident investigations
- Assisting safety and health professionals in developing accident reports
- Keeping up-to-date on safety issues
- Setting a positive example for employees that says "the safe way is the right way"

Typical Management Failures That Cause Accidents

Management failures represent a major cause of accidents on the job. If management is serious about providing a safe and healthy work environment for employees it must (1) show employees that safe and healthy work practices are expected by including such practices in job descriptions, monitoring employee work practices, and setting an example of safe and healthy work practices; (2) provide training in how to work safely, including orientation training for new employees as well as ongoing updated training for experienced employees; (3) include safe and healthy work practices as criteria in the periodic performance appraisals of employees; and (4) reinforce safe and healthy work practices by rewarding and recognizing employees who use them. Common examples of management failures include the following:

POOR HOUSEKEEPING OR IMPROPER USE OF TOOLS, EQUIPMENT, OR FACILITIES.

Management either has not developed the necessary requirements or has but does not enforce them. The management failure in this case could be lack of safety procedures (failure to let employees know the expectations), lack of training (failure to give employees the knowledge and skills they need to work safely), or failure to properly supervise (failure to monitor employee actions).

PRESSURE TO MEET DEADLINES.

Sometimes management has developed a good safety and health policy, established good safety and health procedures, built safety and health expectations into job descriptions and performance appraisals, and provided the necessary training only to put all this aside when a rush order comes in. This may be the most problematic of the many different types of management failures that can occur because it can undermine all of the organization's safety and health efforts. When management allows safety and health procedures to be ignored or, worse yet, encourages them to be ignored to speed up production in the short run, employees soon get the message that safety and health are important only when there is no rush. This is an example of management failing to set the proper example.

OBESITY AND ACCIDENT CAUSATION

Researchers at Ohio State University found that extremely obese people are more likely than normal-weight people to injure themselves.[17] This is bad news for what an Australian study conducted at Queensland University calls "sedentary workplaces"—those that involve a lot of sitting at desks. It is bad news because the Australian study concluded that the more the people sit at desks during the workday, the more likely they are to be overweight.

Obesity has long been associated with such chronic diseases as high blood pressure, coronary heart disease, diabetes, and certain types of cancer, but these studies now tie it to workplace injuries too. The ramifications for safety and health professionals are profound. The World Health Organization estimates that there are more than 300 million obese people worldwide. In the industrialized nations of the world—nations such as the United States—the number of people considered obese is growing rapidly. These studies used a standard body mass index (BMI) score of 30 or above to define obesity.

In the study conducted by Ohio State University, researchers collected data on more than 2,500 adults. The data show that 26 percent of extremely obese male subjects reported personal injuries. The percentage for extremely obese women was only slightly lower at approximately 22 percent. Researchers compared these percentages with those for normal-weight people and found a noticeable difference in the percentages of males and females who reported injuries (17 percent for males and 12 percent for women). Researchers used a BMI of 18.5 to 24.9 to define "normal weight."

The most common causes of injuries to obese people were the result of overexertion (35.2 percent) and falls (29.9 percent). Underweight people—BMI of 18.5 or less—reported the fewest number of injuries.

SWISS CHEESE MODEL OF ACCIDENT CAUSATION

Originally presented by James Reason and Dante Orlandella in the 1990s, the *Swiss Cheese Model* is now often referred to as the *cumulative act effect*. This model recommends placing numerous defensive layers of preventive/precautionary measure between hazards and the people and property that might be harmed by those hazards. Hence, the model led to the concept of layered security that is now used in computer systems and by the military for planning defense in depth (primary, secondary, and tertiary defensive positions).

Reason posited that the majority of workplace accidents can be attributed to one or more of the following groups of causes: organizational factors, supervisory factors, preconditions, and specific human acts. Because there can be more than one cause or contributing cause of accidents, there must be successive layers of preventive measures. His model is called the "Swiss Cheese Model" because it imagines standing separate layers of Swiss cheese slices (preventive and precautionary measures) between hazards and the harm they might cause. Just as Swiss cheese slices have holes in them, all layers of preventive measures have weaknesses or potential failure points.

The practical application of the concept is that if a hazard makes it past one layer (one slice of Swiss cheese), it will be impeded by the next and the next and the next. These successive layers are the essence of the Swiss Cheese Model. Of course, no system or preventive measure is fool-proof. Consequently, Reason posited that a breakdown in the various layers occurs when the holes in the Swiss cheese slice line up. In other words, when the weaknesses of each precautionary layer line up in a way that allows a hazard to break through all layers.

SUMMARY OF ACCIDENT CAUSATION MODELS AND APPLICATIONS

There is much debate among safety professionals over the viability of the various accident causation models currently available. Is one better than the others? Should one be used and the others discarded? The answer to these questions must be, as is often the case, "it depends." Some models are better than others in some applications. This section summarizes the specific situations in which each model has useful applications.

- *Domino Theory.* Much of what Heinrich posited has since been discounted. However, his theory still has application for the purpose of context and historical understanding of the evolution of thought in accident causation.
- *Human Factors Theory.* Human error is a major cause of workplace accidents. Consequently, when investigating accidents and near misses, the three factors posited in this theory can provide valuable guidance. It is always wise to consider overload, inappropriate response/incompatibility, and inappropriate activities when investigating the cause of a workplace accident.
- *Accident/Incident Theory.* Since this theory is an extension of the human factors theory, with the addition of ergonomic factors, decision to err, and systems failure, its best application is the same as that of the human factors theory: guidance for accident investigations. In fact, the better approach is to combine the two theories and consider all of the factors posited in both.
- *Epidemiological Theory.* This theory posits that the same models used to study relationships between environmental factors and disease can be used for determining relationships between environmental factors and accidents. This is one of the least practical, more academic of the various models. Consequently, its use in the workplace is limited. However, it is useful in conducting studies that make connections between environmental factors and accidents.
- *Systems Theory.* This theory views any situation in which an accident might occur holistically as a system with three components: person (host), machine (agency), and environment. The systems theory has value in reminding safety professionals to look broadly when considering accident causation. When trying to determine the cause of an accident as well as contributing causes, it is easy to get so caught up in the details that the larger picture is ignored or neglected.
- *Combination Theory.* This theory posits that no one model can explain all accidents and their causes. Rather, it is better to select elements of more than one theory when considering accident causation. This theory has validity since all of the others have both strengths and weaknesses. In practical application, this theory is probably the one applied most frequently.
- *Behavioral Theory.* This theory is based on seven principles, all of which have validity. This theory has particular application whenever certain human

behaviors are desirable and other behaviors are not. It is particularly useful for use in accident prevention activities in which the behavior of workers must be modified to prevent accidents.

- **Individual Factors.** This theory is especially applicable for use by front-line supervisors who interact closely with workers—those who are able to observe individual behavior in the moment and how it is affected by such factors as drugs, depression, attitude, obesity, and personality. Supervisors should know what the various individual factors associated with accident causation are and how to intervene when these factors might lead to an accident. For example, a supervisor should never allow an individual who appears to be impaired by drugs, alcohol, or depression to continue working.

- **Management Failures.** Ultimately, it is the responsibility of an organization's management team to provide a safe and healthy work environment, establish expectations of safe work practices, reward working safely and take action against those who do not, and factor safety into performance evaluations. This is so fundamental to safety and health in the workplace that its absence can cancel out the safety professional's efforts to prevent accident and injuries.

- **Swiss Cheese Model.** This model is best used for designing multiple-layered defenses against known hazards. It is more of an accident prevention model than one for determining accident causation. However, before one can develop effective preventive measures, it is first necessary to determine what various hazards/causes might be present or introduced in the work environment. Consequently, accident causation determinations must be part of the Swiss Cheese Model.

Key Terms and Concepts

- Accident/incident theory
- Ancestry
- Axioms of Industrial Safety
- Behavior-based safety (BBS)
- Causal relationship
- Central factor
- Clinical depression
- Combination theory
- Domino theory
- Environment
- Environmental factors
- Epidemiological theory
- Ergonomic traps
- Hazardous condition
- Human error
- Human factors theory
- Inappropriate activities
- Inappropriate responses
- Industrial hygiene
- Industrial place accidents
- Internal factors
- Mechanical hazards
- Neurotransmitters
- Obesity
- Overload
- Physical hazards
- Preceding factors
- Predispositional characteristics
- Situational characteristics
- Situational factors
- Social environment
- Stressors
- Swiss Cheese Model
- Systems theory
- Unsafe acts
- Unsafe behavior

Review Questions

1. Explain the domino theory of accident causation, including its origin and its impact on more modern theories.
2. What were the findings of Herbert W. Heinrich's 1920s study of the causes of industrial accidents?
3. List five of Heinrich's Axioms of Industrial Safety.
4. Explain the following concepts in the domino theory: preceding factor; central factor.

5. What are the three broad factors that lead to human error in the human factors theory? Briefly explain each.
6. Explain the systems failure component of the accident/incident theory.
7. What are the key components of the epidemiological theory? How does their interaction affect accident causation?
8. Explain the systems theory of accident causation.
9. What impact do stressors have in the systems theory?
10. List five factors to consider before making workplace decisions that involve risk.
11. Explain the principles of behavior-based safety.
12. What is the role of the safety and health professional with regard to handling employees who might be drug or alcohol abusers?
13. List the warning signs of clinical depression.
14. What must management do if it is serious about providing a safe and healthy work environment for employees?
15. Explain the connection between obesity and injuries.
16. Explain the Swiss Cheese Model from your perspective of accident prevention.

Endnotes

1. National Safety Council, *Accident Facts* (Chicago: National Safety Council, 2017), 23.
2. Ibid., 24.
3. Ibid., 26.
4. Industrial Foundation for Accident Prevention. Retrieved from ifap.asn.au on March 11, 2013.
5. Ibid.
6. Ibid.
7. Ibid.
8. Ibid.
9. M. A. Topf, "Chicken/Egg/Chegg!" *Occupational Health Safety* 68, no. 6: 60–66. 2002.
10. D. L. Goetsch, *Implementing Total Safety Management* (Upper Saddle River, NJ: Prentice Hall, 1998), 227.
11. Ibid.
12. Ibid.
13. E. S. Geller, "Behavior-Based Safety: Confusion, Controversy, and Clarification," *Occupational Health & Safety* 68, no. 1: 40–49.
14. Stephen G. Minter, "The Safety Threat from Within," *Occupational Hazards* 64, no. 4: 8. 2001.
15. Todd Nighswonger, "Depression: The Unseen Safety Risk," *Occupational Hazards* 64, no. 4: 38–42. 2001.
16. Retrieved from online.misu.kodak.edu/19577/AccCautrac.htm on March 5, 2017.
17. Occupational Health & Safety Online, *Obesity Studies Focus on Injuries, Sedentary Workplaces*. Retrieved from ohsonline.com/stevens/ohspub.nsf/d3d5b4f93822b6e8625670c006dbc58/d9 on April 5, 2017.

CHAPTER

4

ROLES AND PROFESSIONAL CERTIFICATIONS FOR SAFETY AND HEALTH PROFESSIONALS

Learning Objectives

- Summarize the composition of modern workplace safety and health teams
- Explain the role of the safety and health manager
- Explain the role engineers can play in workplace safety and health
- Describe the role of the industrial hygienist in workplace safety and health
- Describe the role of the health physicist in safety and health
- Explain the role of the occupational physician in workplace safety and health
- Explain the role of the occupational health nurse in workplace safety and health
- Describe the role of the risk manager as it relates to workplace safety and health
- Explain the various certifications that are available to safety and health professionals
- Explain the role of employees in workplace safety and health
- Describe the emerging role of safety and health professionals

This book was designed for use by prospective and practicing safety and health managers. People with such titles are typically responsible to higher management for the safety and health of a company's workforce. Modern safety and health managers seldom work alone. Rather, they usually head a team of specialists that may include engineers, physicists, **industrial hygienists**, occupational physicians, and occupational health nurses.

It is important for safety and health managers today to understand not only their roles but also the roles of all members of the safety and health team. This chapter provides the information that prospective and practicing safety and health managers need to know about the roles of safety personnel in the age of high technology and the certifications that they need.

MODERN SAFETY AND HEALTH TEAMS

The issues that concern modern safety and health managers are multifaceted and complex. They include such diverse issues as stress; explosives; laws, standards, and codes; radiation; acquired immunodeficiency syndrome (AIDS); product safety and liability; ergonomics; ethics; automation; workers' compensation; and an ever-changing multitude of others.

It would be unreasonable to expect one person to be an expert in all the many complex and diverse issues faced in the modern workplace. For this reason, the practice of safety and health management in the age of high technology has become a team sport. Figure 4–1 illustrates the types of positions that may comprise a safety and health team. In the remaining sections of this chapter, the roles, duties, responsibilities, and relationships of members of the safety and health team are described. In small companies, one person may have to wear several of these hats and contract for nursing services.

SAFETY AND HEALTH MANAGER

The most important member of the safety and health team is its manager. Companies that are committed to providing a safe and healthy workplace employ a **safety and health manager** at an appropriate level in the corporate hierarchy. The manager's position in the hierarchy is an indication of the company's commitment and priorities. This, more than anything else, sets the tone for a company's safety and health program.

In times past, companies with a highly placed safety and health manager were rare. However, the passage of the Occupational Safety and Health Act (OSH Act) in 1970 (see Chapter 5) began to change this. The OSH Act, more than any other single factor, put teeth in the job descriptions of safety and health professionals. Occupational Safety

FIGURE 4–1 A modern safety and health team.

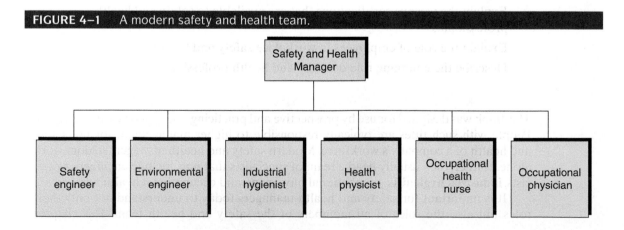

and Health Administration (OSHA) standards, on-site inspections, and penalties have encouraged a greater commitment to safety and health. Environmental, liability, and workers' compensation issues have also had an impact, as has the growing awareness that providing a safe and healthy workplace is the right thing to do from both an ethical and a business perspective.

Job of the Safety and Health Manager

The job of the safety and health manager is complex and diverse. Figure 4–2 is an example of a job description for such a position. The description attests to the diverse nature of the job. Duties range from hazard analysis, accident reporting, standards and compliance, record keeping, training to emergency planning, and so on.

The minimum educational requirement set by Poultry Processing Inc. (PPI) is an occupational certificate from a community college, with the full associate of science or applied science degree preferred. Preference is given to applicants with a bachelor's degree in specifically identified fields.

Role in the Company Hierarchy

The safety and health manager described in Figure 4–2 reports to PPI's local plant manager and has line authority over all other members of the safety and health team. This and the duties set forth in the job description are evidence of the company's commitment to safety and health and show that PPI is large enough to have a dedicated safety and health manager.

In some companies, the safety and health manager may also have other duties, such as those of a production manager or personnel manager. In these cases, the other members of the safety and health team, like those shown in Figure 4–1, are not normally company employees. Rather, they are available to the company on a part-time or consultative basis as needed. The safety and health manager's role in a company depends in part on whether his or her safety and health duties are full-time or are in addition to other duties.

Another role determinant is the issue of authority. Does the safety and health manager have line or staff authority? **Line authority** means that the safety and health manager has authority over and supervises certain employees (i.e., other safety and health personnel). **Staff authority** means that the safety and health manager is the staff person responsible for a certain function, but he or she has no line authority over others involved with that function.

Those occupying staff positions operate like internal consultants—that is, they may recommend, suggest, and promote, but they do not have the authority to order or mandate. This is typically the case with safety and health managers. Even managers with line authority over other safety and health personnel typically have a staff relationship with other functional managers (e.g., personnel, production, or purchasing). For example, consider the following safety and health-related situations:

- A machine operator continually creates unsafe conditions by refusing to practice good housekeeping.
- A certain process is associated with an inordinately high number of accidents.
- A new machine is being purchased that has been proven to be unsafe at other companies.

FIGURE 4–2 Safety and health manager job description.

POULTRY PROCESSING, INC.
Highway 90 West
Crestview, FL 32536
Vacancy Announcement

Position Title: Safety and Health Manager

Position Description: The Safety and Health Manager for PPI is responsible for establishing, implementing, and managing the company's overall safety and health program. The position reports to the local plant manager. Specific duties include the following:

- Establish and maintain a comprehensive companywide safety and health program.
- Assess and analyze all departments, processes, and materials for potential hazards.
- Work with appropriate personnel to develop, implement, monitor, and evaluate accident prevention and hazard control strategies.
- Ensure companywide compliance with all applicable laws, standards, and codes.
- Coordinate the activities of all members of the company's safety and health program.
- Plan, implement, and broker, as appropriate, safety and health-related training.
- Maintain all required safety and health-related records and reports.
- Conduct accident investigations as necessary.
- Develop and maintain a companywide emergency action plan (EAP).
- Establish and maintain an ongoing safety promotion effort.
- Analyze the company's products from the perspectives of safety, health, and liability.

Qualifications Required: The following qualifications have been established by the PPI management team with input from all levels and all departments.

- *Minimum Education.* Applicants must have at least a one-year community college certificate or an associate of science or applied science degree in industrial safety or a closely related degree (AS holders will begin work at a salary 15 percent higher than a certificate graduate).
- *Preferred Education.* Applicants with a bachelor's degree in any of the following major fields of study will be given first priority: industrial safety and health, industrial technology, industrial management, manufacturing technology, engineering technology, and related. Degree programs in these fields must include at least one three-semester or five-quarter-hour course in industrial or occupational safety and health.

In the first example, the safety and health manager could recommend that the employee be disciplined but could not normally undertake or administer disciplinary measures. In the second example, the safety and health manager could recommend that the process be shut down until a thorough analysis could be conducted, hazards identified, and corrective measures taken. However, the manager would rarely have the authority to order the process to be shut down. In the final example, the safety and health manager could recommend that an alternative machine be purchased, but he or she would not normally have the authority to stop the purchase.

Maintaining a safe and healthy workplace while playing the role of internal consultant is often the greatest challenge of safety and health managers. It requires managers to be resourceful, clever, astute with regard to corporate politics, good at building relationships, persuasive, adept at trading for favors, credible, and talented in the development and use of influence.

Problems Safety and Health Managers Face

As if the diversity and complexity of the job were not enough, there are a number of predictable problems that safety and health managers are likely to face. These problems are discussed in the following paragraphs.

Lack of Commitment

Top management may go along with having a company-wide safety and health program because they see it as a necessary evil. The less enthusiastic may even see safety and health as a collection of government regulations that interfere with **profits**. Although this is less often true now than it has been in the past, safety and health professionals should be prepared to confront a less than wholehearted commitment in some companies.

Production versus Safety

Industrial firms are in business to make a profit. They do this by producing or processing products. Therefore, anything that interferes with production or processing is likely to be looked on unfavorably. At times, a health or safety measure will be viewed by some as interfering with **productivity**. A common example is removal of safety devices from machines as a way to speed production. Another is running machines until the last possible moment before a shift change rather than shutting down with enough time left to perform routine maintenance and housekeeping tasks.

The modern marketplace has expanded globally and, therefore, become intensely competitive. To survive and succeed, today's industrial firm must continually improve its productivity, quality, cost, image, response time, and service. This sometimes puts professionals who are responsible for safety and health at odds with others who are responsible for productivity, quality, cost, and response time.

Sometimes, this cannot be avoided. At other times, it is the fault of a management team that is less than fully committed to safety and health. However, sometimes the fault rests squarely on the shoulders of the safety and health manager. This is because one of the most important responsibilities of this person is to convince higher management, middle management, supervisors, and employees that, in the long run, the safe and healthy way of doing business is also the competitive, profitable way of doing business. The next section explains several strategies for making this point.

Company-Wide Commitment to Safety and Health

In many cases, safety and health managers have been their own worst enemy when it comes to gaining a company-wide commitment. The most successful are those who understand the goals of improved productivity, quality, cost, image, service, and response time and are able to convey the message that a safe and healthy workplace is the best way to accomplish these goals. The least successful are those who earn a reputation for being grumpy in-house bureaucrats who quote government regulations chapter and verse but know little and care even less about profits. Unfortunately, in the past, there have been too many safety and health managers who fall into the latter category. This is not the way to gain a company-wide commitment to safety and health, but it is a sure way to engender resentment.

Lack of Resources

Safety and health managers are like other managers in an organization in that they must compete for the resources needed to do the job. Often they find that their departments rank lower in priority than the production and operations departments (at least until a disaster occurs). Safety and health managers need to become proficient in showing the financial benefits of a safe workplace.

Today's safety and health manager must understand the bottom-line concerns of management, supervisors, and employees and be able to use these concerns to gain a commitment to safety and health. Figure 4–3 illustrates the essential message that **competitiveness** comes from continually improving a company's productivity, quality, cost, image, service, and response time. These continual improvements can be achieved and maintained best in a safe and healthy work environment.

Safety and health managers should use this message to gain a commitment from management and employees. Following are some strategies that can be used to get the point across.

FIGURE 4–3   Factors that produce competitiveness.

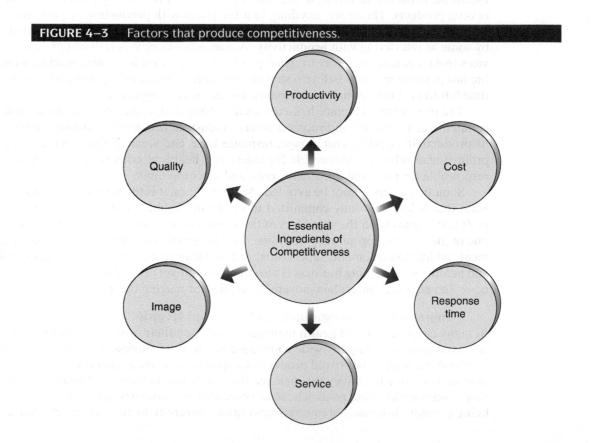

Productivity, Quality, Cost, and Response Time

These four factors, taken together, are the key to productivity in the age of high technology and global competitiveness. The most productive company is the one that generates the most output with the least input. Output is the company's product. Input is any resource—time, talent, money, technology, and so on—needed to produce the product. **Quality** is a measure of reliability and customer satisfaction. **Cost** is the amount of money required to purchase the item. If all other factors are equal, customers will select the product that costs less. **Response time** is the amount of time that elapses between an order being placed and the product being delivered.

To compete in the global marketplace, industrial companies must continually improve these four factors. At the most fundamental level, successfully competing in the global marketplace means having the best people and the best technology and getting the most out of both by applying the best management strategies.

Safety and health managers who understand this can use their knowledge to gain a commitment to their programs. In attempting to do so, the following five points are helpful:

- If it is important to attract and keep the best people, a safe and healthy workplace will help.
- If it is important to get the most out of talented people, it must be important to keep them safe and healthy so that they are functioning at peak performance levels (e.g., the best technician in the world can't help when he or she is out of action as the result of an injury or illness).
- Employees cannot concentrate fully on quality when they are concerned for their safety and health.
- Keeping industrial technologies up-to-date requires the continual investment of funds. Profits that are siphoned off to pay the costs associated with accidents, emergencies, and health problems cannot be reinvested in the latest technologies needed to stay competitive.
- With the skyrocketing costs of medical care, workers' compensation, and litigation, it costs less to prevent accidents than to pay for them.

Image and Service

Image and **service** are also important factors in the competitiveness equation. Of the two, image relates more directly to safety and health.

In today's intensely competitive marketplace, a company's image, internal and external, can be a deciding factor in its ability to succeed. Companies that establish a solid internal image in terms of safety and health will find it easier to attract and keep the best employees. Companies that establish a solid external image with regard to environmental and product safety issues will find it easier to attract and retain customers.

Gaining a full and real commitment to safety and health in the workplace is one of the most important roles of the safety and health manager. Traditionally, safety and health managers have argued their cases from the perspectives of ethics or government mandates. The ethical argument is as valid now as it has always been and should continue to be used.

However, in today's competition-driven workplace, managers responsible for the bottom line may resent arguments that are based on government mandates and regulations. On the other hand, these same managers may respond positively if they can be shown that resources invested in safety and health can actually improve a company's competitiveness. Using the points made earlier about productivity, quality, cost, image, and response time can go a long way in helping to gain management commitment to safety and health.

Education and Training for Safety and Health Managers

Advances in technology, new federal legislation, the potential for costly litigation, and a proliferation of standards have combined to make the job of safety and health professionals more complex than ever before. These factors have correspondingly increased the importance of education and training for safety and health managers. The ideal formula for safety and health professionals is **formal education** prior to entering the profession supplemented by **in-service training** on a lifelong basis afterward.

Universities, colleges, and community colleges across the country have responded to the need for formal education for safety and health managers as well as other safety and health personnel. Many community colleges offer occupational certificates and **associate degrees** in applied science or science degrees with such program titles as industrial safety, occupational safety, environmental technology, safety and health management, and industrial hygiene.

Universities have responded to the need for formal education by making safety and health-related courses either optional or required parts of such **bachelor's** (or baccalaureate) **degree** programs as industrial technology, manufacturing technology, engineering technology, **industrial engineering** technology, industrial management, and industrial engineering. Some universities offer bachelor's degrees in industrial safety and health, occupational safety management, and industrial hygiene.

Formal education provides the foundation of knowledge needed to enter the profession. Having begun a career as a safety and health manager, the next challenge is keeping up as the laws, regulations, standards, and overall body of knowledge relating to safety and health grow, change, and evolve.

In-service training, ongoing interaction with professional colleagues, and continued reading of professional literature are effective ways to stay current. New safety and health managers should move immediately to get themselves "plugged into" the profession. This means joining the appropriate professional organizations, becoming familiar with related government agencies, and establishing links with relevant standards organizations. The next section covers agencies and organizations that can be particularly helpful to safety and health managers.

Helpful Agencies and Organizations

Numerous agencies and organizations are available to help the safety and health manager keep up-to-date. These agencies and organizations provide databases, training, and professional literature. There are professional societies, trade associations, scientific organizations, certification boards, service organizations, and emergency service organizations. Those listed here represent only a portion of those available.

- **Certification boards.** Professional certification is an excellent way to establish one's status in the field of safety and health. To qualify to take a certification

FIGURE 4–4   Professional certification boards.

- **Board of Certified Safety Professionals of America**
  208 Burwash Ave.
  Savoy, IL 61874
  217-359-9265
  http://www.bcsp.org/

- **American Board of Industrial Hygiene**
  6015 W. St. Joseph, Suite 102
  Lansing, MI 48917-3980
  517-321-2638
  http://www.abih.org/

- **Board of Certification in Professional Engineering**
  PO Box 2811
  Bellingham, WA 98227-2811
  360-671-7601
  http://www.bcpe.org/

- **The Institute of Industrial Engineers**
  25 Technology Park
  Norcross, GA 30092
  770-449-0460
  http://www.iienet.org

- **American Board for Occupational Health Nurses**
  201 East Ogden Ave., Suite 114
  Hinsdale, IL 60521
  630-789-5799
  http://www.abohn.org

examination, safety and health managers must have the required education and experience and submit letters of recommendation as specified by the **certification board**. Figure 4–4 contains the names and addresses of certification boards of interest to safety and health managers. Certification is covered in greater detail later in this chapter.

- **Professional societies. Professional societies** are typically formed to promote professionalism, adding to the body of knowledge, and forming networks among colleagues in a given field. Numerous professional societies focus on various safety and health issues. Figure 4–5 summarizes some of these.
- **Scientific standards and testing organizations.** Scientific **standards and testing organizations** conduct research, run tests, and establish standards that identify the acceptable levels for materials, substances, conditions, and mechanisms to which people might be exposed in the modern workplace. Figure 4–6 summarizes those of critical importance to safety and health managers.

FIGURE 4–5   Professional societies.

American Academy of Industrial Hygiene
302 S. Waverly Rd.
Lansing, MI 48917

American Industrial Hygiene Association
475 Wolf Ledges Pkwy.
Akron, OH 44311

American Occupational Medical Association
2340 S. Arlington Heights Rd.
Arlington Heights, IL 60005

American Society of Safety Engineers
1900 E. Oakton St.
Des Plaines, IL 60016

National Safety Council
1121 Spring Lake Dr.
Itasca, IL 60143

Society of Toxicology
1133 I St. NW, Suite 800
Washington, DC 20005

FIGURE 4–6   Scientific standards and testing organizations.

American National Standards Institute (ANSI)
1430 Broadway
New York, NY 10018

International Atomic Energy Agency
Wagramstrasse 5
A01400 Vienna, Austria

National Fire Protection Association (NFPA)
Batterymarch Park
Quincy, MA 02269

Underwriters Laboratories, Inc. (UL)
333 Pfingsten Rd.
New York, NY 10017

American Society of Mechanical Engineers
(ASME)
345 E. 47th St.
New York, NY 10017

American Society for Testing and Materials
(ASTM)
1916 Race St.
Philadelphia, PA 19103

FIGURE 4–7 Government agencies.

Bureau of Mines
Department of the Interior
2401 E St. NW
Washington, DC 20241

National Institute for Occupational Safety and Health
Department of Health and Human Services
Parklawn Bldg.
5600 Fishers La. Rd. 1401
Rockville, MD 20857

Center for Devices and Radiological Health
Food and Drug Administration
8757 Georgia Ave.
Silver Springs, MD 20910

Consumer Product Safety Commission
1111 18th St. NW
Washington, DC 20207

Department of Labor
200 Constitution Ave. NW
Washington, DC 20210

Environmental Protection Agency
401 M St. SW
Washington, DC 20460

Mine Safety and Health Administration
Department of Labor
4015 Wilson Blvd.
Arlington, VA 22203

National Bureau of Standards Headquarters
Route I-270 and Quince Orchard Rd.
Gaithersburg, MD 20899

National Center for Toxicological Research
Food and Drug Administration
5600 Fishers La.
Rockville, MD 20857

National Technical Information Service
Department of Commerce
5285 Port Royal Rd.
Springfield, VA 22161

Nuclear Regulatory Commission
1717 H St. NW
Washington, DC 20555

Occupational Safety and Health Commission
Department of Labor
200 Constitution Ave. NW
Washington, DC 20210

Occupational Safety and Health Review Committee
1825 K St. NW
Washington, DC 20006

Office of Energy Research
Department of Energy
1000 Independence Ave. SW
Washington, DC 20585

Office of Hazardous Materials Transportation
400 7th St. SW
Washington, DC 20590

U.S. Fire Administration
Federal Emergency Management Agency
16825 S. Seton Ave.
Emmitsburg, MD 21727

- **Government agencies.** Many **government agencies** are concerned with various aspects of workplace-related safety and health. Some of the most helpful agencies for the safety and health manager are listed in Figure 4–7.
- **Trade associations.** The purpose of a **trade association** is to promote the trade that it represents. Consequently, material produced by trade associations can be somewhat self-serving. Even so, trade associations can be valuable sources of information and training for safety and health managers. Figure 4–8 lists some of the trade associations that can be particularly helpful.

FIGURE 4–8 Trade associations.

Alliance of American Insurers

American Foundrymen's Association

American Insurance Association

American Iron and Steel Institute

American Metal Stamping Association

American Petroleum Institute

American Welding Society

Associated General Contractors of America

Compressed Gas Association

Industrial Safety Equipment Association

Institute of Makers of Explosives

Lead Industrial Association

National Electrical Manufacturers Association

National LP-Gas Association

National Machine Tool Builders Association

Scaffolding, Shoring, and Forming Institute

Soap and Detergent Association

Technical Association of the Pulp and Paper Industry

Note: Complete addresses are available in the reference section of most college and university libraries. Most have websites that can be found on the Internet.

ENGINEERS AND SAFETY

Engineers can make a significant contribution to safety. Correspondingly, they can cause, inadvertently or through incompetence, accidents that result in serious injury and property damage. The engineer has more potential to affect safety in the workplace than any other person does. The following example illustrates this:

> A car-pooler transports himself and three fellow employees to work each day. He is not a particularly safe driver and does not insist that his passengers use seatbelts. After running a red light, he crashes into the side of a building while swerving to avoid another vehicle. The two passengers wearing seatbelts are not hurt, but the driver and one other passenger, neither of whom were wearing seatbelts, are critically injured.

This brief story illustrates an accident that has two things in common with many workplace accidents. The first is a careless worker—in this case, the driver. The second is other workers who do not follow prescribed safety rules—in this case, failing to use seat belts. Employees such as these can and do cause many workplace accidents, but even the most careless employee cannot cause a fraction of the problems caused by a careless engineer. The following example illustrates this point:

> An engineer is charged with the responsibility for designing a new seatbelt that is comfortable, functional, inexpensive, and easy for factory workers to install. He designs a belt that meets all of these requirements, and it is installed in 10,000 new cars. As the cars are bought and accidents begin to occur, it becomes apparent that the new seatbelt fails in crashes involving speeds over 36 miles per hour. The engineer who designed the belt took all factors into consideration except one: safety.

This brief story illustrates how far-reaching an engineer's impact can be. With a poorly designed seat belt installed in 10,000 automobiles, the engineer has inadvertently endangered the lives of as many as 40,000 people (estimating a maximum of four passengers per automobile).

The engineer's opportunity for both good and bad comes during the design process. The process is basically the same regardless of whether the product being designed is a small toy, an industrial machine, an automobile, a nuclear power plant, a ship, a jumbo jetliner, or a space vehicle. Safety and health professionals should be familiar with the design process so that they can more fully understand the role of engineers concerning workplace safety.

Not all engineers are design engineers. However, engineers involved in design are usually in the aerospace, electrical, mechanical, and nuclear fields. The following paragraphs give an overview of these design-oriented engineering fields as seen in the course descriptions of a college catalog.

AEROSPACE ENGINEERING

The Bachelor of Science in Engineering (Aerospace Engineering) program incorporates a solid foundation of physical and mathematical fundamentals which provides the basis for the development of the engineering principles essential to the understanding of both atmospheric and extra-atmospheric flight. Aerodynamics, lightweight structures, flight propulsion, and related subjects typical of aeronautical engineering are included. Other courses introduce problems associated with space flight and its requirements. Integration of fundamental principles with useful applications is made in design work in the junior and senior years. Thus, the program prepares the student to contribute to future technological growth, which promises exciting and demanding careers in aerospace engineering.… Examples of concentration areas are: aerodynamics; design; flight propulsion; flight structures; space technology; stability, control, and guidance.[1]

ELECTRICAL ENGINEERING

Electrical Engineering is a science-oriented branch of engineering primarily concerned with all phases and development of the transmission and utilization of electric power and intelligence. The study of electrical engineering can be conveniently divided into the academic areas of circuits, electronics, electromagnetics, electric energy systems, communications, control, and computer engineering.[2]

MECHANICAL ENGINEERING

Mechanical Engineering is the professional field that is concerned with motion and the processes whereby other energy forms are converted into motion. Mechanical engineers are the people who are responsible for conceiving, designing, manufacturing, testing, and marketing devices and systems that alter, transfer, transform, and utilize the energy forms that ultimately cause motion. Thus mechanical engineers … are the people who make the engines that power ships, trains, automobiles, and spacecraft; they design the power plants which convert the energy in fuels, atoms, waterfalls, and sunlight into useful mechanical forms; and they construct intelligent machines and robots as well as the gears, cams, bearings, and couplings that facilitate and control all kinds of mechanical motion.[3]

INDUSTRIAL ENGINEERING

Industrial Engineering is concerned with the design, improvement, and installation of integrated systems which include people, material, equipment, and energy. This field of engineering draws upon specialized knowledge and skills in the mathematical, physical, and social sciences in concert with the principles and methods of engineering analysis and design to specify, predict, and evaluate the results to be obtained from such systems.

NUCLEAR ENGINEERING

Nuclear Engineering Sciences comprises those fields of engineering and science directly concerned with the release, control, and safe utilization of nuclear energy. Applications range over such broad topics as the design, development, and operation of nuclear reactor power systems to the applications of radiation in medicine, space, industry, and other related areas. The nuclear engineer, by virtue of his/her engineering and science-based training, is in a unique position to contribute to the many diverse aspects of this major component of the energy radiation field.[4]

Design Process

The **design process** is a plan for reaching a goal that proceeds in five sequential steps:

1. **Problem identification.** Engineers draft a description of the problem. This involves gathering information, considering constraints, reviewing specifications, and combining all of these into a clear and concise description of the problem.
2. **Synthesis.** Engineers combine or synthesize systematic, scientific procedures with creative techniques to develop initial solutions to the problem identified in Step 1. At this point, several possible solutions may be considered.
3. **Analysis and evaluation.** All potential solutions developed in the previous step are subjected to scientific analysis and careful evaluation. Such questions as the following are asked: Will the proposed solution satisfy the functional requirements? Will it meet all specifications? Can it be produced quickly and economically?
4. **Document and communicate.** Engineering drawings, detailed calculations, and written specifications are prepared. These document the design and communicate its various components to interested parties. It is common to revise the design at this point based on feedback from different reviewers.
5. **Produce and deliver.** Shop or detail drawings are developed, and the design is produced, usually as a prototype. The prototype is analyzed and tested. Design changes are made if necessary. The product is then produced and delivered.

The design process gives engineers unparalleled opportunities to contribute significantly to safety in the workplace and in the marketplace by producing products with safety built into them. However, in too many cases, the design process does not serve this purpose. There are two primary reasons for this:

- In analyzing and evaluating designs, engineers consider such factors as function, cost, life span, and manufacturability. All too often safety is not even considered or is only a secondary consideration.
- Even when engineers do consider safety in analyzing and evaluating designs, many are insufficiently prepared to do so effectively.

FIGURE 4–9 Design students should be required to study workplace safety.

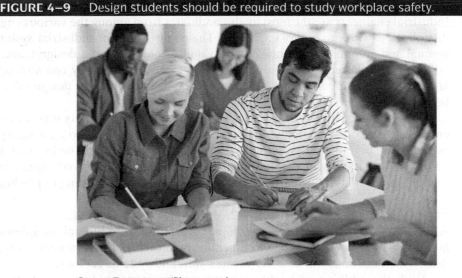

Source: Pressmaster/Shutterstock

Engineers who design products may complete their entire college curriculum without taking even one safety course. Safety courses, when available to engineering students in design-oriented disciplines, often tend to be electives. This limits the contributions that **design engineers** can make to both product and workplace safety, Figure 4–9.

Safety Engineer

The title **safety engineer** is often a misnomer in the modern workplace. It implies that the person filling the position is a degreed engineer with formal education and/or special training in workplace safety. Although this is sometimes the case, typically the title is given to the person who has overall responsibility for the company's safety program (the safety manager) or to a member of the company's safety team. This person is responsible for the traditional aspects of the safety program, such as preventing mechanical injuries; falls, impact, and acceleration injuries; heat and temperature injuries; electrical accidents; fire-related accidents; and so on.

In the former case, the person should be given a title that includes the term "manager." In the latter case, the title "safety engineer" is appropriate. However, persons with academic credentials in areas other than engineering should be encouraged to seek such positions because they are likely to be at least as well prepared and possibly even better prepared than persons with engineering degrees. These other educational disciplines include industrial technology, industrial engineering technology, manufacturing technology, engineering technology, industrial management, and industrial safety technology (bachelor's or associate degree).

There are signs that engineering schools are becoming more sensitive to safety and health issues. Graduate degrees in such areas as nuclear physics and **nuclear engineering** now often require safety courses. The federal government sponsors postgraduate studies in safety.

Industrial Engineers and Safety

Industrial engineers are the most likely candidates from among the various engineering disciplines to work as safety engineers. Their knowledge of industrial systems, both manual and automated, can make them valuable members of a design team, particularly one that designs industrial systems and technologies. They can also contribute after the fact as a member of a company's safety team by helping design job and plant layouts for both efficiency and safety.

Although industrial engineers are more likely to work as safety engineers than are engineers from other disciplines, they are not much more likely to have safety courses as a required part of their program of study. However, their focus on industrial systems and the integration of people and technology does give industrial engineers a solid foundation for additional learning through either in-service training or graduate work.

Environmental Engineers and Safety

A relatively new discipline (when compared with more traditional disciplines such as mechanical engineering) is environmental engineering. Environmental engineering deals with hazardous waste, water, wastewater, and systems ecology.

With the addition of health concerns to the more traditional safety concerns, **environmental engineers** will be sought as members of corporate safety and health teams. The course work they take is particularly relevant since all of it relates either directly or indirectly to health, Figure 4–10.

A person with the type of formal education shown in Figure 4–10 would be a valuable addition to the safety and health team of any modern industrial firm. Environmental engineers typically report to the overall safety and health manager and are responsible for those elements of the program relating to hazardous waste management, atmospheric pollution, indoor air pollution, water pollution, and wastewater management.

FIGURE 4–10   Environmental engineers may now be part of the safety and health team.

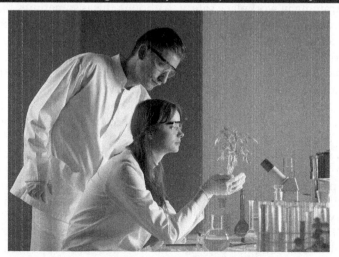

Source: Comaniciu Dan/Shutterstock

Chemical Engineers and Safety

Increasingly, industrial companies are seeking **chemical engineers** to fill the industrial hygiene role on the safety and health team. Their formal education makes people in this discipline well equipped to serve in this capacity. Chemical engineers are concerned with physical changes of matter in processing operations.

INDUSTRIAL HYGIENIST

Industrial hygiene is science of the recognition, evaluation, and control of those environmental factors or stresses, arising in and from the workplace, which may cause sickness, impaired health and well-being, or significant discomfort and inefficiency among workers or among citizens of the community.[5]

Industrial hygienists are primarily concerned about the following types of hazards: solvents, particulates, noise, dermatoses, radiation, temperature, ergonomics, toxic substances, biological substances, ventilation, gas, and vapors. In a safety and health team, the industrial hygienist typically reports to the safety and health manager.

HEALTH PHYSICIST

Health physicists are concerned primarily with radiation in the workplace. Consequently, they are employed by companies that generate or use nuclear power. Their primary duties include the following: monitoring radiation inside and outside the facility, measuring the radioactivity levels of biological samples, developing the radiation components of the company's emergency action plan, and supervising the decontamination of workers and the workplace when necessary.

Nuclear engineering and nuclear physics are the two most widely pursued fields of study for health physicists. A study conducted by Moeller and Eliassen gave the following breakdown of the academic preparation of practicing health physicists:

Associate degree	5.41 percent
Baccalaureate degree	28.38 percent
Master's degree	42.43 percent
Doctorate degree	19.46 percent[6]

The remaining practitioners are nondegreed personnel who have completed various types of noncollege credit training. This breakdown shows that graduate study is particularly important for health physicists. Professionals in this field may be certified by the American Board of Health Physics (ABHP).

OCCUPATIONAL PHYSICIAN

Occupational medicine as a specialized field dates back to World War II, when the United States experienced unprecedented industrial expansion. Production of manufactured goods skyrocketed, and workplace-related medical needs followed suit. Occupational medicine was not classified as a medical specialty by the American

Board of Preventive Medicine until 1955, however. The National Safety Council (NSC) describes the main concerns of the **occupational physician** as follows:

- Appraisal, maintenance, restoration, and improvement of the workers' health through application of the principles of preventive medicine, emergency medical care, rehabilitation, and environmental medicine.
- Promotion of a productive and fulfilling interaction of the worker and the job, via application of principles of human behavior.
- Active appreciation of the social, economic, and administrative needs and responsibilities of both the worker and work community.
- Team approach to safety and health, involving cooperation of the physician with occupational or industrial hygienists, occupational health nurses, safety personnel, and other specialties.[7]

Occupational physicians are fully degreed and licensed medical doctors. In addition, they must have completed postgraduate work in specific areas of safety.

Bernardino Ramazzini is widely thought of as being the first occupational physician. This is primarily as a result of his study of the work-related problems of workers in Modena, Italy, and a subsequent book he authored titled *The Diseases of Workers* (1700).[8] The first leading occupational physician in the United States was Alice Hamilton, MD.

Pioneers such as Ramazzini and Hamilton paved the way for the approximately 3,000 occupational physicians practicing today.[9] Whereas in the past the primary role of the occupational physician was treatment, today the primary role is prevention. This means more analytical, diagnostic, and intervention-related work than in the past and a much more proactive rather than passive approach.

One major difference in the role of modern occupational physicians compared with those in the past is in their relationship with employers. In the past, the occupational physician tended to be an in-house physician. More than 80 percent of the Fortune 100 companies still have at least one in-house occupational physician. However, the trend is away from this approach to the contracted approach in all but the largest companies. Companies contract with a private physician, clinic, or hospital to provide specific medical services. The healthcare provider is typically placed on a retainer in much the same manner as an attorney.

This contracted approach presents a new challenge for the safety and health manager. Including the occupational physician in planning, analyzing, assessing, monitoring, and other prevention-related duties can be more difficult when the physician is not in-house and readily available. For this reason, the safety and health manager must work with higher management to develop a contract that builds in time for prevention activities and compensates the healthcare provider appropriately.

OCCUPATIONAL HEALTH NURSE

Occupational health nurses have long been important members of corporate safety and health programs.

Like occupational physicians, today's occupational health nurses have seen their profession evolve over the years. The shift in emphasis is away from after-the-fact treatment to prevention-related activities such as analysis, monitoring, counseling, and education.

The American Association of Occupational Health Nurses (AAOHN) defines the objectives of occupational nurses as follows:

- To adapt the nursing program to meet the specific needs of the individual company
- To give competent nursing care for all employees
- To ensure that adequate resources are available to support the nursing program
- To seek out competent medical direction if it is not available on-site
- To establish and maintain an adequate system of records relating to workplace healthcare
- To plan, prepare, promote, present, and broker educational activities for employees
- To establish and maintain positive working relationships with all departments within the company
- To maintain positive working relationships with all components of the local healthcare community
- To monitor and evaluate the nursing program on a continual basis and adjust accordingly[10]

Like all members of the safety and health team, occupational nurses are concerned about cost containment. Cost containment is the driving force behind the emphasis on activities promoting health and accident prevention.

Occupational nurses typically report to an occupational physician when he or she is part of the on-site safety and health team. In companies that contract for off-site physician services, occupational nurses report to the overall safety and health manager.

RISK MANAGER

Organizations are at risk every time they open their doors for business. On any given day, an employee may be injured, a customer may have an accident, or a consumer may be injured using the organization's product. *Risk* is defined as a specific contingency or peril. Because the situations that put organizations at risk can be so expensive when they do occur, many organizations employ risk managers.

Risk management consists of the various activities and strategies that an organization can use to protect itself from situations, circumstances, or events that may undermine its security. You are, yourself, a risk manager. You take action every day to protect your personal and economic security. Take, for instance, the act of driving an automobile—a risky undertaking. Every time you drive, you put yourself at risk of incurring injuries, medical bills, lawsuits, and property damage.

You manage the risk associated with driving using two broad strategies: **reduction** and **transference**. The risk associated with driving an automobile can be reduced by wearing a seat belt, driving defensively, and obeying traffic laws. The remaining risk is managed by transferring it to an insurance company by purchasing a policy that covers both collision and liability.

The same approach—managing risk by using reduction and transference strategies—applies in the workplace. Risk managers work closely with safety and health personnel to reduce the risk of accidents and injuries on the job. They also work closely with insurance companies to achieve the most effective transference possible.

EMPLOYEES AND SAFETY

Because this book was written to help prepare technologists, engineers, and managers to become competent safety professionals, the emphasis throughout is on the roles and responsibilities of these professionals. However, employees also have important responsibilities in working safely and in helping maintain a safe and healthy workplace. At a minimum, these responsibilities include the following:

- Complying with all safety and health rules, regulations, and policies of their employers.
- Using all equipment, tools, and devices related to their work safely and properly.
- Ensuring all procedures and processes pertaining to their work are followed safely and properly.
- Properly wearing and using required personal protective equipment.
- Refusing to take shortcuts or cut corners on safety.
- Reporting unsafe or unhealthy conditions to superiors as soon as they are observed.
- Encouraging fellow employees to work safely.
- Participating in all safety- and health-related training provided.

CERTIFICATION OF SAFETY AND HEALTH PROFESSIONALS

Professional certification is an excellent way to establish credentials in the safety, health, and environmental profession. The most widely pursued accreditations are

- Certified Safety Professional (CSP), awarded by the Board of Certified Safety Professionals (BCSP)
- Certified Industrial Hygienist (CIE), awarded by the American Board of Industrial Hygiene (ABIH)
- Certified Professional Ergonomist (CPE), awarded by the Board of Certification in Professional Ergonomics (BCPE)
- Certified Occupational Health Nurse (COHN), awarded by the American Board for Occupational Health Nurses (ABOHN)

Certified Safety Professional

To qualify for the CSP title, applicants must follow these steps:

1. Apply to the BCSP
2. Meet an academic requirement
3. Meet a professional safety experience requirement
4. Pass the Safety Fundamentals Examination
5. Pass the Comprehensive Practice Examination[11]

Academic Requirement

The model educational background for a candidate for the **Certified Safety Professional (CSP)** is a bachelor's degree in safety from a program accredited by the Accreditation Board for Engineering and Technology (ABET). Because many people enter the safety profession from other educational backgrounds, candidates for the CSP may substitute other degrees plus professional safety experience for an accredited bachelor's degree in safety. A CSP candidate must meet one of the following minimum educational qualifications:

- Associate degree in safety and health, or
- Bachelor's degree in any field

Experience Requirement

In addition to the academic requirement, CSP candidates must have four years of professional safety experience in addition to any experience used to meet the academic requirement. Professional safety experience must meet all the following criteria to be considered acceptable by the BCSP:

- The professional safety function must be the primary function of the position. Collateral duties in safety are not considered the primary function.
- The position's primary responsibility must be the prevention of harm to people, property, and the environment, rather than responsibility for responding to harmful events.
- The professional safety function must be at least 50 percent of the position duties.
- The position must be full-time (defined by the BCSP as at least 35 hours per week).
- The position must be at the professional level. This is determined by (1) evaluating the degree of responsible charge and reliance by peers, employers, or clients; (2) the person's ability to defend analytical approaches used in professional practice; and (3) the recommendations made for controlling hazards through engineering or administrative approaches.
- The position must have breadth of duties. This is determined by evaluating the variety of hazards about which a candidate must advise and the range of skills involved in recognizing, evaluating, and controlling hazards. Examples of skills are analysis, **synthesis**, design, investigation, planning, administration, and communication.

Examination Requirements

The process to achieve the CSP designation typically involves passing two examinations: Safety Fundamentals and Comprehensive Practice.

- **Safety Fundamentals Examination.** The Safety Fundamentals Examination covers basic knowledge appropriate to professional safety practice. Candidates who meet the academic standard (achieve 48 points through an associate or bachelor's degree plus experience) may sit for the Safety Fundamentals Examination. Upon passing this examination, candidates receive the Associate Safety Professional (ASP) title to denote their progress toward the CSP.
- **Comprehensive Practice Examination.** All CSP candidates must acquire 96 points and pass the Comprehensive Practice Examination. To take this examination, a candidate must meet both the academic and experience requirements and have passed or waived the Safety Fundamentals Examination. The total

credit for academic degrees at all levels plus the months of professional safety experience must equal or exceed 96 points. After passing this examination, a candidate receives the CSP title. The address, telephone numbers, and website for the BCSP are provided in Figure 4–4.

Certified Industrial Hygienist

In the mid-1950s, a group of industrial hygienists from a national organization recommended that a voluntary certification program be established for industrial hygiene practitioners. In 1960, an independent corporation was established from the two national membership organizations—the American Industrial Hygiene Association (AIHA) and the American Council of Governmental Industrial Hygienists (ACGIH)—to establish a national examination process to certify a minimum level of knowledge in industrial hygiene.[12] Because the program was voluntary, it did not restrict the practice of individuals calling themselves "industrial hygienists." Indeed, today there are many competent persons practicing the profession of industrial hygiene who have not even sought certification. However, since its establishment, the program has proven to be a hallmark of achievement that provides an indicator of success in the field. It measures to a defined standard the knowledge of a practicing industrial hygienist in 16 technical areas of practice (called *rubrics*).

The intended purpose of the examination for **Certified Industrial Hygienist (CIH)** is to ensure that professionals working in this field have the skills and knowledge needed in the practice of industrial hygiene. The intent of the board is that the examinations are fair, reasonable, current, and representative of a broad range of industrial hygiene topics. Successful completion of the examination, meeting the educational requirements, and meeting the comprehensive, professional-level industrial hygiene experience requirements are the means by which the board attempts to identify those practitioners who have demonstrated the necessary skills to be an industrial hygienist.

Certification Process

What is the process for becoming certified? First, there is the obvious need for technical knowledge. There are various methods for gathering this information, but no one system has been identified as being the most effective. Review courses are available to prepare aspiring applicants. Conventional wisdom tends to support knowledge gained through experience and the watchful guidance of a competent mentor. Many individuals take the examination to find out their weak areas and study up on those until they manage to pass.

The professional reference questionnaire (PRQ) is important as well. The board's requirement for experience is based on activity on a professional or apprentice level. This causes a sticking point for many who find they are in a position with titles such as "industrial hygienist" or "project manager" when their scope of practice relies on strict adherence to a regulatory interpretation or exercise of very little independent judgment. Others with titles of "industrial hygiene technician" or "specialist" may be acting totally independently and practicing with a scope of extreme variability and many unknowns. This should be portrayed in the PRQ to be fair to the applicant. Details are important to the board in making their evaluation of an applicant.

The ABIH also has a process established for reviewing each application and preparing each examination. Examinations are prepared in minute detail. Each question is evaluated by a group of practicing CIEs to ensure its correctness and relevance to the practice of industrial hygiene. Each item is rated on difficulty for its target audience (CORE or Comprehensive), and this is used to set the passing score for each test. Each question is also rated by professional testers to ensure its validity as a question for an examination. Questions are selected for use on an examination based on the latest survey of the practice of industrial hygiene, in both rubric areas and domains of practice, to achieve a balance indicative of the current practice and some historical knowledge. After each presentation of the examination, questions are again reviewed for validity. The examination itself is the subject of an effort to ensure that it adheres to the standardized evaluation method.

Certification Maintenance

After a candidate has managed to meet the established standard for certification, maintenance becomes the issue. The board requires that all CIEs demonstrate that they are active in the field and have continued to improve their knowledge. Seven categories of practice are noted for the accumulation of "points" toward the 40 required every five years. Some portion of these points is gathered for active practice, technical committee work, publications, education and meetings, teaching, retest, or other work. This requires that some program of continuing education must be pursued. The address, telephone number, and website for the ABIH are provided in Figure 4–4.

Certified Professional Ergonomist

The examination for **Certified Professional Ergonomist (CPE)** is administered by the BCPE. Details concerning the examination are explained in the following paragraphs.[13]

To take the examination, individuals must meet the following requirements:

- **Academic requirements.** Applicants should have graduated from a regionally accredited college or university with a master's degree, or equivalent, in one of the correlative fields of ergonomics, such as biomechanics, human factors/ergonomics, industrial engineering, industrial hygiene, kinesiology, psychology, or systems engineering. Not everyone trained in these fields necessarily has the capabilities required for certification. The board uses other criteria to determine whether an applicant's ergonomics education has been sufficiently broad.
- **Work experience.** Applicants must have completed at least four years of ergonomic work experience. Appropriateness of work experience is determined from the applicant's employment history and evidence of participation in projects requiring ergonomic expertise.
- **Work product.** One work sample must be submitted with the application. This work sample must demonstrate a breadth of ergonomic knowledge and the ability to use ergonomic methods successfully. The work sample must demonstrate, at a minimum, the candidate's capabilities in the application of ergonomic principles to the design of a product, system, or work environment.

Allowable work products are technical reports, design papers, analysis reports, evaluation reports, patent applications, or a thorough written description of the project.

Associate-Level Certification

The Associate Ergonomics Professional (AEP) category is a precursor to the CPE designation and is available to a candidate who satisfies the following criteria:

- Meets the education requirements for BCPE certification
- Has passed Part 1 of the examination (on basic knowledge of human factors ergonomics)
- Is currently working toward fulfilling the BCPE requirement of four years' practical experience as a human factors and ergonomics professional

Bachelor's Degree Certification

A credential is available that recognizes a candidate for achieving the following levels of knowledge, skills, and experience in ergonomics practice:

- A bachelor's degree from a recognized university
- At least 200 hours of ergonomics training
- At least two full years practicing ergonomics
- A satisfactory score on the four-hour, two-part, multiple-choice examination on ergonomics foundations and ergonomics practice methods

This certification may be obtained by candidates with a bachelor's in engineering, healthcare/rehabilitation, industrial hygiene, and psychology. The address, telephone number, and website for the BCPE are provided in Figure 4–4.

Certified Occupational Health Nurse

The ABOHN offers several certifications relating to safety and health:

- **Certified Occupational Health Nurse (COHN)**
- Certified Occupational Health Nurse–Specialist (COHN–S)
- Certified Occupational Health Nurse/Case Manager (COHN/CM)
- Certified Occupational Health Nurse–Specialist/Case Manager (COHN–S/CM)
- Certified Occupational Health Nurse–Safety Manager (COHN–SM)

All the certifications offered by the ABOHN require that individuals first pass either the COHN or COHN–S certification tests. The other certifications are subspecialties.

Academic and Experience Requirements

In order to sit for the COHN certification tests, individuals must be registered nurses holding an associate degree or higher degree or their international equivalents. The focus of this credential is the nurse's role as a clinician, adviser, coordinator, and case manager. To sit for the COHN–S certification test, individuals must be registered nurses who hold a bachelor's degree or higher. The focus of this credential is on the nurse's role in direct care, management, education, consulting, and case management. The bachelor's degree required for this credential need not be in nursing.

In addition to the academic requirements, those who wish to achieve COHN or COHN–S certification must have 4,000 or more hours of work experience in occupational health and 50 or more contact hours of continuing education completed during the past five years. Additional testing is required for the various subspecialty credentials.

The newest certification offered by the ABOHN is the COHN–SM or Safety Manager credential. The focus of this credential is on occupational health nurses who plan, organize, implement, and evaluate hazard control activities to help organizations reduce or eliminate risks in the workplace. In order to sit for this subspecialty test, an individual must hold the COHN or COHN–S credential, hold a position that requires at least 25 percent of the time be spent on safety management activities, have completed 50 or more hours of safety-related continuing education during the past five years, and have 1,000 or more hours of safety management experience earned during the past five years.

OTHER SAFETY AND HEALTH-RELATED CERTIFICATIONS

In addition to the safety and health certifications explained in the previous section, there are several others students of the profession should be familiar with:

- *Graduate Safety Professional (GSP).* This is an alternate route to becoming a CSP. It requires a bachelor's or master's degree from an ABET-ASAC or AABI program.
- *Associate Safety Professional (ASP).* This certification requires a bachelor's degree in any field or an associate degree in safety, health, or the environment or a closely related field.
- *Occupational Health and Safety Technician (OHST).* This certification requires a high school diploma or a GED.
- *Construction Health and Safety Technician (CHST).* This certification requires a high school diploma or GED.
- *Safety Trained Supervisor (STS).* This certification requires 30 hours of safety and health training.
- *Certified Environmental, Safety, and Health Trainer (CET).* This certification requires the delivery of 270 hours of training or teaching in safety, health, and environmental-related areas.

EMERGING ROLE OF SAFETY PROFESSIONALS

As the world gets flatter, as organizations get leaner, and as global competition becomes more intense, the role of safety professionals—whether managers, engineers, technologists, technicians, or healthcare personnel—is changing. This is the message of Don Groover and Jim Spigener in a series of articles in *Occupational Health & Safety*.[14] According to the authors, the core duty of safety professionals has not changed. They are still responsible for preventing accidents and incidents that lead to injuries to personnel and damage to property or the environment. However, the skills they will need to fulfill this duty are changing.

In today's hypercompetitive business climate, fewer people are doing more work with fewer resources, technology continues to change at a rapid pace, employees work with less supervision while being required to make more complex decisions, baby boomers are retiring and being replaced by less experienced personnel, and older people are reentering the workforce. Add to this that the expectations of a safe and healthy workplace are higher than ever and that society is even more litigious than ever and the challenge of the modern safety professional becomes clear.[15]

According to Groover and Spigener, what all this means is that being an expert in a specific safety and health-related discipline—although still necessary and important—is no longer sufficient. Today's safety professional will need to be able to partner with the top leaders in their organizations in ways that promote the establishment and maintenance of a *safety-first corporate culture* that helps enhance the performance of people, processes, and products. To do this, safety professionals will have to become change leaders who are able to effectively challenge, engage, inspire, and influence people at all levels. In other words, safety professionals will have to become transformational leaders in their organizations.[16]

Key Terms and Concepts

- Associate degrees
- Bachelor's degree
- Certification board
- Certified Industrial Hygienist (CIH)
- Certified Occupational Health Nurse (COHN)
- Certified Professional Ergonomist (CPE)
- Certified Safety Professional (CSP)
- Chemical engineers
- Competitiveness
- Cost
- Design engineers
- Design process
- Environmental engineers
- Formal education
- Government agencies
- Health physicists
- Image
- In-service training
- Industrial engineering
- Industrial hygienist
- Line authority
- Mechanical engineering
- Nuclear engineering
- Occupational health nurses
- Occupational physician
- Productivity
- Professional societies
- Profits
- Quality
- Reduction
- Response time
- Risk management
- Safety and health manager
- Safety engineer
- Service
- Staff authority
- Standards and testing organizations
- Synthesis
- Trade association
- Transference

Review Questions

1. What types of positions may be included in a modern safety and health team?
2. Briefly explain the impact that such issues as workers' compensation and the environment have had on the commitment of corporate management to safety and health.
3. What is the difference between a staff and a line position?

4. Explain the types of problems that safety and health managers can expect to confront in attempting to implement their programs.
5. Briefly explain what a company must do to succeed in today's competitive global marketplace.
6. How can safety and health managers use the competitiveness issue to gain a commitment to their programs?
7. List five different college majors that can lead to a career as a safety and health manager.
8. Explain the importance of ongoing in-service training for modern safety and health managers and how to get it.
9. How can safety and health managers become certified in their profession?
10. Name three professional societies that a modern safety and health manager may join.
11. What is meant by the following statement: "If a physician makes an error, he may harm one person, but an engineer who errs may harm hundreds"?
12. Explain how the design process can affect safety.
13. What types of engineers are most likely to work as design engineers?
14. Why is the title "safety engineer" sometimes a misnomer?
15. What specific strengths may industrial engineers bring to bear as safety engineers?
16. What specific strengths may environmental engineers bring to the safety and health team?
17. What specific strengths may chemical engineers bring to the safety and health team?
18. Describe the job of the industrial hygienist.
19. Who is a health physicist?
20. Describe the job of the occupational physician.
21. Describe the job of the occupational health nurse.
22. Explain the concept of risk management.
23. Explain the role of the ergonomist.
24. Explain the responsibilities of employees in safety and health.
25. Explain how to achieve each of the following certifications: Certified Safety Professional, Certified Industrial Hygienist, and Certified Professional Ergonomist.
26. Explain how to achieve the following certification: Certified Occupational Health Nurse–Safety Manager.

Endnotes

1. *The University Record*, Undergraduate Catalog of the University of Florida, 2017.
2. Ibid., 70.
3. Ibid., 74.
4. Ibid., 75.
5. American Industrial Hygiene Association (AIHA), *Engineering Field Reference Manual*. Retrieved from aiha.org on April 4, 2017.
6. Ibid., 572.
7. Barbara A. Plog, *Fundamentals of Industrial Hygiene*, 6th ed., 2012, p. 407.
8. National Safety Council, "Occupational Physicians: What Is Their New Role?" Retrieved from nsc.org on April 9, 2013.
9. Ibid., 38.
10. Ibid., 79.

11. Board of Certified Safety Professionals (BCSP), "The CSP and Specialty Certifications," July 2013, 1–6, retrieved from bcsp.org/.
12. American Board of Industrial Hygiene (ABIH), "Certified Industrial Hygienist (CIH)—What It Means," July 2013, 1–5, retrieved from abih.org/.
13. Board of Certified Professional Ergonomists (BCPE), "BCPE—Frequently Asked Questions," July 2013, 1–5, retrieved from bcpe.org/.
14. D. Groover and J. Spigener, "The Emerging Role of the Safety Professional, Part 1," *Occupational Health & Safety* 77, no. 4: 24. 2005.
15. D. Groover and J. Spigener, "The Emerging Role of the Safety Professional, Part 4," *Occupational Health & Safety* 77, no. 7: 38. 2005.
16. Ibid.

5

THE OSH ACT, STANDARDS, AND LIABILITY

Learning Objectives

- Explain the rationale for the OSH Act
- Summarize OSHA's mission and purpose
- Summarize what is covered in the OSH Act
- Explain the concept of the OSHA standard
- Explain OSHA's record-keeping requirements
- Describe the OSHA requirement to keep employees informed
- Explain how OSHA enforces compliance
- Summarize the main tenets of OSHA's Whistleblower program
- Explain OSHA's enhanced enforcement policy
- List and explain the various citations/penalties OSHA can assess
- Describe the cost of OSHA fines: direct and indirect
- Explain how OSHA's appeal process works
- Explain the concept of the state-level OSHA program and how it relates to the federal program
- Summarize the various services available from OSHA
- Summarize employer rights and responsibilities under the OSH Act
- Summarize employee rights and responsibilities under the OSH Act
- Describe how employers can keep up to date on OSHA requirements/ services
- Explain the most commonly stated problems with OSHA
- List the other organizations and agencies concerned with workplace safety and health
- Summarize the contents of OSHA's General Industry Standards
- Summarize the contents of OSHA's Maritime Standards
- Summarize the contents of OSHA's Construction Standards
- Explain the concepts of standards and codes
- Explain the laws and liabilities relating to workplace safety and health
- Explain OSHA's position on awarding safety incentives

Since the early 1970s, the number of legislations passed—and the number of subsequent regulations—concerning workplace safety and health has increased markedly. Of all the legislation, by far the most significant has been the **Occupational Safety and Health Act (OSH Act)** of 1970. Prospective and practicing safety and health professionals must be knowledgeable about the OSH Act and the agency established by it—the **Occupational Safety and Health Administration (OSHA)**. This chapter provides students with the information they need about the OSH Act, OSHA, and other pertinent federal legislation and agencies.

RATIONALE FOR THE OSH ACT

Perhaps the most debilitating experience one can have on the job is to be involved in, or exposed to, a work-related accident or illness. Such an occurrence can be physically and psychologically incapacitating for the victim, psychologically stressful for the victim's fellow workers, and extraordinarily expensive for the victim's employer. In spite of this, until 1970, laws governing workplace safety were limited and sporadic. Finally, in 1970, Congress passed the OSH Act with the following stated purpose: "To assure so far as possible every working man and woman in the nation safe and healthful working conditions and to preserve our human resources."[1]

According to the U.S. Department of Labor, in developing this comprehensive and far-reaching piece of legislation, Congress considered the following statistics that applied at that time:

- An average of 14,000 deaths was caused by **workplace accidents** (this figure has since dropped to 4,500).
- Approximately 2.5 million workers were disabled in workplace accidents.
- Approximately 300,000 new cases of **occupational diseases** were reported.[2]

Clearly, a comprehensive, uniform law was needed to help reduce the incidence of work-related injuries, illnesses, and deaths. The OSH Act of 1970 addressed this need. It is contained in Title 29 of the *Code of Federal Regulations* (*CFR*), Parts 1900 through 2200. The act also established OSHA, which is part of the U.S. Department of Labor and is responsible for administering the OSH Act.

OSHA'S MISSION AND PURPOSE

OSHA was created by Congress to ensure safe and healthy working conditions by establishing standards, enforcing those standards, and providing helpful services including training, outreach, education, and compliance assistance. Under the OSH Act, employers are responsible for providing a safe and healthy workplace for their personnel. According to the U.S. Department of Labor, OSHA's mission and purpose can be summarized as follows:

- Encourage employers and employees to reduce workplace hazards.
- Implement new safety and health programs.
- Improve existing safety and health programs.
- Encourage research that will lead to innovative ways of dealing with workplace safety and health problems.

- Establish the rights of both employees and employers regarding the improvement of workplace safety and health.
- Monitor job-related illnesses and injuries through a system of reporting and record keeping.
- Establish training programs to increase the number of safety and health professionals and to improve their competence continually.
- Establish mandatory workplace safety and health standards and enforce those standards.
- Provide for the development and approval of state-level workplace safety and health programs.
- Monitor, analyze, and evaluate **state-level safety and health programs**.[3]

OSH ACT COVERAGE

The OSH Act applies to most employers. If an organization has even one employee, it is considered an employer and must comply with applicable sections of the act. This includes all types of employers from manufacturing and construction to retail and service organizations. There is no exemption for small businesses, although organizations with 10 or fewer employees are exempted from OSHA inspections and the requirement to maintain injury and illness records.

Although the OSH Act is the most comprehensive and far-reaching piece of safety and health legislation ever passed in this country, it does not cover all employers. In general, the OSH Act covers employers in all 50 states, the District of Columbia, Puerto Rico, and all other territories that fall under the jurisdiction of the U.S. government. Exempted employers are

- Persons who are self-employed
- Family farms that employ only immediate members of the family
- Federal agencies covered by other federal statutes (in cases where these other federal statutes do not cover working conditions in a specific area or areas, OSHA standards apply)
- State and local governments (except to gain OSHA's approval of a state-level safety and health plan, states must provide a program for state and local government employees that is at least equal to its private sector plan)
- Coal mines (coal mines are regulated by mining-specific laws)

Employers must adhere to various OSHA requirements. Some apply to all employers—except those exempted—whereas others apply only to specific types of employers. These requirements cover areas of concern such as the following:

- Fire protection
- Electricity
- Sanitation
- Air quality
- Machine use, maintenance, and repair
- Posting of notices and warnings
- Reporting of accidents and illnesses

- Maintaining written compliance programs
- Employee training

In addition to these, other more important and widely applicable requirements are explained later in this chapter.

Coverage of Federal, State, and Local Government Personnel

OSHA's protection applies to all federal government agencies. Section 19 of the OSH Act holds agency heads responsible for providing safe and healthy working conditions for their personnel. OSHA monitors federal agencies in the same way it monitors private sector employers, and it conducts workplace inspections in response to reports received from workers about potentially unsafe conditions. However, OSHA does not fine federal agencies since to do so would be to indirectly punish taxpayers.

Personnel who work in local and state government organizations are not covered by Federal OSHA standards. However, they do have OSH Act protection in states that have OHSA-approved state safety programs. OSHA also allows states and territories to establish plans for covering local and state employees. At this time, four states and one U.S. territory have OSHA-approved state plans that cover local and state government personnel only:

- Connecticut
- Illinois
- New Jersey
- New York
- Virgin Islands

OSHA STANDARDS

OSHA establishes standards in the following areas: General Industry, Construction, Maritime, and Agriculture. These standards are intended to protect workers from a wide range of workplace hazards. For example, OSHA has established standards to

- Provide fall protection.
- Prevent trenching cave-ins.
- Prevent exposure to certain infectious diseases.
- Ensure that confined spaces are safe to work in.
- Prevent exposure to dangerous chemicals.
- Put guards on dangerous machines.
- Provide personal protective gear including respirators.
- Provide training for selected dangerous jobs in a language and vocabulary level that workers can understand.

In carrying out its duties, OSHA is responsible for promulgating legally enforceable standards. OSHA standards may require conditions, or the adoption or use of one or more practices, means, methods, or processes reasonably necessary and appropriate to protect workers on the job. It is the responsibility of employers to become familiar with standards applicable to their establishments and to ensure that employees have and use personal protective equipment when required for safety.[4]

The general duty clause of the OSH Act requires that employers provide a workplace that is free from hazards that are likely to harm employees. This is important because the general duty clause applies when there is no specific OSHA standard for a given situation. Where OSHA standards do exist, employers are required to comply with them as written.

How OSHA Standards Are Developed

OSHA has the authority to develop, adopt, amend, and revoke occupational safety and health standards. The OSHA standard-setting process is comprehensive and provides many opportunities for input from all stakeholders, from individuals to private sector employers to other government agencies. OSHA is authorized to begin the standard-setting process on its own or in response to requests or recommendations from other stakeholders including

- The **National Institute for Occupational Safety and Health (NIOSH)** (NIOSH is the research agency for occupational safety and health)
- State and local government agencies
- Nationally recognized standards producing organizations
- Employers
- Labor organizations
- Interested parties in general

When considering the development or revision of a standard, OSHA typically publishes a request for information (RFI) or an advance notice of proposed rulemaking (ANPRM) in the *Federal Register* to ensure that all stakeholders are aware and have opportunities to provide input. Further, OSHA often holds meetings with stakeholder organizations that have a specific interest in the area covered by the proposed new or revised standard. All responses to an RFI or ANPRM are made public.

When OSHA decides to proceed with issuing a new or revised standard, it must publish a notice of proposed rulemaking (NPRM) in the *Federal Register* and solicit input from stakeholders and the public in general. An NPRM will contain the draft standard and an explanation of why it is needed. Stakeholders may submit their feedback in writing at www.regulations.gov. In addition, OSHA typically holds public hearings to solicit testimony and information from stakeholders. Once input from stakeholders has been considered and any changes based on it have been made, OSHA issues a final standard that then becomes an enforceable document.

Each Spring and Fall, the Department of Labor publishes a list of all regulatory projects that are currently in process. The list is published in the *Federal Register*. OSHA provides a projected schedule for its in-process projects so that stakeholders know the timing of new and revised standards and can participate in the development process. Current and past issues of the Regulatory Agenda are available at http://osha.gov/law-regs.html.

Input from and Help for Small Business in Standards Development and Compliance

OSHA is sensitive to the needs of small business both in developing new and revised standards and in enforcing their compliance. Further, the Small Business Regulatory Enforcement Act (SBREFA) provides help to small businesses that struggle with understanding and complying with OSHA regulations. SBREFA also gives small businesses a voice in the development of new and revised standards by requiring that OSHA

- Produces small entity compliance guides for selected agency rules.
- Is responsive to small business inquiries about complying with OSHA regulations.
- Has a penalty reduction policy for small businesses.
- Involves small businesses in developing proposed rules that are expected to have a significant effect on a large number of small businesses. This is accomplished through Small Business Advocacy Review Panels.
- Gives small businesses opportunities to undertake court challenges to OSHA rules and regulations they believe will adversely affect them.

OSHA Standards versus OSHA Regulations

OSHA issues both standards and regulations. Safety and health professionals need to know the difference between the two. OSHA standards address specific hazards such as working in confined spaces, handling hazardous waste, and working with dangerous chemicals. Regulations are more generic in some cases than standards and more specific in others. However, even when they are specific, regulations do not apply to specific hazards. Regulations do not require the rigorous review process that standards go through. This process is explained in the next section.

How Standards Are Adopted, Amended, or Revoked

OSHA can adopt, amend, or revoke standards. Before any of these actions can be undertaken, OSHA must publish its intentions in the *Federal Register* in either a notice of proposed rule making or an advance notice of proposed rule making. The **notice of proposed rule making** must explain the terms of the new rule, delineate proposed changes to existing rules, or list rules that are to be revoked. The advance notice of proposed rule making may be used instead of the regular notice when it is necessary to solicit input before drafting a rule.

After publishing notice, OSHA must conduct a public hearing if one is requested. Any interested party may ask for a public hearing on a proposed rule or rule change. When this happens, OSHA must schedule the hearing and announce the time and place in the *Federal Register*.

The final step, according to the U.S. Department of Labor, is as follows:

> After the close of the comment period and public hearing, if one is held, OSHA must publish in the Federal Register the full, final text of any standard amended or adopted and the date it becomes effective, along with an explanation of the standard and the reasons for implementing it. OSHA may also publish a determination that no standard or amendment needs to be issued.[5]

How to Read an OSHA Standard

OSHA standards are typically long and complex and are written in the language of lawyers and bureaucrats, making them difficult to read. However, reading OSHA standards can be simplified somewhat if one understands the system.

OSHA standards are part of the *CFR*, published by the Office of the *Federal Register*. The regulations of all federal government agencies are published in the *CFR*. Title 29 contains all the standards assigned to OSHA. Title 29 is divided into several parts, each carrying a four-number designator (such as Part 1901, Part 1910). These parts are divided into sections, each carrying a numerical designation. For example, 29 CFR 1910.1 means *Title 29, Part 1910, Section 1, Code of Federal Regulations.*

The sections are divided into four different levels of subsections, each with a particular type of designator as follows:

First Level:	Alphabetically using lowercase letters in parentheses: (a) (b) (c) (d)
Second Level:	Numerically using numerals in parentheses: (1) (2) (3) (4)
Third Level:	Numerically using roman numerals in parentheses: (i) (ii) (iii) (iv)
Fourth Level:	Alphabetically using uppercase letters in parentheses: (A) (B) (C) (D)

Occasionally, the standards go beyond the fourth level of subsection. In these cases, the sequence just described is repeated with the designator shown in parentheses underlined. For example: (a), (1), (i), (A).

Understanding the system used for designating sections and subsections of OSHA standards can guide readers more quickly to the specific information needed. This helps reduce the amount of cumbersome reading needed to comply with the standards.

Temporary Emergency Standards

The procedures described in the previous section apply in all cases. However, OSHA is empowered to pass **temporary emergency standards** on an emergency basis without undergoing normal adoption procedures. Such standards remain in effect only until permanent standards can be developed.

To justify passing temporary standards on an emergency basis, OSHA must determine that workers are in imminent danger from exposure to a hazard not covered by existing standards. Once a temporary standard has been developed, it is published in the *Federal Register.* This step serves as the notification step in the permanent adoption process. At this point, the standard is subjected to all the other adoption steps outlined in the preceding section.

How to Appeal a Standard

After a standard has been passed, it becomes effective on the date prescribed. This is not necessarily the final step in the **appeals process**, however. A standard, either permanent or temporary, may be appealed by any person who is against it.

An appeal must be filed with the U.S. Court of Appeals serving the geographic region in which the complainant lives or does business. Appeal paperwork must be initiated within 60 days of a standard's approval. However, the filing of one or more appeals does not delay the enforcement of a standard unless the court of appeals handling the matter mandates a delay. Typically, the new standard is enforced as passed until a ruling on the appeal is handed down.

Requesting a Variance

Occasionally, an employer may be unable to comply with a new standard by the effective date of enforcement. In such cases, the employer may petition OSHA at the state or federal level for a variance. Following are the different types of variances that can be granted.

Temporary Variance

When an employer advises that it is unable to comply with a new standard immediately but may be able to if given additional time, a **temporary variance** may be requested. OSHA may grant such a variance for up to a maximum of one year. To be granted a temporary variance, employers must demonstrate that they are making a concerted effort to comply and are taking the steps necessary to protect employees while working toward compliance.

Application procedures are very specific. Prominent among the requirements are the following: (1) identifying the parts of the standard that cannot be complied with; (2) explaining the reasons why compliance is not possible; (3) explaining in detail the steps that have been taken so far to comply with the standard; and (4) explaining the steps that will be taken to comply fully.

According to the U.S. Department of Labor, employers are required to keep their employees informed. They must "certify that workers have been informed of the variance application, that a copy has been given to the employees' authorized representative, and that a summary of the application has been posted wherever notices are normally posted. Employees also must be informed that they have the right to request a hearing on the application."[6]

Variances are not granted simply because an employer cannot afford to comply. For example, if a new standard requires employers to hire a particular type of specialist and there is a shortage of people with the requisite qualifications, a temporary variance might be granted. However, if the employer simply cannot afford to hire such a specialist, the variance will probably be denied. Once a temporary variance is granted, it may be renewed twice. The maximum period of each extension is six months.

Permanent Variance

Employers who feel they already provide a workplace that exceeds the requirements of a new standard may request a **permanent variance**. They present their evidence, which is inspected by OSHA. Employees must be informed of the application for a variance and notified of their right to request a hearing. Having reviewed the evidence and heard testimony (if a hearing has been held), OSHA can award or deny the variance. If a permanent variance is awarded, it comes with a detailed explanation of the employer's ongoing responsibilities regarding the variance. If, at any time, the company does not meet these responsibilities, the variance can be revoked.

Other Variances

In addition to temporary and permanent variances, an experimental variance may be awarded to companies that participate in OSHA-sponsored experiments to test the effectiveness of new health and safety procedures. Variances also may be awarded in cases where the secretary of labor determines that a variance is in the best interest of the country's national defense.

When applying for a variance, employers are required to comply with the standard until a decision has been made. If this is a problem, the employer may petition OSHA for an interim order. If granted, the employer is released from the obligation to comply until a decision is made. In such cases, employees must be informed of the order.

Typical of OSHA standards are the confined space and hazardous waste standards. Brief profiles of these standards provide an instructive look at how OSHA standards are structured and the extent of their coverage.

Confined Space Standard

This standard was developed in response to the approximately 300 work-related deaths that occur in confined spaces each year.[7] The standard applies to a broad cross section of industries that have employees working in spaces with the following characteristics: limited openings for entry or exit, poor natural ventilation, and a design not intended to accommodate continuous human occupancy. Such spaces as manholes, storage tanks, underground vaults, pipelines, silos, vats, exhaust ducts, boilers, and degreasers are typically considered confined spaces.

The key component in the standard is the *permit requirement.* Employers are required to develop an in-house program under which employees must have a permit to enter confined spaces. Through such programs, employers must do the following:

- Identify spaces that can be entered only by permit.
- Restrict access to identified spaces to ensure that only authorized personnel may enter.
- Control hazards in the identified spaces through engineering, revised work practices, and other methods.
- Continually monitor the identified spaces to ensure that any known hazards remain under control.

The standard applies to approximately 60 percent of the workers in the United States. Excluded from coverage are federal, state, and local government employees; agricultural workers; maritime and construction workers; and employees of companies with 10 or fewer workers.

Hazardous Waste Standard

This standard specifically addresses the safety of the estimated 1.75 million workers who deal with hazardous waste: hazardous waste workers in all situations, including treatment, storage, handling, and disposal; firefighters; police officers; ambulance personnel; and hazardous materials response team personnel.[8]

The requirements of this standard are as follows:

- Each hazardous waste site employer must develop a safety and health program designed to identify, evaluate, control safety and health hazards, and provide for emergency response.
- There must be preliminary evaluation of the site's characteristics prior to entry by a trained person to identify potential site hazards and to aid in the selection of appropriate employee protection methods.
- The employer must implement a site control program to prevent contamination of employees. At a minimum, the program must identify a site map, site work

zones, site communications, safe work practices, and the location of the nearest medical assistance. Also required in particularly hazardous situations is the use of the buddy system so that employees can keep an eye on one another and provide quick aid if needed.

- Employees must be trained before they are allowed to engage in hazardous waste operations or emergency response that could expose them to safety and health hazards.
- The employer must provide medical surveillance at least annually and at the end of employment for all employees exposed to any particular hazardous substance at or above established exposure levels or those who wear approved respirators for 30 days or more on-site.
- Engineering controls, work practices, and personal protective equipment (PPE), or a combination of these methods, must be implemented to reduce exposure below established exposure levels for the hazardous substances involved.
- There must be periodic air monitoring to identify and quantify levels of hazardous substances and to ensure that proper protective equipment is being used.
- The employer must set up an informational program with the names of key personnel and their alternates responsible for site safety and health, and the requirements of the standard.
- The employer must implement a decontamination procedure before any employee or equipment leaves an area of potential hazardous exposure; establish operating procedures to minimize exposure through contact with exposed equipment, other employees, or used clothing; and provide showers and change rooms where needed.
- There must be an emergency response plan to handle possible on-site emergencies prior to beginning hazardous waste operations. Such plans must address personnel roles; lines of authority, training, and communications; emergency recognition and prevention; safe places of refuge; site security; evacuation routes and procedures; emergency medical treatment; and emergency alerting.
- There must be an off-site emergency response plan to better coordinate emergency action by local services and to implement appropriate control actions.[9]

OSHA'S RECORD KEEPING AND REPORTING

One of the breakthroughs of the OSH Act was the centralization and systematization of **record keeping**. This has simplified the process of collecting health and safety statistics for the purpose of monitoring problems and taking the appropriate steps to solve them.

Over the years, OSHA has made substantial changes to its record-keeping and **reporting** requirements. Employers have complained for years about the mandated injury and illness record-keeping system. Their complaints can be summarized as follows:

- Original system was cumbersome and complicated.
- OSHA record-keeping rule had not kept up with new and emerging issues.
- There were too many interpretations in many of the record-keeping documents.

- Record keeping forms were too complex.
- Guidelines for record keeping were too long and difficult to understand.

In response to these complaints, OSHA initiated a dialogue among stakeholders to improve the record-keeping and reporting process. Input was solicited and received from employers, unions, trade associations, record keepers, OSHA staff, state occupational safety and health personnel, and state consultation program personnel. OSHA's goals for the new record-keeping and reporting system were as follows:

- Simplify all aspects of the process.
- Improve the quality of records.
- Meet the needs of a broad base of stakeholders.
- Improve access for employees.
- Minimize the regulatory burden.
- Reduce vagueness—give clear guidance.
- Promote the use of data from the new system in local safety and health programs.

In recording and reporting occupational illnesses and injuries, it is important to have common definitions. The U.S. Department of Labor uses the following definitions for record-keeping and reporting purposes:

> An occupational injury is any injury such as a cut, fracture, sprain, or amputation that results from a work-related accident or from exposure involving a single incident in the work environment. An occupational illness is any abnormal condition or disorder other than one resulting from an occupational injury caused by exposure to environmental factors associated with employment. Included are acute and chronic illnesses or diseases which may be caused by inhalation, absorption, ingestion, or direct contact with toxic substances or harmful agents.[10]

Reporting Requirements

All occupational illnesses and injuries must be reported if they result in one or more of the following:

- Death of one or more workers
- In-patient hospitalization of any employee
- One or more days away from work
- Restricted motion or restrictions to the work that an employee can do
- Loss of consciousness of one or more workers
- Transfer of an employee to another job
- Medical treatment beyond in-house first aid (if it is not on the first-aid list, it is considered medical treatment)
- Work-related amputations
- Any other condition listed in Appendix B of the rule[11]

Electronic Reporting Requirements

In May 2016, OSHA published its final rule requiring certain employers to file reports on workplace injuries and illnesses electronically. The rule went into effect on January 1, 2017. Information from electronically filed reports will, in turn, be posted on OSHA's website (although personal information about injured employees will be redacted). However, the employer will be identified on the website. The electronic reporting requirement applies to all employers with 250 employees or more. In addition, certain employers with between 20 and 250 personnel but must also file their reports electronically. These are employers that fall into traditionally high-injury-rate industries including construction and manufacturing. There are more than 50 other high-injury-rate industries on OSHA's list. The entire list of industries can be found at: www.osha.gov/recordkeeping/NAICScodesforelectronicsubmission.pdf

Critics of the rule claim that OSHA is using "public shaming" as a strategy to reduce workplace injuries and illnesses. Others criticize the rule claiming posting of reports along with the company names will discourage the filing of complete and accurate reports. OSHA does admit that the public aspect of the rule is intended to give employers an incentive to work harder to provide employees a safe and healthy work environment. The rule as contains the following elements:

- **Anti-retaliation measures.** OSHA's anti-retaliation measures are strengthened to protect employees who report unsafe conditions. Employees who report unsafe conditions are protected from retaliation or discrimination by termination, reduced pay, assignment to less favorable jobs, or other adverse actions.
- **Employee reporting measures.** Employers must establish reasonable requirements for employees to report injuries and illnesses (although OSHA does not define what is considered "reasonable").
- **Safety incentives.** OSHA began discouraged safety incentives in 2012 with a guidance memorandum believing they may inadvertently lead to underreporting of injuries and illnesses. The electronic reporting rule does not prohibit the use of incentives, but the language in the rule makes it clear that OSHA does not encourage them and will scrutinize their use closely.

Record-Keeping Requirements

Employers are required to keep injury and illness records for each location where they do business.[12] For example, an automobile manufacturer with plants in several states must keep records at each individual plant for that plant. Records must be maintained on an annual basis using special forms prescribed by OSHA. Computer or electronic copies can replace paper copies. Records are not sent to OSHA. Rather, they must be maintained locally for a minimum of three years. However, they must be available for inspection by OSHA at any time.

All records required by OSHA must be maintained on the following forms or forms based on them:

OSHA's Form 300. Form 300 is used to record information about every work-related death and every work-related injury or illness that involves loss of consciousness, restricted work activity, job transfer, days away from work, or medical treatment beyond first aid. First aid is defined as follows:

- Using nonprescription medication at nonprescription doses
- Administering tetanus immunizations
- Cleaning, flushing, or soaking wounds on the surface of the skin
- Using wound coverings such as bandages and gauze pads

- Using hot or cold therapy
- Using totally nonrigid means of support, such as elastic bandages, wraps, and nonrigid backbelts
- Using temporary immobilization devices such as splints, slings, neck collars, and back boards while transporting an accident victim
- Drilling a fingernail or toenail to relieve pressure, or draining fluids from blisters
- Using eye patches
- Using irrigation, tweezers, cotton swab, or other simple means to remove splinters or foreign material from areas other than the eye
- Using finger guards
- Using massages
- Drinking fluids to relieve heat stress

Injuries that require no more than these first-aid procedures do not have to be recorded. Form 300 is also used to record significant work-related injuries and illnesses that are diagnosed by a physician or licensed healthcare professional as well as those that meet any of the specific recording criteria set forth in 29 CFR Part 1904.12. Figure 5–1A is OSHA's Form 300.

OSHA's Form 300A. All organizations covered by 29 CFR Part 1904 must complete Form 300A, even if there have been no work-related injuries or illnesses during the year in question. This form is used to summarize all injuries and illnesses that appear on OSHA's Form 300 (Figure 5–1B).

OSHA's Form 301. Form 301 is used for every incidence of a recordable injury or illness. Form 301 must be completed within seven calendar days of learning that a recordable injury or illness has occurred. Figure 5–2 is OSHA's Form 301.

Reporting and Record-Keeping Summary

Reporting and record-keeping requirements appear as part of several different OSHA standards. Not all of them apply in all cases. Following is a summary of the most widely applicable OSHA reporting and record-keeping requirements.

- **29 CFR 1903.2(a) OSHA Poster. OSHA Poster 2203**, which advises employees of the various provisions of the OSH Act, must be conspicuously posted in all facilities subject to OSHA regulations.
- **29 CFR 1903.16(a) Posting of OSHA Citations.** Citations issued by OSHA must be clearly posted for the information of employees in a location as close as possible to the site of the violation.
- **29 CFR 1904.2 Injury/Illness Log.** Employers are required to maintain a log and summary of all *recordable* (Form 300) injuries and illnesses of their employees.
- **29 CFR 1904.4 Supplementary Records.** Employers are required to maintain supplementary records (**OSHA Form 301**) that give complete details relating to all recordable injuries and illnesses of their employees.
- **29 CFR 1904.5 Annual Summary.** Employers must complete and post an annual summary (**OSHA Form 300A**) of all recordable illnesses and injuries. The summary must be posted from February 1 to April 30 every year.

FIGURE 5–1A   OSHA's Form 300.

OSHA's Form 300 (Rev. 01/2004)

Log of Work-Related Injuries and Illnesses

Attention: This form contains information relating to employee health and must be used in a manner that protects the confidentiality of employees to the extent possible while the information is being used for occupational safety and health purposes.

Year 20 ____

U.S. Department of Labor
Occupational Safety and Health Administration

Form approved OMB no. 1218-0176

You must record information about every work-related death and about every work-related injury or illness that involves loss of consciousness, restricted work activity or job transfer, days away from work, or medical treatment beyond first aid. You must also record significant work-related injuries and illnesses that are diagnosed by a physician or licensed health care professional. You must also record work-related injuries and illnesses that meet any of the specific recording criteria listed in 29 CFR Part 1904.8 through 1904.12. Feel free to use two lines for a single case if you need to. You must complete an Injury and Illness Incident Report (OSHA Form 301) or equivalent form for each injury or illness recorded on this form. If you're not sure whether a case is recordable, call your local OSHA office for help.

Establishment name ____

City ____   State ____

Identify the person

(A) Case no.	(B) Employee's name	(C) Job title (e.g., Welder)	(D) Date of injury or onset of illness

Describe the case

(E) Where the event occurred (e.g., Loading dock north end)	(F) Describe injury or illness, parts of body affected, and object/substance that directly injured or made person ill (e.g., Second degree burns on right forearm from acetylene torch)

Classify the case

CHECK ONLY ONE box for each case based on the most serious outcome for that case:

Death (G)	Days away from work (H)	Remained at Work Job transfer or restriction (I)	Other recordable cases (J)

Enter the number of days the injured or ill worker was:

Away from work (K)	On job transfer or restriction (L)
____ days	____ days

Check the "Injury" column or choose one type of illness: (M)

Injury (1)	Skin disorder (2)	Respiratory condition (3)	Poisoning (4)	Hearing loss (5)	All other illnesses (6)

Page totals ▶

Be sure to transfer these totals to the Summary page (Form 300A) before you post it.

Page ____ of ____

Public reporting burden for this collection of information is estimated to average 14 minutes per response, including time to review the instructions, search and gather the data needed, and complete and review the collection of information. Persons are not required to respond to the collection of information unless it displays a currently valid OMB control number. If you have any comments about these estimates or any other aspects of this data collection, contact: US Department of Labor, OSHA Office of Statistical Analysis, Room N-3644, 200 Constitution Avenue, NW, Washington, DC 20210. Do not send the completed forms to this office.

Source: OSHA.

FIGURE 5–1B   OSHA's Form 300A.

OSHA's Form 300A (Rev. 01/2004)

Summary of Work-Related Injuries and Illnesses

U.S. Department of Labor
Occupational Safety and Health Administration

Year 20____

Form approved OMB no. 1218-0176

All establishments covered by Part 1904 must complete this Summary page, even if no work-related injuries or illnesses occurred during the year. Remember to review the Log to verify that the entries are complete and accurate before completing this summary.

Using the Log, count the individual entries you made for each category. Then write the totals below, making sure you've added the entries from every page of the Log. If you had no cases, write "0."

Employees, former employees, and their representatives have the right to review the OSHA Form 300 in its entirety. They also have limited access to the OSHA Form 301 or its equivalent. See 29 CFR Part 1904.35, in OSHA's recordkeeping rule, for further details on the access provisions for these forms.

Number of Cases

Total number of deaths

_____
(G)

Total number of cases with days away from work

_____
(H)

Total number of cases with job transfer or restriction

_____
(I)

Total number of other recordable cases

_____
(J)

Number of Days

Total number of days away from work

_____
(K)

Total number of days of job transfer or restriction

_____
(L)

Injury and Illness Types

Total number of . . .
(M)

(1) Injuries   _____

(2) Skin disorders   _____

(3) Respiratory conditions _____

(4) Poisonings   _____

(5) Hearing loss   _____

(6) All other illnesses _____

Establishment Information

Your establishment name _____

Street _____

City _____ State _____ ZIP _____

Industry description (e.g., Manufacture of motor truck trailers)
_____

Standard Industrial Classification (SIC), if known (e.g., 3715)
___ ___ ___ ___

OR

North American Industrial Classification (NAICS), if known (e.g., 336212)
___ ___ ___ ___ ___ ___

Employment Information (If you don't have these figures, see the Worksheet on the back of this page to estimate.)

Annual average number of employees _____

Total hours worked by all employees last year _____

Sign here

Knowingly falsifying this document may result in a fine.

I certify that I have examined this document and that to the best of my knowledge the entries are true, accurate, and complete.

_____
Company executive   Title

(___) ___ - ___   ___/___/___
Phone   Date

Post this Summary page from February 1 to April 30 of the year following the year covered by the form.

Source: OSHA.

FIGURE 5–2	OSHA's Form 301.

OSHA's Form 301
Injury and Illness Incident Report

U.S. Department of Labor
Occupational Safety and Health Administration

Form approved OMB no. 1218-0176

Attention: This form contains information relating to employee health and must be used in a manner that protects the confidentiality of employees to the extent possible while the information is being used for occupational safety and health purposes.

This *Injury and Illness Incident Report* is one of the first forms you must fill out when a recordable work-related injury or illness has occurred. Together with the *Log of Work-Related Injuries and Illnesses* and the accompanying *Summary*, these forms help the employer and OSHA develop a picture of the extent and severity of work-related incidents.

Within 7 calendar days after you receive information that a recordable work-related injury or illness has occurred, you must fill out this form or an equivalent. Some state workers' compensation, insurance, or other reports may be acceptable substitutes. To be considered an equivalent form, any substitute must contain all the information asked for on this form.

According to Public Law 91-596 and 29 CFR 1904, OSHA's recordkeeping rule, you must keep this form on file for 5 years following the year to which it pertains.

If you need additional copies of this form, you may photocopy and use as many as you need.

Completed by _____

Title _____

Phone (_____) _____ – _____   Date ___ / ___ / ___

Information about the employee

1) Full name _____

2) Street _____

   City _____ State _____ ZIP _____

3) Date of birth ___ / ___ / ___

4) Date hired ___ / ___ / ___

5) ☐ Male
   ☐ Female

Information about the physician or other health care professional

6) Name of physician or other health care professional _____

7) If treatment was given away from the worksite, where was it given?

   Facility _____

   Street _____

   City _____ State _____ ZIP _____

8) Was employee treated in an emergency room?
   ☐ Yes
   ☐ No

9) Was employee hospitalized overnight as an in-patient?
   ☐ Yes
   ☐ No

Information about the case

10) Case number from the Log _____ *(Transfer the case number from the Log after you record the case.)*

11) Date of injury or illness ___ / ___ / ___

12) Time employee began work _____ AM / PM

13) Time of event _____ AM / PM   ☐ Check if time cannot be determined

14) **What was the employee doing just before the incident occurred?** Describe the activity, as well as the tools, equipment, or material the employee was using. Be specific. *Examples:* "climbing a ladder while carrying roofing materials"; "spraying chlorine from hand sprayer"; "daily computer key-entry."

15) **What happened?** Tell us how the injury occurred. *Examples:* "When ladder slipped on wet floor, worker fell 20 feet"; "Worker was sprayed with chlorine when gasket broke during replacement"; "Worker developed soreness in wrist over time."

16) **What was the injury or illness?** Tell us the part of the body that was affected and how it was affected; be more specific than "hurt," "pain," or sore." *Example:* "strained back"; "chemical burn, hand"; "carpal tunnel syndrome."

17) **What object or substance directly harmed the employee?** *Examples:* "concrete floor"; "chlorine"; "radial arm saw." *If this question does not apply to the incident, leave it blank.*

18) **If the employee died, when did death occur?** Date of death ___ / ___ / ___

Public reporting burden for this collection of information is estimated to average 22 minutes per response, including time for reviewing instructions, searching existing data sources, gathering and maintaining the data needed, and completing and reviewing the collection of information. Persons are not required to respond to the collection of information unless it displays a current valid OMB control number. If you have any comments about this estimate or any other aspects of this data collection, including suggestions for reducing this burden, contact: US Department of Labor, OSHA Office of Statistical Analysis, Room N-3644, 200 Constitution Avenue, NW, Washington, DC 20210. Do not send the completed forms to this office.

Source: OSHA.

- **29 CFR 1904.6 Lifetime of Records.** Employers are required to keep injury and illness records on file for three years.
- **29 CFR 1904.7 Access to Records.** Employers are required to give employees, government representatives, and former employees and their designated representatives access to their own individual injury and illness records.
- **29 CFR 1904.8 Major Incident Report.** A major incident is the death of one employee or the hospitalization of five or more employees in one incident. All such incidents must be reported to OSHA.
- **29 CFR 1904.11 Change of Ownership.** When a business changes ownership, the new owner is required to maintain the OSHA-related records of the previous owner.
- **29 CFR 1910.1020(d) Medical/Exposure Records.** Employers are required to maintain medical and/or exposure records for the duration of employment plus 30 years (unless the requirement is superseded by another OSHA standard).
- **29 CFR 1910.1020(e) Access to Medical/Exposure Records.** Employers that keep medical and/or exposure records are required to give employees access to their own individual records.
- **29 CFR 1910.1020(g)(1) Toxic Exposure.** Employees who will be exposed to toxic substances or other harmful agents in the course of their work must be notified when first hired and reminded continually on at least an annual basis thereafter of their right to access their own individual medical records.
- **29 CFR 1910.1020(g)(2) Distribution of Materials.** Employers are required to make a copy of OSHA's *Records Access Standard* (29 CFR 1910.1020) available to employees. They are also required to distribute to employees any informational materials provided by OSHA.

OSHA makes provisions for awarding record-keeping variances to companies that wish to establish their own record-keeping systems (see variance regulation 1905). Application procedures are similar to those described earlier in this chapter for standard variances. To be awarded a variance, employers must show that their record-keeping system meets or exceeds OSHA's requirements.

Incidence Rates

Two concepts can be important when completing OSHA 300 forms: *incidence rates* and *severity rates.* On occasion, it is necessary to calculate the total injury and illness incident rate of an organization in order to complete an OSHA Form 300. This calculation must include fatalities and all injuries requiring medical treatment beyond mere first aid.

The formula for determining the total injury and illness incident rate follows:

$$IR = N \times 200,000 \div T$$

IR = Total injury and illness incidence rate
N = Number of injuries, illnesses, and fatalities
T = Total hours worked by all employees during the period in question

The number 200,000 in the formula represents the number of hours that 100 employees work in a year (40 hours per week × 50 weeks = 2,000 hours per year per employee). Using the same basic formula with only minor substitutions, safety managers can calculate the following types of incidence rates:

- Injury rate
- Illness rate
- Fatality rate
- Lost workday cases rate
- Number of lost workdays rate
- Specific hazard rate
- Lost workday injuries rate

The *number of lost workdays rate*, which does not include holidays, weekends, or any other days that employees would not have worked anyway, takes the place of the old severity rate calculation.

Record-Keeping and Reporting Exceptions

Among the exceptions to OSHA's record-keeping and reporting requirements, the two most prominent are as follows:

- Employers with 10 or fewer employees (full- or part-time in any combination)
- Employers in one of the following categories: real estate, finance, retail trade, or insurance

There are also partial exceptions to OSHA's record-keeping and reporting requirements. Most businesses that fall into Standard Industrial Classifications (SIC) codes 52–89 are exempt from all record-keeping and reporting requirements except in the case of a fatality or an incident in which five or more employees are hospitalized. The following types of organizations are partially exempt. They are not required to maintain OSHA injury and illness records unless asked by OSHA to do so. OSHA's request must be in writing.

Advertising services	Membership organizations
Apparel and accessory stores	Miscellaneous business services
Barber shops	Miscellaneous food stores
Beauty shops	Miscellaneous personnel services
Bowling centers	Miscellaneous shopping goods stores
Candy, nut, and confectionary stores	Motion picture
Child day care services	Motorcycle dealers
Computer and data processing services	Museums and art galleries
Credit reporting and collection services	New and used car dealers
Dairy products stores	Nondepository institutions
Dance studios, schools, and halls	Offices and clinics of dentists
Depository institutions	Offices of osteopathic physicians
Drug and proprietary stores	Offices and clinics of medical doctors
Eating and drinking establishments	Offices of other health practitioners
Educational institutions and services	Photographic studios, portrait
Engineering, accounting, and related	Producers, orchestras, entertainers
Funeral service and crematories	Radio, television, and computer stores
Gasoline service stations	Real estate agents and managers
Hardware stores	Retail stores, not elsewhere classified
Health services, not elsewhere classified	Research, management, and related

Holding and other investment offices
Individual and family services
Insurance agents, brokers, services
Insurance carriers
Legal services
Liquor stores
Meat and fish markets
Mailing, reproduction, and stenographic
Medical and dental laboratories

Retail bakeries
Re-upholstery and furniture repair
Security and commodity brokers
Services, not elsewhere classified
Shoe repair and shoeshine parlors
Social services, not elsewhere classified
Title abstract offices
Used car dealers

KEEPING EMPLOYEES INFORMED

One of the most important requirements of the OSH Act is *communication*. Employers are required to keep employees informed about safety and health issues that concern them. Most of OSHA's requirements in this area concern the posting of material. Employers are required to post the following material at locations where employee information is normally displayed:

- OSHA Poster 2203, which explains employee rights and responsibilities as prescribed in the OSH Act. The state version of this poster may be used as a substitute.
- Summaries of variance requests of all types.
- Copies of all OSHA citations received for failure to meet standards. Unlike other informational material, citations must be posted near the site of the violation. They must remain until the violation is corrected or for a minimum of three days, whichever period is longer.
- OSHA Form 300A (Summary of Workplace Injuries and Illnesses). Each year the new summary must be posted by February 1 and must remain posted until April 30.

In addition to the posting requirements, employers must also provide employees who request them with copies of the OSH Act and any OSHA rules that may concern them. Employees must be given access to records of exposure to hazardous materials and medical surveillance that has been conducted.

WORKPLACE INSPECTIONS AND ENFORCEMENT

One of the methods OSHA uses for enforcing its rules is the **workplace inspection**. OSHA personnel may conduct workplace inspections unannounced, and except under special circumstances, giving an employer prior notice is a crime punishable by fine, imprisonment, or both. OSHA's priorities for carrying out workplace inspections are as follows:

- Imminent danger
- Catastrophes—fatalities or hospitalizations
- Worker complaints and referrals
- Targeted inspections—particular hazards or high injury rates
- Follow-up inspections

When OSHA compliance officers arrive to conduct an inspection, they are required to present their credentials to the person in charge. Having done so, they are authorized to enter, at reasonable times, any site, location, or facility where work is taking place. They may inspect, at reasonable times, any condition, facility, machine, equipment, materials, and so on. Finally, they may question, in private, any employee or other person formally associated with the company.

Under special circumstances, employers may be given up to a maximum of 24 hours' notice of an inspection. These circumstances are as follows:

- When imminent danger conditions exist
- When special preparation on the part of the employer is required
- When inspection must take place at times other than during regular business hours
- When it is necessary to ensure that the employer, employee representative, and other pertinent personnel will be present
- When the local area director for OSHA advises that advance notice will result in a more effective inspection

Employers may require that OSHA have a judicially authorized warrant before conducting an inspection. The U.S. Supreme Court handed down this ruling in 1978 (*Marshall v. Barlow's Inc.*).[13] However, having obtained a legal warrant, OSHA personnel must be allowed to proceed without interference or impediment.

The OSH Act applies to approximately 6 million worksites in the United States. Sheer volume dictates that OSHA establishes priorities for conducting inspections. These priorities are as follows: imminent danger situations, catastrophic fatal accidents, employee complaints, planned high-hazard inspections, and follow-up inspections.

After being scheduled, the inspection proceeds in the following steps:

1. The OSHA compliance officer presents his or her credentials to a company official.
2. The compliance officer conducts an **opening conference** with pertinent company officials and employee representatives. The following information is explained during the conference: why the plant was selected for inspection, the purpose of the inspection, its scope, and applicable standards.
3. After choosing the route and duration, the compliance officer makes the **inspection tour**. During the tour, the compliance officer may observe, interview pertinent personnel, examine records, take readings, and take photographs.
4. The compliance officer holds a **closing conference**, which involves open discussion between the officer and company and employee representatives. OSHA personnel advise company representatives of problems noted, actions planned as a result, and assistance available from OSHA.

OSHA'S WHISTLEBLOWER PROGRAM

OSHA provides avenues for workers who observe hazardous conditions in the workplace to file complaints and protections for those who do. Workers should first attempt to have hazardous conditions resolved by approaching their supervisors. However, if supervisors are not responsive or if there is reason to fear retribution, workers may file a complaint directly with OSHA by calling 1-800-321-OSHA (6742) or by contacting the nearest regional office:

- Region I: (CT, ME, MA, NH, RI, VT): (617) 565-9860
- Region II: (NJ, NY, PR, VI): (212) 337-2378

- Region III: (DE, DC, MD, PA, VA, WVA): (215) 861-4900
- Region IV: (AL, FL, GA, KY, MS, NC, SC, TN): (678) 237-0400
- Region V: (IL, IN, MI, MN, OH, WI): (312) 353-2220
- Region VI: (AR, LA, NM, OK, TX): (972) 850-4150
- Region VII: (IA, KS, MO, NE): (816) 283-8745
- Region VIII: (CO, MT, ND, SD, UT, WY): (720) 264-6550
- Region IX: (AZ, CA, NI, NV): (415) 625-2547
- Region X: (AK, ID, OR, WA): (206) 757-6700

To protect workers who file complaints, Section 11(c) of the OSH Act prohibits any manner of retaliation or discrimination against a worker for exercising his or her rights under the act. The OSH Act gives workers the right to: (1) raise safety and health concerns with an employer, (2) report a work-related injury or illness, (3) file a complaint with OSHA, (4) seek an OSHA inspection, (5) participate in an OSHA inspection, and (6) testify in proceedings relating to an OSHA inspection. Protection from retaliation or discrimination means that employers cannot take a type of "adverse action" against an employee including the following types of actions: firing, laying off, blacklisting, demoting, denying overtime, denying promotions, disciplining, denying benefits, failing to hire, failing to re-hire, intimidation, making threats, reassignment affecting prospects for promotion, reducing pay, and reducing hours. Workers who feel they are being discriminated against because they filed a complaint have 30 days to contact the nearest Regional Office of OSHA or to call OSHA directly at 1-800-321-OSHA (6742).

OSHA'S ENHANCED ENFORCEMENT POLICY

Organizations that receive OSHA citations for high-gravity violations are subject to enhanced enforcement measures. Components of the enhanced enforcement policy include follow-up inspections, programmed inspections, public awareness, settlements, and Section 11(b) summary enforcement orders.

OSHA's Severe Violator Enforcement Program

OSHA's Severe Violator Enforcement Program (SVEP) took effect in June 2010. SVEP focuses on employers who willfully and repeatedly endanger workers by exposing them to seriously hazardous conditions without the proper precautions or protections. Procedures for dealing with severe violators include mandatory follow-up inspections or worksites found to be in violation and other worksites of the same company where similar unacceptable practices and hazardous conditions might be present.

Follow-Up Inspections

High-gravity violations include high-gravity willful violations, multiple high-gravity serious violations, repeat violations at the originating establishment, failure-to-abate notices, and serious or willful violations related to a workplace fatality. Organizations that commit any of these types of violations will receive on-site follow-up inspections. In addition, OSHA's area directors are empowered to conduct follow-up inspections to verify compliance if there is reason to suspect that a required abatement has not occurred.

Programmed Inspections

OSHA's *site-specific targeting process* uses objective selection criteria to schedule programmed inspections. As part of this process, OSHA records the name of the overall corporate entity during inspections and prioritizes within its site-specific targeting list all branch and affiliated facilities under the entity's broad corporate umbrella that have received high-gravity violations.

Public Awareness

When an organization receives a high-gravity violation of the type listed above under "Follow-up Inspections," OSHA makes the public aware of the violation and all applicable enforcement actions taken by issuing press releases through local and national media. In addition, high-gravity violations at branch and affiliated facilities are made known to the company's corporate headquarters through official notification by OSHA, of the citation and corresponding penalties.

Settlements

Provisions for high-gravity violation settlement agreements

1. Require the organization in question to hire consultants to develop a feasible process for changing the health and safety culture in the facility where the violations occurred.
2. Apply the violation settlement not just to the facility in question but also corporate-wide.
3. Include information about other job sites of the organization in question.
4. Require the organization to consent in advance to report all serious injuries and illnesses that require outside medical care and to receive OSHA inspections based on the report.
5. Agree to consent to entry of a court enforcement order under Section 11(b) of the OSH Act.

Section 11(b) Summary Enforcement Orders

In cases of the kinds of high-gravity violations listed earlier in this section, OSHA applies to the appropriate federal court of appeal for orders that summarily enforce the citations in question under Section 11(b) of the OSH Act. This means that once a Section 11(b) order has been entered, organizations that fail to comply may be held in contempt of court—an action OSHA pursues.

CITATIONS AND PENALTIES

Based on the findings of the compliance officer's workplace inspections, OSHA is empowered to issue citations and/or assess penalties. A citation informs the employer of OSHA violations. Penalties are typically fines assessed as the result of citations. The types of citations and their corresponding penalties are as follows:

- **Other-than-serious violation.** A violation that has a direct relationship to job safety and health, but probably would not cause death or serious physical harm. A **proposed penalty** for each violation is discretionary. A penalty for an **other-than-serious violation** may be adjusted downward by as much as 95 percent, depending on the employer's good faith (demonstrated efforts to comply with the act), history of previous violations, and size of business.

- **Serious violation.** A violation in which there is a high probability that death or serious physical injury may result and that the employer knew or should have known of the hazard. OSHA proposes a mandatory penalty for each **serious violation**. The actual penalty may be adjusted downward depending on the employer's good faith, history of prior violations, and the gravity of the alleged violation.
- **Willful violation.** A violation that the employer intentionally and knowingly commits. The employer either knows that what he or she is doing constitutes a violation or is aware that a hazardous condition exists and has made no reasonable effort to eliminate it. Penalties may be proposed for each **willful violation**, with a minimum penalty of $5,000 for each violation. A proposed penalty for a willful violation may be adjusted downward, depending on the size of the business and its history of previous violations. Usually, no credit is given for good faith. If an employer is convicted of a willful violation of a standard that has resulted in the death of an employee, the offense is punishable by a court-imposed fine or by imprisonment for up to six months, or both. A fine of up to $250,000 for an individual or $500,000 for a corporation may be imposed for a criminal conviction.
- **Repeat violation.** A violation of any standard, regulation, rule, or order where, upon reinspection, a substantially similar violation is found. **Repeat violations** can result in a fine for each such violation. To be the basis of a repeat citation, the original citation must be final; a citation under contest may not serve as the basis for a subsequent repeat citation.
- **Failure to abate prior violation.** Failure to correct a prior violation may bring a civil penalty for each day that the violation continues beyond the prescribed abatement date.
- **De minimis violation.** Violations of standards that have no direct bearing on safety and health. **De minimis violations** are documented like any other violation, but they are not included in citations.[14]

In addition to the citations and penalties described in the preceding paragraphs, employers may also be penalized by additional fines and/or imprisoned if convicted of any of the following offenses: (1) falsifying records or any other information given to OSHA personnel; (2) failing to comply with posting requirements; and (3) interfering in any way with OSHA compliance officers in the performance of their duties.

OSHA FINES: HOW MUCH AND WHERE DOES THE MONEY GO?

OSHA fines can be expensive, especially when the company in question is a repeat violator or is guilty of willful violations. In August, 2016 OSHA increased its fines for other than serious, serious, willful, and repeat violations for the first time in 25 years. The standard penalties are now as follows:

- Other than serious violation: $12,471
- Serious violation: $12,471
- Willful violation: $126,000
- Repeat violation: $126,000

This standard fines list can be misleading though. In reality, OSHA can assess much larger fines depending on the circumstances. For example, a company that had a history of OSHA violations was fined $1.9 million when an accident resulted in an amputation. Another company was fined $1.2 million by OSHA for failing to provide its employees the required safety training relating to their jobs. A company was fined $354,000 when a worker was trapped in a trench during an OSHA inspection. Another company that violated the process safety management standard was fined $917,000 when four workers were injured.

Penalties can, of course, be contested. Contested penalties can be reduced by as much as 60 percent for small employers, 25 percent for good faith efforts, and 10 percent for having a good record when it comes to citations. However, as in the cases cited above, repeat and willful violators should not expect their fines to be reduced. In fact, there is one case that is well known in OSHA circles in which a business agreed to shut down and never again employ people to work with explosives to avoid paying a $1.2 million fine. The appeal process that is part of the OSH Act attempts to ensure balance and fairness, but it is not intended to allow negligent employers to avoid accountability.

The fines collected by OSHA annually can be in the billions. So what happens to all of that money? There are those who complain that OSHA awards large fines for the purpose of filling its coffers. This is an unfortunate but common misperception. The fines assessed and collected by OSHA go the general fund of the U.S. Treasury as specified in the OSH Act. There is no monetary gain to OSHA from the fines it assesses. OSHA fines have one overarching purpose: to hold employers accountable for their actions in ways that prevent accidents, injuries, and deaths.

APPEALS PROCESS

Employee Appeals

Employees may not contest the fact that citations were or were not awarded or the amounts of the penalties assessed. However, they may appeal the following aspects of OSHA's decisions regarding their workplace: (1) the amount of time (**abatement period**) given to an employer to correct a hazardous condition that has been cited and (2) an employer's request for an extension of an abatement period. Such appeals must be filed within 10 working days of a posting. Although opportunities for formal appeals by employees are unlimited, employees may request an informal conference with OSHA officials to discuss any issue relating to the findings and results of a workplace inspection.

Employer Appeals

Employers may appeal a citation, an abatement period, or the amount of a proposed penalty. Before actually filing an appeal, however, an employer may ask for an informal meeting with OSHA's area director. The area director is empowered to revise citations, abatement periods, and penalties in order to settle disputed claims. If the situation is not resolved through this step, an employer may formalize the appeal. Formal appeals are of two types: (1) a petition for modification of abatement or (2) a notice of contest. The specifics of both are explained in the following paragraphs.

Petition for Modification of Abatement

The **Petition for Modification of Abatement (PMA)** is available to employers who intend to correct the situation for which a citation was issued, but who need more time. As a first step, the employer must make a good-faith effort to correct the problem within the prescribed time frame. Having done so, the employer may file a petition for modification of abatement. The petition must contain the following information:

- Descriptions of steps taken so far to comply
- Length of additional time needed for compliance and why
- Descriptions of the steps being taken to protect employees during the interim
- Verification that the PMA has been posted for employee information and that the employee representative has been given a copy

Notice of Contest

An employer who does not wish to comply may contest a citation, an abatement period, and/or a penalty. The first step is to notify OSHA's area director in writing. This is known as filing a **notice of contest**. It must be done within 15 federal working days of receipt of a citation or penalty notice. The notice of contest must clearly describe the basis for the employer's challenge and contain all of the information about what is being challenged (i.e., amount of proposed penalty or abatement period).

Once OSHA receives a notice of contest, the area director forwards it and all pertinent materials to the **Occupational Safety and Health Review Commission (OSHRC)**. OSHRC is an independent agency that is associated with neither OSHA nor the U.S. Department of Labor. The Department of Labor describes how OSHRC handles an employer's claim:

> The commission assigns the case to an administrative law judge. The judge may disallow the contest if it is found to be legally invalid, or a hearing may be scheduled for a public place near the employer's workplace. The employer and the employees have the right to participate in the hearing; the OSHRC does not require that they be represented by attorneys. Once the administrative law judge has ruled, any party to the case may request further review by OSHRC. Any of the three OSHRC commissioners also may, at his or her own motion, bring a case before the Commission for review. Commission rulings may be appealed to the appropriate U.S. Court of Appeals.[15]

Employer appeals are common. In fact, each issue of *Occupational Hazards* carries a special section titled "Contested Cases." A review of examples of contested cases can be instructive in both process and outcomes.

STATE-LEVEL OSHA PROGRAMS

States are allowed to develop their own safety and health programs.[16] In fact, the OSH Act encourages it. As an incentive, OSHA will fund up to 50 percent of the cost of operating a state program for states with approved plans. States may develop comprehensive plans covering public and private employers or limit their plans to coverage of public employers only. In such cases, OSHA covers employers not included in the state plan.

To develop an OSHA-approved safety and health plan, a state must have adequate legislative authority and must demonstrate the ability to develop standards-setting, enforcement, and appeals procedures within three years; public employee protection; a sufficient number of qualified enforcement personnel; and education, training, and technical assistance programs. When a state satisfies all these requirements and accomplishes all developmental steps,

> OSHA then certifies that a state has the legal, administrative, and enforcement means necessary to operate effectively. This action renders no judgement on how well or poorly a state is actually operating its program, but merely attests to the structural completeness of its program. After this certification, there is a period of at least one year to determine if a state is effectively providing safety and health protection.[17]

Figure 5–3 lists the states that currently have OSHA-approved safety and health plans. Connecticut and New York cover the public sector only. The Virgin Islands and Puerto Rico account for 2 of the 23 plans.

FIGURE 5–3 States and territories with approved safety and health plans.

States and Territories with Approved Safety and Health Plans

- Alaska
- Arizona
- California
- Connecticut (public sector only)
- Hawaii
- Indiana
- Iowa
- Kentucky
- Maryland
- Michigan
- Minnesota
- Nevada
- New Mexico
- New York (public sector only)
- North Carolina
- Oregon
- Puerto Rico
- South Carolina
- Tennessee
- Utah
- Vermont
- Virginia
- Virgin Islands
- Washington
- Wyoming

SERVICES AVAILABLE FROM OSHA

In addition to setting standards and inspecting for compliance, OSHA provides services to help employers meet the latest safety and health standards. The services are typically offered at no cost. The categories of services available from OSHA are consultation, safety and health achievement recognition program, cooperative programs, alliance program, OSHA strategic partnership program, challenge program, voluntary protection programs, Susan Harwood Training Grants, and training and education services.

Consultation Services

Consultation services provided by OSHA include assistance in (1) identifying hazardous conditions; (2) correcting identified hazards; and (3) developing and implementing programs to prevent injuries and illnesses. To arrange consultation services, employers contact the consultation provider in their state (see Figure 5–4).

The actual services are provided by professional safety and health consultants, who are not OSHA employees. They typically work for state agencies or universities and provide consultation services on a contract basis; OSHA provides the funding.

Consultation services that are available in the various states and Puerto Rico can be found by downloading OSHA's Consultation Project Directory located at the following website: www.osha.gov/Publications/consultation_kit/projDir_Ins.pdf

FIGURE 5–4 OSHA provides on-site consultations.

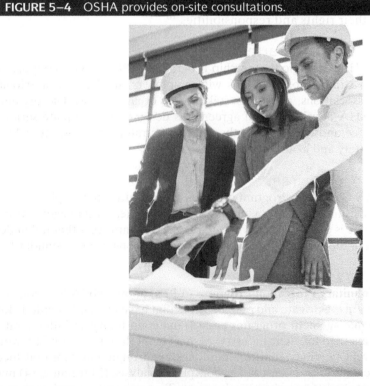

Source: Wavebreak Media ltd/Alamy Stock Photo

Safety and Health Achievement Recognition Program

Under OSHA's consultation program, selected exemplary employers may be allowed—upon request—to participate in OSHA's Safety and Health Achievement Recognition Program (SHARP) and be recognized with SHARP certification. Eligibility for participation includes—but is not limited to—receiving a full-service consultation visit, correcting all identified hazardous conditions, and developing an effective safety and health management program. Worksites that received SHARP recognition are exempt from programmed inspections during the period of the SHARP certification.

Cooperative Programs

Cooperative programs are partnerships of businesses, labor groups, and other organizations. As the name of these programs implies, their aim is to foster cooperation between and among stakeholders in workplace safety. Information on OSHA's cooperative programs is available at osha.gov/dcsp/compliance_assistance/index _programs.html.

Alliance Program

These are programs through which OSHA forms alliances with other groups that are committed to workplace safety and health. Such groups include trade associations, professional organizations employers, unions, consulates, faith-based organizations, community-based organizations, and educational institutions. These groups work with OSHA to develop effective compliance assistance tools and resources, share information with employers and workers, and educate workers about their rights and responsibilities.[18]

OSHA Strategic Partnership Program

The OSHA Strategic Partnership Program (OSPP) provides opportunities for OSHA to form strategic partnerships with employers, workers, professional organizations, trade associations, and other stakeholder organizations. The organizations involved and OSHA develop formal agreements designed to eliminate serious hazards in the workplace and to recognize the efforts of strategic partners to achieve model workplace safety and health practices.

Challenge Program

OSHA's Challenge Program provides opportunities for employers and workers to receive assistance in improving their safety and health management programs. OSHA Challenge is open to private and public sector employers that fall under OSHA's jurisdiction in the General Industry, Maritime, or Construction Standards.

Voluntary Protection Programs

The **Voluntary Protection Programs (VPPs)** allow OSHA to recognize employers and workers in the private and public sectors that have (1) implemented effective safety and health management programs and (2) maintain injury and illness rates below the national average for their industry. Management, labor, and OSHA work cooperatively to prevent fatalities, injuries, and illnesses through a system that focuses on (1) hazard prevention and control, (2) workplace analysis, (3) training, (4) management commitment, and (5) worker involvement. To participate, employers must first undergo a comprehensive on-site inspection conducted by safety and health professionals. If the workforce is represented by a collective bargaining unit, a union representative must participate. VPP participants are exempted from OSHA programmed inspections as long as they maintain their VPP status—which must be re-evaluated every three to five years. Popular VPP programs include the Star Program, Merit Program, and Demonstration Program.[19]

Susan Harwood Training Grants

OSHA awards grants to nonprofit organizations to provide training and education on hazard identification and prevention to workers and small employers. The focus of the grants is small business and hard-to-reach workers in high-hazard industries.[20]

Training and Education Services

Training and education services available from OSHA take several forms. OSHA operates a training institute in Des Plaines, Illinois, that offers a wide variety of services to safety and health personnel from the public and private sectors. The institute has a full range of facilities, including classrooms and laboratories in which it offers more than 60 courses.

To promote training and education in locations other than the institute, OSHA awards grants to nonprofit organizations. Colleges, universities, and other nonprofit organizations apply for funding to cover the costs of providing workshops, seminars, or short courses on safety and health topics currently high on OSHA's list of priorities. Grant funds must be used to plan, develop, and present instruction. Grants are awarded annually and require a match of at least 20 percent of the total grant amount.

EMPLOYER RIGHTS AND RESPONSIBILITIES

OSHA is very specific in delineating the rights and responsibilities of employers regarding safety and health. These rights and responsibilities, as set forth in OSHA publication 2056, are summarized in this section.

Employer Rights

The following is a list of **employer rights** under the OSH Act. Employers have the right to

- Seek advice and consultation as needed by contacting or visiting the nearest OSHA office.
- Request proper identification of the OSHA compliance officer prior to an inspection.
- Be advised by the compliance officer of the reason for an inspection.
- Have an opening and closing conference with the compliance officer in conjunction with an inspection
- Accompany the compliance officer on the inspection
- File a notice of contest with the OSHA area director within 15 working days of receipt of a notice of citation and proposed penalty.
- Apply for a temporary variance from a standard if unable to comply because the materials, equipment, or personnel needed to make necessary changes within the required time are not available.

- Apply for a permanent variance from a standard if able to furnish proof that the facilities or methods of operation provide employee protection at least as effective as that required by the standard.
- Take an active role in developing safety and health standards through participation in OSHA Standards Advisory Committees, through nationally recognized standards-setting organizations, and through evidence and views presented in writing or at hearings.
- Be assured of the confidentiality of any trade secrets observed by an OSHA compliance officer during an inspection.
- Ask NIOSH for information concerning whether any substance in the workplace has potentially toxic effects.[21]

Employer Responsibilities

In addition to the rights set forth in the previous subsection, employers have prescribed responsibilities. The following is a list of **employer responsibilities** under the OSH Act. Employers must

- Meet the general duty responsibility to provide a workplace free from hazards that are causing or are likely to cause death or serious physical harm to employees and to comply with standards, rules, and regulations issued under the OSH Act.
- Be knowledgeable of mandatory standards and make copies available to employees for review upon request.
- Keep employees informed about OSHA.
- Continually examine workplace conditions to ensure that they conform to standards.
- Minimize or reduce hazards.
- Make sure employees have and use safe tools and equipment (including appropriate PPE) that is properly maintained.
- Use color codes, posters, labels, or signs as appropriate to warn employees of potential hazards.
- Establish or update operating procedures and communicate them so that employees follow safety and health requirements.
- Provide medical examinations when required by OSHA standards.
- Provide the training required by OSHA standards.
- Report to the nearest OSHA office within eight hours any fatal accident or one that results in the hospitalization of three or more employees.
- Keep OSHA-required records of injuries and illnesses and post a copy of **OSHA Form 300** from February 1 through April 30 each year (this applies to employers with 11 or more employees).
- At a prominent location within the workplace, post OSHA Poster 2203 informing employees of their rights and responsibilities.
- Provide employees, former employees, and their representatives access to the Log of Work-Related Injuries and Illnesses (OSHA Form 300) at a reasonable time and in a reasonable manner.
- Give employees access to medical and exposure records.
- Give the OSHA compliance officer the names of authorized employee representatives who may be asked to accompany the compliance officer during an inspection.

- Not discriminate against employees who properly exercise their rights under the act.
- Post OSHA citations at or near the worksite involved (each citation or copy must remain posted until the violation has been abated or for three working days, whichever is longer).
- Abate cited violations within the prescribed period.[22]

EMPLOYEE RIGHTS AND RESPONSIBILITIES

Employee Rights

Section 11(c) of the OSH Act delineates **employee rights**. These rights are actually protection against punishment for employees who exercise their right to pursue any of the following courses of action:

- Complain to an employer, union, OSHA, or any other government agency about job safety and health hazards.
- File safety or health grievances.
- Participate in a workplace safety and health committee or in union activities concerning job safety and health.
- Participate in OSHA inspections, conferences, hearings, or other OSHA-related activities.[23]

Employees who feel they are being treated unfairly because of actions they have taken in the interest of safety and health have 30 days in which to contact the nearest OSHA office. Upon receipt of a complaint, OSHA conducts an investigation and makes recommendations based on its findings. If an employer refuses to comply, OSHA is empowered to pursue legal remedies at no cost to the employee who filed the original complaint.

In addition to those just set forth, employees have a number of other rights. Employees may

- Expect employers to make review copies available of OSHA standards and requirements.
- Ask employers for information about hazards that may be present in the workplace.
- Ask employers for information on emergency procedures.
- Receive safety and health training.
- Be kept informed about safety and health issues.
- Anonymously ask OSHA to conduct an investigation of hazardous conditions at the worksite.
- Be informed of actions taken by OSHA as a result of a complaint.
- Observe during an OSHA inspection and respond to the questions of a compliance officer.
- See records of hazardous materials in the workplace.
- See their medical record.
- Review the annual Log of Work-Related Injuries and Illnesses (OSHA Form 300).

- Have an exit briefing with the OSHA compliance officer following an OSHA inspection.
- Anonymously ask NIOSH to provide information about toxicity levels of substances used in the workplace.
- Challenge the abatement period given to employers to correct hazards discovered in an OSHA inspection.
- Participate in hearings conducted by the Occupational Safety and Health Review Commission.
- Be advised when an employer requests a variance to a citation or any OSHA standard.
- Testify at variance hearings.
- Appeal decisions handed down at OSHA variance hearings.
- Give OSHA input concerning the development, implementation, modification, and/or revocation of standards.[24]

Employee Responsibilities

Employees have a number of specific responsibilities. The following list of **employee responsibilities** is adapted from OSHA 2056, 1991 (Revised). Employees must

- Read the OSHA poster at the job site and be familiar with its contents.
- Comply with all applicable OSHA standards.
- Follow safety and health rules and regulations prescribed by the employer and promptly use PPE while engaged in work.
- Report hazardous conditions to the supervisor.
- Report any job-related injury or illness to the employer and seek treatment promptly.
- Cooperate with the OSHA compliance officer conducting an inspection.
- Exercise their rights under the OSH Act in a responsible manner.[25]

KEEPING UP-TO-DATE ON OSHA

OSHA's standards, rules, and regulations are always subject to change. The development, modification, and revocation of standards are an ongoing process. It is important for prospective and practicing safety and health professionals to stay up-to-date with the latest actions and activities of OSHA. Following is an annotated list of strategies that can be used to keep current:

- Establish contact with the nearest regional or area OSHA office and periodically request copies of new publications or contact the OSHA Publications Office at the following website: osha.gov/pls/publications/pubindex.list
- Review professional literature in the safety and health field. Numerous periodicals carry OSHA updates that are helpful.
- Establish and maintain relationships with other safety and health professionals for the purpose of sharing information, and do so frequently.
- Join professional organizations, review their literature, and attend their conferences.

PROBLEMS WITH OSHA

Federal agencies are seldom without their detractors. Resentment of the federal bureaucracy is intrinsic in the American mind-set. Consequently, complaints about OSHA are common. Even supporters occasionally join the ranks of the critics. Often, the criticisms leveled against OSHA are valid.

Criticisms of OSHA take many different forms. Some characterize OSHA as an overbearing bureaucracy with little or no sensitivity to the needs of employers who are struggling to survive in a competitive marketplace. At the same time, others label OSHA as timid and claim it does not do enough. At different times and in different cases, both points of view have probably been at least partially accurate.

Most criticism of OSHA comes in the aftermath of major accidents or a workplace disaster. Such events typically attract a great deal of media attention, which, in turn, draws the attention of politicians. In such cases, the criticism tends to focus on the question, "Why didn't OSHA prevent this disaster?" At congressional hearings, detractors will typically answer this question by claiming that OSHA spends too much time and too many resources dealing with matters of little consequence while ignoring real problems. Supporters of OSHA will typically answer the question by claiming that lack of resources prevents the agency from being everywhere at once. There is a measure of validity in both answers.

On the one hand, OSHA has made a significant difference in the condition of the workplace in this country. On the other hand, large, centralized bureaucratic agencies rarely achieve a high level of efficiency. This is compounded in OSHA's case by the fact that the agency is subject to the ebb and flow of congressional support, particularly in the area of funding. Consequently, OSHA is likely to continue to be an imperfect organization subject to ongoing criticism.

OSHA's "Guide to Restroom Access for Transgender Workers"

One of the more controversial acts of OSHA has been publication of its *"Guide to Restroom Access for Transgender Workers."* Transgender restroom access was already a controversial issue before OSHA entered the fray on June 1, 2016 with its "Restroom Guide." However, the complaints against OSHA's guidelines have been more about bureaucratic overreach and politics than the advisability of allowing transgendered individuals to select the restroom of their choice.

The issue becomes even more clouded because OSHA developed its "Restroom Guide" at the behest of the National Center for Transgender Equality rather than a recognized safety and health organization. The questions detractors ask most frequently about OSHA's new "Restroom Guide" can be summarized as follows: (1) What does an individual's sexual identity have to do with workplace safety and health? (2) Was OSHA pressured by politicians to get involved in an issue that falls outside the scope of its mission?

OSHA has responded that employees can suffer health problems if restrooms are not available to them when they need them. Critics have countered that this justification argues for more restrooms and not restroom selection by gender identity. OSHA counters this argument by pointing out that the Equal Employment Opportunity Commission (EEOC) ruled in April 2015 that transgendered individuals must be allowed access to the restroom of the gender they identify with.

Critics have countered this argument by claiming that OSHA should have stayed out of the debate and simply allowed the EEOC's rule to provide any necessary guidance for employers.

Even some of its supporters would rather OSHA had stayed out of the transgender restroom debate. These OSHA proponents fear that getting involved in debates such as this one will undermine the organization's credibility when it comes to workplace safety and health. Said another way, these supporters fear that by joining in the transgender restroom debate OSHA has given ammunition to its critics and detractors. OSHA counters this argument by stating that employees who are uncomfortable using a restroom that does not fit their gender identity may avoid using the restroom completely. Avoiding the restroom for an entire working shift, OSHA adds, is likely to lead to health problems for the workers in question. This being the case, OSHA has determined that it will stand behind its "Restroom Guide" regardless of controversy.

Where does OSHA's stand on sexual identity and restroom choice leave safety and health professionals? The answer to this question is simple. Once OSHA passes a requirement, it becomes the duty of safety and health professionals to ensure that their organizations comply with it. Many of OSHA's requirements over the years have been controversial when first enacted, and some still are. The OSH Act itself was controversial when first passed, and still is in some quarters. But safety and health professionals do not get to choose which OSHA requirements they comply with. One's personal views—pro or con—concerning OSHA requirements are not the issue. Therefore, safety and health professionals should be familiar with the contents of OSHA's "Restroom Guide." The entire document can be downloaded at: www.dol.gov/asp/policy-development/TransgenderBathroomAccessBestPractices.pdf

OTHER AGENCIES AND ORGANIZATIONS

Although OSHA is the most widely known safety and health organization in the federal government, it is not the only one. Figure 5–5 lists associations, agencies, and organizations (including OSHA) with safety and health as part of their mission.

Of those listed, the most important to modern safety and health professionals are NIOSH and OSHRC. The missions of these organizations are summarized in the following paragraphs.

NIOSH

NIOSH is part of the Department of Health and Human Services (HHS). (Recall that OSHA is part of the U.S. Department of Labor.) NIOSH has two broad functions: research and education. The main focus of the agency's research is on toxicity levels and human tolerance levels of hazardous substances. NIOSH prepares recommendations along these lines for OSHA standards dealing with hazardous substances. NIOSH studies are also published and made available to employers. Each year, NIOSH publishes updated lists of toxic materials and recommended tolerance levels. These publications represent the educational component of NIOSH's mission.

FIGURE 5–5 Agencies, organizations, and associations dealing with safety and health.

American Public Health Association
www.apha.org
Bureau of Labor Statistics
www.bls.gov
Environmental Protection Agency
www.epa.gov
National Institute for Occupational Safety
 and Health (NIOSH)
www.cdc.gov/NIOSH
Occupational Safety and Health
 Administration U.S. Department of Labor
www.osha.gov
Occupational Safety and Health Review
 Committee (OSHRC)
www.oshrc.gov
U.S. Consumer Product Safety
 Commission
www.cpsc.gov

The Department of Health and Human Services describes NIOSH as follows:

In 1973, NIOSH became a part of the Centers for Disease Control (CDC), an arm of the Public Health Service in the Department of Health and Human Services. NIOSH is unique among federal research institutions because it has the authority to conduct research in the workplace, and to respond to requests for assistance from employers and employees.

NIOSH also consults with the Department of Labor and other federal, state, and local government agencies to promote occupational safety and health and makes recommendations to the Department of Labor about worker exposure limits.

NIOSH estimates that more than 7,000 people are killed at work each year, and nearly 12 million nonfatal injuries occur in the workplace. In addition to death and injury, it is estimated that more than 10 million men and women are exposed to hazardous substances in their jobs that can eventually cause fatal or debilitating diseases. To help establish priorities in developing research and control of these hazards, NIOSH developed a list of the 10 leading work-related diseases and injuries:

1. Occupational lung diseases
2. Musculoskeletal injuries
3. Occupational cancers
4. Occupational cardiovascular disease
5. Severe occupational traumatic injuries
6. Disorders of reproduction

7. Neurotoxic disorders
8. Noise-induced hearing loss
9. Dermatological conditions
10. Psychological disorders[26]

Figure 5–6 is an organizational chart showing the major divisions in NIOSH. The four principal divisions are described in the following paragraphs.

Division of Biomedical and Behavioral Science

The **Division of Biomedical and Behavioral Science (DBBS)** conducts research in the areas of toxicology, behavioral science, ergonomics, and the health consequences of various physical agents.[27] DBBS investigates problems created by new technologies and develops biological monitoring and diagnostic aids to ensure that the workplace is not responsible for diminished health, functional capacity, or life expectancy of workers. Consultation and data are furnished for developing the criteria for workplace exposure standards.

FIGURE 5–6   The major divisions of NIOSH.

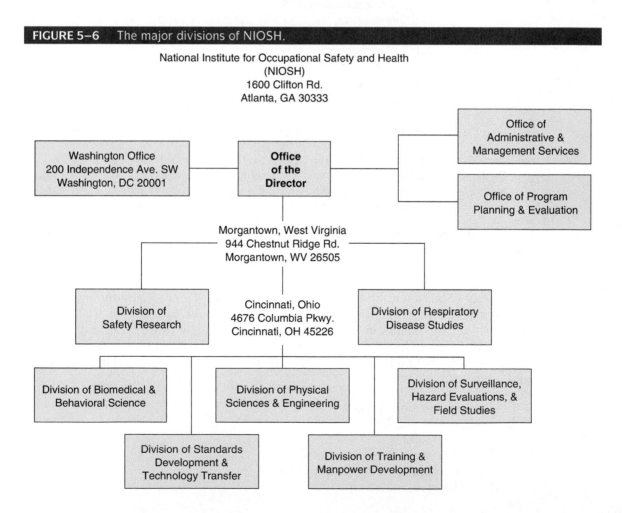

DBBS conducts laboratory and worksite research into the psychological, behavioral, physiological, and motivational factors relating to job stress as well as those induced by chemical and physical agents. The division assesses physical work capacity and tolerance for environmental conditions as influenced by age, gender, body type, and physical fitness. Interventions and control procedures using various approaches are developed and tested for their ability to reduce undue job stress.

The DBBS toxicology research involves dose-response and methods development studies to evaluate the effects of toxic agents and to develop techniques that utilize biomarkers as indicators of toxic exposures and early indicators of pathological change. The studies include applications of cellular biology, immunochemistry, pharmacokinetics, and neurophysiology. Through laboratory analysis of biological samples from animals exposed experimentally and humans exposed occupationally to workplace hazards, DBBS provides clinical and biochemical consultations for ascertaining the extent of exposure and for diagnosing occupational diseases.

DBBS conducts laboratory and worksite research on hazards from physical agents such as noise, vibration, and nonionizing energy sources. Studies seek to identify exposure factors that are significant to the health and well-being of the workforce. Investigations, including instrumentation and methods development, for characterizing and relating exposure factors to biological and performance changes in animal and human populations are also undertaken.

Division of Respiratory Disease Studies

The **Division of Respiratory Disease Studies (DRDS)** is the focal point for the clinical and epidemiological research that NIOSH conducts on occupational respiratory diseases.[28] The division provides legislatively mandated medical and autopsy services and conducts medical research to fulfill NIOSH's responsibilities under the Federal Mine Safety and Health Amendments Act of 1977.

The division conducts field studies of occupational respiratory diseases, in addition to designing and interpreting cross-sectional and prospective morbidity and mortality studies relating to occupational respiratory disease. Field studies are conducted at mines, mills, and other industrial plants where occupational respiratory diseases occur among workers. The division uses epidemiological techniques, including studies of morbidity and mortality, to detect common characteristics related to occupational respiratory diseases.

To formulate and implement programs that will identify factors involved in the early detection and differential rates of susceptibility to occupational respiratory diseases, DRDS conducts cell biology research to determine the role of microorganisms and environmental exposure in these diseases. The division also provides autopsy evaluations and a pathology research program. Research is conducted on immunological mechanisms and cell physiology to determine the effects of environmental exposure as it relates to occupational respiratory diseases.

DRDS provides for planning, coordinating, and processing medical examinations mandated under the Federal Mine Safety and Health Amendments Act of 1977 and operates a certification program for medical facilities and physicians who participate in the examination program. DRDS also evaluates and approves employer programs for the examination of employees in accordance with published regulations as well as arranging for the examination of employees who work at locations that lack an approved examination program. The division conducts the National Coal Workers' Autopsy

Program and performs research into the postmortem identification and quantification of occupational respiratory exposures.

Division of Surveillance, Hazard Evaluations, and Field Studies

The **Division of Surveillance, Hazard Evaluations, and Field Studies (DSHEFS)** conducts surveillance of the nation's workforce and workplaces to assess the magnitude and extent of job-related illnesses, exposures, and hazardous agents.[29] DSHEFS conducts legislatively mandated health hazard evaluations and industry-wide epidemiological research programs, including longitudinal studies of records and clinical and environmental field studies and surveys. It also provides technical assistance on occupational safety and health problems to other federal, state, and local agencies; other technical groups; unions; employers; and employees.

Surveillance efforts are designed for the early detection and continuous assessment of the magnitude and extent of occupational illnesses, disabilities, deaths, and exposures to hazardous agents, using new and existing data sources from federal, state, and local agencies; labor; industry; tumor registries; physicians; and medical centers. DSHEFS also conducts evaluation and validation studies of reporting systems covering occupational illnesses, with the intent of improving methods for measuring the magnitude of occupational health problems nationwide.

Division of Training and Manpower Development

The **Division of Training and Manpower Development (DTMD)** implements Section 21 of the OSH Act, which sets forth training and education requirements.[30] DTMD develops programs to increase the numbers and competence of safety and health professionals. This continuing education program provides short-term technical training courses, including seminars, independent study packages, and specialized workshops to federal, state, and local government; private industry; labor unions; and other organizations in the field of safety and health. The curriculum development component develops courseware and other training materials for NIOSH-sponsored training programs, including those represented by in-house faculty as well as those conducted by universities and other outside training organizations.

The educational resource development program continually assesses human resource needs for safety and health practitioners and researchers on a nationwide basis. To help meet the demand, DTMD administers a major training grant program to foster the development of academically based training programs for occupational physicians, occupational health nurses, industrial hygienists, toxicologists, epidemiologists, and safety professionals. DTMD also develops specific criteria for the selection of qualified organizations to conduct research training, graduate education, continuing education, and outreach programs to expand the network of knowledgeable professionals in occupational safety and health.

DTMD initiates special-emphasis projects targeted to physicians, engineers, managers, vocational education students, and science teachers to facilitate the inclusion of occupational safety and health knowledge in their formal program of study. The division establishes a collaborating relationship with many professional societies and accrediting bodies to formalize the process of long-term commitment through professional networking.

OSHRC

The OSHRC is not a government agency. Rather, it is an independent board whose members are appointed by the president and given quasi-judicial authority to handle contested OSHA citations. When a citation, proposed penalty, or abatement period issued by an OSHA area director is contested by an employer, OSHRC hears the case. OSHRC is empowered to review the evidence, approve the recommendations of the OSHA area director, reject those recommendations, or revise them by assigning substitute values. For example, if an employer contests the amount of a proposed penalty, OSHRC is empowered to accept the proposed amount, reject it completely, or change it.

Mine Safety and Health Administration (MSHA)

The mining industry is exempt from OSHA regulations. Instead, mining is regulated by the Metal and Nonmetallic Mine Safety Act. OSHA does regulate those aspects of the industry that are not directly involved in actual mining work. There is a formal memorandum of understanding between OSHA and the MSHA, the agency that enforces the Metal and Nonmetallic Mine Safety Act.

In 1977, Congress passed the Mine Safety and Health Act, which established the MSHA as a functional unit within the U.S. Department of Labor. MSHA works with MESA to ensure that the two agencies do not become embroiled in jurisdictional disputes.

Federal Railroad Administration

Railroads, for the most part, fall under the jurisdiction of OSHA. The Federal Railroad Administration (FRA) exercises limited jurisdiction over railroads in situations involving working conditions. Beyond this, railroads must adhere to the standards for *General Industry* in CFR Part 1910. OSHA and FRA personnel coordinate to ensure that jurisdictional disputes do not arise.

OSHA'S GENERAL INDUSTRY STANDARDS

The most widely applicable OSHA standards are the *General Industry Standards*. These standards are found in 29 CFR 1910. Part 1910 consists of Subparts A–Z, each carrying an uppercase-letter designation. Subparts A and B contain no compliance requirements. The remaining subparts are described in the following subsections. Subpart C has been removed and is reserved for future use. Subparts U–Y are also reserved for future use.

Subpart D: Walking–Working Surfaces

Subpart D contains the standards for all surfaces on which employees walk or work. Specific sections of Subpart D are as follows:

1910.21	Definitions
1910.22	General requirements
1910.23	Guarding floor and wall openings and holes
1910.24	Fixed industrial stairs
1910.25	Portable wood ladders
1910.26	Portable metal ladders
1910.27	Fixed ladders

1910.28	Safety requirements for scaffolding
1910.29	Manually propelled mobile ladder stands and scaffolds (towers)
1910.30	Other working surfaces

Subpart E: Exit Routes and Emergency Planning

Subpart E requires employers to ensure that employees have a safe, accessible, and efficient means of escaping a building under emergency circumstances. Specific sections of Subpart E are as follows:

1910.33	Table of Contents
1910.34	Coverage and definitions
1910.35	Compliance with alternate exit route codes
1910.36	General requirements
1910.37	Means of Egress—general
1910.38	Employee emergency plan and fire prevention plans

Emergency circumstances may be caused by fire, explosions, hurricanes, tornadoes, flooding, terrorist acts, earthquakes, nuclear radiation, or other acts of nature not listed here. The appendix to Subpart E covers exit routes, emergency action plans, and fire protection plans.

Subpart F: Powered Platforms, Manlifts, and Vehicle-Mounted Work Platforms

Subpart F applies to powered platforms, mechanical lifts, and vehicle-mounted work platforms. The requirements of this subpart apply only to employers who use this type of equipment in facility maintenance operations. Specific sections of Subpart F are as follows:

1910.66	Powered platforms for building maintenance
1910.67	Vehicle-mounted elevating and rotating work platforms
1910.68	Manlifts

Subpart G: Occupational Health and Environmental Controls

The most widely applicable standard in Subpart G is 1910.95 (occupational noise exposure). Other standards in this subpart pertain to situations where ionizing and/or nonionizing radiation are present. Specific sections of Subpart G are as follows:

1910.94	Ventilation
1910.95	Occupational noise exposure
1910.96	Redesignated as 1910.1096
1910.97	Nonionizing radiation
1910.98	Effective dates

Subpart H: Hazardous Materials

Four of the standards in Subpart H are widely applicable. Section 1901.106 is an extensive standard covering the use, handling, and storage of flammable and combustible liquids. Of particular concern are fire and explosions. Section 1910.107 applies to indoor spray-painting processes and processes in which paint (powder coating) is applied in powder form (e.g., electrostatic powder spray). Sections 1910.112, 1910.113, and 1910.121 are reserved for future use.

Section 1910.119 applies to the management of processes involving specifically named chemicals and flammable liquids and gases. Section 1910.120 contains requirements relating to emergency response operations and hazardous waste. All the standards contained in Subpart H are as follows:

1910.101	Compressed gases (general requirements)
1910.102	Acetylene
1910.103	Hydrogen
1910.104	Oxygen
1910.105	Nitrous oxide
1910.106	Flammable and combustible liquids
1910.107	Spray finishing using flammable and combustible materials
1910.108	Dip tanks containing flammable and combustible materials
1910.109	Explosive and blasting agents
1910.110	Storage and handling of liquefied petroleum gases
1910.111	Storage and handling of anhydrous ammonia
1910.119	Process safety management of highly hazardous chemicals
1910.120	Hazardous waste operations and emergency response
1910.122	Table of contents
1910.123	Dipping and coating operations: coverage and definitions
1910.124	General requirements for dipping and coating operations
1910.125	Additional requirements for dipping and coating operations that use flammable or combustible liquids
1910.126	Additional requirements for special dipping and coating applications

Subpart I: Personal Protective Equipment

Subpart I contains three of the most widely applicable standards: 1910.132 General Requirements; 1910.133 Eye and Face Protection; and 1910.134 Respiratory Protection. The most frequently cited OSHA violations relate to these and the other personal protective equipment standards. All the standards in this subpart are as follows:

1910.132	General requirements
1910.133	Eye and face protection
1910.134	Respiratory protection
1910.135	Head protection
1910.136	Foot protection
1910.137	Electrical protective devices
1910.138	Hand protection

Subpart J: General Environment Controls

This subpart contains standards that are widely applicable because they pertain to general housekeeping requirements. Standard 1910.143, Nonwater carriage disposal systems, is reserved. An especially important standard contained in this subpart is

1910.146: Permit-Required Confined Spaces. A confined space is one that meets any or all of the following criteria:

- Large enough and so configured that a person can enter it and perform assigned work tasks therein
- Continuous employee occupancy is not intended

The *lockout/tagout* standard is also contained in this subpart. The standards in this subpart are as follows:

1910.141	Sanitation
1910.142	Temporary labor camps
1910.144	Safety color code for marking physical hazards
1910.145	Accident prevention signs and tags
1910.146	Permit-required confined spaces
1910.147	Control of hazardous energy (lockout/tagout)

Subpart K: Medical and First Aid

This is a short subpart, the most important section of which pertains to eyeflushing. If employees are exposed to *injurious corrosive materials*, equipment must be provided for quickly flushing the eyes and showering the body. The standard also requires medical personnel to be readily available. "Readily available" can mean that there is a clinic or hospital nearby. If such a facility is not located nearby, employers must have a person on hand who has had first-aid training. The standard in this subpart is

1910.151	Medical services and first aid
1910.152	Reserved

Subpart L: Fire Protection

This subpart contains the bulk of OSHA's fire protection standard. These standards detail the employer's responsibilities concerning fire brigades, portable fire-suppression equipment, fixed fire-suppression equipment, and fire-alarm systems. Employers are not required to form fire brigades, but if they choose to, employers must adhere to the standard set forth in 1910.156. Subpart L also contains five Appendices A–E. The standards in this subpart are as follows:

FIRE PROTECTION

1910.155	Scope, application, and definitions applicable to this subpart
1910.156	Fire brigades

PORTABLE FIRE-SUPPRESSION EQUIPMENT

1910.157	Portable fire extinguishers
1910.158	Standpipe and hose systems

FIXED FIRE-SUPPRESSION EQUIPMENT

1910.159	Automatic sprinkler systems
1910.160	Fixed extinguishing systems, general

1910.161	Fixed extinguishing systems, dry chemical
1910.162	Fixed extinguishing systems, gaseous agent
1910.163	Fixed extinguishing systems, water spray and foam

OTHER FIRE PROTECTION SYSTEMS

| 1910.164 | Fire detection systems |
| 1910.165 | Employee alarm systems |

Subpart M: Compressed Gas/Air

This subpart contains just three sections and only one standard, 1910.169. This standard applies to compressed-air equipment that is used in drilling, cleaning, chipping, and hoisting. (Standards 1910.166, 1910.167, and 1910.168 are reserved for future use.) There are many other uses of compressed air in work settings, but 1910.169 applies only to these applications. The standard in this subpart is

| 1910.169 | Air receivers |

Subpart N: Materials Handling and Storage

This subpart contains one broad standard—1910.176—and several more specific standards relating to 1910.176. Subpart N is actually limited in scope. It applies only to the handling and storage of materials, changing rim wheels on large vehicles, and the proper use of specific equipment identified in the standards' titles. All the standards in this subpart are as follows:

1910.176	Handling material—general
1910.177	Servicing multi-piece and single-piece rim wheels
1910.178	Powered industrial trucks (plus Appendix A)
1910.179	Overhead and gantry cranes
1910.180	Crawler locomotive and truck cranes
1910.181	Derricks
1910.183	Helicopters
1910.184	Slings

Subpart O: Machinery and Machine Guarding

This subpart contains standards relating to specific types of machines. The types of machines covered are identified in the titles of the standards contained in Subpart O. These standards are as follows:

1910.211	Definitions
1910.212	General requirements for all machines
1910.213	Woodworking machinery requirements
1910.214	Cooperage machinery
1910.215	Abrasive wheel machinery
1910.216	Mills and calendars in the rubber and plastics industries
1910.217	Mechanical power presses
1910.218	Forging machines
1910.219	Mechanical power-transmission apparatus

Subpart P: Hand Tools/Portable Power Tools and Other Hand-Held Equipment

This subpart contains standards relating to the use of hand tools, portable power tools, and compressed-air-powered tools. The types of tools covered in this subpart, in addition to typical hand tools, include jacks, saws, drills, sanders, grinders, planers, power lawnmowers, and other tools. The standards contained in this subpart are as follows:

1910.241	Definitions
1910.242	Hand- and portable-powered tools and equipment, general
1910.243	Guarding of portable tools and equipment
1910.244	Other portable tools and equipment

Subpart Q: Welding, Cutting, and Brazing

Welding, cutting, and brazing are widely used processes. This subpart contains the standards relating to these processes in all their various forms. The primary safety and health concerns are fire protection, employee personal protection, and ventilation. The standards contained in this subpart are as follows:

1910.251	Definitions
1910.252	General requirements
1910.253	Oxygen-fuel gas welding and cutting
1910.254	Arc welding and cutting
1910.255	Resistance welding

Subpart R: Special Industries

This subpart is different from others in Part 1910. Whereas other subparts deal with specific processes, machines, and materials, Subpart R deals with specific industries. Each separate standard relates to a different category of industry. The standards contained in this subpart are as follows (Standard 1910.267 is reserved):

1910.261	Pulp, paper, and paperboard mills
1910.262	Textiles
1910.263	Bakery equipment
1910.264	Laundry machinery and operations
1910.265	Sawmills
1910.266	Logging operations
1910.268	Telecommunications
1910.269	Electrical power generation and transmission
1910.272	Grain handling facilities

Subpart S: Electrical

These standards are excerpted directly from the National Electrical Code. Those included in Subpart S are as follows (Standards 1910.330, 1910.336–1910.361, and 1910.381–1910.398 are reserved):

1910.301	Introduction
1910.302	Electric utilization systems

1910.303	General requirements
1910.304	Wiring design and protection
1910.305	Wiring methods, components, and equipment for general use
1910.306	Specific-purpose equipment and installations
1910.307	Hazardous (classified) locations
1910.308	Special systems
1910.331	Scope
1910.332	Training
1910.333	Selection and use of work practices
1910.334	Use of equipment
1910.335	Safeguards for personnel protection
1910.399	Definitions applicable to this subpart

Subpart T: Commercial Diving Operations

This subpart applies only to commercial diving enterprises. The standards contained in Subpart T are divided into six categories: (1) general, (2) personnel requirements, (3) general operations and procedures, (4) specific operations and procedures, (5) equipment procedures and requirements, and (6) record keeping. These standards are as follows:

1910.401	Scope and application
1910.402	Definitions
1910.410	Qualifications of dive teams
1910.420	Safe practices manual
1910.421	Pre-dive procedures
1910.422	Procedures during dive
1910.423	Post-dive procedures
1910.424	SCUBA diving
1910.425	Surface-supplied-air diving
1910.426	Mixed-gas diving
1910.427	Live boating
1910.430	Equipment
1910.440	Record keeping

Subpart Z: Toxic and Hazardous Substances

This is an extensive subpart containing the standards that establish *permissible exposure limits* (PELs) for selected toxic and hazardous substances (those for which PELs have been established). All such substances have an assigned PEL, which is the amount of a given airborne substance to which employees can be exposed during a specified period of time.

The standards relating to these specific toxic and hazardous substances are contained in 1910.1000 through 1910.1450 and are as follows:

1910.1000	Air contaminants
1910.1001	Asbestos
1910.1002	Coal tar pitch volatiles; interpretation of term

1910.1003	13 carcinogens (4-Nitrobiphenyl, etc.)
1910.1004	Alpha-Naphthylamine
1910.1006	Methyl chloromethyl ether
1910.1007	Dichlorobenzidine (and its salts)
1910.1008	Bis-Chloromethyl ether
1910.1009	Beta-Naphthylamine
1910.1010	Benzidine
1910.1011	4-Aminodiphenyl
1910.1012	Ethyleneimine
1910.1013	Beta-Propiolactone
1910.1014	2-Acetylaminofluorene
1910.1015	4-Dimethylaminoazobenzene
1910.1016	N-Nitrosodimethylamine
1910.1017	Vinyl chloride
1910.1018	Inorganic arsenic
1910.1020	Access to employee exposure and medical records
1910.1025	Lead
1910.1026	Chromium (VI)
1910.1027	Cadmium
1910.1028	Benzene
1910.1029	Coke oven emissions
1910.1030	Bloodborne pathogens
1910.1043	Cotton dust
1910.1044	1,2-dibromo-3-chloropropane
1910.1045	Acrylonitrile
1910.1047	Ethylene oxide
1910.1048	Formaldehyde
1910.1050	Methylenedianiline
1910.1051	Butadiene
1910.1052	Methylene chloride
1910.1096	Ionizing radiation
1910.1200	Hazard communication
1910.1201	Retention of DOT markings, placards, and labels
1910.1450	Occupational exposure to hazardous chemicals in laboratories

OSHA's General Industry Standards were covered in some depth in this section because they have the broadest application for students of workplace safety. The CFR also contains standards for the maritime industry. These standards are described in the sections that follow, but in less detail than was devoted to OSHA's General Standards because they are not as widely applicable.

OSHA'S MARITIME STANDARDS

OSHA's Maritime Standards apply to shipbuilding, ship-repairing, and ship-breaking operations not already covered by U.S. Coast Guard regulations. It is important to note that Coast Guard regulations do take precedence over OSHA Maritime Standards and supersede those standards in cases of overlap or conflict. Part 1915 of 29 CFR contains the standards relating to shipbuilding, ship repairing, and ship breaking. Part 1917 contains the standards for marine terminals, and Part 1918 contains longshoring standards. Part 1919 contains the gear-certification standards.

Part 1915: Shipyard Employment

Part 1915 is divided into Subparts A–Z covering the various aspects of shipyard operations. Those subparts are as follows:

Subpart A	General provisions
Subpart B	Confined and enclosed spaces and other dangerous atmospheres in shipyard employment
Subpart C	Surface preparation and preservation
Subpart D	Welding, cutting, and heating
Subpart E	Scaffolds, ladders, and other working surfaces
Subpart F	General working conditions
Subpart G	Gear and equipment for rigging and materials handling
Subpart H	Tools and related equipment
Subpart I	Personal protective equipment (PPE)
Subpart J	Ship's machinery and piping systems
Subpart K	Portable, unfired pressure vessels, drums and containers, other than ship's equipment
Subpart L	Electrical machinery
Subpart M–O	(Reserved)
Subpart P	Fire protection in shipyard equipment
Subpart Q–Y	(Reserved)
Subpart Z	Toxic and hazardous substances

Part 1917: Marine Terminals

Part 1917 is divided into seven subparts covering the various operations associated with marine terminals. Those subparts are as follows:

Subpart A	Scope and definitions
Subpart B	Marine terminal operations
Subpart C	Cargo handling gear and equipment
Subpart D	Specialized terminals
Subpart E	Personal protection
Subpart F	Terminal facilities
Subpart G	Related terminal operations and equipment

Part 1918: Longshoring

Longshoring involves loading and unloading maritime vessels. Part 1918 contains the standards relating to longshoring and is subdivided into the following subparts:

Subpart A	Scope and definitions
Subpart B	Gear certification
Subpart C	Gangways and other means of access

Subpart D	Working surfaces
Subpart E	Opening and closing hatches
Subpart F	Vessels' cargo-handling gear
Subpart G	Cargo-handling gear and equipment other than ship's gear
Subpart H	Handling cargo
Subpart I	General working conditions
Subpart J	Personal protective equipment

Part 1919: Gear Certification

Cargo gear- and material-handling devices must be certified as being safe for use in the handling, loading, unloading, and transport of materials. Persons who certify the gear and devices must be properly accredited. Part 1919 contains the following standards regulating the accreditation process:

Subpart A	General provisions
Subpart B	Procedure governing accreditation
Subpart C	Duties of persons accredited to certificate vessels' cargo gear
Subpart D	Certification of vessels' cargo gear
Subpart E	Certification of vessels: tests and proof loads; heat treatment; competent persons
Subpart F	Accreditation to certificate shore-based equipment
Subpart G	Duties of persons accredited to certificate shore-based material handling devices
Subpart H	Certification of shore-based material handling devices

OSHA'S CONSTRUCTION STANDARDS

These standards apply to employers involved in construction, alteration, or repair activities. To further identify the scope of the applicability of its construction standards, OSHA took the terms *construction*, *alteration*, and *repair* directly from the Davis-Bacon Act. This act provides minimum wage protection for employees working on construction projects. The implication is that if the Davis-Bacon Act applies to an employer, OSHA's Construction Standards also apply.[31]

These standards are contained in Part 1926 of the CFR Subparts A–Z. OSHA does not base citations on material contained in Subparts A and B. Consequently, those subparts have no relevance here. The remaining subparts are as follows:

Subpart C	General safety and health provisions
Subpart D	Occupational health and environmental controls
Subpart E	Personal protective and life-saving equipment
Subpart F	Fire protection and prevention
Subpart G	Signs, signals, and barricades
Subpart H	Materials handling, storage, use, and disposal
Subpart I	Tools—hand and power
Subpart J	Welding and cutting
Subpart K	Electrical

Subpart L	Scaffolds
Subpart M	Fall protection
Subpart N	Hoists, elevators, and conveyors
Subpart O	Motor vehicles, mechanized equipment, and marine operations
Subpart P	Excavations
Subpart Q	Concrete and masonry construction
Subpart R	Steel erection
Subpart S	Tunnels, shafts, caissons, cofferdams, and compressed air
Subpart T	Demolition
Subpart U	Blasting and use of explosives
Subpart V	Power transmission and distribution
Subpart W	Rollover protective structures; overhead protection
Subpart X	Stairways and ladders
Subpart Y	Commercial diving operations
Subpart Z	Toxic and hazardous substances
Subpart AA	Confined spaces in construction
Subpart CC	Cranes and derricks in construction
Subpart DD	Cranes and derricks used in demolition and underground construction

STANDARDS AND CODES

A **standard** is an operational principle, criterion, or requirement—or a combination of these. A **code** is a set of standards, rules, or regulations relating to a specific area. Standards and codes play an important role in modern safety and health management and engineering. These written procedures detail the safe and healthy way to perform job tasks and, consequently, to make the workplace safer and healthier.

Having written standards and codes that employees carefully follow can also decrease a company's exposure to costly litigation. Courts tend to hand down harsher rulings to companies that fail to develop or adapt, implement, and enforce appropriate standards and codes. Consequently, safety and health professionals should be familiar with the standards and codes relating to their company.

Numerous organizations develop standards for different industries. These organizations can be categorized broadly as follows: the government, professional organizations, and technical/trade associations.

Organizations that fall within these broad categories develop standards and codes in a wide variety of areas including, but not limited to, the following: dust hazards, electricity, emergency electricity systems, fire protection, first aid, hazardous chemicals, instrumentation, insulation, lighting, lubrication, materials, noise/vibration, paint, power, wiring, pressure relief, product storage and handling, piping materials, piping systems, radiation exposure, safety equipment, shutdown systems, and ventilation.

Figure 5–7 is a list of organizations that develop and publish standards and codes covering a wide variety of fields and areas of concern. Figure 5–8 is a reference for applicable standards and codes providers in various areas of concern (see Figure 5–7 to decode the abbreviations).

FIGURE 5–7   Organizations that develop standards and codes.

Organization	Abbreviation
U.S. Government	
Department of Transportation	DOT
Environmental Protection Agency	EPA
Federal Aviation Administration	FAA
Hazardous Materials Regulation Board	HMRB
Mine Safety and Health Administration	MSHA
National Institute for Standards and Technology	NIST
Occupational Safety and Health Administration	OSHA
U.S. Coast Guard	USCG
Professional Associations	
American Conference of Governmental Industrial Hygienists	ACGIH
American Industrial Hygiene Association	AIHA
American Institute of Chemical Engineers	AIChE
American Society of Heating, Refrigeration, and Air Conditioning Engineers	ASHRAE
American Society of Mechanical Engineers	ASME
Illumination Engineers Society	IES
Institute of Electrical and Electronic Engineers	IEEE
Instrument Society of America	ISA
Technical and Trade Associations	
Air Conditioning and Refrigeration Institute	ARI
Air Moving and Conditioning Association	AMCA
American Association of Railroads	AAR
American Gas Association	AGA
American Petroleum Institute	API
American Water Works Association	AWWA
Chlorine Institute	CI
Compressed Gas Association	CGA
Cooling Tower Institute	CTI
Manufacturing Chemists Association	MCA
Manufacturers Standardization Society	MSS
National Electrical Manufacturers Association	NEMA
National Fluid Power Association	NFPA
Pipe Fabrication Institute	PFI
Scientific Apparatus Makers Association	SAMA
Society and Plastics Industry	SPI
Steel Structures Painting Council	SSPC
Tubular Exchanger Manufacturers Association	TEMA

(continued)

FIGURE 5–7   Organizations that develop standards and codes. (*continued*)

Organization	Abbreviation
Testing Organizations	
American National Standards Institute	ANSI
American Society for Testing and Materials	ASTM
National Fire Protection Association	NFPA
National Safety Council	NSC
Underwriters Laboratories, Inc.	UL
Insurance Organizations	
American Insurance Association	ATA
Factory Insurance Association	FIA
Factory Mutual Systems	FMS
Industrial Risk Insurers	IRI
Oil Insurance Association	OIA

LAWS AND LIABILITY

The body of law pertaining to workplace safety and health grows continually as a result of a steady stream of liability litigation. Often a company's safety and health professionals are key players in litigation alleging negligence on the part of the company when an accident or health problem has occurred. Because safety and health litigation has become so prevalent today, professionals in the field need to be familiar with certain fundamental legal principles relating to such litigation. These principles are explained in the following section.

Fundamental Legal Principles

The body of law that governs safety and health litigation evolves continually. However, even though cases that set new precedents and establish new principles continue to occur, a number of fundamental legal principles surface frequently in the court. The most important of these—as well as several related legal terms—are summarized in the following paragraphs.

Negligence

Negligence means failure to take reasonable care or failure to perform duties in ways that prevent harm to humans or damage to property. The concept of *gross negligence* means failure to exercise even slight care or intentional failure to perform duties properly, regardless of the potential consequences. *Contributory negligence* means that an injured party contributed in some way to his or her own injury. In the past, this concept was used to protect defendants against negligence charges because the courts awarded no damages to plaintiffs who had contributed in any way to their own injury. Modern court cases have rendered this approach outdated with the introduction of *comparative negligence*. This concept distributes the negligence assigned to each party involved in litigation according to the findings of the court.

FIGURE 5–8   Reference table of standards providers.

Area of Concern	Organizations That Develop and Publish Standards
Dust hazards	ANSI, NFPA, UL, NSC, ACGIH, AIHA, BM, USCG
Electricity	ANSI, NFPA, NSC, ADA, FIA, FM, OIA, API, USCG, OSHA, IEEE, NEMA
Emergency electricity	NFPA, AIA, FM, IEEE, AGA, NEMA, USCG
Fire protection	ANSI, NFPA, UL, NSC, AIA, FIA, OIA, AWWA, API, CGA, MCA, NEMA, BM, USCG, OSHA
First aid	ANSI, NFPA, NSC, AIA, ACGIH, AIHA, CI, MCA, DOT, USCG
Flammability of substances	ANSI, NFPA, UL, NSC, FIA, FM, MCA, SPI, DOT, USCG, NIST
Hazardous chemicals	ANSI, NFPA, UL, NSC, AIA, FIA, FM, ACGIH, AIHA, AIChE, CI, CGA, MCA, DOT, USCG, OSHA
Instrumentation	ANSI, ASTM, NFPA, UL, AIA, FIA, FM, OIA, IEEE, ISA, AWWA, ARI, API, CGA, SAMA, NIST
Insulation	ANSI, ASTM, UL, AIA, FM, OIA, ASHRAE, USCG
Lighting	ASTM, NFPA, UL, AIA, FM, OIA, ASHRAE, USCG
Lubrication	ANSI, NFPA, ASME, AMCA
Materials	ANSI, ASTM, NFPA, UL, NSC, AIA, FM, OIA, ISA, AWWA, CI, CGA, CTI, MCA, TEMA, USCG, HMR
Noise—vibration	ANSI, ASTM, NFPA, UL, NSC, ACGIH, AIHA, ASHRAE, ISA, ARI, AMCA, AGA, NFPA, EPA
Paint/coatings	ANSI, ASTM, UL, AIChE, AWWA, SSPC, HMRB
Power writing	ANSI, NFPA, UL, FIA, FM, OIA, IEEE, API, NEMA, USCG, OSHA
Pressure relief	NFPA, FIA, FM, OIA, API, CI, CGA, USCG, HMRB, OSHA
Product storage/handling	ANSI, NFPA, AIA, FIA, FM, OIA, AIChE, AAR, API, CI, CGA, MCA, USCG, OSHA
Piping materials	ANSI, ASTM, NFPA, UL, AIA, FIA, FM, ASME, ASHRAE, AWWA, ARI, AGA, API, CI, CGA, MSS
Piping systems	ANSI, ASTM, NFPA, UL, AIA, FIA, FM, ASME, ASHRAE, AWWA, ARI, AGA, API, CI, CGA, MSS
Radiation exposure	ANSI, NFPA, NSC, AIA, ACGIH, AIHA, ASME, ISA, DOT
Safety equipment	ANSI, UL, NSC, FM, ACGIH, AIHA, CI, CGA, MCA, BM, USCG, OSHA
Shutdown systems	NFPA, UL, AIA, FIA, OIA, API, USCG
Ventilation	ANSI, NFPA, NSC, AIA, FIA, FM, ACGIH, AIHA, ASHRAE, ISA, CI, DOT, HMRB

Liability

Liability is a duty to compensate as a result of being held responsible for an act or omission. A newer, related concept is *strict liability*. This means that a company is liable for damages caused by a product that it produces, regardless of negligence or fault.

Care

Several related concepts fall under the heading of care. *Reasonable care* is the amount that would be taken by a prudent person in exercising his or her legal obligations toward others. *Great care* means the amount of care that would be taken by an extraordinarily prudent person in exercising his or her legal obligations toward others. *Slight care* represents the other extreme: a measure of care less than what a prudent person would take. A final concept in this category is the *exercise of due care*. This means that all people have a legal obligation to exercise the amount of care necessary to avoid, to the extent possible, bringing harm to others or damage to their property.

Ability to Pay

The concept of **ability to pay** applies when there are a number of defendants in a case, but not all have the ability to pay financial damages. It allows the court to assess all damages against the defendant or defendants who have the ability to pay. For this reason, it is sometimes referred to as the "deep pockets" principle.

Damages

Damages are financial awards assigned to injured parties in a lawsuit. *Compensatory damages* are awarded to compensate for injuries suffered and for those that will be suffered. *Punitive damages* are awarded to ensure that a guilty party will be disinclined to engage in negligent behavior in the future.

Proximate Cause

Proximate cause is the cause of an injury or damage to property. It is that action or lack of action that ties one person's injuries to another's lack of reasonable care.

Willful/Reckless Conduct

Behavior that is even worse than gross negligence is **willful/reckless conduct**. It involves intentionally neglecting one's responsibility to exercise reasonable care.

Tort

A **tort** is an action involving a failure to exercise reasonable care that as a result may lead to civil litigation.

Foreseeability

The concept of **foreseeability** holds that a person can be held liable for actions that result in damages or injury only when risks could have been reasonably foreseen.

The types of questions around which safety and health litigation often revolve are these: Does the company keep employees informed of rules and regulations? Does the company enforce its rules and regulations? Does the company provide its employees with the necessary training? The concepts set forth in this section come into play as both sides in the litigation deal with these questions from their respective points of view.

Health and safety professionals can serve their companies best by (1) making sure that a policy and corresponding rules and regulations are in place; (2) keeping employees informed about rules and regulations; (3) encouraging proper enforcement practices; and (4) ensuring that employees get the education and training they need to perform their jobs safely.

OSHA'S STAND ON SAFETY INCENTIVES

In recent years, organizations have begun offering incentives to employees to work safely. On the surface this sounds like a laudatory practice, and depending on how it is done it can be a laudatory practice. However, if not carefully designed and monitored, safety incentive programs can have the unintended consequence of encouraging under-reporting of accidents and injuries. Consequently, OSHA has come out strongly against not safety incentives per se, but against incentives that discourage employees from reporting accidents and injuries.

An OSHA memorandum titled "Employer Safety incentive and Disincentive Policies and Practices" directs field-compliance officers and whistleblower investigators to be especially vigilant in identifying instances in which accidents, incidents, or injuries have not been reported because of safety incentives. In part this memorandum reads as follows:

> There are several types of workplace policies and practices that could discourage reporting and could constitute unlawful discrimination and a violation of federal statutes ...
> OSHA has received reports of employers who have a policy of taking action against employees who are injured on the job, regardless of the circumstance surrounding the injury. Reporting an injury is always a protected activity ... While OSHA appreciates employers using safety as a key management metric, we cannot condone a program that encourages discrimination against workers who report injuries.[32]

Clearly from OSHA's perspective, organizations that plan to use incentives to enhance safety must be prepared to show that those incentives do not discourage employees from reporting accidents or injuries. OSHA does not oppose safety incentives—in fact it appreciates the intent of their use. However, it does oppose incentives that might lead to underreporting.

Key Terms and Concepts

- Abatement period
- Ability to pay
- Appeals process
- Closing conference
- Code
- Consultation services
- Damages
- De minimis violations
- Division of Biomedical and Behavioral Science (DBBS)
- Division of Respiratory Disease Studies (DRDS)
- Division of Surveillance, Hazard Evaluations, and Field Studies (DSHEFS)
- Division of Training and Manpower Development (DTMD)
- Employee responsibilities
- Employee rights
- Employer appeals
- Employer responsibilities
- Employer rights
- Foreseeability
- Inspection tour
- Liability
- National Institute for Occupational Safety and Health (NIOSH)
- Negligence
- Notice of contest
- Notice of proposed rule making
- Occupational diseases
- Occupational Safety and Health Act (OSH Act)
- Occupational Safety and Health Administration (OSHA)
- Occupational Safety and Health Review Commission (OSHRC)
- Opening conference
- OSHA Form 300
- OSHA Form 300A
- OSHA Form 301
- OSHA Poster 2203
- Other-than-serious violation
- Permanent variance
- Petition for Modification of Abatement (PMA)
- Proposed penalty
- Proximate cause
- Record keeping
- Repeat violations
- Reporting
- Serious violation
- Standard

- State-level safety and health programs
- Temporary emergency standards
- Temporary variance
- Tort
- Voluntary Protection Programs (VPPs)
- Whistleblower program
- Willful/reckless conduct
- Willful violation
- Workplace accidents
- Workplace inspection

Review Questions

1. Briefly explain the rationale for the OSH Act.
2. What is OSHA's mission or purpose?
3. List those who are exempted from coverage by OSHA.
4. Explain the difference between an OSHA standard and an OSHA regulation.
5. Explain how the following processes relating to OSHA standards are accomplished: passage of a new standard; request for a temporary variance; appealing a standard.
6. Briefly describe OSHA's latest record-keeping requirements.
7. What are OSHA's reporting requirements?
8. Explain what employers are required to do to keep employees informed.
9. Explain the various components of OSHA's enhanced enforcement policy.
10. Describe how a hypothetical OSHA workplace inspection would proceed from the first step to the last.
11. List and explain three different types of OSHA citations and the typical penalties that accompany them. Who gets the money from OSHA fines?
12. Describe the process for appealing an OSHA citation.
13. List and briefly explain OSHA's voluntary protection programs.
14. List five employer responsibilities.
15. List five employee rights.
16. Describe the purpose and organization of NIOSH.
17. Define the following legal terms as they relate to workplace safety: *negligence*; *liability*; *ability to pay*; *tort*.
18. Describe OSHA's stand on safety incentives.

Endnotes

1. U.S. Department of Labor, Occupational Safety and Health Administration, *All about OSHA*. Retrieved from osha.gov/Publications/all_about_OSHA.pdf on January 6, 2017.
2. Ibid.
3. Ibid., 2.
4. Ibid., 5.
5. Ibid., 7.
6. Ibid., 9.
7. Retrieved from osha.gov/standard on January 8, 2017.
8. Ibid.
9. Ibid.
10. U.S. Department of Labor, *All about OSHA*, 12.
11. Ibid.
12. Ibid., 12–13.
13. Ibid., 17.
14. Ibid., 24–25.
15. U.S. Department of Labor, *All about OSHA*, 27.

16. Ibid., 28.
17. U.S. Department of Labor, *All about OSHA*, 28.
18. Ibid.
19. Ibid., 29.
20. Ibid, 32.
21. Ibid., 34–35.
22. Ibid., 35–36.
23. Ibid., 37.
24. Ibid., 39–40.
25. Ibid., 37.
26. U.S. Department of Health and Human Services. Retrieved from hhs.gov on January 9, 2017.
27. Ibid.
28. Ibid., 2.
29. Ibid., 3.
30. Ibid., 4.
31. For more in-depth coverage of OSHA's construction standards, refer to D. L. Goetsch, *Construction Safety and the OSHA Standards* (Upper Saddle River, NJ: Prentice Hall, 2009).
32. OSHA Memorandum, "Employer Safety Incentive and Disincentive Policies and Practices," March 12, 2012.

CHAPTER

6 ‖ WORKERS' COMPENSATION

Learning Objectives

- Summarize the basic tenets of workers' compensation
- Explain the historical perspective of workers' compensation
- Summarize the most important workers' compensation legislation
- Explain the basic tenets of workers' compensation in today's business environment
- Summarize the various types of workers' compensation insurance
- Describe how workers' compensation disputes are settled
- Explain how injuries are handled through workers' compensation
- Explain how the various levels of disability are handled through workers' compensation
- Explain how monetary benefits of workers' compensation can vary
- Describe how medical treatment and rehabilitation are handled through workers' compensation
- Describe medical management of workplace injuries in workers' compensation cases
- Explain the concepts of administration and case management in workers' compensation
- Explain the concept of cost allocation in workers' compensation
- Summarize the most common problems with workers' compensation
- List the most effective strategies for spotting workers' compensation abuse
- Describe the future of workers' compensation
- List the most effective cost-reduction strategies for workers' compensation

OVERVIEW OF WORKERS' COMPENSATION

The concept of **workers' compensation** developed as a way to allow injured employees to be compensated appropriately without having to take their employer to court. The underlying rationale for workers' compensation had two aspects: (1) fairness to injured employees, especially those without the resources to undertake legal actions that are often long, drawn out, and expensive and (2) reduction of costs to employers associated with workplace injuries (e.g., legal, image, and morale costs). Workers' compensation is intended to be a no-fault approach to resolving workplace accidents by rehabilitating injured employees and minimizing their personal losses because of their reduced ability to perform and compete in the labor market.[1] Workers' compensation benefits can include medical care, rehabilitation expenses, and disability coverage.

Workers' compensation represents a compromise between the needs of employees and the needs of employers. Employees give up their right to seek unlimited compensation for pain and suffering through legal action. Employers award the prescribed compensation (typically through insurance premiums) regardless of the employee's negligence. The theory is that in the long run both employees and employers will benefit more than either would through legal action. As you will see later in this chapter, workers' compensation has not completely eliminated legal actions.

Objectives of Workers' Compensation

Workers' compensation laws are not uniform from state to state. In fact, there are extreme variations. However, regardless of the language contained in the enabling legislation in a specific state, workers' compensation as a concept has several widely accepted objectives:

- Replacement of income
- Rehabilitation of the injured employee
- Prevention of accidents
- Cost allocation[2]

The basic premises underlying these objectives are described in the following paragraphs.

Replacement of Income

Employees injured on the job lose income if they are unable to work. For this reason, workers' compensation is intended to replace the lost income adequately and promptly. Adequate **income replacement** is viewed as replacement of current and future income (minus taxes) at a ratio of two-thirds (in most states). Workers' compensation benefits are required to continue even if the employer goes out of business.

Rehabilitation of the Injured Employee

A basic premise of workers' compensation is that the injured worker will return to work in every case possible, although not necessarily in the same job or career field. For this reason, a major objective of workers' compensation is to rehabilitate the injured employee. The **rehabilitation** program is to provide the needed medical care at no cost to the injured employee until he or she is pronounced fit to return to work. The program also provides vocational training or retraining as needed. Both components seek to motivate the employee to return to the labor force as soon as possible.

Prevention of Accidents

Preventing future accidents is a major objective of workers' compensation. The theory underlying this objective is that employers will invest in **accident prevention** programs to hold down compensation costs. The payoff to employers comes in the form of lower insurance premiums that result from fewer accidents.

Cost Allocation

The potential risks associated with different occupations vary. For example, working as a miner is generally considered more hazardous than working as an architect. The underlying principle of **cost allocation** is to spread the cost of workers' compensation appropriately and proportionately among industries ranging from the most to the least hazardous. The costs of accidents should be allocated in accordance with the accident history of the industry so that high-risk industries pay higher workers' compensation insurance premiums than do low-risk industries.[3]

Who Is Covered by Workers' Compensation and Who Is Not

Workers' compensation laws are written at the state level, and there are many variations among these laws. As a result, it is difficult to make generalizations. Complicating the issue further is the fact that workers' compensation laws are constantly being amended, revised, and rewritten. Additionally, some states make participation in a workers' compensation program voluntary; others excuse employers with fewer than a specified number of employees.

In spite of the differences among workers' compensation laws in the various states, approximately 80 percent of the employees in the United States are covered by workers' compensation. Those employees who are not covered or whose coverage is limited vary according to the laws. However, they can be categorized in general terms as follows[4]:

- Agricultural employees
- Domestic employees
- Casual employees
- Hazardous work employees
- Charitable or religious employees
- Employees of small organizations
- Railroad and maritime employees
- Contractors and subcontractors
- Minors
- **Extraterritorial employees**

Coverage in these types of employment, to the extent there is coverage, varies from state to state as follows:

- **Agricultural employees** have limited coverage in 38 states, Puerto Rico, and the Virgin Islands. In 15 states, workers' compensation coverage for agricultural employees is voluntary. In these states, employers are allowed to provide coverage if they wish but are not required to do so.
- **Domestic employees** have coverage available in all 50 states and Puerto Rico. However, coverage tends to be limited and subject to minimum requirements regarding hours worked and earnings.

- **Casual employees** are employed in positions in which the work is occasional, incidental, and scattered at irregular intervals. Such employees are not typically afforded workers' compensation coverage.
- **Hazardous employment** is the only type afforded workers' compensation coverage in some states. To qualify, a particular type of employment must be on an approved list of hazardous or especially hazardous jobs. However, the trend in these states is to broaden the list of approved jobs.
- **Charitable or religious employees** are not afforded workers' compensation in most states when this work is irregular, temporary, or short-term.
- **Small organizations** that employ fewer than a stipulated number of employees do not fall under the umbrella of workers' compensation in 26 states.
- **Railroad and maritime workers** are not typically covered by workers' compensation. However, in most cases, they are covered by the Federal Employer's Liability Act. This act disallows the use of common law defenses by employers if sued by an employee for negligence.
- **Contractors and subcontractors** are those who agree to perform a job or service for an agreed amount of money in a nondirected, nonsupervised format. In essence, contract and subcontract employees are viewed as being self-employed. For this reason, they are not covered by workers' compensation.
- **Minors** are afforded regular workers' compensation coverage as long as they are legally employed. In some states, coverage is significantly higher for minors who are working illegally.
- **Extraterritorial employees** are those who work in one state but live in another. In these cases, the employee is usually on temporary duty. Such employees are typically afforded the workers' compensation coverage of their home state.[5]

HISTORICAL PERSPECTIVE

Before workers' compensation laws were enacted in the United States, injured employees had no way to obtain compensation for their injuries except to take their employer to court. Although common law did require employers to provide a safe and healthy work environment, injured employees bore the burden of proof that negligence in the form of unsafe conditions contributed to these injuries. Prior to passage of workers' compensation, employees often had to sue their employer to receive compensation for injuries resulting from a workplace accident or occupational disease, even when the following circumstances prevailed:

- The employee was disabled or died as the result of a **workplace accident** or **occupational disease**.
- The injury might have been expected to occur when the risks and hazards of the job were considered.
- **Worker negligence** on the part of a fellow worker or the injured employee clearly caused the injury.[6]

Proving that an injury was the result of employee negligence was typically too costly, too difficult, and too time-consuming to be a realistic avenue of redress for most injured employees. It took from six months to six years for an injured worker's case to work its way through the legal system.[7] Typically, injured workers, having lost their ability to generate income, could barely afford to get by, much less pay medical expenses, legal fees, and court costs. Another inhibitor was the **fear factor**. Injured employees who hoped to return to work after recovering were often afraid to file suit because they feared retribution by their employer. Employers not only might refuse to give them their jobs back but also might **blackball** them with other employers. Add to this that fellow employees were often afraid to testify to the negligence of the employer, and it is easy to see why few injured workers elected to take their employers to court.

Even with all of these inhibitors, some injured employees still chose to seek redress through the courts in the days before workers' compensation. Those who did faced a difficult challenge because the laws at that time made it easy for employers to defend themselves successfully. All an employer had to do to win a decision denying the injured plaintiff compensation was show that at least one of the following conditions existed at the time of the accident:

1. **Contributory negligence was a factor in the accident. Contributory negligence** meant that the injured worker's own negligence contributed to the accident. Even if the employee's negligence was a very minor factor, it was usually enough to deny compensation in the days before workers' compensation.
2. **There was negligence on the part of a fellow worker.** As with contributory negligence, negligence by a fellow employee, no matter how minor a contributing factor it was, could be sufficient to deny compensation. This defense was known as "the fellow servant rule."
3. **There was assumption of risk on the part of the injured employee.** If an employee knew that the job involved risk, he or she could not expect to be compensated when the risks resulted in accidents and injuries.[8] This defense relied on a long-standing principle of tort law known as "**assumption of risk**."

Because the majority of workplace accidents involve at least some degree of negligence on the part of the injured worker or fellow employees, employers typically won these cases. Because it required little more than a verbal warning by the employer to establish grounds for assumption of risk, the odds against an injured employee being awarded compensation become clear.

In his book *American Social Science*, Gagilardo gives an example of a case that illustrates how difficult it was to win compensation in the days before workers' compensation.[9] He relates the example of an employee who contracted tuberculosis while working under clearly hazardous conditions for a candy-making company. She worked in a wet, drafty basement that admitted no sunlight. Dead rats floated in the overflow of a septic tank that covered the basement floor, and a powerful stench permeated the workplace. Clearly, these were conditions that could contribute to the employee contracting tuberculosis. However, she lost the case and was denied compensation. The ruling judge justified the verdict as follows:

> We think that the plaintiff, as a matter of law, assumed the risk attendant upon her remaining in the employment (*Wager v. White Star Candy Company*, 217 N.Y. Supp. 173).

Situations such as this eventually led to the enactment of workers' compensation laws in the United States.

WORKERS' COMPENSATION LEGISLATION

Today, all 50 states, the District of Columbia, Guam, and Puerto Rico have workers' compensation laws. However, these laws did not exist prior to 1948. Considering that Prussia passed a workers' compensation law in 1838, the United States was obviously slow to adopt the concept. In fact, the first workers' compensation law enacted in the United States did not pass until 1908, and it applied only to federal employees working in especially hazardous jobs. The driving force behind passage of this law was President Theodore Roosevelt, who as governor of New York had seen the results of workplace accidents firsthand. Montana was the first state to pass a compulsory workers' compensation law. However, it was short-lived. Ruling that the law was unconstitutional, the Montana courts overturned it.

In 1911, the New York Court of Appeals dealt proponents of workers' compensation a serious blow. The New York state legislature had passed a compulsory workers' compensation law in 1910. However, in the case of *Ives v. South Buffalo Railway Company* (201 N.Y. 271, 1911), the New York Court of Appeals declared the law unconstitutional based on the contention that it violated the due process clause in the Fourteenth Amendment to the U.S. Constitution.[10]

This ruling had a far-reaching impact. However, even with such precedent-setting cases as *Ives* on the books, pressure for adequate workers' compensation grew as unsafe working conditions continued to result in injuries, diseases, and deaths. In fact, shortly after the New York Court of Appeals released its due process ruling, tragedy struck in a New York City textile factory.

On March 25, 1911, the building that housed the Triangle Shirtwaist Factory on its eighth floor caught fire and burned.[11] As a result of the fire, 149 of the company's 600 workers died, and another 70 were injured. Although the cause of the accident could not be determined, it was clear to investigators and survivors alike that unsafe conditions created by the management of the company prevented those who died or were injured from escaping the fire.

Exit passageways on each floor of the building were unusually narrow (20 inches wide), which made it difficult for employees to carry out bolts of material. A wider exit on each floor was kept locked to force employees to use the narrow exit. The two elevators were slow and able to accommodate only small groups at a time.

As the fire quickly spread, employees jammed into the narrow passageways, crushing each other against the walls and underfoot. With all exits blocked, panic-stricken employees began to jump out of windows and down the elevator shafts. When the pandemonium subsided and the fire was finally brought under control, the harsh realization of why so many had been trapped by the deadly smoke and flames quickly set in.

The owners were brought into court on charges of manslaughter. Although they were not convicted, the tragedy did focus nation-wide attention on the need for a safe workplace and adequate workers' compensation. As a result, new, stricter fire codes were adopted in New York, and in spite of the state court's ruling in *Ives*, the state legislature passed a workers' compensation law.

The next several years saw a flurry of legislation in other states relating to workers' compensation. In response to demands from workers and the general public, several states passed limited or noncompulsory workers' compensation laws. Many such states held back out of fear of being overturned by the courts. Others, particularly Washington, publicly disagreed with the New York Court of Appeals and passed compulsory laws. The constitutionality debate continued until 1917, when the U.S. Supreme Court ruled that workers' compensation laws were acceptable.

MODERN WORKERS' COMPENSATION

Since 1948, all states have had workers' compensation laws. However, the controversy surrounding workers' compensation has not died. As medical costs and insurance premiums have skyrocketed, many small businesses have found it difficult to pay the premiums. Unrealistic workers' compensation rates are being cited more and more frequently as contributing to the demise of small business in America.

The problem has even developed into an economic development issue. Business and industrial firms are closing their doors in those states with the highest workers' compensation rates and moving to states with lower rates. States with lower rates are using this as part of their recruiting package to attract new businesses and industry. Where low-rate states border high-rate states, businesses are beginning to move their offices across the border to the low-rate state while still doing business in the high-rate state.

Critics are now saying that workers' compensation has gotten out of hand and is no longer fulfilling its intended purpose. To understand whether this is the case, one must begin with an examination of the purpose of workers' compensation:

1. To provide an appropriate level of income and medical benefits to injured workers or to provide income to the worker's dependents regardless of fault
2. To provide a vehicle for reducing the amount of personal injury litigation in the court system
3. To relieve public and private charities of the financial strain created by workplace injuries that go uncompensated
4. To eliminate time-consuming and expensive trials and appeals
5. To promote employer interest and involvement in maintaining a safe work environment through the application of an experience-rating system
6. To prevent accidents by encouraging frank, objective, and open investigations of the causes of accidents[12]

Early proponents of workers' compensation envisioned a system in which both injured workers and their employers would win. Injured workers would receive prompt compensation, adequate medical benefits, and appropriate rehabilitation to allow them to reenter the workforce and be productive again. Employers would avoid time-consuming, expensive trials and appeals and would improve relations with employees and the public in general.

What proponents of workers' compensation did not anticipate were the following factors: (1) employees who would see workers' compensation as a way to ensure themselves a lifelong income without the necessity of work; (2) enormous increases in the costs of medical care with corresponding increases in workers' compensation insurance premiums; and (3) the radical differences among workers' compensation laws passed by the various states.

Not all employees abide by the spirit of workers' compensation (i.e., rehabilitation in a reasonable amount of time). Attempted abuse of the system was perhaps inevitable.

Unfortunately, such attempts result in a return to what workers' compensation was enacted to eliminate: time-consuming, drawn-out, expensive legal battles, and the inevitable appeals.

Proponents of workers' compensation **reform** can cite a long list of cases that illustrate their point. The city of Pittsfield, Massachusetts, was once overwhelmed by workers' compensation claims. One of the more remarkable cases concerned a city worker who was receiving workers' compensation benefits as the result of a back injury. While collecting benefits, he was a star player for a local softball team.[13] He eventually agreed to waive his right to compensation for a lump sum settlement of $12,000 plus $3,000 for his lawyer, and city officials considered themselves cheaply rid of him.[14]

Another Pittsfield city employee in the Department of Public Works was injured and began collecting workers' compensation at a rate of $295.50 per week. In addition to his job with the city, this worker owned a small diesel oil company. When his workers' compensation benefits were called into question because he owned a business that produced an income, the injured employee sold the business to his son.[15]

Medical costs have skyrocketed in the United States since the 1960s. There are many reasons for this. During this same period, the costs associated with other basic human needs, including food, clothing, transportation, shelter, and education, have also increased markedly. Increases in medical costs can be explained, at least partially, as the normal cost-of-living increases experienced in other sectors of the economy. However, the costs associated with medical care have increased much faster and much more than the costs in these other areas. The unprecedented increases can be attributed to two factors: (1) technological developments that have resulted in extraordinary but costly advances in medical care and (2) a proliferation of **litigation** that has driven the cost of malpractice insurance steadily up. Each of these factors has contributed to higher medical costs. **Malpractice** suits that once might not even have gone to court now result in multimillion-dollar settlements. Such costs are, of course, passed on to whoever pays the medical bill—in this case, employers who must carry workers' compensation insurance.

WORKERS' COMPENSATION INSURANCE

The costs associated with workers' compensation must be borne by employers as part of their overhead. In addition, employers must also ensure that the costs will be paid even if they go out of business. The answer for most employers is workers' compensation insurance.

In most states, though not all, workers' compensation insurance is compulsory.

A common thread woven through all the various compensation laws is the requirement that employers carry workers' compensation insurance. There are three types: **state funds**, **private insurance**, and **self-insurance**. Regardless of the method of coverage chosen, rates can vary greatly from company to company and state to state. The laws vary from state to state, Figure 6–1. To view the specific information for a given state go to **www.cutcomp.com/depts.htm**. Rates are affected by a number of different factors, including the following:

- Number of employees
- Types of work performed (risk involved)
- Accident experience of the employer

FIGURE 6–1   Workers' compensation coverage methods allowed for selected states.*

* Workers compensation laws vary from state to state.
Source: Lane V. Erickson/Shutterstock

- Potential future losses
- Overhead and profits of the employer
- Quality of the employer's safety program
- Estimates by actuaries

Insurance companies use one of the following six methods in determining the premium rates of employers:

1. **Schedule rating.** Insurance companies establish baseline safety conditions and evaluate the employer's conditions against the baselines. Credits are awarded for conditions that are better than the baseline, and debits are assessed for conditions that are worse. Insurance rates are adjusted accordingly.
2. **Manual rating.** A **manual of rates** is developed that establishes rates for various occupations. Each occupation may have a different rate based on its perceived level of hazard. The overall rate for the employer is a pro-rata combination of all the individual rates.
3. **Experience rating.** Employers are classified by type. Premium rates are assigned based on predictions of average losses for a given type of employer. Rates are then adjusted either up or down according to the employer's actual experience over the past three years.
4. **Retrospective rating.** Employees pay an established rate for a set period. At the end of the period, the actual experience is assessed and an appropriate monetary adjustment is made.
5. **Premium discounting.** Large employers receive discounts on their premiums based on their size. The theory behind this method is that it takes the same amount of time to service a small company's account as it does for a large company, but the large company produces significantly more income for the insurer. **Premium discounts** reward the larger company for its size.
6. **Combination method.** The insurer combines two or more of the other methods to arrive at premium rates.[16]

The trend nationwide for the past decade has been for premiums to increase markedly. For example, over the past 10 years, some states experienced increases of more than 60 percent. This trend will ensure that workers' compensation remains a controversial issue in the state legislatures.

RESOLUTION OF WORKERS' COMPENSATION DISPUTES

One of the fundamental objectives of workers' compensation is to avoid costly, time-consuming litigation. Whether this objective is being accomplished is questionable. If a worker is injured because of reckless or intentional action by an employer, that worker can bypass the workers' compensation system and sue in court for various damages including punitive, pain and suffering, and mental anguish. When an injured employee and the employer's insurance company disagree on some aspect of the compensation owed (e.g., weekly pay, length of benefits, and degree of disability), the disagreement must be resolved. Most states have an arbitration board for this purpose. Neither the insurance company nor the injured employee is required to hire an attorney. However, many employees do. There are a number of reasons for this. Some don't feel they can adequately represent themselves. Others are fearful of the "big business running over the little guy" syndrome. In any case, workers' compensation litigation is still very common and expensive.

Allowable attorney fees are set by statute, administrative rule, or policy in most states. In some states, attorney fees can be added to the injured employee's award. In others, the fee is a percentage of the award.

INJURIES AND WORKERS' COMPENSATION

The original workers' compensation concept envisioned compensation for workers who were injured in on-the-job accidents. What constituted an accident varied from state to state. However, all original definitions had in common the characteristics of being *sudden* and *unexpected*. Over the years, the definition of an accident has undergone continual change. The major change has been a trend toward the elimination of the "sudden" characteristic. In many states, the gradual onset of a disease as a result of prolonged exposure to harmful substances or a harmful environment can now be considered an accident for workers' compensation purposes.

A **harmful environment** does not have to be limited to its physical components. Psychological factors (such as stress) can also be considered. In fact, the highest rate of growth in workers' compensation claims over the past two decades has been in the area of stress-related injuries.

The National Safety Council maintains statistical records of the numbers and types of injuries suffered in various industries in the United States. Industries are divided into the following categories: agriculture, mining, construction, manufacturing, transportation/public utilities, trade, services, and public sector. Injuries in these industrial sectors are classified according to the type of accident that caused them. Accident types include overexertion, being struck by or against an object, falls, bodily reactions, caught in or between objects, motor vehicle accident, coming in contact with radiation or other caustics, being rubbed or abraded, and coming in contact with temperature extremes.

More than 30 percent of all disabling work injuries are the result of overexertion when all industry categories are viewed in composite. The next most frequent cause of injuries is struck by/struck against objects at 24 percent. Falls account for just more than 17 percent. The remainder is fairly evenly distributed among the other accident types listed above.[17]

AOE and COE Injuries

Workers' compensation benefits are owed only when the **injury arises out of employment (AOE)** or **injury occurs in the course of employment (COE)**. When employees are injured undertaking work prescribed in their job description, work assigned by a supervisor, or work normally expected of employees, they fall into the AOE category. Sometimes, however, different circumstances determine whether the same type of accident is considered to be AOE. For example, say a soldering technician burns her hand while repairing a printed circuit board that had been rejected by a quality control inspector. This injury would be classified as AOE. Now suppose the same technician brings a damaged printed circuit board from her home stereo to work and burns her hand while trying to repair it. This injury would not be covered because the accident did not arise from her employment. Determining whether an injury should be classified as AOE or COE is often a point of contention in workers' compensation litigation.

Who Is an Employee?

Another point of contention in workers' compensation cases is the definition of the term *employee*. This is an important definition because it is used to determine AOE and COE. A person who is on the company's payroll, receives benefits, and has a supervisor is clearly an **employee**. However, a person who accepts a service contract to perform a specific task or set of tasks and is not directly supervised by the company is not considered an employee. Although definitions vary from state to state, there are common characteristics. In all definitions, the workers must receive some form of remuneration for work done, and the employer must benefit from this work. Also, the employer must supervise and direct the work, both process and result. These factors—supervision and direction—are what set **independent contractors** apart from employees and exclude them from coverage. Employers who use independent contractors sometimes require the contractors to show proof of having their own workers' compensation insurance.

Unless an employer provides transportation, employees are not generally covered by workers' compensation when traveling to and from work. However, in some circumstances, they can be covered.

DISABILITIES AND WORKERS' COMPENSATION

Injuries that are compensable typically fall into one of four categories: (1) temporary partial disability, (2) temporary total disability, (3) permanent partial disability, and (4) permanent total disability (Figure 6–2). Determining the extent of disability is often a contentious issue. In fact, it accounts for more workers' compensation litigation than any other issue. Further, when a disability question is litigated, the case tends to be complicated because the evidence is typically subjective, and it requires hearing officers, judges, or juries to determine the future.

FIGURE 6–2 Disability claims are often litigated.

Source: CoraMax/Shutterstock

Temporary Disability

Temporary disability is the state that exists when it is probable that an injured worker, who is currently unable to work, will be able to resume gainful employment with no or only partial disability. Temporary disability assumes that the employee's condition will substantially improve. Determining whether an employee is temporarily disabled is not normally difficult. Competent professionals can usually determine the extent of the employee's injuries, prescribe the appropriate treatment, and establish a timeline for recovery. They can then determine if the employee will be able to return to work and when the return might take place.

There is an important point to remember when considering a temporary disability case: The ability to return to work relates only to work with the company that employed the worker at the time of the accident.

Temporary disability can be classified as either *temporary total disability* or *temporary partial disability*. A **temporary total disability** classification means the injured worker is incapable of any work for a period of time but is expected to recover fully. Most workers' compensation cases fall in this classification. A **temporary partial disability** means the injured worker is capable of light or part-time duties. Depending on the extent of the injury, temporary partial disabilities sometimes go unreported. This practice is allowable in some states. It helps employers hold down the cost of their workers' compensation premium. This is similar to not reporting a minor fender bender to your automobile insurance agent.

Most states prescribe in law the benefits owed in temporary total disability cases. Factors prescribed typically include a set percentage of an employee's wage that must

SAFETY MYTH

WORKERS' COMPENSATION CLAIMS AND CHARACTER

A leopard doesn't change its spots. A dishonest employee will always abuse the workers' compensation system with fraudulent claims. Right? Maybe not. A company in Oklahoma reduced its workers' compensation costs in just two years from $486,000 to $47,000. How did this company accomplish a reduction of this magnitude? With character training based on biblical principles. The program focuses on such character traits as honesty, attentiveness, gratitude, and dependability. At monthly training sessions, employees discuss one character trait and how it applies on the job.

be paid and a maximum period during which benefits can be collected. Figure 6–3 shows this information for a geographically distributed selection of states. Because workers' compensation legislation changes continually, this figure is provided only as an illustration of how benefits are prescribed in the laws of the various states. Actual rates are subject to change.

FIGURE 6–3 Temporary total disability benefits for selected states.*

State	Percentage of Employee's Wage	Maximum Period
Alabama	66⅔	Duration of disability
Arkansas	66⅔	450 weeks
California	66⅔	Duration of disability
Florida	66⅔	104 weeks
Indiana	66⅔	500 weeks
Kansas	66⅔	Duration of disability
Maryland	66⅔	Duration of disability
Montana	66⅔	Duration of disability*
Nebraska	66⅔	Duration of disability
New Jersey	70	400 weeks
New Mexico	66⅔	Duration of disability
Vermont	66⅔	Duration of disability
Washington	60–75	Duration of temporary disability
Wisconsin	60–75	Duration of disability

*Or until worker is released to preinjury job or similar employment.

* Contents of this chart are subject to change.
Source: U.S. Department of Labor.

Permanent Partial Disability

Permanent partial disability is the condition that exists when an injured employee is not expected to recover fully. In such cases, the employee will be able to work again but not at full capacity. Often employees who are partially disabled must be retrained for another occupation.

Permanent partial disabilities can be classified as *schedule* or *nonschedule* disabilities. **Schedule disabilities** are typically the result of nonambiguous injuries such as the loss of a critical but duplicated body part (e.g., arm, ear, hand, finger, or toe). Because such injuries are relatively straightforward, the amount of compensation that they generate and the period of time that it will be paid can be set forth in a standard schedule. A compilation of information from such schedules for a geographically distributed list of states is shown in Figure 6–4. Workers' compensation legislation changes continually; therefore, this figure is provided only as an example. Actual rates are subject to change continually.

Nonschedule injuries are less straightforward and must be dealt with on a case-by-case basis. Disabilities in this category tend to be the result of head injuries, the effects of which can be more difficult to determine. The amount of compensation awarded and the period over which it is awarded must be determined by studying the evidence. Awards are typically made based on a determination of disability percentage. For example, if it is determined that an employee has a 25 percent disability, the employee

FIGURE 6–4 Permanent partial disability benefits for selected states in 2005. Based on actual wages lost and not subject to minimums.*

State	Percentage of Employee's Wage	Maximum Period
Alabama	66⅔	300 weeks
Arkansas	66⅔	450 weeks
California	66⅔	619.25 weeks
Florida	*	364 weeks
Indiana	66⅔	500 weeks
Kansas	66⅔	415 weeks
Maryland	66⅔	Duration of disability
Montana	66⅔	350 weeks
Nebraska	66⅔	300 weeks
New Jersey	70	600 weeks
New Mexico	66⅔	500 weeks
Vermont	66⅔	330 weeks
Washington	—	—
Wisconsin	66⅔	1,000 weeks

*Shall not exceed 66⅔.

* Contents of this chart are subject to change.
Source: U.S. Department of Labor.

might be entitled to 25 percent of the income he or she could have earned before the injury with normal career progression factored in.

Four approaches to handling permanent partial disability cases have evolved. Three are based on specific theories, and the fourth is based on a combination of two or more of these theories. The three theories are (1) whole-person theory, (2) wage-loss theory, and (3) loss of wage-earning-capacity theory.

Whole-Person Theory

The **whole-person theory** is the simplest and most straightforward of the theories for dealing with permanent partial disability cases. Once it has been determined that an injured worker's capabilities have been permanently impaired to some extent, this theory is applied like a subtraction problem. What the worker can do after recuperating from the injury is determined and subtracted from what he or she could do before the accident. Factors such as age, education, and occupation are not considered.

Wage-Loss Theory

The **wage-loss theory** requires a determination of how much the employee could have earned had the injury not occurred. The wages actually being earned are subtracted from what could have been earned, and the employee is awarded a percentage of the difference. No consideration is given to the extent or degree of disability. The only consideration is loss of actual wages.

Loss of Wage-Earning-Capacity Theory

The most complex of the theories for handling permanent partial disability cases is the **loss of wage-earning-capacity theory** because it is based not just on what the employee earned at the time of the accident, but also on what he or she might have earned in the future. Making such a determination is obviously a subjective undertaking. Factors considered include past job performance, education, age, gender, and advancement potential at the time of the accident, among others. Once future earning capacity has been determined, the extent to which it has been impaired is estimated, and the employee is awarded a percentage of the difference. Some states prescribe maximum amounts of compensation and maximum periods within which it can be collected. For example, in Figure 6–4, Alabama sets 300 weeks as the maximum period for collecting compensation on a nonschedule injury. Maryland, on the other hand, awards compensation for the duration of the disability. Scheduled disabilities are typically compensated for the duration in all states.

The use of schedules has reduced the amount of litigation and controversy surrounding permanent partial disability cases. This is the good news aspect of schedules. The bad news aspect is that they may be inherently unfair. For example, a surgeon who loses his hand would receive the same compensation as a laborer with the same injury if the loss of a hand is scheduled.

Permanent Total Disability

A **permanent total disability** exists when an injured employee's disability is such that he or she cannot compete in the job market. This does not necessarily mean that the employee is helpless. Rather, it means an inability to compete reasonably. Handling permanent total disability cases is similar to handling permanent partial disability cases except that certain injuries simplify the process. In most states, permanent total disability can be assumed if certain specified injuries have been sustained (i.e., loss of

FIGURE 6–5 Duration of permanent total disability benefits for selected states.

State	Maximum Paid
Alabama	Duration of disability
Arkansas	Duration of disability
California	Life
Florida	Duration of disability
Indiana	500 weeks
Kansas	Duration of disability
Maryland	Duration of disability
Montana	Duration of disability
Nebraska	Duration of disability
New Jersey	450 weeks (life in some cases)
New Mexico	Life
Vermont	Duration of disability*
Washington	Life
Wisconsin	Life

*Minimum of 330 weeks.

Source: U.S. Department of Labor.

both eyes or both arms). In some states, compensation is awarded for life. In others, a time period is specified. Figure 6–5 shows the maximum period that compensation can be collected for a geographically distributed list of states. Notice the range of time periods in Figure 6–5.

MONETARY BENEFITS OF WORKERS' COMPENSATION

The **monetary benefits** accruing from workers' compensation vary markedly from state to state. The actual amounts are of less importance than the differences among them. Of course, the amounts set forth in schedules change frequently. However, for the purpose of comparison, consider that at one time the loss of a hand in Pennsylvania resulted in an award of $116,245. The same injury in Colorado brought only $8,736.

When trying to determine a scheduled award for a specific injury, it is best to locate the latest schedule for the state in question. One way to do this is to contact the following agency:

U.S. Department of Labor
Employment Standards Administration
Office of State Liaison and Legislative Analysis
Division of State Workers' Compensation Programs
200 Constitution Ave. NW
Washington, DC 20210
dol.gov

Death and Burial Benefits

Workers' compensation benefits accrue to the families and dependents of workers who are fatally injured. Typically, the remaining spouse receives benefits for life or until remarriage. However, in some cases, a time period is specified. Dependents typically receive benefits until they reach the legal age of maturity unless they have a condition or circumstance that makes them unable to support themselves even after attaining that age. Figure 6–6 contains the death benefits accruing to surviving spouses and children for a geographically distributed list of states. Because the actual amounts of benefits are subject to change, these are provided for illustration and comparison only.

FIGURE 6–6   Death benefits for surviving spouses and children for selected states.

State	Percentage of Employee's Wage		Maximum Period
	Spouse Only	Spouse and Children	
Alabama	50	66⅔	500 weeks
Arkansas	35	66⅔	Widow/widowerhood; children until 18 or married
California	66⅔	66⅔	—
Florida	50	66⅔	Widow/widowerhood; children until 18
Indiana	66⅔	66⅔	500 weeks
Kansas	66⅔	66⅔	Widow/widowerhood; children until 18
Maryland	66⅔	66⅔	Widow/widowerhood; children until 18
Montana	66⅔	66⅔	Surviving spouse—10 years; children until 18
Nebraska	66⅔	75	Widow/widowerhood; children until 18
New Jersey	50	70	Widow/widowerhood; children until 18
New Mexico	66⅔	66⅔	700 weeks
Vermont	66⅔	76⅔	Widow/widowerhood until 62; children until 18
Washington	60	70	Widow/widowerhood; children until 18
Wisconsin	66⅔	—	300 weeks

Source: U.S. Department of Labor.

FIGURE 6–7   Maximum burial allowances for selected states.

State	Maximum Allowance
Alabama	$3,000
Arkansas	6,000
California	5,000
Florida	2,500
Indiana	6,000
Kansas	5,000
Maryland	5,000
Montana	1,400
Nebraska	6,000
New Jersey	3,500
New Mexico	3,000
Vermont	5,500
Washington	*
Wisconsin	6,000

*200% of state's average monthly wage.

Note: Figures are provided for comparison and illustration only. They are subject to change.
Source: U.S. Department of Labor.

Further expenses are provided in addition to death benefits in all states except Oklahoma. As is the case with all types of workers' compensation, the amount of burial benefits varies from state to state and is subject to change. Figure 6–7 contains the maximum burial benefits for a geographically distributed list of states.

MEDICAL TREATMENT AND REHABILITATION

All workers' compensation laws provide for payment of the medical costs associated with injuries. Most states provide full coverage, but some limit the amount and duration of coverage. For example, in Arkansas, employer liability ceases six months after an injury occurs in those cases in which the employee is able to continue working or six months after he or she returns to work in cases where there is a period of recuperation. In either case, the employer's maximum financial liability is $10,000.[18] In Ohio, medical benefits for silicosis, asbestosis, and coal miner's pneumoconiosis are paid only in the case of temporary total or permanent total disability.[19]

The laws also specify who is allowed or required to select a physician for the injured employee. The options can be summarized as follows:

- **Employee selects the physician of choice.** This option is available in Alaska, Arizona, Delaware, Hawaii, Illinois, Kentucky, Louisiana, Maine, Massachusetts, Mississippi, Nebraska, New Hampshire, North Dakota, Ohio,

Oklahoma, Oregon, Rhode Island, Texas, the Virgin Islands, Washington, West Virginia, Wisconsin, and Wyoming.

- **Employee selects the physician from a list provided by the state agency.** This option applies in Connecticut, Nevada, New York, and the District of Columbia.
- **Employee selects the physician from a list provided by the employer.** This option applies in Georgia, Tennessee, and Virginia.
- **Employer selects the physician.** This option applies in Alabama, Florida, Idaho, Indiana, Iowa, Maryland, Montana, New Jersey, New Mexico, North Carolina, South Carolina, and South Dakota.
- **Employer selects the physician, but the selection may be changed by the state agency.** This option applies in Arkansas, Colorado, Kansas, Minnesota, Missouri, Utah, and Vermont.
- **Employer selects the physician, but after a specified period of time, the employee may choose another.** This option applies only in Puerto Rico.

Rehabilitation and Workers' Compensation

Occasionally, an injured worker will need rehabilitation before he or she can return to work. There are two types of rehabilitation: medical and vocational. Both are available to workers whose ability to make a living is inhibited by physical and/or mental work-related problems.

Medical rehabilitation consists of providing whatever treatment is required to restore to the extent possible any lost ability to function normally. This may include such services as physical therapy or the provision of prosthetic devices. **Vocational rehabilitation** involves providing the education and training needed to prepare the worker for a new occupation. Whether the rehabilitation services are medical or vocational in nature or both, the goal is to restore the injured worker's capabilities to the level that existed before the accident.

MEDICAL MANAGEMENT OF WORKPLACE INJURIES

Out-of-control workers' compensation cases in the 1990s led to the concept of **medical management of workplace injuries**.[20] Through better management of workers' compensation claims, more than 30 states have merged the concepts of workers' compensation and managed care. The goals of these state-level efforts are to (1) speed up the processing of workers' compensation claims; (2) reduce costs; (3) reduce fraud and abuse; and (4) improve medical management of workplace injuries.

Workers' compensation and managed care have been merged through the creation of *Health Partnership Programs* (HPPs). HPPs are partnerships between employers and their state's Bureau of Workers' Compensation (BWC). Employers who choose to participate (some states mandate participation) are required to have a *managed care organization* (MCO) that provides medical management of workplace injuries and illnesses.

The HPPs are effective because an MCO coordinates the paperwork generated by the injured employee, the employer, healthcare providers, and the BWC. As the Ohio example illustrates, the results of this coordination are clear. The average time required to process an injury report prior to implementation of the HPP was more than 66 days. The HPP reduced this to approximately 33 days, effectively cutting reporting time in half. Whether managed medical care will be able to continue this positive trend remains to be seen, but the states that have implemented HPPs have learned the following lessons:

- It is better to mandate HPPs than to make them optional.
- Cost containment is only part of the goal. Managed care programs must also include criteria such as lost wages, ability to return to work, and administrative costs to the employer.
- Employees want choice in selecting healthcare providers.
- Smart return-to-work programs are critical.

ADMINISTRATION AND CASE MANAGEMENT

Even though the Occupational Safety and Health Administration specifies what constitutes a recordable accident, it is not uncommon for minor injuries to go unreported. The employee may be given the rest of the day off or treated with first aid and returned to work. This is done to avoid time-consuming paperwork and to hold down the cost of workers' compensation insurance. However, if an accident results in a serious injury, several agencies must be notified. What constitutes a serious injury, like many workers' compensation issues, can differ from state to state. However, as a rule, an injury is serious if it requires more than 24 hours of active medical treatment (this does not include passive treatment such as observation). Of course, a fatality, a major disfigurement, or the loss of a limb or digit is also considered serious and must be reported.

At a minimum, the company's insurer, the state agency, and the state's federal counterpart must be notified. Individual states may require that additional agencies be notified. All establish a time frame within which notification must be made. Once the notice of injury has been filed, there is typically a short period before the victim or dependents can begin to receive compensation unless inpatient hospital care is required. However, when payments do begin to flow, they are typically retroactive to the date of the injury.

State statutes also provide a maximum time period that can elapse before a compensation claim is filed. The notice of injury does not satisfy the requirement of filing a claim notice. The two are separate processes. The statute of limitations on **claim notices** varies from state to state. However, most limit the period to no more than a year except in cases of work-related diseases in which the exact date of onset cannot be determined.

All such activities—filing injury notices, filing claim notices, arriving at settlements, and handling disputes—fall under the collective heading of administration and case management. Most states have a designated agency that is responsible for administration and case management. In addition, some states have independent boards that conduct hearings and hear appeals when disputes arise.

Once a workers' compensation claim is filed, an appropriate settlement must be reached. Three approaches can be used to settle a claim: (1) **direct settlement**, (2) **agreement settlement**, and (3) **public hearing**. The first two are used in uncontested cases, the third in contested cases.

1. **Direct settlement.** The employer or its insurance company begins making what it thinks are the prescribed payments. The insurer also sets the period over which payments will be made. Both factors are subject to review by the designated

state agency. This approach is used in Arkansas, Michigan, Mississippi, New Hampshire, Wisconsin, and the District of Columbia.

2. **Agreement settlement.** The injured employee and the employer or its insurance company work out an agreement on how much compensation will be paid and for how long. Such an agreement must be reached before compensation payments begin. Typically, the agreement is reviewed by the designated state administrative agency. In cases where the agency disapproves the agreement, the worker continues to collect compensation at the agreed-on rate until a new agreement can be reached. If this is not possible, the case becomes a contested case.

3. **Public hearing.** If an injured worker feels he or she has been inadequately compensated or unfairly treated, a hearing can be requested. Such cases are known as *contested cases*. The hearing commission reviews the facts surrounding the case and renders a judgment concerning the amount and duration of compensation. Should the employee disagree with the decision rendered, civil action through the courts is an option.

COST ALLOCATION

Workers' compensation is a costly concept. From the outset, one of the basic principles has been cost allocation. *Cost allocation* is the process of spreading the cost of workers' compensation across an industry so that no individual company is overly burdened. The cost of workers' compensation includes the costs of premiums, benefits, and administration. These costs have risen steadily over the years.

When workers' compensation costs by industry are examined, there are significant differences. For example, the cost of workers' compensation for a bank is less than one-half of 1 percent of gross payroll. For a ceramics manufacturer, the percentage might be as high as 3 or 4 percent.

Cost allocation is based on the **experience rating** of the industry. In addition to being the fairest method (theoretically) of allocating costs, this approach is also supposed to give employers an incentive to initiate safety programs. Opinions vary as to the fairness and effectiveness of this approach. Arguments against it include the following: (1) small firms do not have a sufficient number of employees to produce a reliable and accurate picture of the experience rating; (2) firms that are too small to produce an experience rating are rated by class of industry, thereby negating the incentive figure; (3) premium rates are more directly sensitive to experience levels in larger firms but are less so in smaller companies; and (4) in order to hold down experience ratings, employers may put their efforts into fighting claims rather than preventing accidents. Not much hard research has been conducted to determine the real effects of cost allocation. Such research is badly needed to determine if the theoretical construct of workers' compensation is, in reality, valid.

PROBLEMS WITH WORKERS' COMPENSATION

There are serious problems with workers' compensation in the United States. On the one hand, there is evidence of abuse of the system. On the other hand, many injured workers who are legitimately collecting benefits suffer a substantial loss of income.

Complaints about workers' compensation are common from all parties involved with the system (employers, employees, and insurance companies).

The examples in this chapter of individuals who claim that job stress has disabled them to the point that workers' compensation is justified are representative of many other similar cases. In 1980, there were so few stress claims that they were not even recorded as a separate category. Today, they represent a major and costly category.

Stress claims are more burdensome than physical claims because they are typically reviewed in an adversarial environment. This leads to the involvement of expert medical witnesses and attorneys. As a result, even though the benefits awarded for stress-related injuries are typically less than those awarded for physical injuries, the cost of stress claims is often higher because of the litigation.

In addition to abuse, a steadily increasing caseload is also a problem. For example, California saw its caseload increase by 40,000 in one year.[21] As this trend continues, the cost of workers' compensation will increase on a parallel track.

Although the cost of workers' compensation is increasing steadily, the amount of compensation going to injured workers is often disturbingly low. In a given year, if workers' compensation payments in the United States amount to around $27 billion (which is typical), $17 billion of this goes to benefits. Almost $10 billion is taken up by medical costs.[22] The amount of wages paid to injured workers in most states is 66 percent. This phenomenon is not new.

> Almost one-half million families each year are faced with getting by on a drastically reduced income because of a disabling injury suffered by the principal income earner. On-the-job accidents are supposed to be covered by workers' compensation and all states have compensation systems. However, the injured worker rarely receives an income that comes close to what he or she was earning before the accident.[23]

The most fundamental problem with workers' compensation is that it is not fulfilling its objectives. Lost income is not being adequately replaced, the number of accidents has not decreased, and the effectiveness of cost allocation is questionable. Clearly, the final chapter on workers' compensation has not yet been written.

SPOTTING WORKERS' COMPENSATION FRAUD AND ABUSE

There is evidence of waste, fraud, and abuse of the system in all states that have passed workers' compensation laws. However, the public outcry against fraudulent claims is making states much less tolerant of, and much more attentive to, abuse. For example, the Ohio legislature passed a statute that allows criminal charges to be brought against employees, physicians, and lawyers who give false information in a workers' compensation case. This is a positive trend. However, even these measures will not completely eliminate abuse.

For this reason, it is important for organizations to know how to spot employees who are trying to abuse the system by filing fraudulent workers' compensation claims. Following are some factors that should cause employers to view claims with suspicion. However, just because one or more of these factors are present does not mean that an employee is attempting to abuse the system. Rather, these are simply factors that should raise cautionary flags:

- The person filing the claim is never home or available by telephone or has an unlisted telephone number.

- The injury in question coincides with a layoff, termination, or plant closing.
- The person filing the claim is active in sports.
- The person filing the claim has another job.
- The person filing the claim is in line for early retirement.
- The rehabilitation report contains evidence that the person filing the claim is maintaining an active lifestyle.
- No organic basis exists for disability. The person filing the claim appears to have made a full recovery.
- The person filing the claim receives all mail at a post office box and will not divulge a home address.
- The person filing the claim is known to have skills such as carpentry, plumbing, or electrical that could be used to work on a cash basis while feigning a disability.
- There are no witnesses to the accident in question.
- The person filing the claim has relocated out of state or out of the country.
- Demands for compensation are excessive.
- The person filing the claim has a history of filing.
- Doctors' reports are contradictory in nature.
- A soft tissue injury is claimed to have produced a long-term disability.
- The injury in question occurred during hunting season.[24]

These factors can help organizations spot employees who may be trying to abuse the workers' compensation system. It is important to do so because legitimate users of the system are hurt, as are their employers, by abusers of the system. If one or more of these factors are present, employers should investigate the claim carefully before proceeding.

FUTURE OF WORKERS' COMPENSATION

The future of workers' compensation will be characterized by an ongoing need for reform as well as higher premiums, higher deductibles, and less coverage in insurance policies. The key to reforming workers' compensation is finding a way to allocate more of the cost to benefits and medical treatment and less to administration and litigation. Key elements in any reform effort are as follows:

- Stabilizing workers' compensation costs over the long term
- Streamlining administration of the system
- Reducing the costs associated with the resolution of medical issues
- Limiting stress-related claims
- Limiting vocational rehabilitation benefits
- Increasing benefits paid for temporary and permanent disabilities
- Reducing the amount that insurers may charge for overhead
- Providing more public input into the setting of rates
- Taking litigation out of the process to the extent possible
- Improving medical treatment management
- Improving overall case management
- Streamlining the claim notification and processing system
- Requiring more sufficient justification from insurance carriers concerning their rates

Controlling Rising Workers' Compensation Costs

The first and best line of defense against escalating workers' compensation costs is a safe and healthy workplace.[25] However, in today's litigious business environment, even this may not be enough. In order to be an effective member of the team responsible for keeping workers' compensation costs in check, safety and health professionals need to cooperate with risk management professionals in implementing the following strategies:

1. Establish an effective safety and health program, and document it clearly and comprehensively for the workers' compensation underwriters.
2. Review workers' compensation claims to ensure that they are accurate before they are submitted to underwriters.
3. Analyze concentrations of risk by location and have comprehensive, up-to-date plans on hand for preventing and responding to catastrophic events.
4. Advise risk management professionals on potential hazards and related risks so they can make informed decisions concerning levels of coverage and deductibles.
5. Communicate frequently with risk management personnel—you and they are on the same team.
6. Develop strategies for dealing with terrorism—many workers' compensation underwriters exclude terrorism from their coverage.[26]

COST-REDUCTION STRATEGIES

Safety and health professionals are responsible for helping their organizations hold down workers' compensation costs. Of course, the best way to accomplish this goal is to maintain a safe and healthy workplace, thereby preventing the injuries that drive up the costs. This section presents numerous other strategies that have proven effective in reducing workers' compensation costs after injuries have occurred, which happen in even the safest environments.

General Strategies

Regardless of the type of organization, there are several rules of thumb that can help reduce workers' compensation claims. These general strategies are as follows:

1. **Stay in touch with the injured employee.** Let injured employees know that they have not been forgotten and that they are not isolated. Answer all their questions and try to maintain their loyalty to the organization.
2. **Have a return-to-work program and use it.** The sooner an injured employee returns to work, even with a reduced workload, the lower workers' compensation costs will be. Reduced costs can, in turn, lower the organization's insurance premium. When using the return-to-work strategy, be cautious. Communicate with the employee and his or her medical treatment team. Have a clear understanding of the tasks that can be done and those that should be avoided, such as how much weight the employee can safely lift.

Colledge and Johnson recommend using the "S.P.I.C.E." model for improving the effectiveness of return-to-work programs. It consists of the following components:

- Simplicity
- Proximity

- Immediacy
- Centrality
- Expectancy[27]

Simplicity means that the medical professionals who treat injured employees should work closely with safety professionals to prevent "system"-induced complications. Such complications occur when employees become convinced their injuries are more serious than they really are because of ominous-sounding diagnostic terminology and complicated tests and treatments. Medical professionals and safety personnel should work together to keep the terminology simple and to explain tests and treatments in easily understood lay terms.

Proximity means keeping the injured employee as close to the job as possible. Employees who are physically separated from their place of employment and their fellow employees also become mentally separated. Within a short time, what used to be "us" can become "them." Giving as much injury care as possible at the work site, providing light-duty assignments, and communicating regularly with employees whose injuries preclude on-site assignments or treatment will keep employees connected and maintain the advantages of proximity.

Immediacy means that the faster an employee's injury claims can be handled, the less likely he or she will be to develop psychosocial issues that can complicate the recovery process. The longer it takes to process a claim, conduct a diagnosis, and begin treatments the more likely the employee is going to worry about the injury and to become accustomed to being off work. Immediate diagnosis, processing, and treatment can decrease the amount of time that elapses before the employee can begin a return-to-work program.

Centrality means getting the employee, his or her family, the medical professionals handling the case, insurance personnel, and the employer to agree on a common vision for successfully returning the injured party to work as soon as possible. It is important for injured employees and their families to know that everyone involved has the same goal and that everyone is working in good faith to achieve that goal.

Expectancy means creating the expectation that getting the employee well and back to work is the goal of all parties involved. It is achieved by communicating this message clearly to all parties and reinforcing it by establishing short-term goals and timelines for actually being back on the job. Achievement of each respective short-term goal should move the employee one step closer to recovery and return to work.

3. **Determine the cause of the accident.** The key to preventing future accidents and incidents is determining the cause of the accident in question, and the key to holding down workers' compensation costs is preventing accidents. Eliminating the root cause of every accident is fundamental to any cost-containment effort.

Specific Strategies

In addition to the general strategies presented above, there are numerous specific cost-containment strategies that have proven to be effective. These specific strategies are presented in this section.

1. **Cultivate job satisfaction.** According to Jon Gice, increasing job satisfaction is just as important as eliminating physical hazards in the workplace.[28] High stress, aggression, alienation, and social maladjustment—all factors associated with, and aggravated by, a lack of job satisfaction—can make employees less attentive while working. An inattentive employee is an accident waiting to happen. Gice recommends the following strategies for improving job satisfaction:

   - Recognize and reward employees.
   - Communicate frequently and openly with employees about job-related problems.
   - Give employees as much control over their work as possible.
   - Encourage employees to talk freely among themselves.
   - Practice conflict management.
   - Provide adequate staffing and expense budgets.
   - Encourage employees to use employee assistance programs.[29]

2. **Make safety part of the culture.** The following steps can be effective in making safety part of the organizational culture as a way to reduce workers' compensation costs:

   - Ensure visible, active leadership, involvement, and commitment from senior management.
   - Involve employees at all levels in the safety program and recognize them for their efforts.
   - Provide comprehensive medical care, part of which is a return-to-work program.
   - Ensure effective communication throughout the organization.
   - Coordinate all safety and health processes.
   - Provide orientation and training for all employees.
   - Have written safe work practices and procedures.
   - Have a comprehensive written safety policy.
   - Keep comprehensive safety records and analyze the data contained in those records.[30]

3. **Have a systematic cost-reduction program.** To reduce costs, an organization should have a systematic program that can be applied continually and consistently. Effective strategies include the following:

   - Insert safety notes and reminders in employees' paycheck envelopes.
   - Call injured employees at home to reassure them that they will have a job when they return.
   - Keep supervisors trained on all applicable safety and health issues, procedures, rules, and so on.
   - Hold monthly meetings to review safety procedures, strategies, and techniques.
   - Reward employees who give suggestions for making the workplace safer.[31]

4. **Use integrated managed care.** Managed care is credited by many with reducing workers' compensation costs nationwide. Others claim that cost reduction has occurred because managed care dangerously restricts the types and amount of healthcare provided to injured employees. Managed care can be fully integrated as follows:

   - A comprehensive safety and hazard-prevention program to keep employees safe and healthy

- On-site medical clinics
- Plant compensation teams that include nurses, rehabilitation technicians, and third-party adjudicators
- Claims management
- Company-wide safety and health database
- Return-to-work program[32]

Key Terms and Concepts

- Accident prevention
- Agreement settlement
- Assumption of risk
- Blackball
- Claim notices
- Contributory negligence
- Cost allocation
- Direct settlement
- Employee
- Experience rating
- Extraterritorial employees
- Fear factor
- Harmful environment
- Income replacement
- Independent contractors
- Injury arises out of employment (AOE)

- Injury occurs in the course of employment (COE)
- Litigation
- Loss of wage-earning-capacity theory
- Malpractice
- Manual of rates
- Medical management of workplace injuries
- Medical rehabilitation
- Monetary benefits
- Occupational disease
- Permanent partial disability
- Permanent total disability
- Premium discounts

- Private insurance
- Public hearing
- Reform
- Rehabilitation
- Schedule disabilities
- Self-insurance
- State funds
- Stress claims
- Temporary partial disability
- Temporary total disability
- Vocational rehabilitation
- Wage-loss theory
- Whole-person theory
- Worker negligence
- Workers' compensation
- Workplace accident

Review Questions

1. Explain the underlying rationale of workers' compensation as a concept.
2. List four objectives of workers' compensation.
3. List five types of employees who may not be covered by workers' compensation.
4. What is meant by the term *contributory negligence*?
5. What is meant by the term *assumption of risk*?
6. Explain the reasons for the unprecedented increases in medical costs in the United States.
7. What are the three types of workers' compensation insurance?
8. Insurance companies use one of six methods for determining the premium rates of employers. Select three and explain them.
9. How can one determine if an injury should be considered serious?
10. Explain the concepts of AOE and COE.
11. Distinguish between an employee and an independent contractor.
12. Define the following terms: *temporary disability* and *permanent disability*.
13. Explain the following theories of handling permanent partial disability cases: *whole-person*, *wage-loss*, and *loss of wage-earning capacity*.
14. Distinguish between medical and vocational rehabilitation.
15. What are the three approaches for settling workers' compensation claims?
16. Explain the concept of medical management of workplace injuries.
17. Explain the theory of cost allocation.
18. Summarize briefly the problems most widely associated with workers' compensation.

19. What types of actions are workers' compensation reform movements likely to recommend in the future?
20. Explain the most common workers' compensation cost-reduction strategies.

Endnotes

1. "Workers' Compensation Beginner's Guide." Retrieved from beginnersguide.com/small_ business/workers_compensation/ on March 15, 2013.
2. Ibid.
3. Ibid., 8.
4. Ibid., 10.
5. Ibid., 14.
6. Ibid.
7. E. Priz and S. Priz, *Worker's Compensation: A Field Guide for Employers* (Riverside, IL: Advanced Insurance Management, 2010), 47–48.
8. Ibid.
9. D. Gagilardo, *American Social Science* (New York: Harper & Row, 1949), 149.
10. E. Collart and D. Siracusa, *Workers Comp Made Easy,* The PEO Show, 2011, p. 7.
11. Ibid.
12. Ibid.
13. D. Goetsch, *Workers' Comp Cases*, Report 2017 (Niceville, FL: Human Resource Solutions, January 2017), 12–16.
14. Ibid.
15. Ibid.
16. Ibid.
17. Society of Manufacturing Engineers, *Accident Facts* (Chicago: Society of Manufacturing Engineers, 2017), 36.
18. U.S. Department of Labor, *State Workers' Compensation Laws.* Retrieved from dol.gov on July 1, 2017.
19. Ibid.
20. C. Cappelle, "Making a Strong Case for Managed Care," *Occupational Hazards* 61, no. 4: 67–71.
21. National Safety Council, *Accident Facts* (Chicago: National Safety Council 2016), 43.
22. Ibid.
23. Priz and Priz, *Worker's Compensation*, 76.
24. Ibid., 78.
25. Ibid., 6.
26. Ibid., 267.
27. A. Colledge and H. Johnson, *S.P.I.C.E.—A Model for Reducing the Incidence and Costs of Occupationally Entitled Claims*. Retrieved from drcolledge.com/articles/spice.pdf on July 1, 2017.
28. Jon Gice, *The Relationship between Job Satisfaction and Workers' Compensation Claims.* Retrieved from connection.ebscohost.com on June 12, 2017.
29. Ibid.
30. Ohio Division of Safety and Hygiene, *Ohio Prompts 10 Steps to Reduced Comp Costs.* Retrieved from ohiowc.com on July 3, 2017.
31. P. Bullard and S. Shields, *New Ideas to Reduce Workers' Compensation Losses*. Retrieved from roughnotes.com on July 6, 2017.
32. Highbeam Research, *Managed Care Can Reduce Workers' Comp Costs*. Retrieved from highbeam.com on July 2, 2017.

7

ACCIDENT INVESTIGATION AND REPORTING

Learning Objectives

- Explain the different types of accident investigations
- Explain when to investigate
- Describe what should be investigated
- Explain who should investigate workplace accidents
- Describe how to conduct an accident investigation
- Summarize the key strategies for conducting interviews with witnesses
- Describe the process for reporting accidents
- Summarize accident investigation mistakes to avoid

An *accident* is any unplanned event that interrupts the completion of activity and that may or may not involve human injury or property damage. The term *incident* refers to that did not cause injury or property damage but had the potential to. The term "near miss" is sometimes used in place of the term "incident."[1] When accidents occur in the workplace, they should be investigated. Reasons for conducting accident investigations include the following:

- Determine the cause so that future accidents can be prevented.
- Fulfill any applicable legal and/or regulatory requirements.
- Determine the cost of the accident.
- Determine compliance with applicable safety regulations.
- Provide information for processing workers' compensation claims.[2]

This chapter provides prospective and practicing safety and health professionals the information they need to conduct thorough, effective accident investigations and prepare comprehensive accident reports.

TYPES OF ACCIDENT INVESTIGATIONS

There are *accident reports* and there are *accident-analysis reports*. An **accident report** is completed when the accident in question represents only a minor incident. It answers the following questions: *who, what, where,* and *when.* However, it does not answer the *why* question.[3] An accident report can be completed by a person with very little formal investigation and reporting training or experience. Supervisors often complete accident reports which, in turn, might be used later as part one of a more in-depth accident report. The Occupational Safety and Health Administration's (OSHA's) Form 301 can be used for accident reports (Figure 7–1).

Figure 7–1 is OSHA's Form 301 — "Injury and Illness Incident Report." Although organizations may develop customized accident report forms (as explained later in this chapter), Form 301 is required any time a recordable work-related injury or illness occurs. Hence, Form 301 is to be used even if an organization develops a more comprehensive accident report form that calls for more comprehensive information.

An **accident-analysis report** is completed when the accident in question is serious. This level of report should answer the same questions as the regular accident report plus one more — *why.* Consequently, it involves a formal accident analysis. The analysis is undertaken for the purpose of determining the root cause of the accident. Accident analysis requires special skills and should be undertaken only by an individual with those skills. There are two reasons for this. First, the accident analysis must identify the actual root cause or the company will expend resources treating only symptoms or, even worse, solving the wrong problem. Second, serious accidents are always accompanied by the potential for litigation. If there might be legal action as a result of an accident, it is important to have a professional conduct the investigation even if it means bringing in an outside consultant.

How can safety and health professionals determine when an accident report is sufficient and when an accident-analysis report is called for? Accident reports are called for when the accident in question is a minor incident that did *not* result in any of the following circumstances: death, loss of consciousness, medical treatment beyond first aid, more than one additional day of lost work beyond the day of the accident, or any kind of modifications to the injured employee's work duties beyond those that might occur on the day of the accident.

Accident-analysis reports are called for when any of the following circumstances result from the accident in question: death, loss of consciousness, professional medical treatment beyond first aid, one or more days of lost work over and above any time lost beyond the day of the accident, or any modifications to the injured employee's work duties beyond those that might occur on the day of the injury.

WHEN TO INVESTIGATE

Of course, the first thing to do when an accident takes place is to implement **emergency procedures**. This involves bringing the situation under control and caring for the injured worker. As soon as all emergency procedures have been accomplished, the accident investigation should begin. Waiting too long to complete an investigation can harm the results. This is an important rule of thumb to remember. Another is that *all* accidents, no matter how small, should be investigated. **Evidence** suggests that the

FIGURE 7-1 OSHA's Form 301.

OSHA's Form 301
Injury and Illness Incident Report

U.S. Department of Labor
Occupational Safety and Health Administration

Form approved OMB no. 1218-0176

This *Injury and Illness Incident Report* is one of the first forms you must fill out when a recordable work-related injury or illness has occurred. Together with the *Log of Work-Related Injuries and Illnesses* and the accompanying *Summary*, these forms help the employer and OSHA develop a picture of the extent and severity of work-related incidents.

Within 7 calendar days after you receive information that a recordable work-related injury or illness has occurred, you must fill out this form or an equivalent form. Some state workers' compensation, insurance, or other reports may be acceptable substitutes. To be considered an equivalent form, any substitute must contain all the information asked for on this form.

According to Public Law 91-596 and 29 CFR 1904, OSHA's recordkeeping rule, you must keep this form on file for 5 years following the year to which it pertains.

If you need additional copies of this form, you may photocopy and use as many as you need.

Attention: This form contains information relating to employee health and must be used in a manner that protects the confidentiality of employees to the extent possible while the information is being used for occupational safety and health purposes.

Information about the employee

1) Full name _____

2) Street _____
   City _____ State ____ ZIP _____

3) Date of birth ___ / ___ / ___

4) Date hired ___ / ___ / ___

5) ☐ Male
   ☐ Female

Information about the physician or other health care professional

6) Name of physician or other health care professional _____

7) If treatment was given away from the worksite, where was it given?
   Facility _____
   Street _____
   City _____ State ____ ZIP _____

8) Was employee treated in an emergency room?
   ☐ Yes
   ☐ No

9) Was employee hospitalized overnight as an in-patient?
   ☐ Yes
   ☐ No

Completed by _____

Title _____

Phone (___) ___ - ___   Date ___ / ___ / ___

Information about the case

10) Case number from the Log _____ *(Transfer the case number from the Log after you record the case.)*

11) Date of injury or illness ___ / ___ / ___

12) Time employee began work _____ AM / PM

13) Time of event _____ AM / PM ☐ Check if time cannot be determined

14) **What was the employee doing just before the incident occurred?** Describe the activity, as well as the tools, equipment, or material the employee was using. Be specific. *Examples:* "climbing a ladder while carrying roofing materials"; "spraying chlorine from hand sprayer"; "daily computer key-entry."

15) **What happened?** Tell us how the injury occurred. *Examples:* "When ladder slipped on wet floor, worker fell 20 feet"; "Worker was sprayed with chlorine when gasket broke during replacement"; "Worker developed soreness in wrist over time."

16) **What was the injury or illness?** Tell us the part of the body that was affected and how it was affected; be more specific than "hurt," "pain," or sore." *Examples:* "strained back"; "chemical burn, hand"; "carpal tunnel syndrome."

17) **What object or substance directly harmed the employee?** *Examples:* "concrete floor"; "chlorine"; "radial arm saw." If this question does not apply to the incident, leave it blank. _____

18) **If the employee died, when did death occur?** Date of death _____

Public reporting burden for this collection of information is estimated to average 22 minutes per response, including time for reviewing instructions, searching existing data sources, gathering and maintaining the data needed, and completing and reviewing the collection of information. Persons are not required to respond to the collection of information unless it displays a current valid OMB control number. If you have any comments about this estimate or any other aspects of this data collection, including suggestions for reducing this burden, contact: US Department of Labor, OSHA Office of Statistical Analysis, Room N-3644, 200 Constitution Avenue, NW, Washington, DC 20210. Do not send the completed forms to this office.

Source: OSHA Form 301, Occupational Safety and Health Administration (OSHA®).

same factors that cause minor accidents cause major accidents.[4] Further, a near miss should be treated like an accident and investigated thoroughly.

There are several reasons why it is important to conduct investigations immediately. First, immediate investigations are more likely to produce accurate information. Conversely, the longer the time span between an accident and an investigation, the greater the likelihood of important facts becoming blurred as memories fade. Second, it is important to collect information before the accident scene is changed and before witnesses begin comparing notes. Human nature encourages people to change their stories to agree with those of other witnesses.[5] Finally, an immediate investigation is evidence of management's commitment to preventing future accidents. An immediate response shows that management cares.[6]

WHAT TO INVESTIGATE

The purpose of an **accident investigation** is to collect facts. It is not to find fault. It is important that safety and health professionals make this distinction known to all involved. **Fault finding** can cause reticence among witnesses who have valuable information to share. **Causes** of the accident should be the primary focus. The investigation should be guided by the following words: **who**, **what**, **when**, **where**, **why**, and **how**.

This does not mean that mistakes and breaches of precautionary procedures by workers go unnoted. Rather, when these things are noted, they are recorded as facts instead of faults. If fault must be assigned, that should come later, after all the facts are in.[7]

In attempting to find the facts and identify causes, certain questions should be asked, regardless of the nature of the accident:

- What type of work was the injured person doing?
- Exactly what was the injured person doing or trying to do at the time of the accident?
- Was the injured person proficient in the task being performed at the time of the accident? Had the worker received proper training?
- Was the injured person authorized to use the equipment or perform the process involved in the accident?
- Were there other workers present at the time of the accident? If so, who are they and what were they doing?
- Was the task in question being performed according to properly approved procedures?
- Was the proper equipment being used, including personal protective equipment?
- Was the injured employee new to the job?
- Was the process, equipment, or system involved new?
- Was the injured person being supervised at the time of the accident?
- Are there any established safety rules or procedures that were clearly not being followed?
- Where did the accident take place?
- What was the condition of the accident site at the time of the accident?

- Has a similar accident occurred before? If so, were corrective measures recommended? Were they implemented?
- Are there obvious solutions that would have prevented the accident?

A Five-Component Approach to Accident Investigation

Another approach to accident investigation that has gained widespread acceptance involved structuring the investigation questions according to five components: task-related questions, material-related questions, environment-related questions, personnel-related questions, and management-related questions.[8] This approach is an excellent way to organize an accident investigation for systematically collecting information. The types of questions an investigator might ask in each of the components include the following:

Task-Related Questions

- Were the proper safe-work procedures used?
- Had conditions changed in ways that made normal procedures unsafe?
- Were the appropriate tools and materials available?
- Were the appropriate tools and materials used?
- Were all safety devices engaged properly?
- Were all safety devices working properly?
- Were lockout or tagout procedures used where necessary?

Material-Related Questions

- Was there an equipment failure? If so, what caused it?
- Was the machinery involved poorly designed?
- Were hazardous substances being used? If so, were they clearly and properly identified? Was there a less hazardous substitute that could have been used? Were workers trained the safe use of the hazardous material in question?
- Was the raw material in use substandard in any way?
- Was personal protective equipment (PPE) used? Was it the right PPE for the situation? Was the PPE properly maintained and in good working order? Was the PPE properly worn? Were workers trained in the proper use of the PPE?

Environment-Related Questions

- What were the weather conditions? Was it too hot or too cold? Was it wet or dry?
- What was the state of housekeeping?
- Was noise a problem?
- Was there sufficient lighting?
- Were toxic dusts, gases, or fumes present?
- Were there obstacles or impediments in the worker's environment?

Personnel-Related Questions

- Were workers properly trained to perform the tasks in question?
- Were workers experienced in performing the tasks in question?

- Were workers physically capable of performing the tasks in question?
- Were workers fatigued?
- Were workers in healthy physical condition?
- Were workers overly stressed from work, personal, or other factors?

Management-Related Questions

- Is there a safety policy with corresponding procedures? If so, were these things followed?
- Were applicable safety procedures clearly communicated to all workers?
- Were the safety procedures enforced?
- Were workers properly trained? Supervised?
- Had any hazards been identified before the work began? If so, had the appropriate steps been taken to eliminate or mitigate the hazards?
- Were tools, equipment, and machinery properly maintained?
- Had safety inspections been completed on a regular and systematic basis?

The questions listed under each of the five components are provided as examples. They are not comprehensive. There are other questions that could be asked under each of the five components.

The answers to these questions should be carefully and copiously recorded. You may find it helpful to dictate your findings into a microcassette recorder. This approach allows you to focus more time and energy on investigating and less on taking written notes.

Regardless of how the findings are recorded, it is important to be thorough. What may seem like a minor unrelated fact at the moment could turn out to be a valuable fact later when all the evidence has been collected and is being analyzed.

Common Causes of Accidents

Common causes of accidents can be categorized as follows: **personal beliefs and feelings**, **decision to work unsafely**, **mismatch or overload**, **systems failures**, **traps**, **unsafe conditions**, and **unsafe acts**.[9] The common causes in each of these categories can help investigators determine the root cause of an accident.

Personal beliefs and feelings. Causes in this category include the following: individual did not believe the accident would happen to him or her; individual was working too fast, showing off, or being a know-it-all; individual ignored the rules out of contempt for authority and rules in general; individual gave in to peer pressure; and individual had personal problems that clouded his or her judgment.

Decision to work unsafely. Some people, for a variety of reasons, feel it is in their best interests or to their benefit to work unsafely. Hence, they make a conscious decision to do so.

Mismatch or overload. Causes in this category include the following: individual is in poor physical condition; individual is fatigued; individual has a high stress level; individual is mentally unfocused or distracted; the task required is too complex or difficult; the task required is boring; the physical environment is stressful (e.g., excessive noise, heat, dust, or other factors); the work in question is very demanding—even for an individual in good physical condition; and individual has a negative attitude (e.g., hostile, uncooperative, and apathetic).

Systems failure. Causes in this category consist of the various errors management makes that are not grossly negligent or *serious and willful*. Common causes in this category include lack of a clear policy; lack of rules, regulations, and procedures; poor hiring procedures; inadequate monitoring and inspections; failure to correct known hazards; insufficient training for employees; rules that are in place but are not enforced; no rewarding or reinforcement of safe behavior; inadequate tools and equipment provided; production requirements set too high; inadequate communication to employees of safety concerns, statistics, and rules; poor safety management; no or insufficient job safety analysis; and insufficient management support for safety.

Traps. Poor design of workstations and processes can create *traps* that, in turn, lead to unsafe behavior. Common causes in this category include defective equipment; failure to provide, maintain, and replace proper personal protective equipment; failure to train employees in the proper use of their personal protective equipment; overly complicated and confusing controls; poorly laid out work area; mechanical lifting equipment that is inadequate for the jobs required of it; uncontrolled hazards that might lead to slips and falls; excessive reaching, bending, stooping, and twisting; excessive contact pressure, vibration, or force; awkward postures that result from poor workstation or tool design; excessive temperature extremes; insufficient lighting; and insufficient ventilation.

Unsafe conditions. Common causes in this category include the following: unsafe condition created by the person injured in the accident; unsafe condition created by a fellow employee; unsafe condition created by a third party; unsafe condition created by management; unsafe condition knowingly overlooked by management; and unsafe condition created by the elements (e.g., rain, sun, snow, ice, wind, and darkness).

Unsafe acts. Common causes in this category include the following: individual chooses to ignore the rules; people are involved in horseplay or fighting; individual uses drugs or alcohol; individual uses unauthorized tools or equipment; individual chooses an improper work method; individual fails to ask for information or other resources needed to do the job safely; individual forgets a rule, regulation, or procedure; individual does not pay proper attention; and individual uses improper body mechanics.

Why Safety Professionals Need to Know the Common Causes of Accidents

The only way to make sure another accident does not happen for the same reason as the one being investigated is to identify the root cause of the accident in question and eliminate it. But finding the root cause of an accident can be like searching through fog. This is the bad news. The good news is that if safety professionals understand the more common causes of workplace accidents, this knowledge can clear away much of the fog and quickly narrow down the range of possibilities. Knowing the more common causes of workplace accidents means that safety professionals do not have to start from scratch every time they investigate. They already know what to look for and what not to look for. Knowing the common causes of workplace accidents can help narrow down the field of possibilities.

However, knowing common causes should never be allowed to lead investigators to make assumptions or to draw unwarranted conclusions. Investigations can take some surprising turns. However, investigating workplace accidents is similar to investigating automobile accidents. Investigators who look into automobile accidents are going to look first for signs of drunk or otherwise impaired driving, speeding, texting, and other common causes of automobile accidents. Even if these common causes are ruled out, the investigator is still closer to finding the actual cause because he or she has ruled out other possibilities. Part of finding what caused an accident is knowing what did not cause it.

Knowing the more common causes of accidents is like knowing the streets of a city. The individual who knows the streets well has an advantage over others who must consult a map, ask for directions, or enter data into their GPS system. The more knowledge and experience people bring to an accident investigation, the better prepared they are to conduct the investigation. Knowing what to look for can be a major asset to an investigator. It can also prevent investigators from honing on what seems to be an obvious cause and accepting it as the cause without doing a thorough investigation.

WHO SHOULD INVESTIGATE

Who should conduct the accident investigation? Should it be the responsible supervisor? The safety and health professional? A higher-level manager? An outside specialist? There is no simple answer to this question, and there is disagreement among professional people of goodwill.

In some companies, the supervisor of the injured worker conducts the investigation. In others, a safety and health professional performs the job. Some companies form an investigative team; others bring in outside specialists. There are several reasons for the various approaches used. Factors considered in deciding how to approach accident investigations include the following:

- Size of the company
- Structure of the company's safety and health program
- Type of accident
- Seriousness of the accident
- Technical complexity
- Number of times that similar accidents have occurred
- Company's management philosophy
- Company's commitment to safety and health

After considering all the variables listed above, it is difficult to envision a scenario in which the safety and health professional would not be involved in conducting an accident investigation. If the accident in question is very minor, the injured employee's supervisor may conduct the investigation, but the safety and health professional should at least study the accident report and be consulted regarding recommendations for corrective action.

If the accident is so serious that it has widespread negative implications in the community and beyond, responsibility for the investigation may be given to a high-level manager or corporate executive. In such cases, the safety and health professional should assist in conducting the investigation. If a company prefers the team approach, the safety and health professional should be a member of the team and, in most cases, should chair it. Other members of the accident investigation team might include the following:

- Employees with knowledge of the work in question
- Member(s) of the safety committee
- Union representative
- Employees with experience in conducting accident investigations
- Outside expert(s)
- Local government representative[10]

Regardless of the approach preferred by a given company, the safety and health professional should play a leadership role in collecting and analyzing the facts and developing recommendations.

CONDUCTING THE INVESTIGATION

The questions in the previous section summarize what to look for when conducting accident investigations (Figure 7–2).

Isolate the Accident Scene

You may have seen a crime scene that was sealed off by the police. The entire area surrounding such a scene is typically blocked off by barriers or heavy yellow tape. This is done to keep curious onlookers from removing, disturbing, or unknowingly destroying vital evidence. The same approach should be used when conducting an accident investigation. As soon as emergency procedures have been completed and the injured worker has been removed, the accident scene should be *isolated* until all pertinent evidence has been collected or observed and recorded. Further, nothing but the injured worker should be removed from the scene. If necessary, a security guard should be

FIGURE 7–2 Accident investigation checklist.

ACCIDENT INVESTIGATION CHECKLIST

✓ Isolate the accident scene

✓ Record all evidence

✓ Photograph or videotape the scene

✓ Identify witnesses

✓ Interview witnesses

✓ Follow up the accident investigation

posted to maintain the integrity of the **accident scene**. The purpose of isolating the scene is to maintain as closely as possible the conditions that existed at the time of the accident.

Record All Evidence

It is important to make a permanent record of all *pertinent evidence* as quickly as possible. There are three reasons for this: (1) certain types of evidence may be perishable; (2) the longer an accident scene must be isolated, the more likely it is that evidence will be disturbed, knowingly or unknowingly; and (3) if the isolated scene contains a critical piece of equipment or a critical component in a larger process, pressure will quickly mount to get it back in operation. Evidence can be recorded in a variety of ways, including written notes, sketches, photography, videotape, dictated observations, and diagrams. In deciding what to record, a good rule of thumb is *if in doubt, record it*. It is better to record too much than to skip evidence that may be needed later, after the accident scene has been disturbed.

Photograph or Videotape the Scene

This step is actually an extension of the previous step. Modern photographic and videotaping technology has simplified the task of observing and recording evidence. Safety and health professionals should be proficient in the operation of a camera, even if it is just an instant camera, and a videotaping camera.

The advent of the digital camera has introduced a new meaning for the concept of "instant photographs." Using a digital camera in conjunction with a computer, photographs of accident scenes can be viewed immediately and transmitted instantly to numerous different locations. Digital camera equipment is especially useful when photographs of accident scenes in remote locations are needed.

Both still and video cameras should be on hand, loaded, and ready to use immediately should an accident occur. As with the previous step, a good rule of thumb in photographing and videotaping is *if in doubt, shoot it*. When recording evidence, it is better to have more shots than necessary than it is to risk missing a vital piece of evidence.

A problem with photographs is that, by themselves, they don't always reveal objects in their proper perspective. To overcome the perspective problem, place a familiar object in the photograph such as a ruler, coin, or even the photographer's finger. The added object will help viewers gain the right perspective on the subject of the photograph.

Identify Witnesses

In **identifying witnesses**, it is important to compile a witness list. Names on the list should be recorded in three categories: (1) **primary witnesses**, (2) **secondary witnesses**, and (3) **tertiary witnesses**. When compiling the witness list, ask employees to provide names of all three types of witnesses.

Interview Witnesses

Every witness on the list should be interviewed, preferably in the following order: primary witnesses, first; secondary, next; and tertiary, last. After all witnesses have been interviewed, it may be necessary to reinterview witnesses for clarification or corroboration. Interviewing witnesses is such a specialized process that the next major section is devoted to it.

Following Up an Accident Investigation

As has already been noted, it is important to begin an accident investigation as soon as possible after an accident occurs. It is also important to follow up quickly once an accident investigation has been completed. Follow-up steps once an accident investigation is complete include the following:

- Write the accident report (explained later in this chapter).
- Develop a plan for corrective action.
- Implement the corrective-action plan.
- Monitor and evaluate the effectiveness of corrective actions.
- Adjust as necessary to ensure the cause of the accident has been eliminated.
- Incorporate changes into standard operating procedures for continual improvement of the workplace environment.[11]

INTERVIEWING WITNESSES

The techniques used for interviewing accident witnesses are designed to ensure that the information is objective, accurate, as untainted by the personal opinions and feelings of witnesses as possible, and able to be corroborated. For this reason, it is important to understand the *when*, *where*, and *how* of interviewing the accident witnesses.

When to Interview

Immediacy is important. Interviews should begin as soon as the witness list has been compiled and, once begun, should proceed expeditiously. There are two main reasons for this. First, a witness's recollections will be best right after the accident. The more time that elapses between the accident and the interview, the more blurred the witness's memory will become. Second, immediacy avoids the possibility of witnesses comparing notes and, as a result, changing their stories. This is just human nature, but it is a tendency that can undermine the value of testimony given and, in turn, the facts collected. Recommendations based on questionable facts are not likely to be valid. Also, witnesses should be interviewed individually and separately, preferably before they have talked to each other.

Where to Interview

The best place to interview is at the accident scene. If this is not possible, interviews should take place in a private setting elsewhere. It is important to ensure that all distractions are removed, interruptions are guarded against, and the witness is not accompanied by other witnesses. All persons interviewed should be allowed to relate their recollections without fear of contradiction or influence by other witnesses or employees. It is also important to select a neutral location in which witnesses will feel comfortable. Avoid the "**principal's office syndrome**" by selecting a location that is not likely to be intimidating to witnesses.

How to Interview

The key to getting at the facts is to put the witness at ease and to listen. Listen to what is said, how it is said, and what is not said. Ask questions that will get at the information listed earlier in this chapter, but phrase them in an **open-ended** format. For example, instead of asking "Did you see the victim pull the red lever?" phrase your question as follows: "Tell me what you saw." Don't lead witnesses with your questions or influence them with gestures, facial expressions, tone of voice, or any other form of nonverbal communication. Interrupt only if absolutely necessary to seek clarification on a critical point. Remain nonjudgmental and objective.

The information being sought in an accident investigation can be summarized as *who*, *what*, *when*, *where*, *why*, and *how* (Figure 7–3). As information is given, it may be necessary to take notes. If you can keep your notetaking to a minimum during the interview, your chances of getting uninhibited information are increased. Notetaking can distract and even frighten a witness.

An effective technique is to listen during the interview and make mental notes of critical information. At the end of the interview, summarize what you have heard and have the witness verify your summary. After the witness leaves, develop your notes immediately.

A question that sometimes arises is, "Why not tape the interview?" Safety and health professionals disagree on the effectiveness and advisability of taping. Those who favor taping claim it allows the interviewer to concentrate on listening without

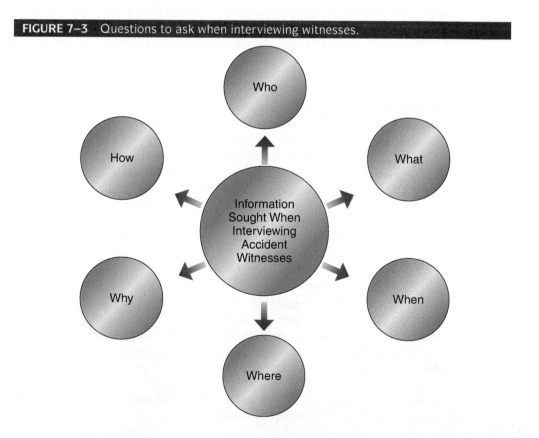

FIGURE 7–3   Questions to ask when interviewing witnesses.

having to worry about forgetting a key point or having to interrupt the witnesses to jot down critical information. It also preserves everything that is said for the record as well as the tone of voice in which it is said. A complete transcript of the interview also ensures that information is not taken out of context.

Those opposed to taping say that taping devices tend to inhibit witnesses so that they are not as forthcoming as they would be without taping. Taping also slows down the investigation while the taped interview is transcribed, and the interviewer wades through voluminous testimony trying to separate critical information from irrelevant information.

In any case, if the interview is to be taped, the following rules of thumb should be applied:

- Use the smallest, most unobtrusive taping device available, such as a microcassette recorder.
- Inform the witness that the interview will be taped.
- Make sure the taping device is working properly and that the tape it contains can run long enough so that you don't have to interrupt the witness to change it.
- Take time at the beginning of the interview to discuss unrelated matters long enough to put the witness at ease and overcome the presence of the taping device.
- Make sure that personnel are available to transcribe the tapes immediately.
- Read the transcripts as soon as they are available and highlight critical information.

An effective technique to use with eyewitnesses is to ask them to reenact the accident for you. Of course, the effectiveness of this technique is enhanced if the reenactment can take place at the accident site. However, even when this is not possible, an **eyewitness** reenactment can yield valuable information.

In using the **reenactment** technique, a word of caution is in order. If an eyewitness does exactly what the victim did, there may be another accident. Have the eyewitnesses explain what they are going to do before letting them do it. Then, have them *simulate* rather than actually perform the steps that led up to the accident.

REPORTING ACCIDENTS

An accident investigation should culminate in a comprehensive *accident report*. The purpose of the report is to record the findings of the accident investigation, the cause or causes of the accident, and recommendations for corrective action.

OSHA has established requirements for reporting and record keeping. According to OSHA document 2056,

> Employers of 11 or more employees must maintain records of occupational injuries and illnesses as they occur. Employers with 10 or fewer employees are exempt from keeping such records unless they are selected by the Bureau of Labor Statistics (BLS) to participate in the Annual Survey of Occupational Injuries and Illnesses.[12]

All injuries and illnesses are supposed to be recorded, regardless of severity, if they result in any of the outcomes shown in Figure 7–4. If an accident results in the death of an employee or hospitalization of three or more employees, a report must be submitted to the nearest OSHA office within eight hours. This rule applies regardless of the size of the company. Reporting locally within an organization for insurance, legal, prevention, and management purposes and reporting for OSHA purposes can

FIGURE 7–4   OSHA record-keeping requirements.

Injuries and illnesses must be recorded if they
result in any of the following:

- Death
- One or more lost work days
- Restriction of motion or work
- Loss of consciousness
- Transfer to another job
- Medical treatment (more than first aid)

be two different tasks. OSHA's reporting requirements were covered in Chapter 5.
Reporting procedures in this section pertain to local in-house reporting.

Accident report forms can vary from company to company. However, the information contained in them is fairly standard.[13] Regardless of the type of form used, an accident report should contain at least the information needed to meet the record-keeping requirements set forth by OSHA. This information includes at least the following:

- Case number of the accident
- Victim's department or unit
- Location and date of the accident or date that an illness was first diagnosed
- Victim's name, social security number, gender, age, home address, and telephone number
- Victim's normal job assignment and length of employment with the company
- Victim's employment status at the time of the accident (i.e., temporary, permanent, full-time, part-time)
- Case numbers and names of others injured in the accident
- Type of injury and body part(s) injured (e.g., burn to right hand; broken bone, lower right leg) and severity of injury (i.e., fatal, first aid only required, hospitalization required)
- Name, address, and telephone number of the physician called
- Name, address, and telephone number of the hospital to which the victim was taken
- Phase of the victim's work day when the accident occurred (e.g., beginning of shift, during break, or end of shift)
- Description of the accident and how it took place, including a step-by-step sequence of events leading up to the accident
- Specific tasks and activities with which the victim was involved at the time of the accident (e.g., task: mixing cleaning solvent; activity: adding detergent to the mixture)
- Employee's posture or proximity related to his or her surroundings at the time of the accident (e.g., standing on a ladder; bent over at the waist inside the robot's work envelope)
- Supervision status at the time of the accident (i.e., unsupervised, directly supervised, indirectly supervised)
- Causes of the accident
- Corrective actions that have been taken so far
- Recommendations for additional corrective action

In addition to these items, you may want to record such additional information as the list of witnesses; dates, times, and places of interviews; historical data relating to similar accidents; information about related corrective actions that were made previously but had not yet been followed up on; and any other information that might be relevant. Figure 7–5 is an example of an accident report form that meets the OSHA record-keeping specifications.

Why Some Accidents Are Not Reported

There are several reasons why accidents go unreported. Be familiar with these reasons so that you can do your part to overcome them. The main reasons are as follows:

1. **Red tape.** Some people see the paperwork involved in accident reporting as red tape and, therefore, don't report accidents just to avoid paperwork.
2. **Ignorance.** Not all managers and supervisors are as knowledgeable as they should be about the reasons for accident reporting. Many are not familiar with OSHA's reporting specifications.
3. **Embarrassment.** Occasionally, people do not report an accident because they are embarrassed by their part in it. A supervisor who did not properly supervise or a manager who has not provided the proper training for employees may be embarrassed to file a report.
4. **Record-spoiling.** Some accidents go unreported just to preserve a safety record, such as the record for days worked without an accident.
5. **Fear of repercussions.** Some accidents go unreported because the people involved are afraid of being found at fault, being labeled accident prone, and being subjected to other negative repercussions.
6. **No feedback.** Some accidents go unreported because those involved feel filing a report is a waste of time. This typically happens when management does not respond to recommendations made in earlier accident reports.[14]

Clearly, these reasons for not reporting accidents present safety and health professionals with a challenge. To overcome these inhibitors, it is necessary to develop a simple reporting system that will not be viewed as too much bureaucratic paperwork to have to do. Safety and health professionals must educate personnel at all levels concerning the purpose of accident reporting and why it is important. An important step is to communicate the fact that fault finding is not the purpose. Another important step is to follow up to ensure that recommendations are acted on or that employees are made aware of why they aren't acted on. This helps ensure the integrity of the process.

Discipline and Accident Reporting

Fault finding is not the purpose of an accident investigation. However, an investigation sometimes reveals that an employee has violated or simply overlooked safety regulations. Should such violations be condoned?

There is a built-in dilemma here that modern safety and health professionals must be prepared to handle. On the one hand, it is important that fault finding not be seen as the purpose of an accident investigation. Such a perception limits the amount of information that can be collected. On the other hand, if those workers whose behavior

FIGURE 7–5 Sample accident report form.

ACCIDENT REPORT FORM

Fairmont Manufacturing Company
1501 Industrial Park Road
Fort Walton Beach, FL 32548
904-725-4041

Victim-Related Information

Person completing report _____ Case no. _____

Victim's name _____

Gender _____ Age _____

Date of accident/illness _____

Victim's home address/telephone _____

Victim's assignment at the time of the accident and length of time in that assignment:
_____
_____
_____

Victim's normal job and length of time in that job: _____
_____
_____

Time of injury/illness and phase of victim's work day: _____
_____

Severity of the injury (e.g., hospitalization required, first aid required, etc.): _____
_____

Type of injury and body part(s) injured: _____
_____

Exact location of the accident (which facility, department, place within the department):
_____

Physician and hospital: _____
_____

Note: Complete one form for each injured worker.

leads to accidents are not disciplined, the credibility of the safety program is undermined. The following procedures are recommended for handling this dilemma: Never discipline an employee because he or she had an accident. Always discipline employees for noncompliance with safety regulations.[15]

Such an approach applied with consistency will help maintain the integrity of both the accident investigation process and the overall safety program.

FIGURE 7–5 *(Continued)*

Accident-Related Information

Accident description with step-by-step sequence of events: _____
_____
_____

Task and specific activity at the time of the accident: _____
_____
_____

Posture/proximity of employee at the time of the accident: _____
_____
_____

Supervision status at the time of the accident: _____
_____
_____

Apparent causes including conditions, actions, events, and activities and other
contributing factors: _____
_____
_____

Recommendations for corrective action: _____
_____
_____

Case numbers and names of other persons injured in the accident: _____
_____
_____

Witnesses to the accident, and dates/places of their interviews:

_____   _____
_____   _____
_____   _____
_____   _____

_____   _____
Reporter name   Date

_____   _____
Employee name   Date

Who Is Responsible for Reporting?

OSHA is clear concerning what must be reported, but who is responsible for completing and filing accident reports? There are two answers to this question; a big-picture answer and a more specific answer. The big picture answer is as follows:

- All employers are required to report to OSHA any work-related death. This must be done within eight hours of the incident.

- All employers are required to report to OSHA any worker amputation, eye loss, or inpatient hospitalization. This must be done within 24 hours of the incident.
- All employers must report to OSHA all fatal heart attacks that occur at work and motor vehicle deaths that occur in a construction work zone.
- Employers with more than ten employees and that are not classified as partially exempt must record serious work-related injuries and illnesses using OSHA forms 300, 300A, and 301 as appropriate.

From these requirements, it is clear that the big-picture answer to the question of "Who is responsible for reporting?" is the employer. OSHA holds the employer responsible for ensuring that its reporting requirements are complied with.

The more specific answer to the question of "Who is responsible for reporting?" is that employers may and often do delegate the job of reporting on accidents, injuries, and deaths to a member of the safety and health team. This individual then becomes responsible to the employer for completing and filing the necessary reports. But it is important to understand that while employers can delegate authority for completing reports to a member of the safety and health team or anyone else they choose, doing so does not relieve them of their responsibility. Ultimately, responsibility for ensuring that OSHA reporting requirements are met rests with the employer. For information about OSHA's reporting requirements, go to:www.osha.gov/recordkeeping

TEN ACCIDENT INVESTIGATION MISTAKES TO AVOID

The amount of information you collect and how you collect it will go a long way toward determining how effective your resultant corrective actions will be following a workplace accident.

- **Failing to investigate near misses.** A near miss is simply an accident that did not happen because of luck. Consequently, investigating near misses can reveal critical accident prevention information.
- **Taking ineffective corrective action.** Ineffective corrective action is often the result of a cursory accident investigation. When investigating, look for the root cause, not the symptoms. Corrective action based on symptoms will not prevent future accidents.
- **Allowing your biases to color the results of the investigation.** Look for facts and be objective when investigating an accident. Do not make assumptions or jump to conclusions. One of the best ways to eliminate bias in accident investigations is to use a standard, structured routine and to skip no steps in the routine.
- **Failing to investigate in a timely manner.** Accident investigations should begin as soon as possible after the accident. The longer you wait to begin, the more likely it is that evidence will be lost, corrupted, or compromised. For example, once people start talking to each other about what they saw, invariably their memories will be shaped by the opinions of their fellow workers and witnesses. People walking through an accident scene can compromise the integrity of the scene by unwittingly destroying evidence.
- **Failing to account for human nature when conducting interviews.** Often what those involved in an accident as well as witnesses to an accident will say during an interview will be shaped by their desire to escape blame, deflect blame

to someone else, or protect a friend. This is why it is important to interview witnesses and others involved privately and individually, and to look for corroborating evidence to support (or refute) their input.

- **Failing to learn investigation techniques.** Before investigating an accident, safety and health professionals should complete specialized training or undertake self-study to learn investigation techniques such as those presented in this chapter. An unskilled investigator is not likely to conduct a valid investigation.

- **Allowing politics to enter into an investigation.** The goal of an investigation is and must be to identify the root cause so that appropriate corrective action can be taken. Personal likes, dislikes, favoritism, and office politics will corrupt an investigation from the outset.

- **Failing to conduct an in-depth investigation.** Everyone is in a hurry and investigating an accident was not on your agenda for the day. In addition, there is sometimes pressure from higher management to "get this thing behind us." Such pressures and circumstances can lead to a rushed investigation in which the goal is to get it over with, not to find the root cause of the accident. Surface-level investigations almost ensure that the same type of accident will happen again.

- **Allowing conflicting goals to enter into an investigation.** The ultimate goal of an accident investigation is to prevent future accidents and injuries. However, even when that is your goal, there may be other people who have different goals. Some may see the investigation as an opportunity to deflect blame, others may see it as an opportunity to protect the organization from litigation, and some may see it as a way to explain not meeting production quotas or performance standards. Safety and health professionals should be aware that other agendas may be in play every time an accident investigation is conducted. For this reason, objectivity, structure, and routine are critical.

- **Failing to account for the effects of uncooperative people.** One would think that employees and management personnel would automatically want to cooperate in accident investigation to ensure that similar accidents are prevented in the future. Unfortunately, this is not always the case. People will not always cooperate for a variety of reasons—all growing out of the concept of perceived self-interest. Further, the lack of cooperation will not always be overt. In fact, often it will be covert (e.g., a person you need to interview keeps putting you off or canceling meetings). Safety and health professionals need to understand that self-interest is one of the most powerful motivators of human beings and factor this into their planning for accident investigations.[16]

These 10 mistakes will probably never be completely eliminated from every accident investigation. However, if safety and health professionals are aware of them, they can at least ensure that such mistakes are minimized. The fewer of these mistakes that are made during an accident investigation, the better the quality of the investigation and the more likely that it will lead to effective corrective action.

Key Terms and Concepts

- Accident-analysis report
- Accident investigation
- Accident report
- Accident scene
- Causes
- Decision to work unsafely
- Emergency procedures
- Evidence
- Eyewitness
- Fault finding
- How

- Identifying witnesses
- Immediacy
- Mismatch or overload
- Open-ended
- Personal beliefs and feelings
- Primary witnesses
- Principal's office syndrome
- Reenactment
- Secondary witnesses
- Systems failures
- Tertiary witnesses
- Traps
- Unsafe acts
- Unsafe conditions
- What
- When
- Where
- Who
- Why

Review Questions

1. Explain the rationale for investigating accidents.
2. When should an investigation be reported? Why?
3. Explain the difference between an accident report and an accident-analysis report.
4. List the categories of the most common causes of accidents.
5. What are the terms that should guide the conduct of an accident investigation?
6. What role should the safety and health professional play in the conduct of an accident investigation?
7. List and explain the steps for conducting an accident investigation.
8. Why is it important to record all pertinent evidence relating to an accident immediately after an accident has occurred?
9. What can you do when taking close-up photographs to put them in the proper perspective?
10. List and differentiate among the three categories of witnesses to an accident.
11. Briefly explain the when and where of interviewing witnesses.
12. Briefly explain the how of interviewing witnesses.
13. What is the purpose of an accident report?
14. Who is responsible for filing an accident with OSHA?

Endnotes

1. Canadian Centre for Occupational Health and Safety, *Accident Investigation.* Retrieved from ccohs.ca/oshanswers/hsprograms/investig.html on July 1, 2017, 1.
2. Ibid.
3. Ibid., 2.
4. Ibid., 6.
5. Ibid., 7.
6. Ibid.
7. Ibid., 3.
8. Ibid., 8–9.
9. Ibid.
10. Ibid., 2.
11. Ibid., 3.
12. U.S. Department of Labor, *OSHA 2056.* Retrieved from osha.gov on June 23, 2017.
13. Canadian Centre for Occupational Health and Safety, 3–6.
14. Ibid., 70–71.
15. Ibid.
16. W. R. Coffee Jr., "Avoid These 10 Mistakes." Retrieved from https://ohsonline.com/articles/2005/05/Avoid-These-10-Mistakes.aspx on January 7, 2017.

8

ERGONOMIC HAZARDS: MUSCULOSKELETAL DISORDERS (MSDS) AND CUMULATIVE TRAUMA DISORDERS (CTDS)

Learning Objectives

- Define the concept of ergonomics
- Explain the concept of human factors in ergonomics
- List the factors associated with physical stress
- Explain why ergonomics is a controversial political issue
- Summarize the key elements of OSHA's voluntary ergonomics guidelines
- Explain how to analyze a worksite for ergonomic hazards
- List several strategies for preventing and controlling ergonomic hazards
- Explain the concept of medical management as it relates to ergonomics
- Explain the key elements of training for ergonomics
- List the common indicators of ergonomic problems
- Explain how to identify specific ergonomic hazards
- List several problem-solving strategies for ergonomic hazards
- Summarize the economics of ergonomics
- Explain the concept of cumulative trauma disorders and soft tissue injuries
- Explain how to avoid back injuries through proper lifting techniques
- Summarize the key elements of participatory ergonomics

The history of workplace development in the Western world is characterized by jobs and technologies designed to improve processes and productivity. All too often in the past, little or no concern was given to the impact of the job process or technology on workers. As a result, work processes and machines have sometimes been unnecessarily dangerous. Another result has been that new technologies have sometimes failed to live up to expectations. This is because, even in the age of high technology, human involvement in work processes is still the key to the most significant and enduring productivity improvements. If a machine or system is uncomfortable, difficult, overly complicated, or dangerous to use, human workers will not be able to derive its full benefit.

The proliferation of uncomfortable and dangerous workplace conditions, whether created by job design or unfriendly technologies, is now widely recognized as harmful to productivity, quality, and worker safety and health. The advent of the science of ergonomics is making the workplace more physically friendly. This, in turn, is making the workplace a safer and healthier place.

ERGONOMICS DEFINED

Minimizing the amount of **physical stress** in the workplace requires continuous study of the ways in which people and technology interact. The insight learned from this study must then be used to improve the interaction. This is a description of the science of **ergonomics**. For the purpose of this book, ergonomics is defined as follows:

Ergonomics is a multidisciplinary science that seeks to conform the workplace and all of its physiological aspects to the worker.

Ergonomics involves the following:

Using special design and evaluation techniques to make tasks, objects, and environments more compatible with human abilities and limitations.

Seeking to improve productivity and quality by reducing workplace stressors, reducing the risk of injuries and illnesses, and increasing efficiency.

The word *ergonomics* is derived from the Greek language. *Ergon* is Greek for *work; nomos* means *laws*. Therefore, in a literal sense, ergonomics means work laws. In practice, it consists of the scientific principles (laws) applied in minimizing the physical stress associated with the workplace (work). Some of the widely accepted benefits of ergonomics include the following:

- Improved health and safety for workers
- Higher morale in the workplace
- Improved quality
- Improved productivity
- Improved competitiveness
- Decreased absenteeism and turnover
- Fewer workplace injuries and health problems

There are benefits to be derived from ergonomics. Both financial and health-related problems can result from giving too little attention to ergonomics. The matter is complicated further because health problems tend to multiply a company's financial problems. Consequently, modern safety and health professionals need to be well versed in ergonomics (see Figure 8–1).

FIGURE 8–1 Ergonomic factors are now an important part of product and work station design.

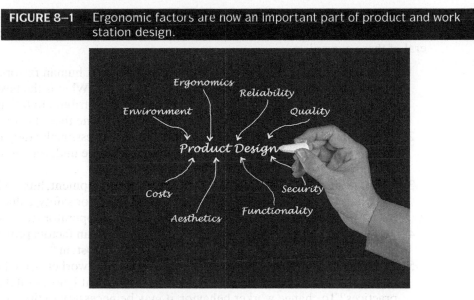

Source: Arka38/Shutterstock

HUMAN FACTORS AND ERGONOMIC HAZARDS

When the topic of ergonomics is discussed, the term *human factors* will usually find its way into the conversation. But what is meant by the term? It can be defined as follows: Consumers are demanding safe and effective products. However, not all people have control over products they use. Therefore, all products must be carefully designed. For example, if a child car seat fails because it does not fit the child or is difficult to install, everyone will lose: the child, the parent, the designer, and the manufacturer. Human factors is a profession to help ensure that equipment and systems are safe and easy to operate by human beings.[1] A human factors researcher gathers and analyzes data on human beings (how they work, their size, their capabilities, and limitations). A human factors engineer works with designers as a team to incorporate data into designs to make sure people can operate and maintain the product or system. Human factors professionals then determine the skills needed to operate or maintain a finished product. Human factors is difficult to define because it is a compilation of many sciences dealing with both humans and machines. Some of the disciplines human factors experts are trained in include psychology, anthropology, engineering, biology, medicine, education, and physiology.[2]

Human Factors Defined

Human factors is a science that combines research with the application of human data.[3] The concept can also be viewed as a science that bridges research about human beings and the application of that research in designing products and systems for human beings. Human factors are human characteristics as well as environmental, organizational, and job factors that influence the behavior of workers in ways that can affect workplace safety and health.

Human Factors in Action

Perhaps the best way to get a feel for the concept of human factors is to consider several of these examples:

1. **Predesign analysis.** In this stage of the design process, human factors professionals conduct research to answer such questions as: What is the best way for humans to interact with computers? What factors contribute to fatigue and stress in an office environment? How can designers overcome these factors?
2. **Preliminary design.** In this stage, human factors professionals study machine and human capabilities to determine which tasks should be undertaken manually and which should be automated.
3. **Detail design and development.** In design and development, human factors professionals define the environment required for operator safety, enhanced operator performance, and the reduction or prevention of operator stress and fatigue.
4. **Test and evaluation.** In this stage of the process, human factors professionals test actual humans in using the prototype equipment or system.[4]
5. **Workplace culture.** Organizational culture can affect worker behavior. Does the organization have a *safety-first* culture or a culture that ignores safety and health practices? To change worker behavior, it may be necessary to first change the culture of the workplace. An organization's top executives set the tone for an organization's culture. If they insist on safety first and make it an organizational priority, workers will adopt a safety-first attitude.
6. **Communication.** If executives, managers, supervisors, and team leaders want safety, they must convey that expectation to employees in every way possible (e.g., verbally, in written policies and procedures, in evaluation criteria, in recognition and reward criteria, and by their examples). Worker behavior is affected by the expectations of those in positions of authority, particularly when authority figures exemplify their expectations in their examples.
7. **Resources.** If executives, managers, supervisors, and team leaders want safety, they must provide workers with the resources necessary to do their jobs in a safe manner. From an ergonomic perspective, this means providing ergonomically correct workstations, tools, and processes.
8. **Employee attitudes and personalities.** Safety and health professionals and other decision makers in an organization must be aware of the effect worker attitudes and personalities can have on safety. Attitudes and personalities are human factors that can help ensure that safe work practices are adhered to or that they are undermined. When attempts to ensuring that a worker follows safe work practices fail because that worker has a negative attitude toward safety or has a risk-taking personality, decision makers should remove the individual in question or transfer him or her to another position that involves less risk.

Human Factors and Safety

Human factors can play an important role in both product safety and workplace safety (where many products are used). What follows is how the science of human factors can help reduce both product and workplace hazards:

1. **Hazard elimination by design.** Human error is frequently the root cause or a contributing cause in accidents on the job. Intelligent design can reduce

human errors by providing controls that are simple to understand and operate and by proving human–machine interaction that is not boring or overly
demanding physically.

2. **Provision and location of safety devices.** The design and location of safety
devices such as emergency cutoff switches can reduce human error on the job,
correspondingly reducing the chances of an accident.

3. **Provision of warning devices.** The color, location, and wording of warning
devices; the pitch and volume of warning signals; and the design of caution
markings on gauges and video displays are all important factors in reducing the
likelihood of human error that might lead to an accident. The science of human
factors can help determine the appropriate way to apply all of these factors in a
given setting.

4. **Establishment of procedures/provision of training.** When hazards cannot be real-
istically designed out of a system, administrative procedures for hazard reduction
must be established, and training relating to those procedures must be provided.
Human factors professionals can help establish appropriate administrative procedures and help develop the necessary training.[5]

FACTORS ASSOCIATED WITH PHYSICAL STRESS

Eight variables that can influence the amount of *physical stress* experienced on the job
are as follows:

Sitting versus standing
Stationary versus moveable/mobile
Large demand for strength/power versus small demand for strength/power
Good horizontal work area versus bad horizontal work area
Good vertical work area versus bad vertical work area
Nonrepetitive motion versus repetitive motion
Low surface versus high surface contact
No negative environmental factors versus negative environmental factors

The following paragraphs summarize the extent of this influence and the forms that
it can take.[6]

Sitting versus Standing

Generally speaking, sitting is less stressful than standing. Standing for extended periods, particularly in one place, can produce unsafe levels of stress on the back, legs, and
feet. Although less strenuous than standing, sitting can be stressful unless the appropriate precautions are taken. These precautions include proper posture, a supportive
back rest, and frequent standing/stretching movement.

Stationary versus Moveable/Mobile

Stationary jobs are those done primarily at one workstation. Of course, even these
jobs involve movement at the primary workstation and occasional movement to
other areas. Mobile jobs, on the other hand, require continual movement from one
station to another. The potential for physical stress increases with stationary jobs

when workers fail to take such precautions as periodically standing/stretching/moving. The potential for physical stress increases with mobile jobs when workers carry materials as they move from station to station.

Large versus Small Demand for Strength/Power

In classifying jobs by these two criteria, it is important to understand that repeatedly moving small amounts of weight over a period of time can have a cumulative effect equal to the amount of stress generated by moving a few heavy weights. Regardless of whether the stress results from lifting a few heavy objects or repeated lifting of lighter objects, jobs that demand larger amounts of strength/power are generally more stressful than those requiring less strength/power.

Good versus Bad Horizontal Work Area

A good **horizontal work area** is one that is designed and positioned so that it does not require the worker to bend forward or to twist the body from side to side. Horizontal work areas that do require these movements are bad. Bad horizontal work surfaces increase the likelihood of physical stress.

Good versus Bad Vertical Work Area

Good **vertical work areas** are designed and positioned so that workers are not required to lift their hands above their shoulders or bend down in order to perform any task. Vertical work areas that do require these movements are bad. Bad vertical work areas increase the likelihood of physical stress.

Nonrepetitive versus Repetitive Motion

Repetitive motion jobs involve short-cycle motion that is repeated continually. Nonrepetitive jobs involve a variety of tasks that are not, or only infrequently, repeated. Repetition can lead to monotony and boredom, thereby increasing the potential for physical stress.

Low versus High Surface Contact

Surface stress can result from contact with hard surfaces such as tools, machines, and equipment. High surface-contact jobs tend to be more stressful in a physical sense than are low surface-contact jobs.

Absence versus Presence of Negative Environmental Factors

Generally, the more **environmental factors** with which a worker has to contend on the job, the more stressful is the job. For example, personal protective equipment, although conducive to reducing environmental hazards, can increase the amount of physical stress associated with the job.

ERGONOMICS: A POLITICAL FOOTBALL

The Occupational Safety and Health Administration's (OSHA's) attempts to develop an ergonomics standard have been the subject of bitter disputes between labor and management in the private sector. These disputes, in turn, have found their way into the political arenas of both the U.S. Congress and the executive branch. OSHA established its voluntary ergonomics guidelines in 1989. Soon thereafter, organized labor began its campaign to have the guidelines made mandatory through passage of an ergonomics standard. During the Clinton administration (1993–2000), Congress steadfastly refused to approve OSHA's proposed ergonomics standard.

Finally, on November 14, 2000, OSHA issued a final Ergonomic Program Standard (29 *CFR* 1901 Subpart W). The final rule became effective on January 16, 2001. However, on March 6, 2001, the United States Senate passed a resolution of disapproval (S.J. Res. 6) of the Ergonomic Program Standard under the Congressional Review Act. The House of Representatives then passed S.J. Res. 6 on March 7, 2001. President George W. Bush signed the resolution into law as Public Law 107-5 on March 20, 2001. At this point, OSHA removed the Ergonomic Program Standard from the Code of Federal Regulations.

The disagreement between business and labor concerning the need for an ergonomic standard continues. There is also opposition from various states which claim that it encroaches on existing workers' compensation programs and statutes. This criticism grows out of the fact that OSHA's Ergonomic Program Standard threatened the exclusive remedy provision of state statutes by requiring a new and entirely separate benefit system for musculoskeletal disorders (MSDs). What further concerned the states was that the eligibility requirements in OSHA's Ergonomic Program Standard were different from those already established in the various state programs for workers' compensation. The standard did contain a provision that allowed an offset of workers' compensation benefits when a worker was entitled to both types of benefits, but state officials complained that this provision would create a bureaucratic nightmare in paperwork and red tape. OSHA has responded to the ongoing controversy in two ways. First, OSHA continues to offer voluntary guidelines beyond those that apply to industry in general for businesses in specific industrial classifications. Second, OSHA has begun to claim that it will use its **general duty clause** to enforce ergonomic safety. However, there is little evidence that this is actually happening. After vowing to make enforcement through the general duty clause part of its four-pronged attack on musculoskeletal hazards, OSHA is, in reality, focusing most of its efforts on the other three prongs: voluntary guidelines, outreach and assistance, and continuing research.

OSHA'S VOLUNTARY ERGONOMICS GUIDELINES

OSHA first published guidelines for general safety and health program management in 1989. OSHA's **ergonomics guidelines** are voluntary and are designed to provide employers with the information and guidance needed to meet their obligations under the Occupational Safety and Health Act (OSH Act) regarding ergonomics.

OSHA has since followed these guidelines with others designed specifically for meatpacking, shipyards, poultry processing, nursing homes, and retail grocery stores. Other guidelines are in the process of being developed. These specific guidelines represent a model for guidelines that are likely to be developed for other specific industries. Meatpacking and poultry processing were singled out because of the high incidence of **cumulative trauma disorders (CTDs)** associated with these industries. CTDs are injuries that result from an accumulation of repetitive motion stress. For example, using scissors continually over time can cause a CTD in the hand and wrist.

OSHA'S ERGONOMICS STANDARD (VOLUNTARY GUIDELINES)

OSHA's voluntary guidelines were well received by employees and labor organizations with one major caveat: the "voluntary" nature of the guidelines. Many people interested in occupational safety and health advocated a mandatory standard for ergonomics. This led to discussions concerning the development by OSHA of a new ergonomics standard. The standard has not yet been developed.

However, OSHA's proposed ergonomics standard failed to win approval from Congress. Consequently, after much political wrangling, OSHA officials decided to convert the proposed ergonomics standard into voluntary ergonomics guidelines. OSHA's current plan for reducing ergonomic hazards in the workplace has four elements:

1. Voluntary guidelines for specific industries.
2. Enforcement of the guidelines under the general duty clause of the OSH Act 5(a)(1). This is a controversial element of the four-part plan because some employers think that OSHA is using the general duty clause to make the "voluntary" guidelines mandatory against the stated will of Congress.
3. Compliance assistance to help employers reduce ergonomic hazards.
4. Research into ergonomic issues to help identify gaps in the body of knowledge surrounding this topic.

Enforcement by OSHA

Although the ergonomics guidelines developed by OSHA are voluntary, at least for the foreseeable future, the agency does claim that it will use the general duty clause of the OSH Act to enforce the guidelines in certain situations. OSHA's criteria for applying the general duty clause are as follows:

1. Is there currently an ergonomic hazard that is causing injuries?
2. Does the employer in question know about the hazard (or should the employer know)?
3. Are the injuries caused by the ergonomic hazard resulting in serious physical harm?
4. Are there feasible alternatives available to the employer for reducing, abating, or minimizing the hazard?

Although the use of the general duty clause to enforce voluntary ergonomics guidelines is controversial and has been questioned by many employers in theory, in practice this is an approach that is rarely used by OSHA.

Rationale for the Voluntary Guidelines

In a typical year, more than 85 percent of the workers' compensation payments made to workers for both lost wages and medical treatment due to disabling injuries can be traced to ten causes. MSDs and CTDs are typically sixth on the list of the top 10 causes accounting for 6.7 percent of disabling injuries. This relatively low figure differs radically from the perceptions of business and labor leaders as well as the general public. Popular perception places MSDs and CTDs at the top of the list. This is probably because these types of injuries have received so much attention in the press and professional journals for more than 30 plus years owing to the ongoing controversy over passage and repeal of OSHA's Ergonomic Program Standard.

Application of the Voluntary Guidelines

OSHA's ergonomics guidelines are geared toward manufacturing and materials handling in the general industry sector. They do not apply to construction, maritime operations, agriculture, or employers that operate a railroad, although they can be applied to other jobs in which the type of work that is fundamental and necessary to perform the job results in MSDs. The ergonomics guidelines apply to more than 1.5 million employees nationwide. Examples of jobs to which the guidelines apply are as follows:

- Patient-handling jobs (nurse assistants and orderlies)
- Shipping, receiving, and delivery (package sorting, handling, delivery, etc.)
- Baggage handlers
- Warehouse work (manual tasks)
- Beverage and water handling and delivery
- Grocery/retail store stocking and bagging
- Garbage and trash collecting
- Assembly-line work
- Piecework assembly
- Product inspection (involving manual tasks such as weighing objectives)
- Meat, poultry, and fish processing
- Machine loading, unloading, and operation
- Textile manufacturing
- Food preparation assembly-line work
- Commercial banking
- Cabinet making
- Tire making

Proposed Requirements of the Voluntary Guidelines

Organizations that fall into the general industry classifications of manufacturing and manual material handling are asked by the guidelines to implement a "basic ergonomics program." This amounts to assigning responsibility for ergonomics to one individual and informing employees about the risks of MSD-related injuries, symptoms of such injuries, and why early reporting of symptoms is important. In addition, the basic program requires employers to establish a system that employees can use to report symptoms of MSD injuries.

The so-called full ergonomics program set forth in the guidelines is not required unless and until an employee's job is determined, by use of a basic screening tool (such as the one shown in Figure 8–2) to have met the "action trigger." The full program consists of the following components:

- Management leadership and employee participation
- Employee participation
- Training
- Record keeping
- Job hazard analysis and control
- Work restriction protection
- MSD management
- Program evaluation

FIGURE 8–2   Basic screening tool for VDT workstations.

Workstation Checklist
for
Video Display Terminals

The workstation should be designed to ensure the following conditions:

A. *Head* and *neck* should be about upright.

B. *Head*, *neck*, and *trunk* should face forward.

C. *Trunk* should be about perpendicular to floor.

D. *Shoulders* and *upper arms* should be about perpendicular to floor and relaxed.

E. *Upper arms* and *elbows* should be close to body.

F. *Forearms*, *wrists*, and *hands* should be straight and parallel to floor.

G. Wrists and hands should be straight.

H. *Thighs* should be about parallel to floor, and lower legs should be about perpendicular to floor.

I. *Feet* should rest flat on floor or be supported by a stable footrest.

J. *VDT tasks* should be organized in a way that allows employee to vary them with other work activities or to take short breaks or recovery pauses while at the VDT workstation.

SEATING (The Chair)

1. *Backrest* should provide support for employee's lower back.

2. *Seat width* and *depth* should accommodate specific employee.

3. *Seat front* should not press against the back of employee's knees and lower legs.

4. *Seat* should have cushioning and be rounded with a "waterfall" front.

5. *Armrests* should support both forearms while the employee performs VDT tasks and not interfere with movement.

KEYBOARD/INPUT DEVICE

The keyboard/input device should be designed for doing VDT tasks so that. . . .

6. *Keyboard/input device platforms* are stable and large enough to hold keyboard and input device.

7. *Input device* (mouse or trackball) is located right next to keyboard so it can be operated without reaching.

8. *Input device* is easy to activate, and shape/size fits hand of specific employee (not too big/small).

9. *Wrists* and *hands* do not rest on sharp or hard edge.

FIGURE 8–2 *(Continued)*

MONITOR

The monitor should be designed for VDT tasks so that . . .

10. *Top line* of screen is at or below eye level so the employee is able to read it without bending head or neck down/back.

11. *Employees with bifocals/trifocals* are able to read the screen without bending their heads or necks backward.

12. *Monitor distance* allows employees to read the screen without leaning their heads, necks, or trunks forward/backward.

13. *Monitor position* is directly in front of the employee so he or she does not have to twist the head or neck.

14. *No glare* (e.g., from windows, lights) is present on the screen that may cause employees to assume an awkward position to read screen.

WORK AREA

The work area is designed for doing VDT tasks so that . . .

15. *Thighs* have clearance space between chair and the VDT table/keyboard platform.

16. *Legs* and *feet* have clearance space under the VDT table so the employee is able to get close enough to the keyboard/input device.

ACCESSORIES

17. *Document holder,* if provided, should be stable and large enough to hold documents that are used.

18. *Document holder,* if provided, should be placed at about the same height and distance as the monitor screen so there is little head movement while the employee looks from document to screen.

19. *Wrist rest,* if provided, should be padded and free of sharp and square edges.

20. *Wrist rest,* if provided, should allow employees to keep forearms, wrists, and hands straight and parallel to ground when using the keyboard/input device.

21. *Telephone* can be used with head upright and shoulders relaxed if the employee does VDT tasks at the same time.

GENERAL

22. Workstation and equipment should have sufficient adjustability so that the employee is able to be in a safe working posture and to make occasional changes in posture while performing VDT tasks.

23. VDT workstation, equipment, and accessories should be maintained in serviceable condition and function properly.

Note to Employers:
Workstation must meet ALL of criteria A–J and all but two of criteria 1–23.

Source: Adapted from Appendix D-2 of OSHA Standard 1910.900.

WORKSITE ANALYSIS PROGRAM FOR ERGONOMICS

Although complex analyses are best performed by a professional ergonomist, the "ergonomics team"—or any qualified person—can use this program to conduct a **worksite analysis** and identify stressors in the workplace. The purpose of the following information is to give a starting point for finding and eliminating those tools, techniques, and conditions that may be the source of ergonomic problems.[7]

In addition to analyzing current workplace conditions, planned changes to existing and new facilities, processes, materials, and equipment should be analyzed to ensure that changes made to enhance production will also reduce or eliminate **ergonomic risk factors**. As emphasized before, this program should be adapted to each individual workplace.

The discussion of the recommended program for worksite analysis is divided into four main parts: (1) gathering information from available sources; (2) conducting baseline screening surveys to determine which jobs need closer analysis; (3) performing ergonomic job hazard analyses of those workstations with identified risk factors; and (4) after implementing control measures, conducting periodic surveys and follow-up studies to evaluate changes.

Information Sources

Records Analysis and Tracking

The essential first step in worksite analysis is **records analysis and tracking** to develop the information necessary to identify ergonomic hazards in the workplace. Existing medical, safety, and insurance records, including OSHA 300 logs, should be analyzed for evidence of injuries or disorders associated with CTDs. Healthcare providers should participate in this process to ensure confidentiality of patient records.

Incidence Rates

Incidence rates for upper extremity disorders and/or back injuries should be calculated by counting the incidence of CTDs and reporting the number for each 100 full-time workers per year by facility:*

$$\text{Incidence rate} = \frac{(\text{Number of new cases/yr}) \times (200{,}000 \text{ work hrs}) \text{ per facility}}{\text{Number of hours worked/facility/yr}}$$

Screening Surveys

The second step in worksite analysis is to conduct baseline screening surveys. Detailed baseline screening surveys identify jobs that put employees at risk of developing CTDs. If the job places employees at risk, an effective program will then require the ergonomic job hazard analysis.**

---

* Adapted from OSHA 3123.

** The same method should be applied to departments, production lines, or job types within the facility.

Checklist

The survey is performed with an ergonomic checklist. This checklist should include components such as posture, materials handling, and upper extremity factors. (The checklist should be tailored to the specific needs and conditions of the workplace.)

Ergonomic Risk Factors

Identification of **ergonomic hazards** is based on ergonomic risk factors: conditions of a job process, workstation, or work method that contribute to the risk of developing CTDs. Not all these risk factors will be present in every CTD-producing job, nor is the existence of one of these factors necessarily sufficient to cause a CTD.

CTD Risk Factors

Some of the risk factors for CTDs of the upper extremities include the following:

- Repetitive and/or prolonged activities
- Forceful exertions, usually with the hands (including pinch grips)
- Prolonged static postures
- Awkward postures of the upper body, including reaching above the shoulders or behind the back, and twisting the wrists and other joints to perform tasks
- Continued physical contact with work surfaces (soft tissue compression)
- Excessive vibration from power tools
- Cold temperatures
- Inappropriate or inadequate tool design
- **High wrist acceleration**
- Fatigue (inadequate recovery time)
- Use of gloves

Back Disorder Risk Factors

Risk factors for back disorders include the following:

- Bad body mechanics such as continued bending over at the waist, continued lifting from below the knees or above the shoulders, and twisting at the waist, especially while lifting
- Lifting or moving objects of excessive weight or asymmetric size
- Prolonged sitting, especially with poor posture
- Lack of adjustable chairs, footrests, body supports, and work surfaces at workstations
- Poor grips on handles
- Slippery footing

Multiple Risk Factors

Jobs, operations, or workstations that have **multiple risk factors** have a higher probability of causing CTDs. The combined effect of several risk factors in the development of CTDs is sometimes referred to as *multiple causation*.

Ergonomic Job Hazard Analyses

At this point, the employer has identified—through the information sources and screening surveys discussed above—jobs that place employees at risk of developing CTDs. As an essential third step in the worksite analysis, an effective ergonomics program requires a job hazard analysis for each job so identified.

The job hazard analysis should be routinely performed by a qualified safety and health professional, preferably an ergonomist, for jobs that put workers at risk of developing CTDs. This type of analysis helps verify lower risk factors at light-duty or restricted activity work positions and to determine if risk factors for a work position have been reduced or eliminated to the extent feasible.

Workstation Analysis

An adequate **workstation** analysis would be expected to identify all risk factors present in each studied job or workstation. A workstation consists of the equipment, furniture, tools, and accessories required to do the job in question. When selecting a workstation for analysis, consider at least the following factors:

- Are repetitive movements required of the worker? (yes)
- Are awkward postures required of the worker? (yes)
- Are long periods of sitting or standing required of the worker?(yes)
- Is the noise level appropriate? (no)
- Is the lighting at the proper level for the tasks in question? (no)
- Has glare been eliminated as a potential hazard? (no)
- Is the temperature appropriate for the tasks in question? (no)
- Is the humidity appropriate for the tasks in question? (no)
- Has static electricity been eliminated as a potential hazard? (no)

The answer in parentheses after each question is a trigger answer. If the answer in parentheses applies to the station in question, the station should be subjected to a thorough assessment. The more trigger responses that apply to a given workstation, the more critical it is to conduct a thorough assessment of that workstation.

For upper extremities, three measures of repetitiveness are the total hand manipulations per cycle, the cycle time, and the total manipulations or cycles per work shift. Force measurements may be noted as an estimated average effort and a peak force. They may be recorded as light, moderate, or heavy.

Tools should be checked for excessive vibration. The tools, personal protective equipment, and dimensions and adjustability of the workstation should be noted for each job hazard analysis. Finally, hand, arm, and shoulder postures and movements should be assessed for levels of risk.

Lifting Hazards

For manual materials handling, the maximum weight-lifting values should be calculated.

DVD

The use of DVDs, where feasible, is suggested as a method for analysis of the work process. Slow-motion videotape or equivalent visual records of workers performing their routine job tasks should be analyzed to determine the demands of the task on the worker and how each worker actually performs each task.

Periodic Ergonomic Surveys

The fourth step in worksite analysis is to conduct periodic review. Periodic surveys should be conducted to identify previously unnoticed factors or failures or deficiencies in work practices or engineering controls. The periodic review process should also include feedback, follow-up, and trend analysis.

Feedback and Follow-Up

A reliable system should be provided for employees to notify management about conditions that appear to be hazardous and to utilize their insight and experience to determine work practice and engineering controls. This may be initiated by an ergonomics questionnaire and maintained through an active safety and health committee or by employee participation with the ergonomics team. Reports of ergonomic hazards or signs and symptoms of potential CTDs should be investigated by ergonomics screening surveys and appropriate ergonomic hazard analysis in order to identify risk factors and controls.

Trend Analysis

Trends of injuries and illnesses related to actual or potential CTDs should be calculated, using several years of data where possible. Trends should be calculated for several departments, process units, job titles, or workstations. These trends may also be used to determine which work positions are most hazardous and need to be analyzed by the qualified person.

Using standardized job descriptions, incidence rates may be calculated for work positions in successive years to identify trends. Using trend information can help determine the priority of screening surveys and/or ergonomic hazard analyses.

HAZARD PREVENTION AND CONTROL

Engineering solutions, where feasible, are the preferred method for ergonomic **hazard prevention and control**. The focus of an ergonomics program is to make the job fit the person—not to make the person fit the job. This is accomplished by redesigning the workstation, work methods, or tools to reduce the demands of the job, including high force, repetitive motion, and awkward postures. A program with this goal requires research into currently available controls and technology. It should also include provisions for utilizing new technologies as they become available and for in-house research and testing. Following are some examples of engineering controls that have proven to be effective and achievable.

Workstation Design

Workstations should be designed to accommodate the persons who actually use them; it is not sufficient to design for the average or typical worker. Workstations should be easily adjustable and should be either designed or selected to fit a specific task, so that they are comfortable for the workers who use them. The workspace should be large enough to allow for the full range of required movements, especially where knives, saws, hooks, and similar tools are used.

Design of Work Methods

Traditional work method analysis considers static postures and repetition rates. This should be supplemented by addressing the force levels and the hand and arm postures involved. The tasks should be altered to reduce these and the other stresses associated with CTDs. The results of such analyses should be shared with the healthcare providers to assist in compiling lists of light-duty and high-risk jobs.

Tool Design and Handles

Tools should be selected and designed to minimize the risks of upper extremity CTDs and back injuries. In any tool design, a variety of sizes should be available. Examples of criteria for selecting tools include the following:

- Designing tools to be used by either hand or providing tools for both left- and right-handed workers.
- Using tools with triggers that depress easily and are activated by two or more fingers.
- Using handles and grips that distribute the pressure over the fleshy part of the palm, so that the tool does not dig into the palm.
- Designing and selecting tools for minimum weight; counterbalancing tools heavier than one or two pounds.
- Selecting pneumatic and power tools that exhibit minimal vibration and maintaining them in accordance with manufacturer's specifications or with an adequate vibration-monitoring program. Wrapping handles and grips with insulation material (other than wraps provided by the manufacturer for this purpose) is normally *not* recommended, as it may interfere with a proper grip and increase stress.

MEDICAL MANAGEMENT PROGRAM

An effective **medical management program** for CTDs is essential to the success of an employer's ergonomic program in industries with a high incidence of CTDs.[8] It is not the purpose of these guidelines to dictate medical practice for an employer's **healthcare providers**. Rather, they describe the elements of a medical management program for CTDs to ensure early identification, evaluation, and treatment of signs and symptoms; to prevent their recurrence; and to aid in their prevention. Medical management of CTDs is a developing field, and healthcare providers should monitor developments on the subject. These guidelines represent the best information currently available.

A physician or occupational health nurse (OHN) with training in the prevention and treatment of CTDs should supervise the program. Each work shift should have access to healthcare providers in order to facilitate treatment, surveillance activities, and recording of information. Where such personnel are not employed full time, the part-time employment of appropriately trained healthcare providers is recommended.

In an effective ergonomics program, healthcare providers should be part of the ergonomics team interacting and exchanging information routinely to prevent and treat CTDs properly. The major components of a medical management program for the prevention and treatment of CTDs are trained first-level healthcare providers, health surveillance, employee training and education, early reporting of symptoms, appropriate medical care, accurate record keeping, and quantitative evaluation of CTD trends throughout the plant.

Trained and Available Healthcare Providers

Appropriately trained healthcare providers should be available at all times and on an ongoing basis as part of the ergonomics program. In an effective medical management program, first-level healthcare providers should be knowledgeable in the prevention, early recognition, evaluation, treatment, and rehabilitation of CTDs, as well as in the principles of ergonomics, physical assessment of employees, and OSHA record-keeping requirements.

Periodic Workplace Walk-Through

In an effective program, healthcare providers should conduct periodic, systematic workplace walk-throughs to remain knowledgeable of operations and work practices, to identify potential light-duty jobs, and to maintain close contact with employees. Healthcare providers should also be involved in identifying risk factors for CTDs in the workplace as part of the ergonomics team.

These walk-through surveys should be conducted every month or whenever a particular job task changes. A record should be kept documenting the date of the walk-through, areas visited, risk factors recognized, and action initiated to correct identified problems. Follow-up should be initiated to correct problems identified and should be documented to ensure that corrective action is taken when indicated.

Symptoms Survey

Those responsible for the medical management program should develop a standardized measurement to determine the extent of work-related disorder symptoms in each area of the plant. This measurement will help determine which jobs are exhibiting problems and to measure progress of the ergonomics program.

Institute a Survey

A **symptoms survey** of employees should be conducted to measure employee awareness of work-related disorders and to report the location, frequency, and duration of discomfort. Body diagrams should be used to facilitate gathering this information. Surveys normally should not include employee's personal identifiers to encourage employee participation.

The survey is one method for identifying areas or jobs where potential CTD problems exist. The major strength of the survey approach is in collecting data on the number of workers who may be experiencing some form of CTD. Reported pain symptoms by several workers on a specific job would indicate the need for further investigation of that job.

Conduct the Survey Annually

Conducting the survey annually should help detect any major change in the prevalence, incidence, and/or location of reported symptoms.

Keep a List of Light-Duty Jobs

The ergonomist or other qualified person should analyze the physical procedures used in the performance of each job, including lifting requirements, postures, hand grips, and frequency of repetitive motion.

The ergonomist and healthcare providers should develop a list of jobs with the lowest ergonomic risk. For such jobs, the ergonomic risk should be described. This information will assist healthcare providers in recommending assignments to light- or restricted-duty jobs. The light-duty job should, therefore, not increase ergonomic stress on the same muscle-tendon groups. Healthcare providers should likewise develop a list of known high-risk jobs. Supervisors should periodically review and update the lists.

Health Surveillance

Baseline

The purpose of baseline health surveillance is to establish a base against which changes in healthcare status can be evaluated, not to prevent people from performing work. Prior to assignment, all new and transferred workers who are to be assigned to positions involving exposure of a particular body part to ergonomic stress should receive baseline health surveillance.

Conditioning Period Follow-Up

New and transferred employees should be given the opportunity during a four- to six-week break-in period to condition their muscle-tendon groups prior to working at full capacity. Unfortunately, this is not always possible; however, when it is possible, conditioning should be done. Healthcare providers should perform a follow-up assessment of these workers after the break-in period (or after one month, if the break-in period is longer than a month) to determine if conditioning of the muscle-tendon groups has been successful; whether any reported soreness or stiffness is transient and consistent with normal adaptation to the job or whether it indicates the onset of a CTD; and if problems are identified, what appropriate action and follow-up are required.

Periodic Health Surveillance

Periodic health surveillance—every two to three years—should be conducted on all workers who are assigned to positions involving exposure of a particular body part to ergonomic stress. The content of this assessment should be similar to that outlined for the baseline. The worker's medical and occupational history should be updated.

Employee Training and Education

Healthcare providers should participate in the training and education of all employees, including supervisors and other plant management personnel, on the different types of CTDs and means of prevention, causes, early symptoms, and treatment of CTDs. This information should be reinforced during workplace walk-throughs and the individual health surveillance appointments. All new employees should be given such education during orientation. This demonstration of concern and the distribution of information should facilitate the early recognition of CTDs prior to the development of more severe and disabling conditions and increase the likelihood of compliance with prevention and treatment.

Encourage Early Report of Symptoms

Employees should be encouraged by healthcare providers and supervisors to report early signs and symptoms of CTDs to the in-plant health facility. This allows for timely and appropriate evaluation and treatment without fear of discrimination or reprisal by employers. It is important to avoid any potential disincentives for employee reporting, such as limits on the number of times that an employee may visit the health unit.

Protocols for Healthcare Providers

Healthcare providers should use written protocols for health surveillance and the evaluation, treatment, and follow-up of workers with signs or symptoms of CTDs. A qualified healthcare provider should prepare the protocols. These protocols should be available in the plant health facility. Additionally, the protocols should be reviewed and updated annually or as state-of-the-art evaluation and treatment of these conditions changes.

Evaluation, Treatment, and Follow-Up of CTDs

If CTDs are recognized and treated appropriately early in their development, a more serious condition can likely be prevented. Therefore, a good medical management program that seeks to identify and treat these disorders early is important.

OSHA Record-Keeping Forms

The OSH Act and record-keeping regulations in Title 29, *Code of Federal Regulations (CFR)* 1904, provide specific recording requirements that comprise the framework of the occupational safety and health recording system. The Bureau of Labor Statistics (BLS) has issued guidelines that provide official agency interpretations concerning the record keeping and reporting of occupational injuries and illnesses. These guidelines— U.S. Department of Labor, BLS: *Record-Keeping Guidelines for Occupational Injuries and Illnesses*, September 1986 (or later editions as published)—provide supplemental instructions for the OSHA record-keeping forms and should be available in every plant healthcare facility. Because healthcare providers often provide information for OSHA logs, they should be aware of record-keeping requirements and participate in fulfilling them.

Monitor Trends

Healthcare providers should periodically (e.g., quarterly) review healthcare facility sign-in logs, OSHA Form 300, and individual employee medical records to monitor trends for CTDs in the plant. This ongoing analysis should be made in addition to the symptoms survey to monitor trends continuously and to substantiate the information obtained in the annual symptoms survey. The analysis should be done by department, job title, work area, and so on.

The information gathered from the annual symptoms survey will help identify areas or jobs where potential CTD problems exist. This information may be shared with anyone in the plant because employees' personal identifiers are not solicited. The analysis of medical records (e.g., sign-in logs and individual employee medical records) may reveal areas or jobs of concern, but it may also identify individual workers who require further follow-up. The information gathered while analyzing medical records is confidential; thus, care must be exercised to protect the individual employee's privacy. The information gained from the CTD trend analysis and symptoms survey will help determine the effectiveness of the various programs initiated to decrease CTDs in the plant.

TRAINING AND EDUCATION

The fourth major program element for an effective ergonomics program is training and education.[9] The purpose of training and education is to ensure that employees are sufficiently informed about the ergonomic hazards to which they may be exposed and thus able to participate actively in their own protection (see Figure 8–3).

FIGURE 8–3 Training a worker in proper lifting techniques.

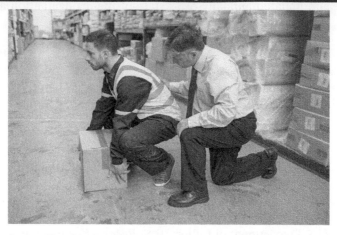

Source: Wavebreakmedia/Shutterstock

Training and education allow managers, supervisors, and employees to understand the hazards associated with a job or process, their prevention and control, and their medical consequences. A training program should include all affected employees, engineers and maintenance personnel, supervisors, and healthcare providers.

The program should be designed and implemented by qualified persons. Appropriate special training should be provided for personnel responsible for administering the program. The program should be presented in language and at a level of understanding appropriate for the individuals being trained. It should provide an overview of the potential risk of illnesses and injuries, their causes and early symptoms, the means of prevention, and treatment.

The program should also include a means for adequately evaluating its effectiveness. This may be achieved by using employee interviews, testing, and observing work practices to determine whether those who received the training understand the material and the work practices to be followed. Also, students and safety professionals are encouraged to take advantage of the training relating to ergonomics as well as other topics from the OSHA Training Institute (OTI). For a complete listing of the courses available from the OTI, go to the following website: www.osha.gov/dte/oti/

In addition to the in-class and online courses provided by the OTI, OSHA maintains a network of Education Centers nationwide. OTI Education Centers are non-profit organizations such as colleges and universities that are authorized by OSHA to deliver safety and health training. To see a list of all OTI Education Centers as well as the courses they provide and accompanying fees, go to the following website: www.osha.gov/otiec

COMMON INDICATORS OF PROBLEMS

Does my company have ergonomic problems? Are injuries and illnesses occurring because too little attention is paid to ergonomic factors? These are questions that modern safety and health professionals should ask themselves. But how do you answer such questions? According to the National Institute for Occupational Safety and Health (NIOSH), the factors discussed in the following paragraphs can be examined to determine whether ergonomic problems exist in a given company.[10]

Apparent Trends in Accidents and Injuries

By examining accident reports, record-keeping documents such as OSHA Form 300 (Log of Work-Related Illnesses and Injuries), first-aid logs, insurance forms, and other available records of illnesses or injuries, safety and health professionals can identify trends if they exist. A pattern or a high incidence rate of a specific type of injury typically indicates that an ergonomic problem exists.

Incidence of CTDs

Factors associated with CTDs include a high level of repetitive work, greater than normal levels of hand force, awkward posture, high levels of vibration, high levels of mechanical stress, extreme temperatures, and repeated hand-grasping or pinch-gripping. By observing the workplace and people at work, safety and health professionals can determine the amount of exposure that employees have to these factors and the potential for ergonomics-related problems.

Absenteeism and High Turnover Rates

High absentee rates and high turnover rates can be indicators of ergonomic problems. People who are uncomfortable on the job to the point of physical stress are more likely to miss work or leave for less stressful conditions.

Employee Complaints

A high incidence of **employee complaints** about physical stress or poor workplace design can indicate the presence of ergonomic problems.

Employee-Generated Changes

Employees tend to adapt the workplace to their needs. The presence of many workplace adaptations, particularly those intended to decrease physical stress, can indicate the presence of ergonomic problems. Have employees added padding, modified personal protective equipment, brought in extra lighting, or made other modifications? Such **employee-generated changes** may be evidence of ergonomic problems.

Poor Quality

Poor quality, although not necessarily caused by ergonomic problems, can be the result of such problems. Poor quality is at least an indicator that there may be ergonomic problems. Certainly, poor quality is an indicator of a need for closer inspection.

Manual Material Handling

The incidence of musculoskeletal injuries is typically higher in situations that involve a lot of **manual material handling**. Musculoskeletal injuries increase significantly when the job involves one or more of the following: lifting large objects, lifting bulky objects, lifting objects from the floor, and lifting frequently. When such conditions exist, the company has ergonomic problems.

IDENTIFYING SPECIFIC ERGONOMIC PROBLEMS

A task analysis of the job in question can identify specific ergonomic problems. Figure 8–4 lists the types of problems that can be identified by a thorough task analysis.[11]

General Observation

General observation of a worker or workers performing the task(s) in question can be an effective task analysis technique. The effectiveness is usually enhanced if the workers are not aware that they are being observed. When observing employees at work, be especially attentive to tasks requiring manual material handling and repetitive movements.

Questionnaires and Interviews

Questionnaires and **interviews** can be used for identifying ergonomic problems. Questionnaires are easier to distribute, tabulate, and analyze, but interviews generally provide more in-depth information.

Videotaping and Photography

Videotaping technology has simplified the process of task analysis considerably. Videotaping records the work being observed as it is done, it is silent so it is not intrusive, and such capabilities as freeze and playback enhance the observer's analysis capabilities significantly. Photography can also enhance the observer's analysis capabilities by recording each motion or step involved in performing a task. If photography is used, be aware that flashes can be disruptive. High-speed film will allow you to make photographs without using a flash.

Drawing or Sketching

Making a neat sketch of a workstation or a drawing showing workflow can help identify problems. Before using a drawing or sketch as part of a task analysis, make sure that it is accurate. Ask an employee who is familiar with the area or process sketched to check the drawing.

FIGURE 8–4   Problems that can be pinpointed by a task analysis.

- Tasks that involve potentially hazardous movements
- Tasks that involve frequent manual lifting
- Tasks that involve excessive wasted motion or energy
- Tasks that are part of a poor operations flow
- Tasks that require unnatural or uncomfortable posture
- Tasks with high potential for psychological stress
- Tasks with a high fatigue factor
- Tasks that could or should be automated
- Tasks that involve or lead to quality control problems

Measuring the Work Environment

Measurements can help identify specific ergonomic problems. How far must a worker carry the material manually? How high does a worker have to lift an object? How much does an object weigh? How often is a given motion repeated? Answers to these and similar questions can enhance the effectiveness of the analysis process.

Understanding the Ergonomics of Aging

When identifying specific ergonomic problems in the workplace, don't overlook the special challenges presented by aging workers. A good ergonomics program adapts the job to the person. Because nearly 30 percent of the workforce is 45 years of age or older, organizations must be prepared to adapt workstations to employees whose physical needs are different from those of their younger counterparts.

In adapting workstations and processes for employees who are 45 or older, keep in mind the following rules of thumb:

- Nerve conduction velocity, hand-grip strength, muscle mass, range of motion, and flexibility all begin to diminish about age 45.
- Weight and mass tend to increase through about the early 50s.
- Height begins to diminish beginning around age 30.
- Lower back pain is more common in people 45 years of age and older.
- Visual acuity at close range diminishes with age.[12]

These rules of thumb mean that safety and health professionals cannot take a "one-size-fits-all" approach to ergonomics. Adaptations for older workers must be individualized and should take aging factors into account.

Anthropometry and Ergonomics

Anthropometry is the science of body size, form, and function. Anthropometric measurements are used to study the interaction of workers with equipment, tools, vehicles, workstations, and personal protective equipment (PPE). Since ergonomics involves fitting the job to the worker, anthropometric measurements can play an important role in identifying potential ergonomic problems in the workplace. Commonly made anthropometric measurements fall into several categories of concern:

- Body composition
- Body size and structure
- Body (circumference)
- Body width
- Body length (height)
- Somatotype (type of physique or combination of body composition and size)

There are three somatotypes: (1) ectomorphs are tall and lean with little body fat or muscle mass; (2) endomorphs have a lot of body fat and muscle mass; and (3) mesomorphs are fit, strong, and solid. In reality, people often have characteristics of two of these body types. These body types can be valuable in quick decisions concerning the fit of PPE, the design of a workstation, or the fit of an individual to a specific task. However, body-type observations should not be allowed to replace actual anthropometric measurements.

Anthropometric measurements include such variables as standing overhead reach, standing height, standing eye height, standing forward reach, sitting height, sitting eye height, sitting knee height, set depth, body mass, girth, handspan, arm length, and recumbent length. All of these measurements, as well as others, can be invaluable in designing products, processes, and tasks so that ergonomic hazards are eliminated or at least minimized. Consider just the example of selecting the right PPE for "fit." Having accurate anthropometric measurements can mean the difference between selecting the right or wrong PPE for an individual, and the right "fit" can mean the difference between working safely in a hazardous environment and being injured.

ERGONOMIC PROBLEM-SOLVING STRATEGIES

The factors that influence stress were explained earlier in this chapter. These factors can be combined in different ways, and the ways in which they are combined determine the type and amount of stress experienced. For the purpose of recommending ergonomic problem-solving strategies, the following combinations can be used:

- Seated repetitive work with light parts
- Seated work with larger parts
- Seated control work
- Standing work
- Standing for heavy lifting and/or carrying work in one place or in motion
- Work with hands above chest height
- Work with hand tools
- Work with video display terminals (VDTs)[13]

Seated Repetitive Work with Light Parts
This type of work can produce more physical stress than you may suspect. Back, neck, shoulder, and lower leg pain are commonly associated with this type of work. Some of the problems associated with seated repetitive work are the result of the nature of the job. The fixed work position and repetitive motion can contribute to ergonomic problems. To solve these problems, it may be necessary to modify both the job and the workstation. Improvement strategies include the following:

- Include other work tasks to break the monotony of repetition (see Figure 8–5).
- Use job rotation, with workers rotating from one or more different jobs.
- Adjust the height of the work surface and/or position.
- Use an adjustable chair equipped with hand, wrist, or arm supports as appropriate.
- Make sure that there is sufficient legroom (height, width, and depth).
- Use ergonomic devices to adjust the height and angle of work.[14]

Seated Work with Larger Parts
This type of work, which involves interacting with objects that may be too large to manipulate manually, is associated with assembly and welding jobs. Problems associated with this type of work are typically related to posture, illumination,

FIGURE 8–5   Adjust the angle of work to reduce stress.

Source: Rikard Stadler/Shutterstock

reach, and lifting. Ergonomic strategies for improving work conditions include the following:

- Use technology to lift and position the work for easy access that does not require bending, twisting, and reaching (see Figure 8–6).
- Use supplemental lighting at the worksite.
- Use adjustable chairs and work surfaces as appropriate.[15]

FIGURE 8–6   Some personal computers have tiltable screens.

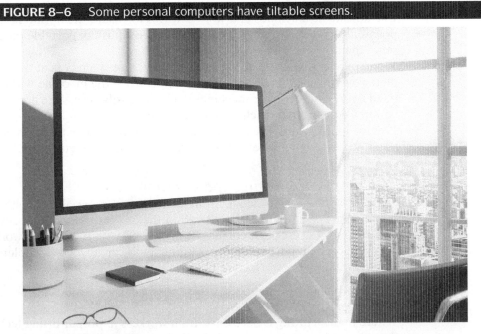

Source: Who is Danny/Shutterstock

Seated Control Work

This type of work involves sitting in one location and using wheels, levers, knobs, handles, and buttons to control a process, system, or piece of equipment. The physical stress associated with seated control work is typically the result of excessive vibration or bending and twisting to achieve better visibility. Ergonomic strategies for improving work conditions include the following:

- Use an adjustable swivel chair with inflatable back and seat support.
- Keep both feet on the floor.
- Provide comfortable and convenient locations for control devices.
- Sit with a straight back and shift positions frequently.
- Use control devices that meet two standards: finger control systems that do not require more than five newtons (1.1 pounds) and hand levers that do not exceed 20 newtons (4.5 pounds).
- Position the control seat so that a clear line of sight exists between the work and the person controlling it.
- Get up and walk around on a regular basis.
- Provide a ladder if the workstation is more than 14 inches above ground.[16]

Standing Work

This category includes most jobs that are performed while standing. Such jobs do not involve a great deal of repetitive motion but do involve handling medium to heavy materials. An example is a machine operator's job (lathe, mill, drill, punch, saw, and so on). Physical stress associated with this type of work includes leg, arm, and back strains. Occasionally, side strains occur when bending and twisting are necessary. Ergonomic strategies for improving work conditions include the following:

- Use adjustable machines and work surfaces to ensure the optimum height and position. When the machine height cannot be adjusted, portable platforms can serve the same purpose.
- When purchasing new machines, make sure there is a recess at the bottom for feet. This will allow the operator to stand close to the machine without having to bend over. Also, look for machines that have easily accessible controls that fall within a comfortable reach zone for operators.
- Provide ample free space around machines for moving material in and out and to allow for ease of movement in servicing machines (Figure 8–7).[17]

Standing for Heavy Lifting and Carrying

This type of work involves heavy lifting and moving material while standing. Lifting and moving may be a relatively small part of the job but are required somewhat regularly. The physical stress most commonly associated with this type of work is back and muscle strains resulting from improper lifting. Falls can also be a problem. Ergonomic strategies for improving work conditions include the following:

- Eliminate manual lifting to the extent possible using various lifting and hoisting technologies.
- Where manual lifting is necessary, train workers in proper lifting techniques.
- Provide sufficient room around all objects to allow lifting without twisting.

FIGURE 8–7 Do not put computers in cramped, crowded locations.

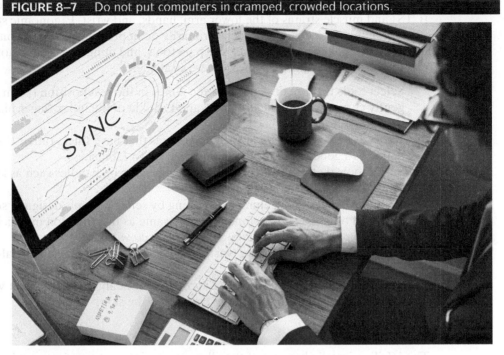

Source: Rawpixel.com/Shutterstock

- Supply the appropriate personal protection equipment such as sure-grip shoes and gloves.
- Keep floors around materials to be lifted clean and dry to prevent slips.
- Do not allow manual carrying of heavy objects upstairs. Stairs increase the physical stress of carrying and, in turn, the potential for injury.[18]

Work with Hands above Chest Height

This type of work can be done in either a standing or sitting position. It may or may not involve material handling. Physical stress associated with this type of work includes neck, upper body, and heart strain. Of these, the most potentially dangerous is heart strain. Prolonged work with the arms above the shoulder level requires the heart to work harder to pump blood to the elevated areas. Ergonomic strategies for improving work conditions include the following:

- Eliminate manual lifting to the extent possible by raising the work floor using lifts and various other technologies.
- Use extension arms or poles when the work floor cannot be raised.
- When purchasing new machines, look for machines with controls that are easily accessible below the horizontal plane of a worker's shoulders.[19]

Work with Hand Tools

All the types of work presented in this section may involve the use of hand tools to some extent. However, because hand tools introduce a variety of potential hazards that are indigenous to their use, they are best examined as a separate work type. Physical stress associated most commonly with the use of hand tools includes carpal tunnel syndrome (CTS) and muscle strains of the lower arm, hands, and wrist. Ergonomic strategies for improving the work conditions focus primarily on improving hand positions during the use of tools, enhancing the worker's grip on tools, and minimizing the amount of twisting involved. Following are some of these strategies:

- Select tools that are designed to keep hands in the rest position (palm down, gently curved, thumb outstretched, and knuckle of the index finger higher than that of the little finger).
- Reduce stress on the hand by selecting tools that have thick, rather than thin, handles (a good range for the diameter is 0.8 to 1.2 inches).
- Select tools that have enhanced gripping surfaces on handles such as knurling, filing, and other enhancements.
- To the extent possible, eliminate twisting by selecting tools designed so that the direction of movement or function is the same as the direction in which force is applied or by using technology (e.g., power screwdriver).
- For tools that do not involve twisting, select handles that have an oval-shaped cross section.
- Select tools with handles made of hard, nonpermeable materials that will not absorb toxic liquids that could be harmful to the skin.[20]

Work with VDTs

The VDT, primarily because of the all-pervasive integration of personal computers in the workplace, is now the most widely used piece of office equipment. This fact, coupled with the ergonomic hazards associated with VDTs, has created a whole new range of concerns for safety and health professionals. Using ergonomics to design a workspace will make it easier, safer, more comfortable, and more efficient to use. Following are some strategies that can be used to reduce the hazards associated with VDTs:

- **Arrange the keyboard properly.** It should be located in front of the user, not to the side. Body posture and the angle formed by the arms are critical factors (see Figure 8–8).
- **Adjust the height of the desk.** Taller employees often have trouble working at average-height desks. Raising the desk with wooden blocks can solve this problem.
- **Adjust the tilt of the keyboard.** The rear portion of the keyboard should be lower than the front.
- **Encourage employees to use a soft touch on the keyboard and when clicking a mouse.** A hard touch increases the likelihood of injury.
- **Encourage employees to avoid wrist resting.** Resting the wrist on any type of edge can increase pressure on the wrist.
- **Place the mouse within easy reach and take your hand off of it when not in use.** Extending the arm to its full reach increases the likelihood of injury.
- **Relax the hand and occasionally change hands.** Also maintain a neutral wrist and float the whole forearm rather than just moving the wrist.
- **Locate the VDT at a proper height and distance.** The VDT's height should be such that the top line on the screen is slightly below eye level. The optimum

distance between the VDT and user will vary from employee to employee, but it will usually be between 16 and 32 inches.

- **Minimize glare.** Glare from a VDT can cause employees to adopt harmful postures. Changing the location of the VDT, using a screen hood, and closing or adjusting blinds and shades can minimize glare.
- **Reduce lighting levels.** Reducing the lighting level in the area immediately around the VDT can eliminate vision strain.
- **Dust the VDT screen.** VDT screens are magnets to dust. Built-up dust can make the screen difficult to read, contributing to eye strain.
- **Eliminate telephone cradling.** Cradling a telephone receiver between an uplifted shoulder and the neck while typing can cause a painful disorder called **cervical radiculopathy** (compression of the cervical vertebrae in the neck). Employees who need to talk on the telephone while typing should wear a headset.
- **Require typing breaks.** Continuous typing for extended periods should be avoided. Repetitive strain injuries are cumulative. Breaking up the repetitive motion in question (typing and clicking) can help prevent the accumulation of strain.[21]

FIGURE 8–8 Ergonomics of VDTs. The left of the diagram highlights optimal postures and positions for the computer user.

Head—Positioned directly over shoulders without straining, about an arm's length from the screen

Neck and Shoulders—Relaxed, without slouching

Back—Upright or inclined slightly forward from the hips; maintain the same spinal curve the back has when standing

Elbows—Relaxed at a 70°–135° angle

Wrist—Relaxed and straight, not angled up or down

Knees—Placed slightly lower than hips

Chair—Sloped slightly forward so knees are in the correct position

0°–60°

70°–135°

Screen—Positioned at eye level or slightly lower

Keyboard—Placed flat, close to elbow-level

Work Surface—Positioned at a height to allow the elbows to be relaxed at a 70°–135° angle; chair may need to be raised to accommodate this

Feet—Planted firmly on the floor. A footrest may be needed for shorter people or high work surfaces

ECONOMICS OF ERGONOMICS

Perhaps the most underresearched subject relating to ergonomics is the cost-effectiveness of safety and health. There are two main reasons for this: (1) such research is often more complex and extensive than the safety and health measures that have been undertaken and (2) many decision makers think such studies are irrelevant because promoting safety and health is the right thing to do, regardless of costs. As with safety and health in general, there are no in-depth studies available that conclusively pin down the cost benefits of specific ergonomic measures.

There is disagreement among well-informed professionals on whether ergonomic improvements should be expected to meet the test of cost-benefit analysis. However, there is growing support for research that will produce reliable data. This point of view is beginning to characterize the outlook of safety and health professionals in government and academe.[22] Consequently, it is important to understand the problems in attempting to undertake or participate in cost-benefit studies.

Inhibitors of Cost-Benefit Studies

A number of factors inhibit hard research into the economics of ergonomics. The most problematic of these are as follows:

- Record-keeping systems in industry are not sufficient to support such studies. As a result, a base of comparison from which to work has not been established.
- Industry does not track injuries and illnesses in ways that provide the controls necessary for true hard research. There are no control groups against which to compare groups of injured workers.
- It is difficult and sometimes impossible to determine what improvements can be attributed directly to specific ergonomic strategies and what improvements should be attributed to other factors.
- Undertaking hard research studies requires a commitment of both time and money. A longitudinal study of the effects of a given ergonomic improvement may take three to five years to produce reliable data.
- Follow-up evaluations of injuries that summarize the direct and indirect costs do not exist to the extent necessary to contribute to hard research.[23]

These inhibitors cannot be overcome unless industrial firms are willing to invest enormous sums of money and time. Although the data produced by such investments would be valuable, they are generally not perceived as adding to a company's bottom line. Therefore, industry is likely to continue to rely on academe and governmental agencies for research. In today's intensely competitive international marketplace, industry is likely to invest all available funds in efforts that are perceived as improving productivity and quality. Consequently, hard research into the economics of ergonomics may remain a low priority for some time to come.

CUMULATIVE TRAUMA DISORDERS/ SOFT TISSUE INJURIES

The personal computer has become an all-pervasive and universal work tool. Jobs from the shop floor to the executive office now involve frequent, repetitive computer use. This means that people in the workplace are typing and clicking at an unprecedented pace. Frequent and, for some, constant computer use have led to an explosion of injuries that until now were seen mostly in the meatpacking industry. Collectively, these injuries are known as CTDs.

Definition

CTD is an umbrella term that covers a number of injuries caused by forceful or awkward movements repeated frequently over time. Other aggravating factors include poor posture, an improperly designed workstation, poor tool design, and job stress. CTDs occur to the muscles, nerves, and tendons of the hands, arms, shoulders, and neck.

FIGURE 8–9   CTD checklist—types of injuries by classification.

Muscle and Tendon Disorders
- ✓ Tendinitis
- ✓ Muscle damage (myofacial)
- ✓ Tenosynovitis
- ✓ Stenosing tenosynovitis
  - • DeQuervain's disease
  - • Trigger finger (flexor tenosynovitis)
- ✓ Shoulder tendinitis
- ✓ Bicipital tendinitis
- ✓ Rotator cuff tendinitis
- ✓ Forearm tendinitis
  - • Flexor carpi radialis tendinitis
  - • Extensor tendinitis
  - • Flexor tendinitis
- ✓ Epicondylitis
- ✓ Ganglion cysts

Cervical Radiculopathy

Tunnel Syndromes
- ✓ Carpal tunnel syndrome
- ✓ Radial tunnel syndrome
- ✓ Sulcus ulnaris syndrome
- ✓ Cubital tunnel syndrome
- ✓ Guyon's canal syndrome

Nerve and Circulation Disorders
- ✓ Thoracic outlet syndrome
- ✓ Raynaud's disease

Classifications of CTDs

For years, CTDs have been incorrectly referred to as CTS, which is actually one type of CTD. This is like referring to all trees as oaks. Figure 8–9 is a checklist of the most common CTDs organized into four broad classifications.

Muscle and Tendon Disorders

Tendons connect muscles to bones. They can accommodate very little in the way of stretching and are prone to injury if overused. Overworking a tendon can cause small tears in it. These tears can become inflamed and cause intense pain. This condition is known as **tendinitis**.

Myofascial muscle damage can also be caused by overexertion. It manifests itself in soreness that persists even when resting. Muscles may burn and be sensitive to the touch. When sore muscles become inflamed and swell, the symptoms are aggravated even further by nerve compression. Tendons that curve around bones are encased in protective coverings called *sheaths*. Sheaths contain a lubricating substance known as synovial fluid. When tendons rub against the sheath too frequently, friction is produced. The body responds by producing additional synovial fluid. Excess build-up of this fluid can cause swelling that, in turn, causes pressure on the surrounding nerves, causing a condition known as **tenosynovitis**.

Chronic tenosynovitis is known as *stenosing tenosynovitis* of which there are two types: DeQuervain's disease and flexor tenosynovitis (trigger finger). DeQuervain's disease affects the tendon at the junction of the wrist and thumb. It causes pain when the thumb is moved or when the wrist is twisted. Flexor tenosynovitis involves the locking of a digit in a bent position, hence the term *trigger finger*. However, it can occur in any finger.

Shoulder tendinitis is of two types: bicipital and rotator cuff tendinitis. Bicipital tendinitis occurs at the shoulder joint where the bicep muscle attaches. The rotator cuff is a group of muscles and tendons in the shoulder that move the arm away from the body and turn it in and out. Pitchers in baseball and quarterbacks in football often experience rotator cuff tendinitis.

Forearm tendinitis is of three types: flexor carpi radialis tendinitis, extensor tendinitis, and flexor tendinitis. Flexor carpi radialis tendinitis causes pain in the wrist at the base of the thumb. Extensor tendinitis causes pain in the muscles in the top of the hand, making it difficult to straighten the hands. Flexor tendinitis causes pain in the fingers, making it difficult to bend them.

Epicondylitis and *ganglion cysts* are two muscle and tendon disorders. Epicondylitis (lateral) affects the outside of the elbow, whereas epicondylitis (medial) affects the inside. The common term for this disorder is "tennis elbow." Ganglion cysts grow on the tendon, tendon sheath, or synovial lining, typically on top of the hand, on the nail bed, above the wrist, or on the inside of the wrist.

Cervical Radiculopathy

This disorder is most commonly associated with holding a telephone receiver on an upraised shoulder while typing. This widely practiced act can cause compression of the cervical discs in the neck, making it painful to turn the head. It is always dangerous to put the body in an unnatural posture while using the hands.

FIGURE 8–10 Tunnel injuries are common in jobs involving repetitive motion.

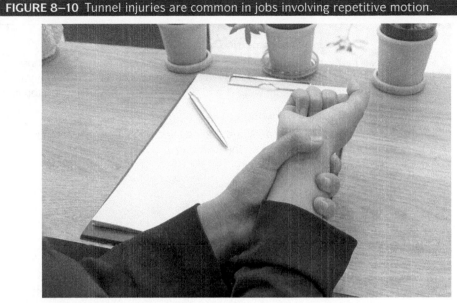

Source: Kittisak Jirasittichai/Shutterstock

Tunnel Syndromes

Tunnels are conduits for nerves that are formed by ligaments and other soft tissues. Damage to the soft tissues can cause swelling that compresses the nerves that pass through the tunnel. These nerves are the median, radial, and ulnar nerves that pass through a tunnel in the forearm and wrist. Pain experienced with tunnel injuries can be constant and intense. In addition to pain, people with a tunnel injury might experience numbness, tingling, and a loss of gripping power. The most common **tunnel syndromes** are CTS, radial tunnel syndrome, sulcus ulnaris syndrome, cubital tunnel syndrome, and Guyon's canal syndrome (see Figure 8–10).

Nerve and Circulation Disorders

When friction or inflammation cause swelling, both nerves and arteries can be compressed, restricting the flow of blood to muscles. This can cause a disorder known as *thoracic outlet syndrome*. The symptoms of this disorder are pain in the entire arm, numbness, coldness, and weakness in the arm, hand, and fingers.

If the blood vessels in the hands are constricted, it can result in *Raynaud's disease*. Symptoms include painful sensitivity, tingling, numbness, coldness, and paleness in the fingers. It can affect one or both hands. This disorder is also known as *vibration syndrome* because it is associated with vibrating tools.

Preventing CTDs

The best way to prevent CTDs is proper work design. In addition, it also helps make employees aware of the hazards that can cause it. These hazards include poor posture at the workstation, inappropriate positioning of the hands and arms, a heavy hand on a keyboard or mouse, and any other act that repeatedly puts the body in an unnatural posture while using the hands. Ergonomically sound workstations can help prevent CTDs, especially when they can be modified to fit the individual employee. However,

FIGURE 8–11 Checklist for safe selection and use of hand tools.

Use Anthropometric Data

Anthropometric data has to do with human body dimensions. Such data can be used to determine the proper handle length, grip span, tool weight, and trigger length when selecting tools.

Reduce Repetition

Repetition is a hazard that can and should be reduced using such strategies as the following:

- Limit overtime.
- Change the process.
- Provide mechanical assists.
- Require breaks.
- Encourage stretching and strengthening exercises.
- Automate where possible.
- Rotate employees regularly.
- Distribute work among more employees.

Reduce the Force Required

The more force required, the more potential for damage to soft tissue. Required force can be reduced using the following strategies:

- Use power tools wherever possible.
- Use the power grip instead of the pinch grip.
- Spread the force over the widest possible area.
- Eliminate slippery, hard, and sharp gripping surfaces.
- Use jigs and fixtures to eliminate the pinch grip.

Minimize Awkward Postures

Awkward postures contribute to CTDs. The following strategies can reduce posture hazards:

- Keep the wrist in a neutral position.
- Keep elbows close to the body (90°–110° where bent).
- Avoid work that requires overhead reaching.
- Minimize forearm rotation.

even the best ergonomic design cannot prevent a heavy hand on the keyboard or mouse. Consequently, ergonomics is only part of the answer. Following are some preventive strategies that can be applied in any organization:

1. **Teach employees the warning signs.** CTDs occur cumulatively over time; they sneak up on people. Employees should be aware of the following warning signs: weakness in the hands or forearms, tingling, numbness, heaviness in the hands, clumsiness, stiffness, lack of control over the fingers, cold hands, and tenderness to the touch.

2. **Teach employees how to stretch.** Employees whose jobs involve repetitive motion work such as typing may help prevent CTDs by using stretching exercises. Limbering up the hands and forearms each day before starting work and again after long breaks such as the lunch hour may help eliminate the stress on muscles and tendons that can lead to CTDs. The term "may" is used because the jury is still out as to the efficacy of stretching. There is no consensus in the ergonomics community on this preventive measure.

3. **Teach employees to start slowly.** Long-distance runners typically start slowly, letting their bodies adjust and their breathing find its rhythm. They pick up the pace steadily, until eventually settling in at a competitive pace. This approach is an excellent example of how employees in CTD-prone jobs should work. Teach employees to limber up, then begin slowly and increase their pace gradually.

4. **Avoid the use of wrist splints.** Teach employees to position their hands properly without using wrist splints. Splints can cause the muscles that they support to atrophy, thereby actually increasing the likelihood of problems.

5. **Exercise regularly.** Exercises that strengthen the hands and forearms coupled with exercises that gently stretch hand and forearm muscles may be a preventive measure. Exercises that strengthen the back can help improve posture, and good posture helps prevent CTDs. As with stretching, the jury is still out on the efficacy of exercise as a preventive measure.

6. **Select tools wisely.** CTS and other CTDs are most frequently associated with the repetitive use of VDTs and hand tools. Selecting and using hand tools properly can help prevent CTDs. Figure 8–11 is a checklist for the proper selection and use of hand tools. Note that ergonomically designed hand tools will not overcome poor job design. Good job design and proper tool selection together are the best strategy.

LIFTING HAZARDS

Back injuries that result from improper lifting are among the most common in the workplace. In fact, back injuries account for approximately $12 billion in workers' compensation costs annually. The following statistics concerning workplace back injuries illustrate the scope and seriousness of this problem:

- Lower back injuries account for 20 to 25 percent of all workers' compensation claims.
- Thirty-three to 40 percent of all workers' compensation costs are related to lower back injuries.
- Each year, there are approximately 46,000 back injuries in the workplace.
- Back injuries cause approximately 100 million lost workdays each year.
- Approximately 80 percent of the population will experience lower back pain at some point in their lives.

Typical causes of back injuries in the workplace include: improper lifting, reaching, sitting, and bending. **Lifting hazards** such as poor posture, ergonomic factors, and personal lifestyles also contribute to back problems. Consequently, a company's overall safety and health program should include a **back safety/lifting** component.

Back Safety/Lifting Program

Prevention is critical in back safety. Consequently, safety and health professionals need to know how to establish back safety programs that overcome the hazards of lifting and other activities. What follows is an example of a comprehensive back safety/ lifting program:

1. **Make proper lifting techniques visible.** Posters and worksheets that illustrate proper lifting, reaching, sitting, and bending techniques should be displayed strategically throughout the workplace and in workstations. This is as important in offices as in manufacturing settings. Clerical and office personnel actually sustain a higher proportion of back injuries than employees in general. Sitting too long without standing, stretching, and walking can put as much pressure on the back as lifting.

2. **Screen applicants. Preemployment screening** can identify people who already have back problems when they apply for jobs. This is important because more than 40 percent of back injuries occur in the first year of employment and the majority of these injuries are related to pre-existing back problems.

3. **Conduct safety inspections on a regular basis.** Regular inspections of the workplace can identify potential problems so that corrective action can be taken immediately. Occasionally, bringing a workers' compensation consultant in to assist with an inspection can help identify hazards that company personnel may miss.

4. **Provide education, training, and mentoring.** Education, training, and mentoring designed to help employees understand how to lift, bend, reach, stand, walk, and sit safely can be the most effective preventive measure undertaken. Companies that provide back safety training report a significant decrease in back injuries.

5. **Make use of external services.** A variety of external healthcare agencies can help companies develop, maintain, and evaluate their back safety/lifting programs. Identify local healthcare providing agencies and organizations, what services they can provide, and a contact person in each. Maintaining a positive relationship with these external service contact people can increase the services available to employers. Also, make use of the services available from OSHA at: www.osha.gov/dcsp/compliance_assistance/cas.html. Small businesses can get help from OSHA that is geared specifically for their needs at: www. osha.gov/dcsp/smallbusiness/benefits.html.

6. **Document the program in writing.** The back safety/lifting program should be written down and incorporated in the company's overall safety and health program. The written program should be reviewed periodically and updated as appropriate.

Despite a company's best efforts, back injuries will still occur; see Figure 8–12. Consequently, safety and health professionals should be familiar with the treatment and therapy that injured employees are likely to receive. Treatment for reconditioning addresses five goals: restoring function, reducing pain, minimizing deficits in strength,

FIGURE 8–12   Back injuries can occur in even the most safety conscious organizations

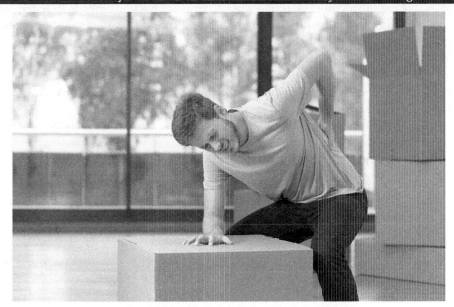

Source: Antonio Guillem/Shutterstock

reducing lost time, and returning the body to preinjury fitness levels. A concept that is gaining acceptance in bridging the gap between treatment or therapy and a safe return to work is known as **work hardening**. A work hardening program involves assigning recovering workers are to specially designed work centers that contain various types of work stations, exercise equipment, and aggressive protocols that allow them to begin replicating actual work conditions. The objectives of a work hardening program are:

- A return to maximum physical abilities as soon as possible
- Improvement of general body fitness
- Reducing the likelihood of re-injury
- Work simulation that duplicates real work conditions

The work centers referred to above replicate in as much detail as possible the injured worker's actual work environment. In addition to undergoing carefully controlled and monitored therapy in the work center, the employee is encouraged to use exercise equipment. Employees who undergo work center therapy should have already completed a program of acute physical therapy and pain management, and they should be medically stable.

Health and safety professionals can help facilitate the fastest possible safe resumption of duties by injured employees by identifying local healthcare providers that use the work hardening approach. Such services and local providers of them should be made known to higher management so that the company can take advantage of them. Some organizations choose to have in-house work hardening programs operated by properly trained professionals in the field of physical rehabilitation.

Proper Lifting Techniques

One of the most effective ways to prevent back injuries is to teach employees proper lifting techniques. Following are lifting techniques that should be taught as part of an organization's safety program.

PLAN AHEAD

- Determine if you can lift the load. Is it too heavy or too awkward?
- Decide if you need assistance.
- Check your route to see whether it has obstructions and slippery surfaces.

LIFT WITH YOUR LEGS, NOT YOUR BACK

- Bend at your knees, keeping your back straight (Figure 8–13).
- Position your feet close to the object.
- Center your body over the load.
- Lift straight up smoothly; don't jerk (Figure 8–14).
- Keep your torso straight; don't twist while lifting or after the load is lifted.
- Set the load down slowly and smoothly with a straight back and bent knees; don't let go until the object is on the floor.

PUSH, DON'T PULL

- Pushing puts less strain on your back; don't pull objects.
- Use rollers under the object whenever possible.

FIGURE 8–13   Improper lifting technique leads to back injuries.

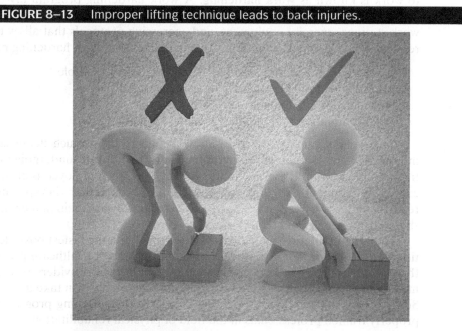

Source: Hakandogu/Shutterstock

FIGURE 8–14   Keeping the back straight while lifting is critical.

Source: Mia Caruana/Alamy Stock Photo

HELPFUL ASSESSMENT TOOLS: NIOSH, RULA, REBA, AND HAL

It is important for safety and health professionals who may not be ergonomic experts to be able to conduct comprehensive assessments of ergonomic hazards in the workplace. To make this challenge easier, several assessment tools have been developed. NIOSH developed the ***Ergonomic Guidelines for Manual Material Handling (EGMMH).*** A private firm, Ergonomics Plus developed the **RULA** and **REBA** assessment tools. The American Conference of Governmental Industrial Hygienists (ACGIH) developed the TLV for Hand Level Activity or **HAL** and the **Strain Index**. All of these assessment tools are available for use by safety and health professionals.

NIOSH Ergonomic Guidelines for Manual Material Handling (EGMMH)

NIOSH teamed with the California Occupational Safety and Health Administration and the Ergonomic Assist and Systems Equipment Council of the Material Handling Industry of America to develop the **Ergonomic Guidelines for Manual Material Handling (EGMMH)**. The guidelines were developed primarily for smaller companies that cannot afford to hire a staff of safety and health professionals. They are divided into four parts:

- The ergonomic process
- A matrix of common material handling tasks
- Several chapters on how to approach common material handling tasks
- A resource index

The guidelines provide more than one safe way to approach the various material handling tasks covered, and they incorporate photographs and pictures to illustrate these approaches. Safety and health professionals will find the assessment checklists provided in Appendix B helpful and easy to use. Appendix B contains a checklist for manual material handling and another for hazard evaluation. Organizations or safety and health professionals who are interested in obtaining a copy of the EGMMH should visit the NIOSH website at: www.cdc.gov/niosh.

RULA Assessment Tool

The Rapid Upper Limb Assessment or RULA was developed to allow safety and health professionals to quickly assess worker exposure to ergonomic hazards relating to the upper body. It assesses the requirements/hazards of work tasks relating to the neck, trunk, and upper extremities. RULA assessments use a specially designed worksheet to assess posture, force, and repetition relating to work tasks. Risk is rated as follows: (1) negligible, requiring no changes to how the work task is performed; (2) low risk, requiring some changes to how the work task is performed; (3) medium risk requiring additional investigation and definite changes to how the work task is performed and how soon; and (4) very high risk requiring immediate changes to how the work task is performed. The guide for administering the RULA and the instrument itself are available at: http://ergo-plus.com/rula-assessment-tool-guide/

REBA Assessment Tool

The Rapid Entire Body Assessment or REBA is similar to the RULA except that, at the name implies, it covers the entire body. It was developed to allow safety and health professionals to quickly assess worker exposure to ergonomic hazards relating to the entire body. REBA, like RULA, uses a one-page worksheet to assess whole body postural risks associated with work tasks. REBA uses the same rating scale for risks as RULA: (1) negligible, requiring no changes to how the work task is performed; (2) low risk, requiring some changes to how the work task is performed; (3) medium risk, requiring additional investigation and definite changes to how the work task is performed and how soon; and (4) very high risk, requiring immediate changes to how the work task is performed. The guide for administering the REBA and the instrument itself are available at: http://ergo-plus.com/reba-assessment-tool-guide/

ACGIH HAL and the Strain Index Assessment Tools

The HAL assessment tool evaluates the risk of developing disorders of the hand, wrist, and forearm in monotask jobs that involve four or more hours of repetitive hand work. Such factors as posture, movements, external force, vibration, and contact forces are assessed according to magnitude/amplitude, duration, and frequency. The Strain Index is similar, but is used for assessing the risk of workers developing distal upper extremity disorders. Also, it may be used for multi-task jobs. Special training is required before safety and health professionals can use HAL or the Strain Index effectively. For additional information about HAL or the Strain Index, go to www.acgih.org and search the site for "HAL" or the "Strain Index" depending on your interest.

PARTICIPATORY ERGONOMICS

Participatory ergonomics (PE) is the involvement of people at work in planning for and controlling the ergonomic aspects of their work environment. There is one caveat to be observed here: Before people are involved in planning and controlling the ergonomic aspects of their work environment, they must first be equipped with prerequisite knowledge. This means that employees will typically require training before they can become involved in participatory ergonomics.[24]

Participatory ergonomics is an approach to intervention that combines the best of two worlds: outside expertise and inside experience. When undertaking an ergonomics intervention in the workplace, the most common approach is to form a PE team. This team should consist of the outside expert(s) in the field of ergonomics, an internal safety and health professional, management personnel, and employees. Before beginning the intervention process, the team is trained—typically by the outside expert(s) in the field of ergonomics. The goal of the PE team is to design ergonomic interventions that are tailored specifically to the workplace in question.

Nine of ten studies reviewed by the Institute for Work & Health showed that PE interventions can have a positive effect on safety and health in the workplace. These studies showed that PE interventions had a positive effect on the musculoskeletal symptoms of workers and on reducing workplace injuries as well as workers' compensation claims.[25]

Making PE Interventions Effective

The Institute for Work & Health found that there are several factors that, when taken together, serve to make PE interventions effective. These factors are as follows:

- Acceptance of the PE team members by other employees and unions (where applicable)
- Acceptance of the PE team members by management
- Availability of an ergonomics expert to lead the team and provide the prerequisite training
- Sufficient support from management in the form of resources

When these factors are present, an organization can reduce injuries and workers' compensation claims by applying the concept of participatory ergonomics when making workplace interventions.

Key Terms and Concepts

- Back safety/lifting
- Cervical radiculopathy
- Cumulative trauma disorders (CTDs)
- Employee complaints
- Employee-generated changes
- Environmental factors
- Ergonomic hazards
- Ergonomic risk factors
- Ergonomics
- Ergonomics guidelines
- General duty clause
- General observation
- Hazard prevention and control

- HAL
- Healthcare providers
- High wrist acceleration
- Horizontal work area
- Incidence rates
- Interviews
- Lifting hazards
- Manual material handling
- Medical management program
- Multiple risk factors
- Participatory ergonomics
- Physical stress
- Preemployment screening

- Questionnaires
- REBA
- Records analysis and tracking
- Repetitive motion
- RULA
- Strain Index
- Symptoms survey
- Tendinitis
- Tenosynovitis
- Tunnel syndromes
- Vertical work areas
- Work hardening
- Worksite analysis
- Workstation

Review Questions

1. Define the term *ergonomics*. Explain its origins.
2. Explain how the following opposing factors can influence the amount of physical stress associated with a job: sitting versus standing, large versus small demand for strength/power, nonrepetitive versus repetitive motion.
3. Explain the concept of human factors and how it relates to ergonomics.

4. Name the elements of OSHA's voluntary ergonomics guidelines.
5. What are the four main parts of OSHA's recommended worksite analysis program for ergonomics?
6. List five risk factors associated with CTDs.
7. Briefly explain the steps in conducting an ergonomic job hazard analysis.
8. Briefly explain the components of a hazard prevention and control program.
9. Who should participate in an ergonomics training program?
10. List and briefly explain three common indicators of the existence of ergonomic problems.
11. Describe three approaches that can be used to pinpoint specific ergonomic problems.
12. Describe an ergonomic problem-solving strategy for each of the following types of work: seated repetitive work with light parts, work with hands above the chest height,
and work with hand tools.
13. Explain why so little hard research has been done concerning the economics of ergonomics.
14. Define the term *CTD*. Describe the most common types.
15. List and briefly explain the assessments that are available relating to ergonomic hazards.

Endnotes

1. G. Salvendy, *Handbook of Human Factors and Ergonomics* (New York: Wiley, 2012), 6.
2. Ibid., 10–11.
3. Ibid., 12.
4. Ibid., 13.
5. Ibid., 23.
6. *Ergonomics Report.* Retrieved from ergonomicsreport.com/publish on July 1, 2017.
7. Adapted from OSHA 3123.
8. Ibid.
9. Ibid.
10. National Institute for Occupational Safety and Health (NIOSH), *Elements of Ergonomics Programs*. Retrieved from cdc.gov/niosh on July 1, 2017.
11. Ibid., 12.
12. S. Minter, "Ergonomic Challenge: The Aging Workforce," *Occupational Hazards* 64, no. 9: 6. 2002.
13. NIOSH, *Elements of Ergonomics Programs*, 13–14.
14. Ibid., 14.
15. Ibid., 15.
16. Ibid., 16.
17. Ibid., 17.
18. Ibid., 18.
19. Ibid., 19.
20. Ibid., 20.
21. Roberta Carson, "Ergonomic Innovations: Free to a Good Company," *Occupational Hazards*, January 1996, 61–64.
22. Ibid., 38.
23. Ibid., 39.
24. D. Cole, I. Rivilis, D. Van Eerd, K. Cullen, E. Irvin, and D. Kramer, "Effectiveness of Participatory Ergonomic Interventions: A Systematic Review," *Institute for Work & Health*. Retrieved from www.iwh.on.ca on July 1, 2017.
25. Ibid.

CHAPTER
9

STRESS AND SAFETY

Learning Objectives

■ Define the concept workplace stress
■ Explain the most common sources of workplace stress
■ Describe the types of human reactions to stress
■ Demonstrate how workplace stress can be measured
■ Explain how shift work can increase worker stress
■ List effective stress-reduction strategies
■ Explain how safety managers are vulnerable to stress
■ Summarize the relationship between stress and workers compensation

WORKPLACE STRESS DEFINED

Our emotions are affected in the workplace by social, occupational, environmental, and psychological factors that we perceive as threats. These perceived threats are external stimuli, and **stress** is the reaction of the human body to these stimuli. The amount of stress felt depends as much on the individual's ability to deal with the external stimuli as the relative intensity of the stimuli. For example, the threat of falling out of a boat into the water would cause more stress for a person who cannot swim than it would for a person who is an accomplished swimmer.

In this example, the water is just as deep and the boat is just as far from shore for both individuals, but the threat and the amount of stress it causes are less for the one who can swim than for the nonswimmer. This is because of the good swimmer's ability to deal with the threatening situation. The same rule of thumb can be applied to all perceived threats in the workplace.

As they often do, management personnel and line employees often view stress from two different perspectives. Managers tend to view stress as an individual problem tied to the personality and emotional makeup of the employee. Employees often view stress as

227

a problem induced by poor supervision, unrealistic expectations, and other management shortcomings. Practically speaking, both managers and employees are partially right in their opposing perspectives. On the one hand, both the personality and emotional makeup of employees are factors in how they respond to the stimuli that produce stress. On the other hand, management shortcomings are also factors in how the employees respond to these stimuli.

However it is defined, stress is a serious problem in the modern workplace. Stress-related medical bills and the corresponding absentee rates cost employers more than $150 billion annually. Almost 15 percent of all occupational disease claims are stress related.

Workplace stress involves the emotional state resulting from a perceived difference between the level of occupational demand and a person's ability to cope with this demand. Because preparations and emotions are involved, workplace stress is considered a subjective state. An environment that a worker finds stressful may generate feelings of tension, anger, fatigue, confusion, and anxiety.

Workplace stress is primarily a matter of person-workload fit. The status of the person-workload fit can influence the acceptance of the work and the level of acceptable performance of that work. The perception of workload may be affected by the worker's needs and his or her level of job satisfaction. The relation between job demands and the worker's ability to meet those demands further influences workplace stress. Because workplace stress may be felt differently by people in similar situations, it must be concluded that there are many causes of workplace stress.

SOURCES OF WORKPLACE STRESS

The sources of on-the-job stress may involve physical working conditions, work overload, **role ambiguity**, lack of **feedback**, personality, personal and family problems, or role conflict. Other sources of workplace stress are discussed in the following paragraphs:

- **Task complexity** relates to the number of different demands made on the worker. A job perceived as being too complex may cause feelings of inadequacy and result in emotional stress. Repetitive and monotonous work may lack complexity so that the worker becomes bored and dissatisfied with the job and possibly experiences some stress associated with the boredom.

- **Control** of employees over their job responsibilities can also be a source of workplace stress. Being responsible for work without being in control of it is a formula for stress. The more work can be organized and structured to allow for maximum control by those who do it, the less stressful the work will be. In addition, even when workers do feel as if they control the work they are responsible for, if they are monitored electronically their stress levels tend to rise. In the eyes of a line employee, electronic monitoring is the same as having a supervisor standing over your shoulder constantly. Job control has also been tied to cardiovascular disease and heart attacks. Workers with high stress related to a lack of job control experience elevated levels of plasma fibrinogen—a blood-clotting factor that has been linked to cardiovascular events such as heart attacks. A study by Belgian scientists at Ghent University tied not just job stress, but also the specific type of stress associated with a lack of job control, to plasma fibrinogen—a specific medical marker for cardiovascular problems.[1]

- A **feeling of responsibility** for the welfare or safety of family members may produce on-the-job stress. Being responsible for the welfare of his or her family may cause a worker to feel that options to take employment risks are limited. A worker may then perceive that he or she is "trapped in the job." Overly constrained employment options may lead to anxiety and stress. The feeling of being responsible for the safety of the general public has also been shown to be a stressor. Air traffic controllers are known to experience intense stress when their responsibility for public safety is tested by a near-accident event. A feeling of great responsibility associatcd with a job can transform a routine activity into a stress-inducing task.

- **Job security** involves the risk of unemployment. A worker who believes that his or her job is in jeopardy will experience anxiety and stress. The ready availability of other rewarding employment and a feeling that one's professional skills are needed reduce the stress associated with job security issues.

- An organizational culture leaves the employee feeling left out, out of the loop, and ill-informed. Organizations in which managers fail to communicatc frcquently and effectively with employees are creating high-stress environments for workers.

- Work schedules that are unpredictable, never-changing, and ever-changing can induce stress in employees. Employees have lives outside of their jobs. Consequently, the ability to predict their work schedules is important. When work schedules are unpredictable, stress increases. It also increases when work schedules are inflexible—they cannot be changed no matter what other obligation the employee might have. On the other hand, ever-changing work schedules such as those associated with shift work can also increase the level of employee stress. The big-picture issue with regard to work schedules is the level of control employees have over their lives. The less control, the more stress.

- Home and family problems can create added stress for workers. There was a time when employees were expected to "leave their problems at the front door" when they came to work. This, of course, is a practical impossibility. The demands of raising children, working out home and job schedules, dealing with the conflicting agendas inherent in dual-career households, and otherwise handling the everyday work and home conflicts that inevitably arise can markedly increase an employee's stress level.

- Work relationships can lead to on-the-job stress. People are social beings by nature. They like to get along with the people they spend a lot of time with. However, office politics, turf battles, and internal competition for recognition and rewards can be hard on work relationships. When employees do not get along with their fellow workers, stress levels increase.

- Human resource management (HRM) issues can be a source of workplace stress. People who work have a vested interest in their wages, salaries, working conditions, and benefits. If even one of these factors is negative, employee stress levels can increase significantly. In addition to these factors, other stress-inducing HRM issues include being underemployed, failing to get promoted, and working in a position that is clearly not valued by management.

- **Workload demands** can stimulate stress when they are perceived as being overwhelming. These demands may involve time constraints and cognitive constraints such as speed of decision making and mandates for attention. Workload demands may also be physically overwhelming if the worker is poorly matched to the physical requirements of the job or is fatigued. Whenever the worker believes the workload to be too demanding, it can result in stress.

- **Psychological support** from managers and coworkers gives a feeling of acceptance and belonging and helps defuse stress. A lack of such support may increase the perception of a burdensome workload and result in stress.

- The lack of **environmental safety**, including the potential for workplace violence, can also be a cause of stress. Feeling that one is in danger can be a stressor. Workers need to feel safe from environmental hazards such as extreme temperatures, pressure, electricity, fire, explosives, toxic materials, ionizing radiation, noises, and dangerous machinery. To reduce the potential for stress due to environmental hazards, workers should feel that their managers are committed to safety and that their company has an effective safety program.

- **Technology that is 24/7/365.** Computers and handheld technologies have made work for some a 24-hours-a-day, 7-days-a-week, 365-days-a-year enterprise. Rather than relieve people of work-related stress, technology has actually increased the level of stress. Because of technology, some people literally never get away from their work. For example, a smartphone will give people immediate access to their work any time of day from any location within the zone of a cell tower. Further, to recommend that people just turn their smartphones and computers off is unrealistic because when important work issues are pending, the temptation to check for messages from work can be overwhelming.

- **Multitasking.** Much has been made of the concept of multitasking. It was once believed that an individual who was incapable of multitasking was incapable of climbing the corporate ladder. However, the more that is learned about the concept, the more it is being recognized as a stress producer that can actually undermine both productivity and safety. Anytime the demands of work exceed an individual's ability to meet them, stress is the result. The more important the tasks in question, the higher the stress level. Trying to accomplish several important tasks simultaneously—multitasking—just multiplies the stress level of workers. An emerging definition of a multitasker is an unfocused individual whose attention is divided among several tasks such that no task gets sufficient attention. In a work setting, this lack of attention can be dangerous.

COMMON CAUSES OF STRESS IN THE WORKPLACE

- The company was recently purchased by another company.
- Downsizing or layoffs have occurred in the past year.
- Employee benefits were significantly cut recently.

- Mandatory overtime is frequently required.
- Employees have little control over how they do their work.
- The consequences of making a mistake on the job are severe.
- Workloads vary greatly.
- Most work is machine paced or fast paced.
- Workers must react quickly and accurately to changing conditions.
- Personal conflicts on the job are common.
- Few opportunities for advancement are available.
- Workers cope with a great deal of bureaucracy in getting work done.
- Staffing, money, or technology is inadequate.
- Pay is below the going rate.
- Employees are rotated among shifts.[2]

Making Use of the List of Common Causes of Stress

The maxim about being prepared certainly applies in the workplace. To be prepared to eliminate or, at least, mitigate the more common causes of stress in workers, decision makers in organizations must know the common causes. A list of common causes of stress in the workplace was presented in the previous section. It is important for safety professionals and decision makers in organizations to know these common causes of stress. However, just knowing them is not enough. Knowing the causes of stress will do little to reduce stress unless this knowledge is effectively put to use.

Consequently, it is important for employers to use items in the list of common causes to inform their actions when any item on the list applies to their organizations. The list of common causes—if understood by key management personnel—can be used to plan and implement appropriate preventive measures before the cause(s) in question leads to problems. For example, if management personnel understand that conflict on the job is a stress inducer and that conflict does, in fact, exist in the organization, they can take the appropriate steps to (1) resolve existing conflicts and (2) institute conflict management strategies including training all employees in conflict resolution and management. For another example, if layoffs are a possibility, rather than just let fears about job security fester and grow, management personnel can take positive steps to (1) inform workers of the facts and (2) institute re-employment assistance measures so that employees know that anyone who is laid off will at least have help finding a new job.

It has long been known among safety and health professionals and practitioners of occupational medicine that stress can have a detrimental effect on attendance, productivity, employee retention, and morale. The sources of stress contained in this list are just some of the more common ones. There are many others. For example, technological developments have increased stress levels on the job.

The fact that cellular telephones can be obnoxiously intrusive and distracting has tended to increase stress levels at work. Continual change in the field of computer technology—change that forces workers to upgrade their skills almost as soon as they become comfortable with a given software package—has become a source of stress in the workplace. Stress has become such an all-pervasive source of safety and health-related problems in the workplace that professionals in the field must acknowledge it and work to reduce its harmful effects like any other workplace hazard.

HUMAN REACTIONS TO WORKPLACE STRESS

Human reactions to workplace stress may be grouped into the following categories: subjective or emotional (anxiety, aggression, guilt); behavioral (accident proneness, trembling); cognitive (inability to concentrate or make decisions); physiological (increased heart rate and blood pressure); and organizational (absenteeism and poor productivity). Continual or persistent stress has been linked to many physiological problems. Initially, the effects may be psychosomatic, but with continued stress, the symptoms show up as actual organic dysfunction. The most common forms of stress-related diseases are gastrointestinal, particularly gastric or duodenal ulcers. Research has linked some autoimmune diseases with increased long-term workplace stress.[3]

The human response to workplace stress can be compared to a rubber band being stretched. As the stress continues to be applied, the rubber band stretches until a limit is reached when the rubber band breaks. For humans, various physical and psychological changes are observed with the repetitive stimuli of stress. Until the limit is reached, the harmful effects can be reversed. With an increase in intensity or duration of the stress beyond the individual's limit, the effects on the human become pathological.[4]

There are three stages of the human stress response: (1) alarm, (2) resistance, and (3) exhaustion.[5] The alarm reaction occurs when the stress of a threat is sensed. The **stage of alarm** is characterized by pallor, sweating, and an increased heart rate. This stage is usually short. It prepares the body for whatever action is necessary.

When the stress is maintained, the **stage of resistance** initiates a greater physical response. The alarm symptoms dissipate, and the body develops an adaptation to the stress. The capacity for adaptation during this stage is limited.

Eventually, with sustained stress, the **stage of exhaustion** is reached. This stage is demonstrated by the body's failure to adapt to the continued stress. Psychosomatic diseases such as gastric ulcers, colitis, rashes, and autoimmune disorders may begin during this stage. The tendency to develop a specific stress-related disease may be partially predetermined by heredity, personal habits such as smoking, and personality.

From an evolutionary viewpoint, the adverse effects of stress on health may be considered a maladaptation of humans to stress. What does this tell us? Either we (1) learn to do away with all stress (unlikely), (2) avoid all stressful situations (equally unlikely), (3) learn to adapt to being sick because of stress (undesirable), or (4) learn to adapt to workplace stress (the optimal choice). The first step in learning to adapt to stress is understanding the amount of stress to which we are subjected.

MEASUREMENT OF WORKPLACE STRESS

Workplace stress can be seen as an individual's psychological reaction to the work environment. Although psychological response cannot be directly measured in physical terms, one method commonly employed uses a measurement of mental workload. Mental workload can be measured in one of three ways:

1. With **subjective ratings**, the workers are asked to rate their perceived level of workload. The perceived workload is then viewed as a direct reflection of workplace stress. The workers may be asked to rate their mood in relation to the work situation. The data gathered by this method are obviously subjective and state

dependent. **State-dependent** data are directly related to the circumstances or state under which they are collected and, therefore, have a built-in bias.

2. **Behavioral time-sharing** techniques require the simultaneous performance of two tasks. One of the tasks is considered to be primary or the most important; the other is of secondary importance. The decrease in performance efficiency of the secondary task is considered an index of workload for behavioral time-sharing or human multitasking. Workplace stress is thought to increase as behavioral time-sharing increases.

3. **Psychophysiological techniques** require simultaneous measurement of heart rate and brain waves, which are then interpreted as indexes of mental workload and workplace stress.

4. **Physiological measures**. The effects of stress are not limited to the emotional and psychological. Stress can also affect a person physically, and some of the physical responses can be measured. For example stress is known to increase blood pressure and heart rate, both of which can be easily measured. Pupil dilation is another physiological response to stress that can, at least, be observed. Other physiological responses are more difficult to measure outside of a controlled laboratory environment. These responses include increased perspiration, inability to sleep, difficulty thinking clearly, and the inability to remember. Much work has been done in controlled laboratory studies in which physiological responses to stress are measured based on analysis of saliva, for example. However, these are not measures that safety professionals can make in a live workplace setting.

Behavioral time-sharing and psychophysiological techniques are related to theoretical models, making data easier to interpret. These two techniques also require sophisticated equipment and data collection methods.

Subjective ratings may be collected using questionnaires or survey instruments. These instruments may ask about the physical working conditions, the individual's health and mental well-being, and perceived overall satisfaction with the job. The data may then be compared to standardized scales developed by various researchers.

Psychosocial questionnaires evaluate workers' emotions about their jobs. Workers may be asked about job satisfaction, workload, pace, opportunities for advancement, management style, and organizational climate. Psychosocial questionnaires are another form of subjective rating and are also subject to state-dependent bias in the data. Regardless of the measurement method, because workplace stress is dependent on personal awareness, no direct means of measuring workplace stress are now available.

SHIFT WORK, STRESS, AND SAFETY

Shift work can require some employees to work when the majority of people are resting. In some cases, shift work requires rotating between two or three different starting times, which may vary by eight hours or more. Shift work has traditionally been required by the medical community, the transportation industry, utilities, security, and, increasingly, by retail sales.[6]

Basic physiological functions are scheduled by the biological clock called the **circadian rhythm**. Most children in the United States grow up on the day shift, going to school during the day and sleeping at night. After a life of being on the day shift, the body perceives a change in work shift as being stressful. If the person takes a job starting at midnight, his or her body will still expect to be sleeping at night and active during the day.

Many physical and psychological functions are affected by circadian rhythm. Blood pressure, heart rate, body temperature, and urine production are measurably slower at night. The same functions are normally faster during the day (active time).

Behavioral patterns also follow the circadian slower-at-night and more-active-during-the-day pattern. Sleep demand and sleep capacity for people aged 14 or older are greatest at night and least during daylight hours. Alertness has been determined to be decreased at night.

Workers surveyed have consistently reported lower job satisfaction with rotating shifts. Day-shift workers with the same task definitions report higher job satisfaction and less stress than their second- or third-shift counterparts. Rotating shifts over several weeks can result in desensitization to the circadian rhythms. With this desensitization comes a measurable loss in productivity, increased numbers of accidents, and reported subjective discomfort. After returning to a predictable shift, workers regained their biological clock and circadian rhythm.

Not working the normal day-shift hours results in an increase in workplace stress, with rotating shifts being the most stressful. From a safety viewpoint, shift workers are subjected to more workplace stress in terms of weariness, irritability, depression, and a lack of interest in work. Shift work increases workplace stress and may lead to a less safe worker.

Reducing the Stress Associated with Shift Work

Shift work is and will probably always be a fact of life for employees in certain occupations. More than 15 million people in the United States have jobs that require shift work. To reduce the stress associated with shift work, safety and health professionals can apply the following strategies:

1. Encourage shift workers to exercise regularly. Regular exercise can have the double benefit of improving the quality of an individual's sleep and relieving pent-up stress.
2. Encourage shift workers to avoid caffeine, alcohol, or other drugs that can inhibit their ability to sleep.
3. If shift workers cannot sleep without some type of sleep aid, the food supplement melatonin or other natural sleep inducers should be recommended rather than sleeping pills that contain synthetic chemicals and can have side effects that might contribute to other stress-inducing effects.[7]

IMPROVING SAFETY BY REDUCING WORKPLACE STRESS

Not all sources of stress on the job can be eliminated, and employment screening is unlikely to identify all those who are sensitive to stress. People can learn to adapt to stress, however. **Training** can help people recognize and deal with stress effectively. Employees need to know what is expected of them at any given time and to receive recognition when it is deserved. Managers can reduce role ambiguity and stress caused by lack of feedback by providing frequent feedback.

Stress can result from low participation or lack of **job autonomy**. A manager can help employees realize their full potential by helping them match their career goals with the company's goals and giving them more control over their jobs.

Managers can help design jobs in ways that lead to worker satisfaction, thereby lessening work stress. **Physical stress** can be reduced by improving the work environment and establishing a sound safety and health program. Managers can also assist in the effort to provide varied and independent work with good possibilities for contact and collaboration with fellow workers and for personal development.

Organizational approaches to coping with work stress include avoiding a monotonous, mechanically **controlled pace**, standardized motion patterns, and constant repetition of **short-cycle operations**. Other stress-inducing work design features to avoid include jobs that do not make use of a worker's knowledge and initiative, that lack human contact, and that have authoritarian-type supervision.

There are also several individual approaches to coping with stress. One of the most important factors in dealing with stress is learning to recognize its symptoms and taking them seriously. Handling stress effectively should be a lifelong activity that gets easier with practice. Keeping a positive mental attitude can help defuse some otherwise stressful situations.

People can analyze stress-producing situations and decide what is worth worrying about. Individuals can effectively respond to a stressful workload by **delegating responsibility** instead of carrying the entire load. Relaxation techniques can also help reduce the effects of stress. Some common relaxation methods include meditation, biofeedback, music, and exercise.

The following strategies are recommended for reducing workplace stress:

- *Management recognizes workplace stress and takes steps regularly to reduce this stress.* If this strategy is not in place, other strategies will not work. There must be management recognition of stress as a detriment to safety, quality, and productivity before other strategies can be effective.[8]
- *Mental health and Employee Assistance Program (EAP) benefits are provided in the worker's health insurance coverage.* Mental health is just as important as physical health when it comes to the safety and health of workers. Consequently, it is important that insurance policies cover mental health issues such as stress-related problems. It is also important for organizations to provide an Employee Assistance Program (EAP) that covers stress-related problems.
- *The employer has a formal program for communicating with workers.* One of the more stressful situations for workers is to feel that no one is listening. Consequently, it is important to establish a formal communication program that not only keeps workers informed but allows them to give feedback concerning stress-related problems and other safety and health issues.
- *Workers are regularly given up-to-date information on how to cope with stress.* Executives, managers, supervisors, and safety professionals should never assume that workers know how to cope effectively with stress. In fact, it is better

to assume they don't. Management should provide up-to-date information on coping strategies to all personnel on a regular basis.

- *Workers have current, accurate, and clear job descriptions.* One of the most stressful things workers can confront is ambiguity concerning their job description. Workers need to know what their duties and responsibilities are and what is expected of them. Communicating this information to workers begins with the job description.
- *Management and workers talk openly with one another.* Some executives and managers are afraid to talk openly with workers about stress and other safety and health issues for fear of "opening a can of worms." They believe the less said about these issues the fewer problems there will be. Their fear is an increase in workers compensation claims based on the power of suggestion. This attitude toward talking with workers about stress and other safety and health issues is outmoded and inaccurate. The better informed workers are, the better able they are to prevent problems.
- *Workers are free to talk with each other during work.* Open communication helps improve an organization's safety and health record. Consequently, workers should be encouraged to share what they know, lessons they have learned, and information they have accumulated with other workers, particularly less-experienced workers.
- *Employers offer wellness programs, exercise facilities, and other stress-reduction classes.* Experts on stress emphasize the importance of exercise and other wellness-related stress-reduction strategies for workers. But workers may not know how to eat right, exercise properly, or apply other stress-reducing strategies. Consequently, offering classes—especially on-site classes—can be an effective management strategy for reducing stress.
- *Workers are recognized and rewarded with nonmonetary prizes for their contributions.* Organizations demonstrate what is important by what they recognize and reward. If decision makers want workers to take stress-reduction seriously, they need to reward and recognize workers who take positive steps to make it happen.
- *Work rules are published and are the same for everyone.* One of the more stressful conditions that can exist in the workplace is ambiguity concerning the rules and expectations of management. Further, when the rules and expectations appear to be different for some workers, the issue of favoritism arises. Both situations are stress-inducing. They are also indicators of poor management and leadership.
- *Child care programs are available.* One of the most stressful situations parents face is having access to affordable child care for their children. One way to relieve the stress from this source is to provide either on-site childcare or cash subsidies for off-site childcare.
- *Workers can work flexible hours.* Life can get complicated for workers who must juggle children, school, college, and other personal responsibilities with work. This juggling act is a major source of stress for some workers. Allowing them to opt for flexible hours can be an effective way to reduce stress from this source.

- *Perks are granted fairly based on a person's level in the organization.* In most organizations workers earn certain perquisites as they climb the organizational ladder. Perquisites are provided by organizations as incentives, morale boosters, and performance enhancers. However, if they are not fairly and properly allocated, they can have the opposite effect and cause workers to feel they are being treated unfairly. Such a feeling can cause the stress level in workers to skyrocket. Consequently, it is important for organizations to be clear concerning who can receive what perquisites.

- *Workers have the training, equipment, support, and technology access that they need.* Few things can be more stressful in the workplace than being expected to perform at a certain level but not being provided the resources needed to perform at that level. Organizations should never be guilty of expecting peak performance of workers unless they have provided them with the resources needed to perform at that level.

- *Employers encourage work and personal support groups.* Sometimes the best support for workers is other workers who face the same problems, challenges, situations, and circumstances every day. By encouraging workers to form their own support groups, organizations can go a long way toward helping them reduce work-related stress.

- *Workers have a place and time to relax during the workday.* It helps to get away from the everyday grind of work during the workday, even for a short time. Workers need to occasionally catch their breath so to speak and they need a place that is comfortable, accessible, and theirs to do this. Providing a comfortable place for workers to take a break and relax during the workday can be an effective stress-reduction strategy.

- *Elder care programs are available.* Just when the children have grown up and are finally on their own, some workers are faced with the reality of having to take care of their aging parents. Some workers are still taking care of children while also taking care of their aging parents. Few situations could be more demanding and stressful. Consequently, in organizations that employ a significant number of workers who find themselves in this situation, providing access to elder care is just as important as child care.

- *Employees' work spaces are not crowded.* Feeling crowded is a stress-inducing phenomenon. People who live in large cities understand this. People who visit large cities soon learn it. Workers do not like to feel cramped or crowded as they try to do their jobs. Making sure that workers have sufficient room to do what is expected of them is an effective stress reliever.

- *Workers can put up personal items in their work areas.* Familiarity and comfort are effective stress reducers. Consequently, allowing workers to personalize their work areas, within reason, can be an effective way to help them cope with job stress.

- *Management appreciates humor in the workplace.* One of the most effective stress reducers known is laughter. Consequently, by encouraging humor in the workplace decision makers can help workers cope with stress. Caution is urged though with this strategy. In a diverse workplace, what is funny to one person may not be funny to others.

DISCUSSION CASE

WHAT IS YOUR OPINION?

"We are beginning to see more and more stress-related problems. I'm afraid that if we don't deal with the issue, stress is going to cause even more serious injuries." "Nonsense. We all have stress. There has always been stress on the job. It goes with the territory. A few employees go to a seminar on workplace stress and, all of a sudden, everyone is complaining about stress." This discussion took place between the safety director and the CEO of Gulf Coast Electric Company. Is stress really a legitimate workplace hazard? What is your opinion?

There is no one clear answer to workplace stress. The suggestions given above are a good starting place for management and employees to begin the process of being aware of and dealing effectively with workplace stress.

STRESS IN SAFETY MANAGERS

Safety and health management can be a stressful profession. As safety managers assess the workplace for stress problems, they should remember that they too can become victims of stress. Specific stressors for safety managers include the following:

- Overload
- Ever-changing safety regulations
- Communication problems with employees, managers, and supervisors
- Competing loyalties[9]

Safety and health professionals are sometimes overloaded when corporate *downsizing* results and they are delegated more and more responsibilities. Trying to keep up with the ever-changing multitude of regulations is a stress-inducing challenge. Communication relating to safety and health is always a challenge. However, when economic forces focus an organization's attention on other matters, it can be even more difficult than usual to get the safety and health message across. This increased difficulty can lead to increased stress. Line managers who are more concerned with meeting production quotas than with the safety of their employees sometimes try to influence safety and health managers—their colleagues and sometimes their friends—to look the other way. This subjects safety and health managers to the pressures of competing loyalties.[10]

Safety and health managers can cope with these four common triggers of stress by applying the following strategies: (1) prioritize activities by focusing on those that present the most risk to the organization, (2) work closely with the organization's legal staff and subscribe to an online updating service, (3) formalize communication and hold regularly scheduled safety and health meetings for all operating employees, and (4) focus on the risks to the organization and refuse to take sides.[11]

SAFETY MYTH

MANAGERS CANNOT REDUCE EMPLOYEE STRESS

Of course, the workplace is stressful! This fact cannot be helped. Some people can handle stress, and some cannot. There is nothing managers can do. Right? Wrong. There is a lot that managers can do. Managers can provide training to help employees learn to recognize stress and deal with it. Managers can empower employees, giving them as much control as possible over their jobs. Managers can improve communication with employees. Managers can provide clear, understandable job descriptions that eliminate ambiguity. Managers can provide child care or child care assistance. Managers can provide exercise and stress-reduction programs. There are many strategies that managers can employ to reduce stress on the job.

STRESS AND WORKERS' COMPENSATION

There are serious problems with workers' compensation in the United States. On the one hand, there is evidence of abuse of the system. On the other hand, many injured workers who are legitimately collecting benefits suffer a substantial loss of income. Complaints about workers' compensation are common from all parties involved with the system (employers, employees, and insurance companies).

Stress claims are more burdensome than physical claims because they are typically reviewed in an adversarial environment. This leads to the involvement of expert medical witnesses and attorneys. As a result, even though the benefits awarded for stress-related injuries are typically less than those awarded for physical injuries, the cost of stress claims is often higher because of litigation.

Key Terms and Concepts

- Behavioral time-sharing
- Circadian rhythm
- Control
- Controlled pace
- Delegating responsibility
- Environmental safety
- Feedback
- Feeling of responsibility
- Human reactions
- Job autonomy
- Job security
- Physical stress
- Psychological support
- Psychophysiological techniques
- Psychosocial questionnaires
- Role ambiguity
- Shift work
- Short-cycle operations
- Stage of alarm
- Stage of exhaustion
- Stage of resistance
- State-dependent
- Stress
- Subjective ratings
- Task complexity
- Training
- Workload demands
- Workplace stress

Review Questions

1. Define *stress*.
2. How is workplace stress different from general stress?
3. List five sources of workplace stress and give an on-the-job example for each source.

4. Explain why lack of job autonomy may cause workplace stress.
5. Give five categories of human reaction to workplace stress.
6. How are psychosomatic reactions to stress and actual physiological illness related?
7. How are some autoimmune diseases and workplace stress related?
8. Explain three stages of human reaction to stress.
9. Explain three ways in which mental workload can be measured.
10. Discuss efforts to rid the workplace of all causes of workplace stress.
11. What type of data do psychosocial questionnaires provide? Discuss the bias in this type of data.
12. Discuss how shift work causes workplace stress. Give suggestions for minimizing workplace stress from shift work.
13. Give specific steps that can be taken by managers to help reduce workplace stress.
14. Discuss at least five methods to reduce workplace stress, according to the life insurance company research.
15. Explain how individuals can reduce workplace stress.

Endnotes

1. Occupational Health & Safety Online, *Study: Job Stress Linked to Increased Inflammation*. Retrieved from ohsonline.com/stevens/ohspub.nsf/d3d5b4f-93862266e8625670c006dbc58/c16 on July 3, 2017.
2. Health Advocate, Inc., "Stress in the Workplace." Retrieved from healthadvocate.com/downloads/webinars/stress-workplace.pdf on January 7, 2017.
3. Ibid.
4. Ibid.
5. Ibid.
6. D. Morshead, "Stress and Shiftwork," *Occupational Health & Safety* 71, no. 4: 36–37. 2003.
7. Ibid.
8. Ibid.
9. *Roles in Tackling Stress*. Retrieved from www.hse.gov.uk/stress/roles/hsmanagers.htm on July 6, 2017.
10. Ibid.
11. Ibid.

10 MECHANICAL HAZARDS AND MACHINE SAFEGUARDING

Learning Objectives

- List the more common mechanical injuries
- Define the term *safeguarding*
- Summarize OSHA's requirements for machine safeguarding
- Explain the concept of risk assessment in machine operation
- Describe the design requirements for safeguards
- Explain the concept of point-of-operation guards
- Explain the concept of point-of-operation devices
- Summarize the key elements of a machine guarding self-assessment
- Describe the various kinds of feeding and ejection systems
- Describe the types of safeguards used with robots
- Explain how *lockout/tagout* systems are used to control hazardous energy in machines
- Explain how permanent electrical safety devices are used in lockout/tagout programs
- List the general precautions to be observed in machine safeguarding
- Explain the basic content for a machine safeguarding program
- Summarize how to take corrective action when a mechanical hazard is observed

Failure to provide proper machine guards and enforce their use can be costly for companies. A manufacturing firm in Syracuse, New York, learned this fact the hard way when it was cited by the Occupational Safety and Health Administration (OSHA) for failure to provide appropriate machine guards and require their proper use by employees. Because the company had been issued a similar citation earlier, it was fined $119,000. Mechanical hazards that are not properly guarded are implicated in thousands of workplace injuries every year.

As another example, a windshield manufacturing firm was fined more than $105,000 by OSHA when it allowed employees to be exposed to electrical hazards and knowingly failed to meet lockout/tagout standards. The small gains in productivity that might be obtained by willfully bypassing mechanical safeguards on machines can cost companies huge fines and even expensive medical bills.

Mechanical hazards are those associated with power-driven machines, whether automated or manually operated. Concerns about mechanical hazards date back to the Industrial Revolution and the earliest days of mechanization. Machines driven by steam, hydraulic, or electric power introduced new hazards into the workplace. In spite of advances in safeguarding technologies and techniques, mechanical hazards are still a major concern today. In addition, automated machines have introduced new concerns.

COMMON MECHANICAL INJURIES

In an industrial setting, people interact with machines that are designed to drill, cut, shear, punch, chip, staple, stitch, abrade, shape, stamp, and slit such materials as metals, composites, plastics, and elastomers. If appropriate safeguards are not in place or if workers fail to follow safety precautions, these machines can apply the same procedures to humans. When this happens, the types of **mechanical injuries** that result are typically the result of cutting, tearing, shearing, crushing, **breaking**, straining, or puncturing (see Figure 10–1). Information about each of these hazards is provided in the following paragraphs.

Cutting and Tearing
A cut occurs when a body part comes in contact with a sharp edge. The human body's outer layer consists of the following, starting from the outside: *epidermis*, the tough outer covering of the skin; *dermis,* the greatest part of the skin's thickness; *capillaries*,

FIGURE 10–1 Some common mechanical hazards.

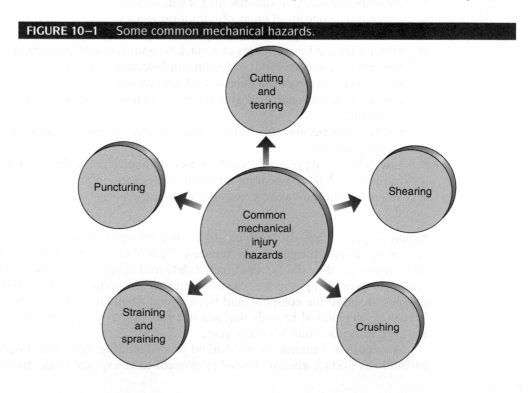

the tiny blood vessels that branch off the small arteries and veins in the dermis; *veins*, the blood vessels that collect blood from the capillaries and return it to the heart; and *arteries*, the larger vessels that carry blood from the heart to the capillaries in the skin. The seriousness of **cutting** or **tearing** the skin depends on how much damage is done to the skin, veins, arteries, muscles, and even bones.

Shearing

To understand what **shearing** is, think of a paper cutter. It shears the paper. Power-driven shears for severing paper, metal, plastic, elastomers, and composite materials are widely used in manufacturing. In times past, such machines often amputated fingers and hands. These tragedies typically occurred when operators reached under the shearing blade to make an adjustment or placed materials there and activated the blade before fully removing their hand. Safeguards against shearing accidents are explained later in this chapter.

Crushing

Injuries from **crushing** can be particularly debilitating, painful, and difficult to heal. They occur when a part of the body is caught between two hard surfaces that progressively move together, thereby crushing anything between them. Crushing hazards can be divided into two categories: *squeeze-point* types and *run-in points*.

Squeeze-point hazards exist where two hard surfaces, at least one of which must be in motion, push close enough together to crush any object that may be between them. The process can be slow, as in a manually operated vise, or fast, as with a metal-stamping machine.

Run-in point hazards exist where two objects, at least one of which is rotating, come progressively closer together. Any gap between them need not become completely closed. It need only be smaller than the object or body part lodged in it. Meshing gears and belt pulleys are examples of run-in point hazards (see Figures 10–2, 10–3, and 10–4).

FIGURE 10–2   This nip point can pull hands, feet, or articles of clothing.

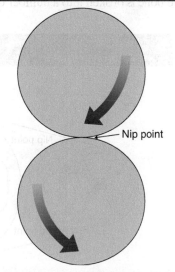

Nip point

FIGURE 10–3   Fingers might get caught between the bench top and the revolving wheel.

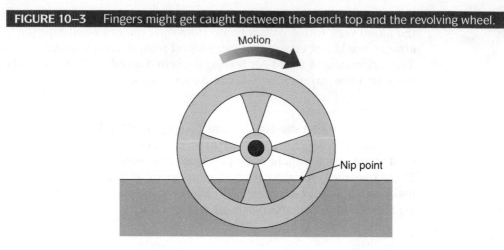

Body parts can also be crushed in other ways—for example, a heavy object falling on a foot or a hammer hitting a finger. However, these are impact hazards, which are covered in Chapter 11.

Breaking

Machines used to deform engineering materials in a variety of ways can also cause broken bones. A break in a bone is known as a *fracture*. Fractures are classified as simple, compound, complete, and incomplete.

A simple fracture is a break in a bone that does not pierce the skin. A compound fracture is a break that has broken through the surrounding tissue and skin. A complete fracture divides the affected bone into two or more separate pieces. An incomplete fracture leaves the affected bone in one piece but cracked.

Fractures are also classified as transverse, oblique, and comminuted. A transverse fracture is a break straight across the bone. An oblique fracture is diagonal. A comminuted fracture exists when the bone is broken into a number of small pieces at the point of fracture.

FIGURE 10–4   Nip points can catch fingers, hands, hair, clothing, etc., with dangerous results.

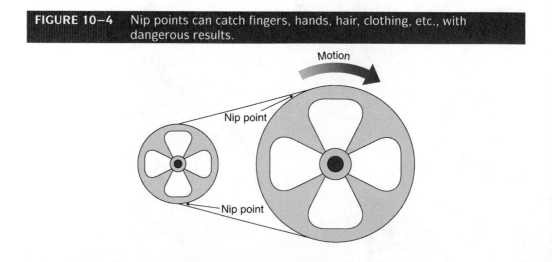

Straining and Spraining

There are numerous situations in an industrial setting when **straining** of muscles or **spraining** of ligaments is possible. A strain results when muscles are overstretched or torn. A sprain is the result of torn ligaments in a joint. Strains and sprains can cause swelling and intense pain.

Puncturing

Punching machines that have sharp tools can puncture a body part if safety precautions are not observed or if appropriate safeguards are not in place. **Puncturing** results when an object penetrates straight into the body and pulls straight out, creating a wound in the shape of the penetrating object. The greatest hazard with puncture wounds is the potential for damage to internal organs.

SAFEGUARDING DEFINED

All the hazards explained in the previous section can be reduced by the application of appropriate safeguards. CFR 1910 Subpart O contains the OSHA standards for machinery and machine guarding (1910.211–1910.222). **Safeguarding** can be defined as follows:

Machine safeguarding is to minimize the risk of accidents of machine-operator contact. The contact can be:

1. An individual making the contact with the machine—usually the moving part—because of inattention caused by fatigue, distraction, curiosity, or deliberate chance taking;
2. From the machine via flying metal chips, chemical and hot metal splashes, and circular saw kickbacks, to name a few;
3. Caused by the direct result of a machine malfunction, including mechanical and electrical failure.[1]

Safeguards can be broadly categorized as point-of-operation guards, point-of-operation devices, and feeding/ejection methods. The various types of safeguards in these categories are explained later in this chapter.

OSHA'S REQUIREMENTS FOR MACHINE GUARDING

The OSHA standard containing the general requirements for machine guarding is 29 CFR 1910.212. A more specific standard (29 CFR 1926.300) exists for the construction industry. This section focuses on 29 CFR 1910.212—OSHA's requirements for all industries. Those requirements are summarized as follows:

Types of guarding. One or more methods of machine guarding must be provided to protect people from such point of operation hazards as nip points, rotating parts, flying chips, and sparks. "Point of operation" refers to the area on the machine where work is performed on the material being processed. Examples of point-of-operation machine guards are barriers, two-hand switches and tripping devices, and electronic sensors.

General requirements for machine guards. Where possible, guards should be affixed to the machine in question. When this is not possible, guards should be secured in the most feasible location and method away from the machine. Guards must be affixed in such a way that they do not create a hazard themselves.

Guarding the point of operation. Any point of operation that might expose a person to injury must be guarded. Guarding devices must comply with all applicable standards. In the absence of applicable standards, the guard must be designed, constructed, and installed in such a way as to prevent the machine operator from having any part of his body (including clothing and hair) in the danger zone during the operating cycle of the machine.

Machines requiring point of operation guards. The following are examples of machines that require point of operation guards: guillotine cutters, shears, alligator shears, power presses, milling machines, power saws, jointers, portable power tools, forming rolls, and calenders.

Exposure of blades. Fans must be guarded in any case in which the periphery of the fan blades is less than seven feet above the floor or working level. Guards for fans shall have no openings that exceed one-half inch.

Anchoring fixed machinery. Machines that are designed to be fixed in one location must be securely anchored to prevent movement.[2]

Minimum General Requirements for Safeguards

OSHA Standard 29 CFR 1910.212 requires that machine safeguards for all industries meet the following minimum requirements:[3]

- **Prevent contact.** Must prevent any part of the worker's body from making contact with dangerous moving parts.
- **Be secure.** Must be secure enough to prevent workers from easily removing them, durable enough to withstand normal use, and firmly secured to the machine where possible (or secured elsewhere if it is not possible to secure them to the machine).
- **Protect from falling objects.** Must ensure that falling objects cannot fall into moving parts.
- **Create no interference.** Must not impede the worker's ability to perform required tasks quickly and comfortably.
- **Allow safe lubrication.** Must be able to be lubricated without removal from the machine if possible.

RISK ASSESSMENT IN MACHINE OPERATION

Risk assessment in this context is the process of quantifying the level of risk associated with the operation of a given machine. It should be a structured and systematic process that answers the following four specific questions:

- How *severe* are potential injuries?
- How *frequently* are employees exposed to the potential hazards?
- What is the *possibility* of avoiding the hazard if it does occur?
- What is the *likelihood* of an injury should a safety control system fail?

A widely used risk-assessment technique is the decision tree, coupled with codes representing these four questions and defined levels of risk. Figure 10–5 is an example

FIGURE 10–5 Risk-assessment decision tree.

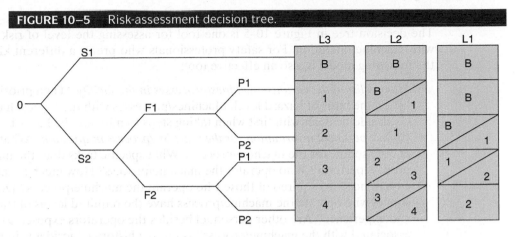

of a risk-assessment decision tree. In this example, the codes and their associated levels of risk are as follows:

S = Severity

QUESTION 1: SEVERITY OF POTENTIAL INJURIES

S1   Slight injury (bruise, abrasion)

S2   Severe injury (amputation or death)

F = Frequency

QUESTION 2: FREQUENCY OF EXPOSURE TO POTENTIAL HAZARDS

F1   Infrequent exposure

F2   From frequent to continuous exposure

P = Possibility

QUESTION 3: POSSIBILITY OF AVOIDING THE HAZARD IF IT DOES OCCUR

P1   Possible

P2   Less possible to not impossible

L = Likelihood

QUESTION 4: LIKELIHOOD THAT THE HAZARD WILL OCCUR

L1   Highly unlikely

L2   Unlikely

L3   Highly likely

RL = Risk Levels

Associated risk factors ranging from lowest (B) to highest (4) By applying the decision tree in Figure 10–5 or a similar device, the risk associated with the operation of a given machine can be quantified. This allows safety personnel to assign logical priorities for machine safeguarding and hazard prevention.

Risk Assessment Model

The decision tree in Figure 10–5 is one tool for assessing the level of risk associated with machine operation. For safety professionals who prefer a different kind of tool, the following model is also an effective tool:

- *Make a list of the various machines/processes in the facility.* Then prioritize the list on the basis of hazard level. Machines/processes with the highest hazard levels should be dealt with first when taking steps to mitigate the hazards.
- *Collect pertinent information for the machine/process in question.* What is the energy source for the machine/process? What specific tasks does the machine/process perform? Who operates the machine/process? How much training and experience are required of those who operate the machine/process? Do all personnel who operate the machine/process have the required levels of training and experience? Are other personnel besides the operators exposed to the risk associated with the machine/process? Is there a history of accidents/injuries associated with the machine/process?
- *Remove all safeguards.* In the next step the machine/process will be observed with a mind to identifying potentially hazardous conditions. This should be done without safeguards in place so that a hazard is not overlooked simply because it is mitigated by a safeguard.
- *Observe the machine/process in operation without safeguards and make a list of potential hazards.* Observe each step that is necessary to operate the machine/process and make note of any potential hazards. These operational steps should include at least the following: set-up, start-up, normal operation, tool installation, tool changes, housekeeping and clean-up, and regular maintenance tasks. Look for such potential hazards as contact with live parts, cutting, crushing, puncturing, slip and fall, trip and fall, contact with ejected parts, contact with ejected fluids, and electrical hazards.
- *Choose a method for labeling risk levels.* There are several widely recognized methods that can be used for labeling risk levels on machines/processes. One of the more widely used methods is contained in ANSI B11.TR3. Section 7.2 of ANSI B11.TR3 addresses the potential severity of harm and labels these levels as *catastrophic, serious, moderate,* or *minor.* Section 7.3 of ANSI B11.TR3 addresses the probability of the occurrence of harm and labels the probability as *very likely, likely, unlikely,* or *remote.*
- *Choose safeguards on the basis of the level of risk associated with the machine/process.* Safeguards should be chosen on the basis of risk level as assigned in the previous step. For example, a machine or process that has the potential to produce a catastrophic injury should be fitted with safeguards that provide the highest level of risk reduction. Safeguards providing the highest level of risk reduction are barrier guards or devices that prevent intentional exposure of a body part to the hazard. Safeguards that provide a high to intermediate level of risk reduction are barrier guards or devices that prevent unintended exposure of a body part to the hazard. Safeguards that provide a low to intermediate level of risk reduction are barrier guards or devices that prevent inadvertent exposure of a body part to the hazard. Safeguards that provide a low level of risk reduction are physical barriers that provide only tactile or visual awareness of the hazard.

DESIGN REQUIREMENTS FOR SAFEGUARDS

The various machine motions present in modern industry involve mechanisms that rotate, reciprocate, or do both. This equipment includes tools, bits, chucks, blades, spokes, screws, gears, shafts, belts, and a variety of different types of stock. Safeguards can be devised to protect workers from harmful contact with such mechanisms while at the same time allowing work to progress at a productive rate. The following require-ments should be met when designing safeguards:

1. **Prevent contact.** Safeguards should prevent human contact with any potentially harmful machine part. The prevention extends to machine operators and any other person who might come in contact with the hazard.
2. **Be secure and durable.** Safeguards should be attached so that they are secure. This means that workers cannot render them ineffective by tampering with or disabling them. This is critical because removing safeguards in an attempt to speed production is a common practice. Safeguards must also be durable enough to withstand the rigors of the workplace. Worn-out safeguards won't protect workers properly.
3. **Protect against falling objects.** Objects falling onto moving machine mecha-nisms increase the risk of accidents, property damage, and injury. Objects that fall on a moving part can be quickly hurled out, creating a dangerous projectile. Therefore, safeguards must do more than just prevent human contact. They must also shield the moving parts of machines from falling objects.
4. **Create no new hazard.** Safeguards should overcome the hazards in question without creating new ones. For example, a safeguard with a sharp edge, un-finished surface, or protruding bolts introduces new hazards while protecting against the old.
5. **Create no interference.** Safeguards can interfere with the progress of work if they are not properly designed. Such safeguards are likely to be disregarded or disabled by workers feeling the pressure of production deadlines.
6. **Allow safe maintenance.** Safeguards should be designed to allow the more frequently performed maintenance tasks (e.g., lubrication) to be accomplished without the removal of guards. For example, locating the oil reservoir outside the guard with a line running to the lubrication point will allow for daily mainte-nance without removing the guard.[4]

Design and construction of safeguards are highly specialized activities requiring a strong working knowledge of machines, production techniques, and safety. However, it is critical that all the factors explained in this section be considered and accommo-dated during the design process.

POINT-OF-OPERATION GUARDS

Guards are most effective when used at the point of operation, which is where haz-ards to humans exist. Point-of-operation hazards are those caused by the shearing, cutting, or bending motions of a machine. Pinch-point hazards result from guiding material into a machine or transferring motion (e.g., from gears, pressure rollers, or chains and sprockets). Single-purpose safeguards, because they guard against only

FIGURE 10–6   Point-of-operation guards may be used on machines such as this CNC machining center.

Source: Aumm graphixphoto/Shutterstock

one hazard, typically are permanently fixed and nonadjustable. Multiple-purpose safeguards, which guard against more than one hazard, typically are adjustable.[5]

Point-of-operation guards are of three types, each with its own advantages and limitations: fixed, interlocked, and adjustable (see Figure 10–6).

- **Fixed guards** provide a permanent barrier between workers and the point of operation. They offer the following advantages: They are suitable for many specific applications, can be constructed in-plant, require little maintenance, and are suitable for high-production, repetitive operations. Limitations include the following: They sometimes limit visibility, are often limited to specific operations, and sometimes inhibit normal cleaning and maintenance.
- **Interlocked guards** shut down the machine when the guard is not securely in place or is disengaged. The main advantage of this type of guard is that it allows safe access to the machine for removing jams or conducting routine maintenance without the need for taking off the guard. There are also limitations. Interlocked guards require careful adjustment and maintenance and, in some cases, can be easily disengaged.
- **Adjustable guards** provide a barrier against a variety of different hazards associated with different production operations. They have the advantage of flexibility. However, they do not provide as dependable a barrier as other guards do, and they require frequent maintenance and careful adjustment. Some guards in this category are self-adjusting.

Figures 10–7 through 10–10 show various guards used in modern manufacturing settings.

POINT-OF-OPERATION DEVICES

A number of different **point-of-operation devices** can be used to protect workers. The most widely used are explained in the following paragraphs:

- **Photoelectric devices** are optical devices that shut down the machine whenever the light field is broken. These devices allow operators relatively free movement. They do have limitations including the following: They do not protect

FIGURE 10–7   When the doors are opened, the milling tool stops automatically.

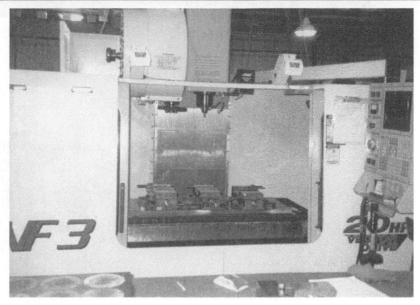

Courtesy of David Goetsch

against mechanical failure, they require frequent calibration, they can be used only with machines that can be stopped, and they do not protect workers from parts that might fly out of the point-of-operation area.

- **Radio-frequency devices** are capacitance devices that brake the machine if the capacitance field is interrupted by a worker's body or another object. These devices have the same limitations as photoelectric devices.

FIGURE 10–8   In order for this shearing machine to cut, both the foot pedal and the hand button must be engaged.

Courtesy of David Goetsch

FIGURE 10–9   This door protects the operator in the event of an exploding or shattering grinding wheel.

Courtesy of David Goetsch

- **Electromechanical devices** are contact bars that allow only a specified amount of movement between the worker and the hazard. If the worker moves the contact bar beyond the specified point, the machine will not cycle. These devices have the limitation of requiring frequent maintenance and careful adjustment.
- **Pullback devices** pull the operator's hands out of the danger zone when the machine starts to cycle. These devices eliminate the need for auxiliary

FIGURE 10–10   The safety door on this drilling machine must be closed or the drill will not operate.

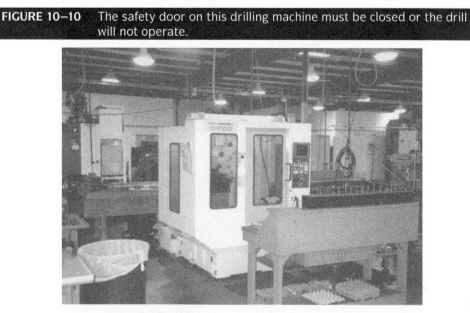

Courtesy of David Goetsch

barriers. However, they also have limitations. They limit operator movement, must be adjusted for each individual operator, and require close supervision to ensure proper use.

- **Restraint devices** hold the operator back from the danger zone. They work well, with little risk of mechanical failure. However, they do limit the operator's movement, must be adjusted for each individual operator, and require close supervision to ensure proper use.
- **Safety trip devices** include trip wires, trip rods, and body bars. All these devices stop the machine when tripped. They have the advantage of simplicity. However, they are limited in that all controls must be activated manually. They protect only the operator and may require the machine to be fitted with special fixtures for holding work.
- **Two-hand controls** require the operator to use both hands concurrently to activate the machine (e.g., a paper cutter or metal-shearing machine). This ensures that hands cannot stray into the danger zone. Although these controls do an excellent job of protecting the operator, they do not protect onlookers or passers-by. In addition, some two-hand controls can be tampered with and made operable using only one hand.
- **Gates** provide a barrier between the danger zone and workers. Although they are effective at protecting operators from machine hazards, they can obscure the work, making it difficult for the operator to see.[6]
- **Wireless control devices**. Wireless communication is now so widely used that it is only natural that there would be safety applications for this technology. Wireless machine control devices are being used increasingly for risk reduction in machine operation. The most widely used wireless technologies for machine safety applications are Bluetooth and WLAN or Wi-Fi. Bluetooth technology uses radio waves in the 2.4 GHz frequency to transmit signals. It is more popular than Wi-Fi in some cases because it can transmit through walls and solid objects using less power than Wi-Fi. WLAN (Wireless Local Area Network) or Wi-Fi devices operate in one of two modes: peer-to-peer or infrastructure. In peer-to-peer mode (sometimes called ad hoc mode), wireless devices communicate directly with each other without the need for central access points. In infrastructure mode, a wireless network is connected to a wired Ethernet network and central access points are needed. Regardless of the type of wireless technology used, risk reduction is accomplished by a signal that is sent from the sensor device that, in turn, brakes or otherwise incapacitates the machine to prevent intentional or inadvertent contact between the hazard and a body part.

HOW TO CHOOSE A MACHINE GUARD OR DEVICE

Choosing the right machine guard or device is a critical step in the risk reduction process. Earlier in this chapter, a model was presented for assessing the level of hazards relating to specific machines/processes. Such an assessment should be completed and the result known before attempting to choose the right machine guard or device. The type of machine guard chosen for a given situation depends on the level of risk reduction desired. When choosing a machine guard or device, apply the following criteria and read the specification of guards and devices to determine if they can meet the criteria:

- For the highest level of risk reduction, choose a guard or device that (1) is secured with a lock or fasteners the operator cannot remove, (2) prevents intentional contact with the hazard, and (3) has built-in redundancy with continuous self-checking.
- For a high level of risk reduction, choose a guard or device that (1) is not removable or adjustable by unauthorized personnel, (2) does not require adjustments by the operator, (3) prevents unintentional contact with the hazard, and (4) has redundancy with self-checking built in.
- For an intermediate level of risk reduction, choose a guard or device that (1) is movable, (2) can be adjusted by the operator, (3) prevents inadvertent contact with the hazard, and (4) checks redundancy manually.
- For a low level of risk reduction, choose a guard or device that (1) provides only minimal protection (tactile or visual awareness) against inadvertent contact with the hazard and (2) is moveable by the operator.

MACHINE GUARDING SELF-ASSESSMENT

One of the most effective ways to ensure that machines are properly guarded is to conduct periodic self-assessments.[7] These self-assessments can be conducted by safety personnel, supervisors, or employees. Developing self-assessment criteria and encouraging supervisors and employees to use them daily is an excellent strategy for safety and health professionals. The following questions can be used for conducting machine guarding self-assessments:

1. Are all machines that might expose people to rotating parts, nip points, flying chips, sparks, flying particles, or other similar hazards properly guarded?
2. Are all mechanical power transmission belts and the nip points they create properly guarded?
3. Are all exposed power shafts located less than seven feet above the working level properly guarded?
4. Are all hand tools and other hand-operated equipment inspected regularly for hazardous conditions?
5. Is compressed air used to clean tools, machines, equipment, and parts reduced to less than 30 pounds per square inch (psi)?
6. Are power saws and similar types of equipment properly guarded?
7. Are the tool rests for grinding wheels set to within one-eighth or less of the grinding wheel?
8. Are hand tools regularly inspected on a systematic basis for burred ends, cracked handles, and other potentially hazardous conditions?
9. Are all compressed gas cylinders inspected regularly and systematically for obvious signs of defects, deep rusting, or leakage?
10. Do all employees who handle and store gas cylinders and valves know how to do so without causing damage?

11. Are all air receivers periodically and systematically inspected, including safety valves?
12. Are all safety valves tested regularly, systematically, and frequently?

FEEDING AND EJECTION SYSTEMS

Feeding and ejection systems can be effective safeguards if properly designed and used. The various types of feeding and ejection systems available for use with modern industrial machines are summarized as follows:

- **Automatic feed** systems feed stock to the machine from rolls. Automatic feeds eliminate the need for operators to enter the danger zone. Such systems are limited in the types and variations of stock that they can feed. They also typically require an auxiliary barrier guard and frequent maintenance.
- **Semiautomatic feed** systems use a variety of approaches for feeding stock to the machine. Prominent among these are chutes, moveable dies, dial feeds, plungers, and sliding bolsters. They have the same advantages and limitations as automatic feed systems.
- **Automatic ejection** systems eject the work pneumatically or mechanically. The advantage of either approach is that operators don't have to reach into the danger zone to retrieve workpieces. However, these systems are restricted to use with relatively small stock. Potential hazards include blown chips or debris and noise. Pneumatic ejectors can be quite loud.[8]
- **Semiautomatic ejection** systems eject the work using mechanisms that are activated by the operator. Consequently, the operator does not have to reach into the danger zone to retrieve workpieces. These systems do require auxiliary barriers and can be used with a limited variety of stock.

ROBOT SAFEGUARDS

Robots have become commonplace in modern industry. Only the guarding aspects of robot safety are covered in this section. The main hazards associated with robots are (1) entrapment of a worker between a robot and a solid surface, (2) impact with a moving robot arm, and (3) impact with objects ejected or dropped by the robot.

The best guard against these hazards is to erect a physical barrier around the entire perimeter of a robot's **work envelope** (the three-dimensional area established by the robot's full range of motion). This physical barrier should be able to withstand the force of the heaviest object that a robot could eject.

Various types of shutdown guards can also be used. A guard containing a sensing device that automatically shuts down the robot if any person or object enters its work envelope can be effective. Another approach is to put sensitized doors or gates in the perimeter barrier that automatically shut down the robot as soon as they are opened.

These types of safeguards are especially important because robots can be deceptive. A robot that is not moving at the moment may simply be at a stage between cycles. Without warning, it might make sudden and rapid movements that could endanger any person inside the work envelope.

CONTROL OF HAZARDOUS ENERGY (LOCKOUT/TAGOUT SYSTEMS)

OSHA's standard for the control of hazardous energy, often referred to as the "lockout/ tagout" standard, is 29 CFR 1910.147. The purpose of this standard is to protect people in the workplace from hazardous energy while they are performing service or maintenance on machines, tools, and equipment. A key element of the standard is to prevent the accidental or inadvertent activation of a machine while it is being serviced or repaired. The lockout/tagout standard identifies the proper procedures for shutting down machines and equipment and locking or tagging it out so that accidental or inadvertent activation does not occur. The standard also calls for employee training and periodic inspections. The overall requirement of 29 CFR 1910.147 is that before service or maintenance operations are performed, the machines or equipment in question must be disconnected from their **energy source**, and the energy source must be either locked out or tagged out to prevent accidental or inadvertent activation.[9] In addition, 29 CFR 1910.333 establishes requirements to protect employees working on electrical circuits and equipment.

Lockout/Tagout Language

The following terms and phrases are frequently used in the language of lockout/tagout. Safety and health professionals should be knowledgeable of these terms:

- **Affected employee.** Employees who perform their jobs in areas in which the procedure in question is implemented and in which service or maintenance operations are performed. **Affected employees** do not implement energy control procedures unless they are *authorized*.
- **Authorized employee.** Employees who perform service or maintenance on a machine and use lockout/tagout procedures for their own protection.
- **Energized.** Machines, equipment, and tools are **energized** if they are connected to an energy source or when they still contain stored or residual energy even after being disconnected.
- **Capable of being locked out.** A device is considered to be capable of being locked out if it meets one of the following requirements: (1) it has a hasp to which a lock can be attached; (2) it has another appropriate integral part through which a lock can be attached; (3) it has a built-in locking mechanism; or (4) it can be locked without permanently dismantling, rebuilding, or replacing the **energy-isolating device**.
- **Energy-isolating device.** Any mechanical device that physically prevents the release or transmission of energy (e.g., circuit breakers, disconnect switches, or blocks).
- **Energy source.** Any source of power that can activate a machine or piece of equipment (e.g., electrical, mechanical, hydraulic, pneumatic, chemical, or thermal).
- **Energy control procedure.** A written document containing all the information an authorized person needs to know in order to properly control hazardous energy when shutting down a machine or equipment for maintenance or service.
- **Energy control program.** A systematic program for preventing the accidental or inadvertent energizing of machines or equipment during maintenance or servicing. This is sometimes called the organization's lockout/tagout program.
- **Lockout.** Placing a **lockout device** such as a padlock on an energy-isolating device to prevent the accidental or inadvertent energizing of a machine during maintenance or servicing.

FIGURE 10–11   Lockout system.

- **Lockout device.** Any device (see Figure 10–11) that uses a positive means to keep an energy-isolation device in the *safe* position to prevent the accidental or inadvertent energizing of a machine or piece of equipment.
- **Tagout.** Placing a tag (see Figure 10–12) on an energy-isolation device to warn people so that they do not accidentally or inadvertently energize a machine or piece of equipment.
- **Tagout device.** Any prominent warning device such as a tag that can be affixed to an energy-isolation device to prevent the accidental or inadvertent energizing of a machine or piece of equipment.

FIGURE 10–12   Tagout system.

DANGER

This tag may only
be removed by:

Simpson McGraw
Maintenance Department
Ext. 218

Provisions of the Standard

OSHA's standard for control of hazardous energy contains provisions in the following areas: **energy control program**, **energy control procedure**, energy-isolating devices, lockout/tagout devices, periodic inspections, application of controls and lockout/tagout devices, removal of locks and tags, testing or positioning of machines, outside personnel, group lockout or **tagout**, and shift or personnel changes. Those provisions are as follows:

- **Energy control program.** Organizations must establish energy control programs that have fully documented energy control procedures, provide employee training, and ensure periodic inspections.
- **Energy control procedure.** Organizations must develop, document, and use energy control procedures that contain at least the following elements: (a) a statement on how the procedure will be used; (b) procedural steps used to shut down, isolate, block, and secure machines or equipment; (c) steps designating the safe placement, removal, and transfer of lockout/tagout devices and who has responsibility for them; and (d) specific requirements for testing machines or equipment to verify the effectiveness of energy control measures.
- **Energy-isolating devices.** Organizations must have appropriate energy-isolating devices for preventing the accidental or inadvertent release of energy on all machines and equipment. The preferred type of device is one that can be locked out. However, when this type of device is not feasible, tagout procedures may be used. When a tagout system is used, employees must receive training on the following limitations of tags: (a) tags are just warning devices and do not provide the safety margin of locks; (b) tags may be removed only by the person who affixes them and should never be bypassed, ignored, or otherwise overcome; (c) tags must be legible and understandable by all employees (this can mean providing labels in more than one language); (d) tags and their means of attachment must be made of material that is durable enough to withstand the environment to which they will be subjected; (e) tags can evoke a false sense of security; and (f) tags must be attached securely enough that they will not come off during servicing or maintenance operations.
- **Requirements for lockout/tagout devices.** Lockout and **tagout devices** must meet the following requirements: (a) durable enough to withstand the environment to which they will be exposed; (b) standardized in terms of color or size and print and format; (c) substantial enough to minimize the likelihood of accidental or unauthorized removal; and (d) identifiable in terms of the employee who affixed them and the warning message (e.g., *Do Not Start, Do Not Close, Do Not Energize, Do Not Open*).
- **Employee training.** Organizations must provide both initial training and retraining as necessary and certify that the necessary training has been given to all employees covered in the standard (e.g., authorized, affected, and other). Training for **authorized employees** must cover the following topics at a minimum: (a) details about the types and magnitude of hazardous energy sources present in the workplace and (b) methods, means, and procedures for isolating and controlling these sources. Training for affected employees (usually machine operators or users) and other employees must cover the following topics at a minimum: (a) how to recognize when the energy control procedure is implemented and (b) the importance of never attempting to start up a locked-out or tagged-out machine. All training must

ensure that employees understand the purpose, function, and restrictions of the energy control program, and that authorized employees have the knowledge and skills necessary to properly apply, use, and remove energy controls.

- **Periodic inspections.** Inspections must be performed at least annually to ensure that the energy control program is up-to-date and being properly implemented. In addition, the organization must certify that the periodic inspections have actually taken place.

- **Application of controls and lockout/tagout devices.** Controls and lockout/ tagout devices must be applied properly. The appropriate procedure consists of the following steps that must be implemented in sequence: (a) prepare for shutdown; (b) shut down the machine or equipment in question; (c) affix the lockout or tagout device; (d) render safe any residual or stored energy that might remain in the machine; and (e) verify that the energy source has been effectively isolated and that the machine or equipment has been effectively de-energized.

- **Removal of locks or tags.** Before locks or tags are removed, the following procedures must be completed: (a) inspect the work area to ensure that nonessential items have been removed and that the machine is capable of operating properly; (b) check the area around the machine to ensure that all employees are safely back or removed from the area; (c) notify affected employees immediately after removing the energy control devices and before energizing the machine; and (d) ensure that energy control devices are removed by the individual who affixed them (if this is not possible, make sure that the person who does remove them follows the proper procedures).

- **Additional safety requirements.** OSHA allows for special circumstances as set forth in this subsection. When a machine must be energized in order to test or position it, energy control devices may be removed only as follows: (a) clear the machine of tools and materials; (b) remove employees from the area; (c) remove the devices as set forth in the standard; (d) energize the machine and conduct the test or proceed with the positioning procedure; and (e) de-energize the machine, isolate the energy source, and reapply the energy control devices. Organizations must ensure that outside personnel such as contractors are fully informed about energy control procedures. In group lockout or tagout situations, each individual employee performing maintenance or service tasks must be protected by his or her own personal energy control device. Organizations must have specific procedures for ensuring continuity in spite of personnel and shift changes.

PERMANENT ELECTRICAL SAFETY DEVICES IN LOCKOUT/TAGOUT PROGRAMS

An emerging technology in electrical safety is the permanent electrical safety device (PESD). PESDs have excellent potential to help workers safely isolate electrical energy, especially when used as part of an organization's Lockout/Tagout Procedures.

With PESDs incorporated into safety procedures, installed correctly into electrical enclosures, and validated before and after each use, workers can transition the once-risky endeavor of verifying voltage into a less precarious undertaking that never exposes them to voltage … every electrical incident has one required ingredient: voltage. Electrical safety is radically improved by eliminating exposure to voltage while still validating zero energy from outside the panel.[10]

Traditionally electrical energy that might be present is detected using a voltage detector. To ensure safety, NFPA 70E 120.1(5) requires that voltage detectors be checked before and after every use to ensure that they are operating properly. This requirement means that voltage detectors must be checked against an independent source before being used to detect the presence of electrical energy in a machine a worker needs to use. Then after checking the machine for electrical energy, the voltage detector must be validated again to an independent source.

This same principle applies to PESDs, but because they are permanently mounted PESDs cannot be moved between two power sources as voltage meters can. With PESDs, the validation technique is different. The PESD validation technique is based on the detection of a small amount of current flowing between two sources. A voltage detector is used to relate this small current flow to actual voltage. The worker is then given an appropriate visual or audible signal to indicate whether it is safe to proceed.

PESDs are designed for one purpose and one purpose only: to ensure electrical safety. Consequently, when used properly they can be even more reliable than the traditional method of using voltage meters. Further, PESDs can be used by the workers themselves without the need for having an electrician make the determination of the presence or absence of electrical energy. PESDs improve electrical safety by eliminating exposure to voltage while still ensuring zero voltage from outside the panel. It has been estimated that using PESDs can allow as much as 75 percent of locked out and tagged out equipment to be put back into service.

A PESD becomes a real safety device only after it is included as part of a written LOTO procedure. Without this, it is nothing more than another electrical component. The LOTO procedure must explain to the worker each step that involves the PESD. At a minimum, workers will need to verify proper operation of the PESD before and after performing a LOTO procedure. Mechanical maintenance workers receive a huge benefit when these devices are used in mechanical LOTO procedures. Workers performing mechanical LOTO procedures (work involving no contact with conductors or circuit parts) still must isolate electrical energy. PESDs provide a means of checking voltage inside an electrical panel without exposure to that same voltage. Without these devices, a mechanic performing mechanical LOTO would be required to work in tandem with an electrician using a voltmeter to physically verify zero voltage inside an electrical panel before work begins. With PESDs, the mechanic can single-handedly check for zero electrical energy without any exposure to voltage, thereby making the LOTO procedure safer and more productive.[11]

Evaluating Lockout/Tagout Programs

Lockout/tagout violations are frequently cited by OSHA during on-site inspections. The following questions can be used to evaluate an organization's lockout/tagout program:

- Are all machinery or equipment capable of movement required to be de-energized or disengaged and blocked or locked out during cleaning, servicing, adjusting, or setting up operations?
- Where the power disconnect equipment does not disconnect the electrical control circuit, are the appropriate electrical enclosures identified?
- If the power disconnect for equipment does not disconnect the electrical control circuit, is a means provided to ensure that the control circuit can be disconnected and locked out?

- Is it required to lock out main power disconnects instead of locking out control circuits?
- Are all equipment control valve handles equipped with a means for locking out?
- Does the lockout procedure require that stored energy—whether it is mechanical, hydraulic, or air—be released or blocked before the equipment is locked out for repairs?
- Are appropriate employees provided with individually keyed personal safety locks?
- Are these employees required to keep personal control of their keys while they have safety locks in use?
- Is only the employee exposed to the hazard required to install or remove the safety lock?
- Are employees required to check the safety lockout by attempting a startup after making sure no one is exposed?
- After the safety is checked, does the employee again place the switch in the "off" position?
- Are employees instructed always to push the control circuit stop button before reenergizing the main power switch?
- Are all employees who are working on locked-out equipment identified by their locks or accompanying tags?
- Are enough accident prevention signs, tags, and safety padlocks provided for any reasonably foreseeable repair emergency?
- When machine operations, configuration, or size require the operator to leave his or her control station to install tools or perform other operations, is he or she required to lock or tag out separately any parts of the machine that could move if accidentally activated?
- If the equipment or lines cannot be shut down, locked out, and tagged, is a safe job procedure established and rigidly followed?
- Have employees been trained not to start machinery or equipment if it is locked out or tagged out?
- Are all workers notified when the machinery or equipment they usually use is shut down and locked out for maintenance or servicing purposes?
- After maintenance is completed, is the machinery checked to ensure that nonessential items have been removed and the machine is operationally intact?
- Before the machinery is activated, are employees removed from possible danger?
- When the machinery is fully operational, are employees notified?[12]

GENERAL PRECAUTIONS

The types of safeguards explained in this chapter are critical. In addition to these specific safeguards, there are also a number of general precautions that apply across the board in settings where machines are used. Some of the more important general precautions are as follows:

- All operators should be trained in the safe operation and maintenance of their machines.
- All machine operators should be trained in the emergency procedures to take when accidents occur.

- All employees should know how to activate emergency shutdown controls. This means knowing where the controls are and how to activate them.
- Inspection, maintenance, adjustment, repair, and calibration of safeguards should be carried out regularly.
- Supervisors should ensure that safeguards are properly in place when machines are in use. Employees who disable or remove safeguards should be disciplined appropriately.
- Operator teams (two or more operators) of the same system should be trained in coordination techniques and proper use of devices that prevent premature activation by a team member.
- Operators should be trained and supervised to ensure that they dress properly for the job. Long hair, loose clothing, neckties, rings, watches, necklaces, chains, and earrings can become caught in equipment and, in turn, pull the employee into the hazard zone.
- Shortcuts that violate safety principles and practices should be avoided. The pressures of deadlines should never be the cause of unsafe work practices.
- Other employees who work around machines but do not operate them should be made aware of the emergency procedures to take when an accident occurs.

BASIC PROGRAM CONTENT

Machine safeguarding should be organized, systematic, and comprehensive. A company's safeguarding program should have at least the following elements:

- Safeguarding policy that is part of a broader company-wide safety and health policy
- Machine hazard analysis
- Lockout/tagout (materials and procedures)
- Employee training
- Comprehensive documentation
- Periodic safeguarding audits (at least annually)

TAKING CORRECTIVE ACTION

What should be done when a mechanical hazard is observed? The only acceptable answer to this question is, take *immediate corrective action*. The specific action indicated will depend on what the problem is. Figure 10–13 shows selected examples of problems and corresponding corrective actions.

These are only a few of the many different types of problems that require corresponding corrective action. Regardless of the type of problem, the key to responding is immediacy. As shown in the examples given earlier in this chapter, waiting to take corrective action can be fatal. It is important to note that it is often prudent to exceed some OSHA guarding requirements (i.e., the seven-foot rule).

FIGURE 10–13 Selected examples of problems and corresponding actions.

Problem	Action
Machine is operating without the safety guard.	Stop machine immediately and activate the safety guard.
Maintenance worker is cleaning a machine that is operating.	Stop machine immediately and lock or tag it out.
Visitor to the shop is wearing a necktie as he observes a lathe in operation.	Immediately pull the visitor back and have him remove the tie.
An operator is observed disabling a guard.	Stop the operator, secure the guard, and take disciplinary action.
A robot is operating without a protective barrier.	Stop the robot and erect a barrier immediately.
A machine guard has a sharp, ragged edge.	Stop the machine and eliminate the sharp edge and ragged burrs by rounding it off.

Key Terms and Concepts

- Adjustable guards
- Affected employees
- Authorized employees
- Automatic ejection
- Automatic feed
- Breaking
- Crushing
- Cutting
- Electromechanical devices
- Energized
- Energy control procedure
- Energy control program
- Energy source
- Energy-isolating device
- Fixed guards
- Gates
- Interlocked guards
- Lockout device
- Mechanical hazards
- Mechanical injuries
- Photoelectric devices
- Point-of-operation devices
- Point-of-operation guards
- Pullback devices
- Puncturing
- Radio-frequency devices
- Restraint devices
- Risk assessment
- Safeguarding
- Safety trip devices
- Semiautomatic ejection
- Semiautomatic feed
- Shearing
- Spraining
- Straining
- Tagout
- Tagout devices
- Tearing
- Two-hand controls
- Wireless control devices
- Work envelope

Review Questions

1. List and briefly explain the common types of mechanical injury hazards.
2. Explain the concept of safeguarding.
3. Summarize OSHA's requirements for machine guarding.
4. What are the requirements all safeguards should meet?
5. Describe the three types of point-of-operation guards.
6. Describe four types of point-of-operation devices.
7. What are the relative advantages and disadvantages of feeding and ejection systems?
8. Describe the primary hazards associated with robots.
9. Explain how to guard against the hazards associated with robots.
10. What is a lockout system?
11. What is a tagout system?

12. What impact may a lockout/tagout system have if carefully followed nationwide?
13. Summarize the main provisions of OSHA's lockout/tagout standard.
14. Explain the concept of risk assessment as it relates to machine operation.
15. Explain how to evaluate lockout/tagout programs.

Endnotes

1. OHSA, *Methods of Machine Guarding.* Retrieved from osha.gov/Publications/Mach_Safeguard/Chapt2.html on July 16, 2017, 1.
2. Retrieved from osha.gov/pls/oshaweb on July 6, 2017.
3. Grainger, *General OSHA Machine Guarding Requirements.* Retrieved from grainger.com on July 3, 2017, 5.
4. OHSA, *Methods of Machine Guarding,* 2–3.
5. Ibid.
6. Ibid.
7. Ibid.
8. Ibid., 44.
9. Retrieved from osha.gov/SLTC/smallbusiness/sec11.html on July 17, 2017.
10. P. Allen. "Using Permanent Electrical Safety Devices," *Occupational Health and Safety* 81, no. 1: 36.
11. Ibid. 37.
12. L. Johnson, *The "Red Flags" of LOTC.* Retrieved from oshonline.com on July 6, 2017, 55.

CHAPTER

11

FALLING, IMPACT, ACCELERATION, AND VISION HAZARDS WITH APPROPRIATE PPE

Learning Objectives

- List the most common causes of falls in the workplace
- Summarize the various kinds of falls
- Describe slip and fall hazards
- Summarize the basic elements of a slip and fall prevention program
- Summarize the key elements of OSHA's Fall Protection Standards
- Explain the key elements of ladder safety
- Describe what to do after a fall has occurred
- Explain how to monitor fall protection equipment and why it fails
- Summarize the most common impact/acceleration hazards
- Explain the hazards of standing
- Explain the basic elements of hand protection
- Explain key concepts associated with Personal Protective Equipment (PPF)
- Summarize the key elements of forklift safety (powered industrial trucks)

Some of the most common accidents in the workplace happen as the result of slipping, falling, and improper lifting. Impact from a falling object is also a common cause of accidents. This chapter provides the information needed by modern safety and health professionals to prevent such accidents. It also provides specific information about head, hand, back, eye, face, and **foot protection** as well as forklift safety.

CAUSES OF FALLS

More than 16 percent of all disabling work-related injuries are the result of falls. Clearly, falls are a major concern of safety and health professionals. The primary causes of falls are as follows:

- A foreign object on the walking surface
- A design flaw in the walking surface
- Slippery surfaces
- An individual's impaired physical condition[1]

A **foreign object** is any object that is out of place or in a position to trip someone or to cause a slip. There is an almost limitless number of **design flaws** that may cause a fall. A poorly designed floor covering, a ladder that does not seat properly, and a catwalk that gives way are all examples of design flaws that may cause falls. Slippery surfaces are particularly prevalent in industrial plants where numerous different lubricants and cleaning solvents are used.

Automobile accidents are often caused when a driver's attention is temporarily drawn away from the road by a visual distraction. This is also true in the workplace. Anything that distracts workers visually can cause a fall. When a person's physical condition is impaired for any reason, the potential for falls increases. This is a particularly common problem among aging workers. Understanding these causes is the first step in developing fall prevention techniques.

KINDS OF FALLS

Falling from ladders and other elevated situations is covered later in this chapter. This section deals with the more common surface falls. Such falls can be divided into the following four categories:

- **Trip and fall** accidents occur when workers encounter an unseen foreign object in their path. When the employee's foot strikes the object, he or she trips and falls.
- **Stump and fall** accidents occur when a worker's foot suddenly meets a sticky surface or a defect in the walking surface. Expecting to continue at the established pace, the worker falls when his or her foot is unable to respond properly.
- **Step and fall** accidents occur when a person's foot encounters an unexpected step down (e.g., a hole in the floor or a floorboard that gives way). This can also happen when an employee thinks he or she has reached the bottom of the stairs when, in reality, there is one more step.
- **Slip and fall** accidents occur when the worker's center of gravity is suddenly thrown out of balance (e.g., an oily spot causes a foot to shoot out from under the worker). This is the most common type of fall.

WALKING AND SLIPPING

Judging by the number of injuries that occur each year as the result of slipping, it is clear that walking can be hazardous to a worker's health. This is, in fact, the case when walking on an unstable platform. A stable platform for walking is any surface with a high degree of traction that is free of obstructions. It follows that an unstable platform is one lacking traction, one on which there are obstructions, or both.

0.40*   0.60*

Minimal
traction

Moderate
traction

High traction

*Based on ANSI/NFSI B101.1

Measuring Surface Traction

In order to understand **surface traction**, *you* must have a basis for comparison. An effective way for comparing the relative traction of a given surface is to use the **coefficient of friction**, which is a numerical comparison of the resistance of one surface (shoe or boot) against another surface (the floor) (see Figure 11–1).

The National Floor Safety Institute (NFSI) was accredited by the American National Standards Institute (ANSI) as a Standards Developing Organization in 2006. Since that time, NFSI has done significant work on the prevention of slip and fall accidents. Its most notable work has been the development of B101.1—"Test Method for Measuring Wet SCOF of Common Hard-Surface Floor Materials." This standard provides a method for testing the *Coefficient of Friction* (COF) for walking surfaces, assigning a numerical rating, and tying that numerical rating to the amount of traction the surface provides when wet.

Measuring surface traction when wet was a major step forward in preventing slip and fall accidents since 80 percent of them occur on wet surfaces. Figure 11–1 is based on ANSI/ NFSI B101.1. It shows that walking surfaces (when wet) with a COF of 0.60 or greater have high traction and, in turn, a lower probability of producing slip and fall accidents. Surfaces with a COF between 0.40 and 0.59 provide moderate traction and an increased likelihood of producing a slip and fall accident. Surfaces with a COF of 0.39 and less provide only minimal friction and a higher probability of producing a slip and fall accident.

Factors That Decrease Traction

Good housekeeping can be a major factor in reducing slip and fall hazards. Water, oil, soap, coolant, and cleaning solvents left on a floor can decrease traction and turn an otherwise safe surface into a danger zone. For example, the friction coefficient of concrete (0.43) is reduced by almost 15 percent if the concrete is wet. Rubber-soled shoes can decrease slipping hazards somewhat, but changing the type of shoe is not enough to ensure safety. Additional precautions are needed.

General Strategies for Preventing Slips

Modern safety and health professionals are concerned with preventing slips and falls. Slip prevention should be a part of the company's larger safety and health program. Here are some strategies that can be used to help prevent slipping:

1. **Choose the right material from the outset.** Where the walking surface is to be newly constructed or an existing surface is to be replaced, safety and health professionals should encourage the selection of surface materials that have the highest possible COF. Getting it right from the start is the best way to prevent slipping accidents.

2. **Retrofit an existing surface.** If it is too disruptive or too expensive to replace a **slippery surface** completely, **retrofit** it with friction enhancement devices or materials. Such devices or materials include runners, skid strips, carpet, grooves, abrasive coatings, grills, and textured coverings.

3. **Practice good housekeeping.** Regardless of the type of surface, keep it clean and dry. Spilled water, grease, oil, solvents, and other liquids should be removed immediately. When the surface is wet intentionally, as when cleaning or mopping, rope off the area and erect warning signs.

4. **Require nonskid footwear.** Employees who work in areas where slipping is likely to be a problem should be required to wear shoes with special nonskid soles. This is no different from requiring steel-toed boots to protect against falling objects. **Nonskid footwear** should be a normal part of a worker's PPE.

5. **Inspect surfaces frequently.** Employees who are working to meet production deadlines may be so distracted that they don't notice a wet surface, or they may notice it but feel too rushed to do anything about it. Consequently, safety and health professionals should conduct frequent inspections and act immediately when a hazard is identified.[2]

Effective strategies for preventing slips and falls include the following:

1. Review and analyze accident statistics to determine where slip and fall accidents are happening and why; then take the appropriate corrective measures.
2. Monitor the condition of walking surfaces continually and make appropriate preventive corrections immediately.
3. Make sure that ramps and sloped floors have high-friction surfaces.
4. Use safety mats, nonslip flooring, and **slip-resistant** safety shoes.
5. Make sure that stairs have handrails.
6. Make sure that visibility is good in potentially hazardous areas. Add extra lighting if necessary. Also make sure that the color of paint in these areas is bright and helpful in calling attention to potential hazards.
7. Make sure that spills are cleaned up immediately and that the underlying cause of the spill is corrected.
8. Make sure that employees who work in potentially hazardous areas select and wear the right slip-resistant footwear.
9. Use appropriate technologies such as **vertical incidence tribometers** to measure the slip resistance of floors and take appropriate action based on the results.
10. Conduct periodic audits of walking surfaces throughout the facilities in question, document carefully the findings, and take appropriate action in a systematic way.[3]

Specific Strategies for Preventing Slips

In addition to the general strategies already explained, there are a number of specific strategies that will be helpful for preventing slips and falls. These specific strategies are presented in this section.[4]

Building Lobbies

The lobbies to buildings often have slick, highly polished floor surfaces designed to project a certain image. Unfortunately, such floors represent a real hazard for slipping and falling. The hazard level is often increased when outside moisture from rain, sleet, or snow is brought in by pedestrian traffic. The following prevention strategies will help decrease the hazard level: (1) use large welcome mats that are wide enough to allow several "drying steps" to be taken before reaching the slick floor; (2) provide umbrella holders so that dripping umbrellas are not brought onto the slick floor; (3) monitor the floor surface continually and dry any moisture that makes its way onto the floor immediately; and (4) substitute nonslip surfaces for the slick, highly polished flooring.

Restrooms

Because of the nature of their use, restrooms are likely to have tiled flooring. Certain types of tiles become slippery when water from the sinks, toilets, or urinals splash onto the floor or overflow. The hazards in this situation are multiplied when soap is added in restrooms. The following prevention strategies will help decrease the hazard level in restrooms: (1) monitor restrooms continually and clean up spills immediately; (2) use "wet floors" warning signs; (3) block off any wet areas until they dry; and (4) conduct periodic inspections of public restrooms systematically.

Kitchens

Some business and industrial firms have commercial kitchens for providing meals to their personnel. When this is the case, the most common hazard is polymerized grease. Grease on almost any floor surface, but especially on tile, concrete, and linoleum floors, creates a serious slip and fall hazard. The following strategies will help decrease the level of hazard in commercial kitchens: (1) use a nonslip floor surface; (2) require kitchen employees to wear slip-resistant footwear; and (3) frequently dry clean the floors after-hours using a method other than wet mopping (which typically just distributes the grease).

Processing Areas

Companies that process materials typically experience high hazard levels in the processing areas. Processing often requires the use of fluids that can spill over and onto the floor. For example, companies that process meat have to contend with blood, fats, and meat juices on the floor. Companies that process chemicals must contend with spillage of those chemicals. The following strategies will help decrease the hazard level in processing areas: (1) use nonslip flooring, (2) monitor floor surfaces continually and take immediate action to clean up spills, (3) require processing employees to wear slip-resistant footwear, and (4) inspect and clean floor surfaces regularly.

SLIP AND FALL PREVENTION PROGRAMS

Every year slips, trips, and falls cause more than 1 million workplace injuries and approximately 16,000 deaths. A company's overall safety and health program should include a slip and fall prevention component. Such a component should have the following elements:

1. **A policy statement/commitment.** Statement to convey management's commitment. Areas that should be included in the policy statement are management's intent, scope of activity, responsibility, accountability, the safety professional's role, authority, and standards.

2. **Review and acceptance of walkways.** Establish the criteria that will be used for reviewing all walking surfaces and determining if they are acceptable. For example, a criterion may be a minimum COF value. Regardless of the criteria, the methodology that will be used for applying them to the **review and acceptance** of walkways should also be explained.

3. **Reconditioning and retrofitting.** Include recommendations and timetables for **reconditioning** or retrofitting existing walking surfaces that do not meet review and acceptance criteria.

4. **Maintenance standards and procedures.** State the maintenance standards for walking surfaces (e.g., how often surfaces should be cleaned, resurfaced, replaced). In addition, this section should contain procedures for meeting the standards.

5. **Inspections, audits, tests, and records.** Provide a comprehensive list of inspections, audits, and tests (including the types of tests) that will be done, how frequently, and where. Maintain records of the results.

6. **Employee footwear program.** Specify the type of footwear required of employees who work on different types of walking surfaces.

7. **Defense methods for legal claims.** Outline the company's legal defenses so that aggressive action can be taken immediately should a lawsuit be filed against the company. In such cases, it is important to be able to show that the company has not been negligent (e.g., the company has a slip and fall prevention program that is in effect).

8. **Measurement of results.** Contain the following two parts: (a) an explanation of how the program will be evaluated and how often (e.g., comparison of yearly, quarterly, or monthly slip and fall data) and (b) records of the results of these evaluations.[5]

Practical Prevention Measures

In addition to maintaining the type of comprehensive slip and fall prevention program explained above, it is also important to apply several practical measures that can reduce slip and fall hazards in the workplace. These practical measures include the following:

- **Reduce "tracked-in" hazards.** Every time people enter a building they bring in slip and fall hazards with them. They might track in moisture from rain, sleet, or snow. They might track in mud, dirt, dust, and grime, or they might track in lubricants picked when they walked through a spill in the parking lot or some other source. All of these tracked-in substances can accumulate and become slip and fall hazards. Using entrance mats to absorb substances such as these as well as making a point of cleaning entranceways frequently can help reduce tracked in hazards.

- **Choose antifatigue mats carefully.** Providing antifatigue mats for employees whose work requires long periods of standing is important. However, these mats can solve one problem only to cause another. For example, foam-backed mats tend to absorb oils, lubricants, and solvents. Absorbing liquids such as these can turn antifatigue mats into slip and fall hazards. When they are going to be used in areas that will subject them to oils, lubricants, and solvents, choose antifatigue mats that will not absorb these substances or quickly deteriorate because of them.

- **Locate absorbents strategically.** In work areas where oils, lubricants, and solvents are used, there should be containers of absorbents nearby. Employees should never have to walk long distances to central storage bins for the absorbents they need when there is a spill. Chances are when walking to the central storage bin they are spreading the hazards along the way. Having absorbents stored close by allows employees to respond quickly to spills and avoids the problems caused by spreading the hazard.

- **Choose safety footwear carefully.** When deciding what types of safety footwear will be required for employees, think about more than just the traditional protections built into them. Make sure they also have slip- and fall-resistant soles. The best safety footwear is that which protects the feet from falling objects and slips.

OSHA FALL PROTECTION STANDARDS

Falls are one of the most common causes of accidents, injuries, and deaths in the workplace. Consequently, employers are required to take appropriate measures to prevent fall-related accidents. OSHA fall protection requirements are set forth in the General Industry Standard (29 CFR 1910), Shipyard Standard (29 CFR 1915), Marine Terminals Standard (29 CFR 1917), Longshoring Standard (20 CFR 1918), and Construction Standard (29 CFR 1926). The General Standard contains fall prevention requirements in the following sections: (1) 1910.23—Guarding floor and wall openings and holes; (2) 1910.25—Portable wood ladders; (3) 1910.26—Portable metal ladders; (4) 1910.27—Fixed ladders; (5) 1910.28—Safety requirements for scaffolding; (6) 1910.66—Powered platforms for building maintenance (Appendix C: Personal Fall Arrest System); (7) 1910.67—Aerial lifts; (8) 1910.68—Manlifts; (9) 1910.132—Personal Protective Equipment; and (10) 1910.269—Electric power generation, transmission, and distribution.

In general, OSHA requires employers to take the following precautions against fall-related accidents:

- Guard every floor hole into which a worker might accidentally fall using a guardrail and toeboard or a floor hole cover.
- Provide a guardrail and toeboard around all open-sided platforms, floors, or runways that are four feet or more above the ground or the next level.
- Provide guardrails and toeboards—regardless of height—to prevent workers from falling onto or into machines, equipment, vats, conveyor belts, and so on.
- Provide other means of fall protection as required on certain jobs such as safety harnesses and lines, safety nets, stair railings, and handrails.
- Provide a work environment that is free of known dangers.
- Keep floors in work areas clean and sanitary.
- Select and provide appropriate PPE at no cost to employees.
- Train workers about job hazards in a language they can understand.

Testing Fall Protection Equipment

Testing of fall protection equipment to ensure its viability has long been an issue in occupational safety circles. However, with development of ANSI/ASSE Z359.7—"Certification Testing of Fall Protection Products," much of the concern has been alleviated. This standard requires accreditation as an ISO 17025 laboratory for any laboratory that tests fall protection equipment. Test information from these laboratories is required to be provided to end users upon request by any manufacturer that claims the ANSI standard mark.

In addition to these OSHA standards, the ANSI publishes a Fall Protection Standard (ANSI Z359.1: *Safety Requirements for Personal Fall Arrest Systems,*

Subsystems, and Components). The most comprehensive and most controversial fall protection standard is OSHA's Fall Protection Standard for the construction industry (Subpart M of 29 CFR 1926).

OSHA's Fall Protection Standard for Construction

Fall protection is especially important in the construction industry. OSHA addresses the duty to provide fall protection at construction sites in 29 CFR 1926.501 (b) (13) and specific aspects of fall protection in sections of 29 CFR 1926. In the past, residential contractors had what amounted to an exemption—in practice if not in theory—from the requirements set forth in 29 CFR 1926.501 (13) based on a directive from OSHA (STD 03-00-001). However, a new directive—STD 03-11-002—rescinded the earlier directive so that 29 CFR 1926.501 (b) (13) now applies to commercial, industrial, and residential contractors. Consequently, contractors that have personnel working six feet (1.8 meters) or more above the next lower level must comply with all applicable aspects of OSHA's standard including the provision of such PPE as safety harnesses (see Figure 11–2).[6]

FIGURE 11–2   Body harness for fall protection.

Source: Steve Gorton/Dorling Kindersley

Various additional aspects of fall protection for construction are covered in the following sections of 29 CFR 1926: (1) General requirements (scaffolding)—1926.451; (2) Additional requirements applicable to specific types of scaffolds—1926.452; (3) Training requirements (scaffolding)—1926.454; (4) Fall protection systems criteria and practices; (5) Training requirements (fall protection)—1926.503; (6) Steel erection (fall protection)—1926.760; (7) Underground construction—1926.800; (8) General requirements (stairs and ladders)—1926.1051; (9) Stairways—1926.1052; (10) Ladders—1926.1053; (11) Training requirements (stairways and ladders)—1926.1060; (12) Cranes and derricks in construction (fall protection)—1926.1423; and (13) Cranes and derricks used in demolition and underground construction—1926.1501. The general requirements of 29 CFR 1926.501 (b) (13) are explained in the next section.

OSHA's Recommendations for Effective Fall Protection

Because slip and fall accidents account for approximately one million workplace injuries every year, organizations obviously need to have a strong fall prevention program in place. But what does it take to have an effective fall prevention program? OSHA recommends the following strategies:

- **Have a plan.** An organization should develop a written plan that is part of its larger safety and health plan. The fall protection plan should contain a statement of commitment from both management and employees, rules and regulations relating to fall protection, and an explanation of the training program and training requirements.
- **Establish proper fall protection requirements.** Require the use of fall protection equipment any time an employee works more than 4 feet above the floor in general industry, 6 feet or more in construction, and 10 feet or more when on scaffolding.
- **Provide proper fall protection equipment and procedures and require their use.** Organizations should determine what types of fall protection equipment and procedures are needed, provide them to employees, and require their proper use. As examples, this might include **personal fall arrest systems**, guardrails, safety nets, positioning devices, warning lines, controlled access zones, and safety monitoring.
- **Ensure fall protection device replacement.** Organizations should replace fall protection devices on a regular schedule even if there are no significant signs of wear.
- **Ensure proper use and type of equipment.** Ensure that the fall protection equipment provided to employees is the proper type for the situation in question and that employees inspect it before putting it on, that it fits properly, and that it is properly attached to **anchorage** points. Personal fall limiters (PFL) or self-retracting **lifelines** are beginning to see use in place of traditional shock-absorbing **lanyards**. Employers must stay up to date concerning new developments in fall protection equipment.
- **Provide training.** Provide fall protection training for supervisors and employees, including how to recognize fall-related hazards and how to properly use all applicable fall protection equipment.[7]

ANSI Z359 FALL PROTECTION CODE

The ANSI fall protection standards are not mandatory, but they can be helpful for companies that are establishing comprehensive fall protection programs. The ANSI standards—particularly ANSI Z359—are important enough that safety professionals should be well versed in their requirements. The critical components of ANSI Z359 are as follows:

- ANSI Z359.0: Definitions and nomenclature used for fall protection and fall arrest
- ANSI Z359.1: Safety requirements for personal fall arrest systems, subsystems, and components
- ANSI Z359.2: Minimum requirements for a comprehensive managed fall protection program
- ANSI Z359.3: Safety requirements for positioning and travel restraint systems
- ANSI Z359.4: Safety requirements for assisted-rescue and self-rescue systems, subsystems, and components
- ANSI Z359.6: Specifications and design requirements for active fall protection systems
- ANSI Z359.7: Qualification and verification testing of fall protection products
- ANSI Z359.12: Connecting components for personal fall arrest systems
- ANSI Z359.13: Personal energy absorbers and energy absorbing lanyards
- ANSI Z359.14: Self-retracting devices for personal fall arrest and rescue systems

ANSI Z359.2: Managed Fall Protection Program

Several of the components of ANSI Z359 provide guidance for designers, manufacturers, and testers of fall protection systems, subsystems, and components. However, ANSI Z359.2 applies directly to construction companies in that it provides guidance for establishing a comprehensive fall protection program. ANSI Z359.2 provides guidelines for appropriate policies, responsibilities/duties, training, fall protection procedures, rescue procedures, incident investigations, and program evaluation. What follows is a sample outline for an ANSI Z359 approved fall protection program:

- Develop a fall protection policy that established the scope of the program.
- Develop specific goals for the program.
- Conduct a comprehensive hazard analysis and record all known fall hazards.
- Develop measures for eliminating, preventing, mitigating, or controlling all known fall hazards identified.
- Plan and implement fall protection training.
- Develop job-site specific plans for all job sites (these plans are comprised of the safety measures that are specific to the job site in question).
- Establish and implement procedures maintaining and regularly inspecting fall protection equipment.
- Develop a comprehensive rescue plan and include it in the training provided for workers.
- Monitor the program continually and conduct periodic audits to ensure its viability. Make corrections, adjustments, and improvements the moment the need for them is identified. Revises the written program accordingly.

LADDER SAFETY

Jobs that involve the use of ladders (see Figure 11–3) introduce their own set of safety problems, one of which is an increased potential for falls. OSHA's requirements for ladder safety are contained in 29 CFR 1926.800, 1926.1053, and 1926.1060. The National Safety Council (NSC) recommends that ladders be inspected before every use and that employees who use them follow a set of standard rules.[8]

Inspecting Ladders

Taking a few moments to look over a ladder carefully before using it can prevent a fall:

- See if the ladder has the manufacturer's instruction label on it.
- Determine whether the ladder is strong enough.

FIGURE 11–3 Ladders can be hazardous.

Source: Matthew Ward/Dorling Kindersley

- Read the label specifications about weight capacity and applications.
- Look for the following conditions: cracks on side rails; loose rungs, rails, or braces; or damaged connections between rungs and rails.
- Check for heat damage and corrosion.
- Check wooden ladders for moisture that may cause them to conduct electricity.
- Check metal ladders for burrs and sharp edges.
- Check fiberglass ladders for signs of *blooming*, deterioration of exposed fiberglass.[9]

Dos and Don'ts of Ladder Use

Many accidents involving ladders result from improper use. Following a simple set of rules for the proper use of ladders can reduce the risk of falls and other ladder-related accidents:

- Check for slipperiness on shoes and ladder rungs.
- Secure the ladder firmly at the top and bottom.
- Set the ladder's base on a firm, level surface.
- Apply the **four-to-one ratio** (base one foot away from the wall for every four feet between the base and the support point).
- Face the ladder when climbing up or down.
- Barricade the base of the ladder when working near an entrance.
- Don't lean a ladder against a fragile, slippery, or unstable surface.
- Don't lean too far to either side while working (stop and move the ladder).
- Don't rig a makeshift ladder; use the real thing.
- Don't allow more than one person at a time on a ladder.
- Don't allow your waist to go any higher than the last rung when reaching upward on a ladder.
- Don't separate the individual sections of extension ladders and use them individually.
- Don't carry tools in your hands while climbing a ladder.
- Don't place a ladder on a box, table, or bench to make it reach higher.[10]

OSHA standards for walking and working surfaces and **ladder safety** are set forth in 29 CFR Part 1910 (Subpart D). The standards contained in Subpart D are as follows:

1910.21	Definitions
1910.22	General requirements
1910.23	Guarding floor and wall openings and holes
1910.24	Fixed industrial stairs
1910.25	Portable wood ladders
1910.26	Portable metal ladders
1910.27	Fixed ladders
1910.28	Safety requirements for scaffolding
1910.29	Manually propelled mobile ladder stands and scaffolds (towers)
1910.30	Other working surfaces

WHAT TO DO AFTER A FALL

When using fall arrest systems, employers must have a rescue plan and provide rescue training. If, in spite of your best efforts, a fall occurs on the job, what employees do in the immediate aftermath can mean the difference between life and death for the victim. First, make sure your organization has a fall rescue plan in place that includes the following: (1) training for all personnel in how to carry out a rescue—what to do and what not to do; (2) proper equipment on site and readily available; (3) coordination with local emergency authorities; and (4) assigned responsibilities. Then, in addition, make sure all employees who work at heights or with others who work at heights understand the following basics:[11]

- **Never work alone.** There should always be two or more people working in close proximity when working at heights. It is important to have someone available to initiate the fall rescue plan.
- **Keep legs moving.** When a worker is dangling from his fall arrest gear, it is important for him to keep his legs moving—not frantically, but just rhythmically and regularly. This will help prevent the venous pooling of blood that can lead to shock. If it is possible, the suspended worker should try to move into an upright position.
- **Raise the worker to a seated position.** Once the suspended worker has been brought to the ground, the tendency is to lie him down in a horizontal position. This is a mistake because it can suddenly release pooled blood that can strain the heart and cause death. Instead, move the victim into a seated position.

MONITOR FALL PROTECTION EQUIPMENT AND KNOW WHY IT FAILS

Although more people than ever are using fall protection equipment on the job, the number of injuries and deaths from falls continues to increase. There are several reasons for this, including poor training, deterioration of equipment over time, and selection of the wrong equipment for the job. However, the reason that should concern safety and health professionals most is failure due to lack of monitoring. In many cases, equipment failures could have been prevented by a systematic monitoring process in which all equipment is inspected before being used. Consequently, it is important for safety and health professionals to understand why fall protection equipment fails, and to ensure that equipment is properly monitored to detect potential failure points before it is used.[12]

When inspecting fall protection equipment, look for the following types of potential problems: weld splatter; webbing cuts and abrasions; broken stitching; frayed or burned webbing; chemical damage; discoloration; deformed hardware; loose, distorted, or broken grommets; knotted webbing; and malfunctioning snap hooks. Just making sure that employees know how to monitor their equipment and that they follow through and do it—every time—will save lives.

IMPACT AND ACCELERATION HAZARDS

An employee working on a catwalk drops a wrench. The falling wrench accelerates over the 20-foot drop and strikes an employee below. Had the victim not been wearing a hard hat he might have sustained serious injuries from the impact. A robot loses its grip on a part, slinging it across the plant and striking an employee. The impact from the part breaks one of the employee's ribs. These are examples of accidents involving **acceleration** and **impact**. So is any type of fall because, having fallen, a person's rate of fall accelerates (increases) until striking a surface (impact). Motor vehicle accidents are also acceleration and impact instances.

Because falls were covered in the previous section, this section will focus on hazards relating to the acceleration and impact of objects. Approximately 25 percent of the workplace accidents that occur each year as the result of acceleration and impact involve objects that become projectiles.

Protection from Falling or Accelerating Objects

Objects that fall, are slung from a machine, or otherwise become projectiles pose a serious hazard to the heads, faces, feet, and eyes of workers. Consequently, protecting workers from projectiles requires the use of appropriate PPE and strict adherence to safety rules by all employees.

Head Protection

Approximately 120,000 people sustain head injuries on the job each year.[13] Falling objects are involved in many of these accidents. These injuries occur in spite of the fact that many of the victims were wearing hard hats. Such statistics have been the driving force behind the development of tougher, more durable hard hats.

Originally introduced in 1919, the hard hats first used for **head protection** in an industrial setting were inspired by the helmets worn by soldiers in World War I. Such early versions were made of varnished resin-impregnated canvas. As material technology evolved, hard hats were made of vulcanized fiber, then aluminum, and then fiberglass. Today's hard hats are typically made from the thermoplastic material polyethylene, using the injection-molding process.[14] Basic hard hat design has not changed radically since before World War II. They are designed to provide limited protection from impact primarily to the top of the head, and thereby reduce the amount of impact transmitted to the head, neck, and spine.[15]

It is important not only to wear hard hats, but to wear them properly (e.g., never wear them backwards to make a fashion statement). Some companies adhere double-stick tape or flat magnets to the upper visor area to minimize the amount of dust or iron filings that fall into workers' eyes.

The ANSI standard for hard hats is Z89.1. This standard was revised in 2008 to include voluntary guidelines for reverse donning, high visibility, and temperature extremes. OSHA subsequently adopted this standard as its hard hat standard (29 CFR 1010.135 and 1926.100).

The 2009 revision of ANSI Z89.1 established three nonmandatory tests that manufacturers of hard hats (work helmets) can use to rate their products:

- **Reverse donning.** Helmets rated for reverse donning pass all applicable tests whether worn frontward or backward. This is important since reverse donning of work helmets has become common.
- **Lower temperature.** Helmets rated for lower temperatures (LT) pass all applicable tests at temperatures of −30°C (−22°F).
- **High visibility.** Helmets rated for high visibility (HV) pass all applicable tests for HV colors including tests for chromaticity and luminescence.[16]

Hard hats can help reduce the risk associated with falling or projected objects, but only if they are worn. The use of hard hats in industrial settings in which falling objects are likely has been mandated by federal law since 1971.[17] In addition to making the use of hard hats mandatory when appropriate and supervising to ensure compliance, Feuerstein recommends the use of incentives.[18]

Resources expended promoting the use of hard hats are resources wisely invested. "Work accidents resulting in head injuries cost employers and workers an estimated $2.5 billion per year in workers' compensation insurance, medical expenses and accident investigation as well as associated costs due to lost time on the job and substitute workers. That is an average cost of $22,500 for each worker who received a head injury."[19]

Types and Classes of Hard Hats

To comply with the requirements of ANSI Z89.1, which OSHA has adopted, hard hats must meet or exceed the following requirements:

- *Type I hard hats.* Reduce the force of a blow to the top of the head only. This type of hard hat is designed to protect workers from falling objects.
- *Type II hard hats.* Reduce the force of a blow to the top of the head and the force of a lateral blow or off-center blow. This type of hard hat is designed to protect workers from blows to the top and side of the head.
- *Class E (electrical) hard hats.* Reduce exposure to high-voltage conductors and provide dielectric protection up to 20,000 volts (phase to ground). Note that this protection pertains to the head only.
- *Class G (general) hard hats.* Reduce exposure to low-voltage conductors and offer dielectric protection up to 2,200 volts (phase to ground). Note that this protection pertains to the head only.
- *Class C (conductive) hard hats.* Not designed to protect workers from electricity. They protect from impact and can in some instances provide better breathability than other hard hats.

Eye and Face Protection

Eye and face protection are critical in the modern workplace. Eye injuries are a common and costly phenomenon.

Every day, an estimated 1,000 eye injuries occur in American workplaces, according to the Bureau of Labor Statistics (BLS). What is the result? More than $300 million a year in lost production time, medical expenses, and workers' compensation. Why is this continuing to happen?

First, too many people are not wearing eye protection. BLS found three out of every five workers with eye injuries were not wearing eye protection. Second, they were not wearing the right kind of eye protection. Flying particles, according to the bureau, cause most eye injuries. Almost 70 percent of injuries resulted from flying or falling objects or sparks striking the eye. About 20 percent of the injuries were caused by contact with chemicals. Nearly half of the accidents occurred in manufacturing; just over 20 percent were in construction.[20]

Eye and face protection typically consist of safety glasses, safety goggles, or face shields. The ANSI standard for face and eye protective devices is Z87.1. OSHA has

FIGURE 11–4   ANSI Standard Z87.1: Industrial Eyewear Impact Standard.

High-Mass Impact Test—Purpose

This test is intended to ensure the level of mechanical integrity of a protective device and a level of protection from relatively heavy, pointed objects traveling at low speeds. Frames shall be capable of resisting impact from a 500-gram (17.6-ounce) missile with a 30-degree conical heat-treated tip and a 1-mm (.039-inch) radius dropped from a height of 130 cm (51.2 inches). No parts or fragments shall be ejected from the protector that could contact an eye.

High-Velocity Impact Test—Purpose

This test is intended to ensure a level of protection from high-velocity, low-mass projectiles. Frames shall be capable of resisting impact from 6.35-mm (¼-inch) steel balls weighing 1.06 grams (.04 ounce) at 150 feet per second (fps) from 0 degrees to 90 degrees for frames; 250 fps for goggles, 300 fps for face shields.

Impact Test—Drop Ball

A 25.4-mm (1-inch) steel ball, weighing 68 grams (24 ounces), free fall from 127 cm (50 inches).

Lens Thickness

Thickness is 3.0 mm (.118 inch) except lenses that withstand high velocity impact, then 2.0-mm (.079-inch) thickness is acceptable.

Impact Test—Penetration

Lens shall be capable of resisting penetration from a Singer needle on a holder weighing 1.56 ounces dropped freely from 50 inches.

also adopted this standard. It requires that nonprescription eye and face protective devices pass two impact tests: a high-mass, low-speed test and a low-mass, high-speed test. Figure 11–4 summarizes the purpose of the tests and their individual requirements concerning impact and penetration. Figures 11–5, 11–6, and 11–7, show examples of the types of devices available for eye and face protection.

The high-mass impact test determines the level of protection provided by face and eye protective devices from relatively heavy, pointed objects that are moving at low speeds. The high-velocity impact test determines the level of protection provided from low-mass objects moving at high velocity.

Assessing the Workplace for Eye Hazards

The type of eye protection needed in a given setting depends on the type of work done in that setting and the corresponding hazards. Before establishing a vision protection program, it is necessary to assess the workplace. OSHA recommends using the following questions in making a workplace assessment:

- Do employees perform tasks that may produce airborne dust or flying particles?
- Do employees work near others who perform tasks that may produce airborne dust or flying particles?
- Do employees handle hazardous liquid chemicals or blood?
- Do employees work near others who handle hazardous liquid chemicals or blood?

FIGURE 11–5 Wrap-around safety glasses.

Source: Dorling Kindersley

- Do employees work in conditions in which their lenses may become fogged?
- Do employees work in situations that may expose their eyes to chemical or physical irritants?
- Do employees work in situations that may expose their eyes to intense light or lasers?[21]

Based on the answers to these questions, a vision protection program can be developed to protect employees. That program should meet certain requirements as recommended by OSHA. These requirements are summarized in the next section.

FIGURE 11–6 Hood with facemask.

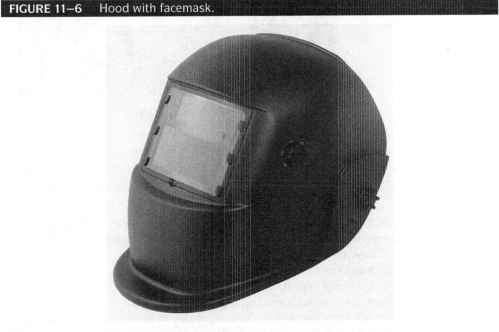

Source: Nikitin Victor/Shutterstock

FIGURE 11–7   Faceshield.

Source: Karin Hildebrand Lau/Shutterstock

Requirements When Choosing Vision-Protection Devices

There are many different types of eye protection devices available that vary in terms of function, style, fit, lens, and other options. OSHA recommends applying the following criteria when selecting vision protection devices:

- Select only those that meet the standards set forth in ANSI Z87.1.
- Select devices that protect against the specific hazard(s) identified in that assessment.
- Select devices that are as comfortable as possible to wear.
- Select devices that do not restrict vision in any way.
- Select devices that have fogging prevention capabilities built in.
- Select devices that are durable, easy to clean, and easy to disinfect.
- Select devices that do not interfere with the functioning of other PPE.[22]

Training

Once the workplace has been assessed and eye protection devices have been selected, it is important to provide employees with training in the proper use of the devices. This accomplishes the following: First, it ensures that the eye protection devices are used properly. Second, it shows employees that they have a critical role to play in the protection of their eyes. OSHA recommends training that covers the following topics:

- Why it is important to use the eye protection devices
- How the devices protect the eyes
- Limitations of the devices
- When the devices should be used
- How the devices are properly worn
- How straps are adjusted for both effectiveness and comfort
- How the employee can identify signs of wear that may lessen the effectiveness of the devices
- How the devices are cleaned and disinfected and how often[23]

First Aid for Eye Injuries

Even with proper eye protection, there is still the risk that an employee may sustain an injury. Even the best vision protection program is not perfect. When this happens, the following guidelines for first aid apply:

- Be gentle with the employee. Don't add to the injury with rough treatment.
- Do not attempt to remove objects embedded in the eyeball.
- Rinse the eyes with a copious amount of water for 15 to 30 minutes to remove the chemicals. Call for professional help. Cover both eyes after the rinsing has been completed.
- Never press on an injured eye or put any pressure on it (as when covering the eyes).
- Do not allow the employee to rub his or her eyes.[24]

Contact Lenses in a Chemical Environment

For years, it was commonly thought among safety and health professionals that workers should not wear contact lenses in a chemical environment. In fact, until 2003 the National Institute for Occupational Safety and Health (NIOSH) recommended that workers in chemical environments not wear contact lenses. However, over time much has been learned about this issue. Contact lenses may, in fact, be worn in chemical environments, and NIOSH has published an "intelligence bulletin" (number 59) explaining how to safely wear contact lenses in chemical environments.[25]

There are still environments in which contacts lens should not be worn. These are environments in which certain chemicals such as the following are present:

- 1,2-dibromo-3-chloropropane (DBCP)
- 4,4′-methylenedianiline (MDA)
- Ethyl alcohol
- Ethylene oxide
- Isopropyl alcohol
- Methylene chloride

This list is neither exhaustive nor comprehensive. Rather, it contains a partial list of the types of chemicals that can make a work environment especially hazardous for contact lens wearers. Before allowing employees who wear contact lenses to work in a chemical environment, it is best to conduct a comprehensive hazard assessment in which the types of chemicals that will be present are identified. NIOSH's Intelligence Bulletin 59: "Contact Lens Use in a Chemical Environment" is a good source of information when conducting hazard assessments of chemical environments. Once the hazard assessment has been completed, provided none of the "contact-lens-prohibited" chemicals are present in the work environment, workers who wear contact lenses should use the same eye protection recommended for other workers.

Foot Protection

The OSHA regulations for foot protection are found in 29 CFR 1910.132 and 126. Foot and toe injuries account for almost 20 percent of all disabling workplace injuries in the United States.[26] There are more than 180,000 foot and toe injuries in the workplace each year.[27] The major kinds of injuries to the foot and toes are from the following:

- Falls or impact from sharp or heavy objects (this type accounts for 60 percent of all injuries)
- Compression when rolled over by or pressed between heavy objects
- Punctures through the sole of the foot
- Conductivity of electricity or heat
- Electrocution from contact with an energized, conducting material
- Slips on unstable walking surfaces
- Hot liquid or metal splashed into shoes or boots
- Temperature extremes[28]

The key to protecting workers' feet and toes is to match the protective measure with the hazard. This involves the following steps: (1) identify the various types of hazards present in the workplace, (2) identify the types of footwear available to counter the hazards, and (3) require that proper footwear be worn. Shoes selected should meet all applicable ANSI standards and have a corresponding ANSI rating. For example, "a typical ANSI rating is Z41PT83M1–75C–25. This rating means that the footwear meets the 1983 ANSI standard and the **steel toe** cap will withstand 75 foot pounds of impact and 2,500 pounds of compression."[29]

Modern safety boots are available that provide comprehensive foot and toe protection. The best safety boots provide all of the following types of protection:

- *Steel* toe for impact protection
- *Rubber* or *vinyl* for chemical protection
- *Puncture-resistant* soles for protection against sharp objects
- *Slip-resistant* soles for protection against slippery surfaces
- *Electricity-resistant* material for protection from electric shock

Employers are not required to provide footwear for employees, but they are required (29 CFR 1910.132 and 136) to provide training on foot protection. The training must cover the following topics as a minimum:

- Conditions when protective footwear should be worn
- Type of footwear needed in a given situation
- Limitations of protective footwear
- Proper use of protective footwear

OSHA Regulations Relating to Footwear

Foot protection is a high priority with OSHA. This can be seen in the number of regulations OSHA has developed and now mandates relating to footwear. The most prominent of these and a summary of the respective requirements follow:

1. **29 CFR 1910.132(d):** Requires hazard assessment in the workplace.
2. **29 CFR 1910.136:** Lists the general requirements for foot protection in the workplace. Requires that employers ensure that employees use the appropriate foot protection when working in areas where there is a danger of foot injuries due to falling or rolling objects, or objects piercing the sole, and where such employees' feet are exposed to electrical hazards.
3. **29 CFR 1910.132(f)(a,iv,v):** Specifies the training mandated for employees and fitting of footwear.[30]

In addition to these regulations, the following additional regulations also deal with specific aspects of footwear and foot protection: 29 CFR 1910.94 (ventilation); 29 CFR 1910.156 (foot, hand, eye, face, head, and body protection for employees who serve in fire brigades); 29 CFR 1910.269 (foot protection when working in electric power generation, transmission, and distribution); and 29 CFR 1910.1029 (foot protection when working with coke ovens).

STANDING HAZARDS

Many jobs require that workers stand or walk for prolonged periods. Prolonged walking and/or standing can cause lower back pain, sore feet, varicose veins, and a variety of other related problems. The following hazard mitigation strategies can help minimize these problems for workers who must stand and walk for prolonged periods.

Antifatigue Mats

Antifatigue mats provide cushioning between the feet and hardworking surfaces such as concrete floors (see Figure 11–8). This cushioning effect can reduce muscle fatigue and lower back pain. However, too much cushioning can be just as bad as too little. Consequently, it is important to test mats on a trial basis before buying a large quantity. Mats that become slippery when wet should be avoided. In areas where chemicals are used, be sure to select mats that will hold up to the degrading effects of chemicals.

FIGURE 11–8   Antifatigue mats.

Source: Hosha/Shutterstock

Shoe Inserts

When antifatigue mats are not feasible because employees must move from area to area and, correspondingly, from surface to surface, shoe inserts may be the answer. Such inserts are worn inside the shoe and provide the same type of cushioning the mats provide. Shoe inserts can help reduce lower back, foot, and leg pain. It is important to ensure proper fit. If inserts make an employee's shoes too tight, they will do more harm than good. In such cases, employees may need to wear a slightly larger shoe size.

Foot Rails

Foot rails added to work stations can help relieve the hazards of prolonged standing. Foot rails allow employees to elevate one foot at a time four or five inches. The elevated foot rounds out the lower back, thereby relieving some of the pressure on the spinal column. Placement of a rail is important. It should not be placed in a position that inhibits movement or becomes a tripping hazard.

Workplace Design

A well-designed workstation can help relieve the hazards of prolonged standing. The key is to design workstations so that employees can move about while they work and can adjust the height of the workstation to match their physical needs.

Sit/Stand Chairs

Sit/stand chairs are higher-than-normal chairs that allow employees who typically stand while working to take quick mini-breaks and return to work without the hazards associated with getting out of lower chairs. They have the advantage of giving the employee's feet, legs, and back an occasional rest without introducing the hazards associated with lower chairs.

Proper Footwear

Proper footwear is critical for employees who stand for prolonged periods. Well-fitting, comfortable shoes that grip the work surface and allow free movement of the toes are best.

HAND PROTECTION

In the United States there are more than 500,000 hand injuries every year. Hand injuries are both serious and costly for employers and for employees. Section 138 of OSHA standard 29 CFR 1910.132 covers PPE for hands. This standard requires employers to base the selection of hand protection (gloves) on a comprehensive assessment of the tasks performed for a given job, hazards present, and the duration of exposure to the hazards. The assessment must be documented in writing.[31]

With the assessment completed, employers are required to review specification information from manufacturers of safety gloves and select the gloves that are best suited for the individual situation. Selecting just the right gloves for the job has historically been one of employers' greatest difficulties in complying with the standard. It is not a simple task. For example, take the issue of fit. A poorly fitted set of gloves cannot offer the degree of protection that a responsible employer or employee wants. Yet, because manufacturers have not developed a consistent set of metrics for sizing gloves, the only way to determine whether a pair fits properly is for the employee to try them on.

Fit is just one of the problems faced when selecting gloves. Other critical features include the protection capability, comfort, and tactile sensitivity of the gloves. Often, greater comfort and tactile sensitivity can mean less protection. Correspondingly, greater protection can mean less comfort and tactile sensitivity.

In an attempt to simplify the process of selecting the right gloves, ANSI and the Industrial Safety Equipment Association (ISEA) developed a joint hand-protection standard, ANSI/ISEA 105–1999. This standard simplifies glove selection by (1) defining characteristics of protection in a variety of critical areas, including cuts, puncture resistance, abrasion, protection from cold and heat, chemical resistance (including both permeation and degradation), viral penetration, dexterity, liquid-tightness, and flame/heat resistance, and (2) standardizing the tests used to measure all of these various characteristics.

Common Glove Materials

Depending on the individual hazards available in a given situation, the right gloves for the application may be made of a variety of different materials (see Figures 11–9, 11–10, and 11–11). The most widely used materials in manufacturing gloves are as follows:

- **Leather.** Offers comfort, excellent abrasion resistance, and minimum cut resistance.
- **Cotton.** Offers comfort, minimal abrasion resistance, and minimum cut resistance.

FIGURE 11–9   Work gloves.

Source: Andy Crawford/Dorling Kindersley

- **Aramids.** Offer comfort, good abrasion resistance, excellent cut resistance, and excellent heat resistance.
- **Polyethylene.** Offers comfort, excellent abrasion resistance, and minimal cut resistance. Gloves made of this material should not be subjected to high temperatures.
- **Stainless steel cord (wrapped in synthetic fiber).** Offers comfort, good abrasion resistance, and optimal cut resistance.
- **Chain link or metal mesh.** Offers very little comfort, but maximum abrasion and cut resistance.
- **Butyl rubber.** Offers little comfort, but has excellent resistance to heat, ozone, tearing, and certain chemicals, including alcohols, aldehydes, ketones, esters, nitriles, gases, amides, acids, and nitro compounds.

FIGURE 11–10 Work gloves as PPE.

Source: James Stevenson/Dorling Kindersley

FIGURE 11–11 Rubber work gloves.

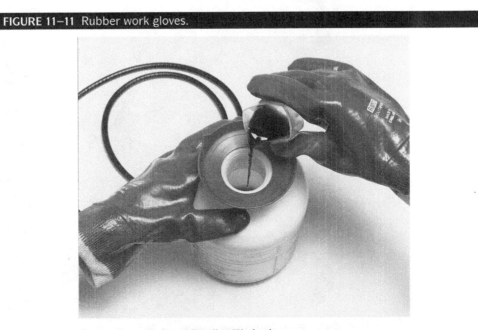

Source: Peter Anderson/Dorling Kindersley

- **Nitrile-based material.** Offers greater comfort and protection. Consequently, there is increased use of this type of material for the substrate coating of glasses.
- **Viton rubber.** Offers little comfort, but performs well with chemicals that butyl rubber cannot protect against, including aliphatics, halogenated, and aromatics. Like butyl gloves, viton gloves also perform well in handling alcohols, gases, and acids.[32]

PERSONAL PROTECTIVE EQUIPMENT

PPE is a critical component in the safety program of most organizations. Head, hand, back, eye, face, foot, skin, and breathing protection all involve the use of PPE. Unfortunately, it can be difficult to convince employees to wear PPE properly or, sometimes, to wear it at all. Employees often balk at the perceived "inconvenience" of PPE. They don't like the way it looks or how it feels. They think it is cumbersome in which to work in or time consuming to put on and take off. Sometimes, they just forget to use it (see Figure 11–12).

FIGURE 11–12   PPE Poster.

Source: Barry Barnes/Shutterstock

Making employees comfortable with PPE is a serious and sometimes difficult challenge for safety and health professionals. The following strategies can be used to meet this challenge:

- **Make maximum use of engineering and administrative controls.** PPE should be the last line of defense in protecting employees from hazardous conditions. Before adopting PPE, organizations should first use every engineering and administrative control available to minimize potential hazards. If employees see that the organization is doing its part by applying these controls, they will be less reluctant to do their part in properly using PPE.
- **Ensure the optimum choice of PPE by using risk assessment.** Employees know when there is a mismatch between the hazards they face and the PPE they are provided. OSHA requires that PPE be selected on the basis of a comprehensive risk assessment. This approach helps employers make the optimum choice when selecting PPE. Employees who know that the PPE provides adequate protection from hazards will be more likely to use it.
- **Involve employees in all aspects of the PPE program.** Always involve employees when making decisions that affect them. This is good policy for two reasons: (1) employees may be able to provide input that will improve the quality of the decisions being made because they understand the work tasks being discussed and (2) employees who are involved in the decision making are more likely to buy into and support that decision.
- **Provide comprehensive education and training programs.** Employees need to understand why PPE is important and how to properly use it. Employers should never assume that employees understand the "why" or "how" of PPE. Training programs should begin with the why aspects, cover them thoroughly, and give employees ample opportunities to ask questions and voice concerns. Once employees understand why PPE is important, they should be given comprehensive training on how to use it properly. No employee should be expected to use PPE without first understanding why and how they should use it.
- **Reinforce the proper use of PPE and challenge its improper use.** Employers should never fall into the trap of taking PPE use for granted. Proper behavior relating to PPE should be reinforced by supervisors and managers. Correspondingly, improper use should be challenged. Employers should use PPE properly themselves and reward employees who follow suit. Rewards need not be formal. Publicly complimenting an employee can be reward enough. Correspondingly, when an employee is seen failing to use the required PPE or using it improperly, that employee should be corrected. However, whereas compliments are given publicly, correction should be done in private.
- **Be clear on who pays for PPE.** The employer is required by OSHA to pay for basic minimal PPE for the hazards present. Employers do not have to pay for PPE that is lost or intentionally damaged. Be clear on who pays for what and when. It is wise for employers to pay for PPE in most situations, even when not required.
- **Be sensitive to fit, comfort, and style issues.** Ill-fitting PPE poses a double problem: (1) it may not provide the necessary protection because of the improper fit and (2) if it does not fit well, employees may be reluctant to wear it because it

FIGURE 11–13   Work boots and other PPE.

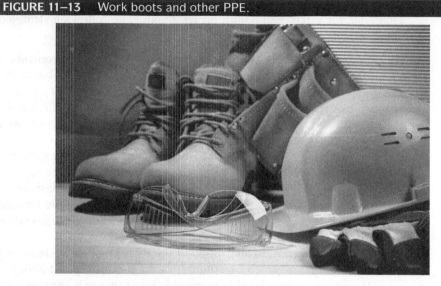

Source: Kuzmafoto/Shutterstock

is uncomfortable. Style can also be a problem in that employees are often self-conscious about their appearance. All these factors should be considered when choosing PPE (see Figure 11–13).

- **Work to make PPE a normal part of the uniform.** By applying these strategies, employers can make PPE a normal part of the uniform. When this happens, using PPE will become the standard operating procedure, and its proper use will cease to be an issue.[33]

FORKLIFT SAFETY (POWERED INDUSTRIAL TRUCKS)

Powered industrial trucks or forklift safety is included here because forklift-related injuries often result from impact or acceleration hazards. OSHA issues its standards for forklift safety in 29 CFR 1910.178 under the heading "Powered Industrial Truck." These regulations apply to forklifts, platform lift trucks, and motorized hand trucks. The latest edition of OSHA's forklift standard is based primarily on a standard produced by the American Society of Mechanical Engineers titled, "Safety for Low Lift and High Lift Trucks" (ASME/ANSI B56.1).

Forklifts are different from cars and trucks in several ways. Employees who drive forklifts should understand how they are different. The primary differences are as follows:

- Forklifts are typically steered by the rear wheels.
- An empty forklift can be more difficult to steer than one with a load.
- Forklifts are frequently driven in reverse.
- Forklifts have three-point suspension so that the center of gravity can move from the rear of the vehicle closer to the front when it is loaded.

Because of these differences, it is important to ensure that only properly trained employees drive forklifts and that these employees follow some basic rules of accident prevention. The rules fall into four categories: (1) general, (2) lifting, (3) traveling, and (4) placing.

General Rules

The rules in this section are general and apply to all phases of forklift operation:

- Keep arms, hands, and legs inside the vehicle at all times.
- Face in the direction of travel at all times.
- If the load blocks your view, drive backward.
- Allow plenty of room for braking—at least three vehicle lengths.
- Make sure there is sufficient overhead clearance before moving a load.

RULES FOR OPERATION OF THE FORKLIFT

- The forklift must be taken out of service if found to be in need of repair, defective, or in any way unsafe (until it has been restored to safe operating condition).
- Fuel tanks are to be filled only when the truck's engine is off and spillage of fuel is to be avoided.
- Spilled fuel should be washed away or completely evaporated and the fuel tank cap in place before the truck's engine is started.
- The forklift may not be operated if there is a leak in the fuel system (until the leak is repaired).
- Open flames may not be used for checking fuel levels or electrolyte levels in batteries.

RULES FOR MAINTENANCE OF THE FORKLIFT

- All necessary repairs to the forklift must be made by authorized personnel.
- No repairs may be made in Class I, II, and III locations.
- Repairs to fuel and ignition systems that involve fire hazards must be made in locations designated specifically for these kinds of repairs.
- The battery must be disconnected before electrical repairs are made.
- Replacement parts must be the equivalent of those they replace from the perspective of safety.
- The forklift must not be altered in any way to cause the relative positions of various parts to be different from they were when originally received from the manufacturer, not shall it be altered by parts not provided by the manufacturer or parts removed. Additional counterweighting of forklifts must be approved by the manufacturer.
- The forklift must be examined daily and shall not be used unless it is determined to be safe for operation.
- Water mufflers must be kept at no less than 75 percent of capacity. Further, muffler screens must not be allowed to become clogged.
- If the temperature of any part of a forklift exceeds normal operating temperature, the forklift must be removed from service until the problem has been eliminated.

- The forklift must be kept in a clean condition. Only noncombustible cleaning agent should be used.
- If the forklift was originally designed for gasoline, it may be converted to lique-fied petroleum gas only if the conversion results in a forklift that embodies the features of LP or LPS forklifts.

RULES FOR PICKING UP A LOAD

- Make sure the load is within the capacity of the forklift.
- Make sure forks are positioned properly.
- Make sure the load is properly balanced.
- Make sure the load is secure.
- Raise the load to the proper height.
- Run the forks all the way into the pallet, and tilt the mast back to stabilize the load before moving.
- Back out and stop completely before lowering the load.

RULES FOR TRAVELING WITH A LOAD

- Always give pedestrians the right-of-way.
- Never allow passengers on the forklift.
- Keep the forks low while moving.
- Keep the load tilted back slightly while moving.
- Drive slowly; a forklift is not a car.
- Slow down at all intersections; stop and sound the horn at blind intersections.
- Drive up and back down ramps and inclines.
- Never lift or lower the load when traveling.
- Keep to the right just as you do when driving a car.
- Watch for oil, grease, and wet spots, which could inhibit traction.
- Cross railroad tracks at a skewed angle, never at a right angle.
- Watch for edges on loading docks and other changes in elevation.
- Maintain at least four seconds of spacing between your forklift and the one in front of you.

RULES FOR PLACING A LOAD

- Stop the forklift completely before raising or lowering the load.
- Move slowly and cautiously with the load raised.
- Never walk or stand under a raised load or allow anyone else to do so.
- Be certain the forks have cleared the pallet before turning and before changing height.
- Stack the load square and straight.
- Check behind and on both sides before backing up.

OSHA's Training Requirements

OSHA estimates that 155,000 injuries and 100 fatalities occur every year as a result of accidents involving powered industrial trucks. Because of this, OSHA revised its training requirements for operators of powered industrial trucks (29 CFR 1910.178).[34] OSHA estimates that employers can save as much as $135 million annually by following these training requirements. Of this sum, $83 million could be saved in direct costs such as medical treatment, and $52 million could be saved in indirect costs.

Initial Training Program Content

Before they are allowed to operate a powered industrial truck (forklift), employees are required by OSHA's 29 CFR 1910.178 to complete initial training in the following topics:

1. Operating instructions, warnings, and precautions for the types of trucks the operator will be authorized to operate.
2. Differences between the truck and the automobile.
3. Truck controls and instrumentation: where they are located, what they do, and how they work.
4. Engine or motor operation.
5. Steering and maneuvering.
6. Visibility (including restrictions due to loading).
7. Fork and attachment adaptation, operation, and use limitations.
8. Vehicle capacity.
9. Vehicle stability.
10. Any vehicle inspection and maintenance the operator will be required to perform.
11. Refueling and/or charging and recharging of batteries.
12. Operating limitations.
13. Any other operating instructions, warnings, or precautions listed in the operator's manual for the types of vehicles the employee is being trained to operate.
14. Workplace-related topics including the following: surface conditions where the vehicle will be operated; composition of loads to be carried and load stability; load manipulation, stacking, and unstacking; pedestrian traffic in areas where the vehicle will be operated; narrow aisles and other restricted places where the vehicle will be operated; hazardous (classified) locations where the vehicle will be operated; ramps and other sloped surfaces that affect the vehicle's stability; closed environments and other areas where insufficient ventilation or poor vehicle maintenance could cause a buildup of carbon monoxide or diesel exhaust; other unique or potentially hazardous environmental conditions in the workplace that could affect safe operation.

Refresher Training Program Content

Refresher training, including an evaluation of the effectiveness of that training, must be conducted to ensure that operators of powered industrial trucks have the up-to-date knowledge and skills to operate them safely. Refresher training is required whenever any one of the following conditions exists: the operator has been observed operating a vehicle in an unsafe manner, the operator has been involved in an accident or a near-hit accident, the operator has received an evaluation that indicates he or she is not operating the powered truck safely, the operator is assigned to drive a different type of truck, or any condition in the workplace changes in such a manner that it could affect the safe operation of the truck. In addition to these requirements, employers are required to conduct an evaluation of the performance of all industrial powered truck operators at least once every three years.

Key Terms and Concepts

- Acceleration
- Anchorage
- Coefficient of friction
- Design flaws
- Eye and face protection
- Foot protection
- Foreign object
- Four-to-one ratio
- Good housekeeping
- Head protection
- Impact

- Ladder safety
- Lanyards
- Lifelines
- Nonskid footwear
- Personal fall arrest systems
- Reconditioning
- Retrofit
- Review and acceptance
- Slip and fall

- Slip-resistant
- Slippery surface
- Steel toe
- Step and fall
- Stump and fall
- Surface traction
- Trip and fall
- Vertical incidence tribometers

Review Questions

1. List the primary causes of falls.
2. Explain briefly the most common kinds of falls.
3. Explain how surface traction is measured.
4. List and briefly explain five strategies for preventing slips.
5. Describe the various components of a slip and fall prevention program.
6. Explain the "trigger height" controversy over OSHA's Fall Protection Standard for construction.
7. What are the requirements for personal fall arrest systems in OSHA regulation 1926.451(d)(16)?
8. What is a lanyard?
9. Explain OSHA's recommendations for effective fall protection, and what action to take when a worker is dangling from his fall arrest gear.
10. Explain how to assess the workplace for eye hazards.
11. What topics should be covered in eye protection training?
12. What should an employee look for when inspecting a ladder?
13. Briefly summarize the evolution of hard hats in this country.
14. List six major kinds of injuries to the foot and toes that occur in the workplace each year, and explain the most widely used footwear terms.
15. Describe the six-step back safety/lifting program, and explain what the EGMMHs are.
16. List four ways to minimize standing hazards.
17. What are the critical factors to consider when selecting gloves?
18. Explain how to get employees more comfortable with using PPE.
19. Explain the training requirement contained in OSHA's standard 29 CFR 1910.178 (powered industrial trucks).

Endnotes

1. J. Rhodes, "A Risk Manager's Roadmap," *Occupational Health & Safety Online,* May 11, 2006, 1–4. Retrieved from 1105media.com on July 2, 2017.
2. Ibid., 2.
3. Ibid., 2–3.
4. Ibid., 3–4.
5. T. Christensen and S. Rupard, "After the Fall: Why Fall Protection Isn't Always Enough," *Occupational Health & Safety* 75, no 3: 48–49.

6. OSHA Regulation Title 24, *Code of Federal Regulations,* Part 1926.

7. James L. Nash, "OSHA's Fight against Fatal Falls," *Occupational Hazards* 65, no. 4: 37.

8. National Safety Council, *Ladder Safety Tips*. Retrieved from nsc.org/pubs/sh/clip1099.htm on July 2, 2017.

9. Ibid.

10. Ibid.

11. Christensen and Rupard, "After the Fall," 49.

12. T. Cox, "Is Your Fall Protection Equipment a Silent Hazard?" *Occupational Health & Safety* 75, no. 5: 69–70.

13. Retrieved from public.ansi.org/ansionline/portal/search on July 2, 2017.

14. Ibid.

15. Ibid.

16. Ibid.

17. P. Feuerstein, "Head Protection Looks Up," *Safety & Health* 144, no. 3: 38.

18. Ibid., 39.

19. M.S. Bross, "Advances Lead to Tougher, More Durable Hard Hat," *Occupational Health & Safety* 74, no. 10: 4.

20. A. Chambers, "Safety Goggles at a Glance," *Occupational Health & Safety* 71, no. 10: 58.

21. J. Hensel, "Setting Up a Vision Program," *Occupational Health & Safety* 68, no. 10: 36.

22. Ibid.

23. Ibid.

24. Ibid.

25. B. Weissman, "Contact Lenses in a Chemical Environment," *Occupational Health & Safety* 74, no. 10: 56–58.

26. J. Goodwin, "A Cure for Common Foot Hazards," *Occupational Health & Safety* 74, no. 7: 84.

27. Ibid.

28. Ibid., 84–86.

29. Ibid., 86.

30. Retrieved from osha.gov/pls/oshaweb on July 2, 2017.

31. Title 29, *Code of Federal Regulations,* Part 1910.132, Section 38.

32. T. Busshart, "A Cut Above," *Occupational Safety & Health* 67, no. 1: 36.

33. T. Andrews, "Getting Employees Comfortable with PPE," *Occupational Hazards* 62, no. 1: 35–38.

34. *Title 29, Code of Federal Regulations, Part 1910.178—Powered Industrial Trucks.*

CHAPTER
12

HAZARDS OF TEMPERATURE EXTREMES

Learning Objectives

- Summarize the most important concepts relating to thermal comfort
- List and define the most important heat stress–related concepts
- Explain the concept of cold stress and how it is prevented
- Compare and contrast first-, second-, and third-degree burns
- Explain the severity and effects of chemical burns
- Summarize OSHA's guidelines relating to hazards of temperature extremes

Part of providing a safe and healthy workplace is appropriately controlling the temperature, humidity, and air distribution in work areas. A work environment in which the temperature is not properly controlled can be uncomfortable. Extremes of either heat or cold can be more than uncomfortable—they can be dangerous. **Heat stress**, **cold stress**, and burns are major concerns of modern safety and health professionals. This chapter provides the information that professionals need to know to overcome the hazards associated with extreme temperatures.

THERMAL COMFORT

Thermal comfort in the workplace is a function of a number of different factors.[1] Temperature, humidity, air distribution, personal preference, and acclimatization are all determinants of comfort in the workplace. However, determining optimum conditions is not a simple process.

To understand fully the hazards posed by temperature extremes, safety and health professionals must be familiar with several basic concepts related to thermal energy. The most important of these are summarized here:

- **Conduction** is the transfer of heat between two bodies that are touching each other or from one location to another within a body. For example, if an employee touches a workpiece that has just been welded and is still hot, heat will be conducted from the workpiece to the hand. Of course, the result of this heat transfer is a burn.
- **Convection** is the transfer of heat from one location to another by way of a moving medium (a gas or a liquid). **Convection** ovens use this principle to transfer heat from an electrode by way of gases in the air to whatever is being baked.
- **Metabolic heat** is produced within a body as a result of activity that burns energy. All humans produce **metabolic heat**. This is why a room that is comfortable when occupied by just a few people may become uncomfortable when it is crowded. Unless the thermostat is lowered to compensate, the metabolic heat of a crowd will cause the temperature of a room to rise to an uncomfortable level.
- **Environmental heat** is produced by external sources. Gas or electric heating systems produce **environmental heat** as do sources of electricity and a number of industrial processes.
- **Radiant heat** is the result of electromagnetic nonionizing energy that is transmitted through space without the movement of matter within that space.

HEAT STRESS AND STRAIN

The key question that must be answered by safety and health professionals concerning employees whose work may subject them to heat stress is as follows:

> What are the conditions to which most adequately hydrated, unmedicated, healthy employees may be exposed without experiencing **heat strain** or any other adverse effects?

The American Conference of Governmental Industrial Hygienists (ACGIH) publishes a comprehensive manual to help safety and health professionals answer this question for the specific situations and conditions that they face. This manual, titled *TLVs and BEIs: Threshold Limit Values for Chemical Substances and Physical Agents and Biological Exposure Indices*, provides reliable guidance and should be in every safety and health professional's library. In addition to using the information contained in this manual, all safety and health professionals should have a comprehensive **heat stress management** program in place and apply sound professional judgment.[2]

Heat Stress Defined

Heat stress is the net heat load to which a worker may be exposed from the combined contributions of metabolic effect of work, environmental factors (i.e., air temperature, humidity, air movement, and radiant heat exchange), and clothing requirements.

A mild or moderate heat stress may cause discomfort and may adversely affect performance and safety, but it is not harmful to health. As the heat stress approaches human tolerance limits, the risk of heat-related disorders increases.[3]

What follows are some widely used heat stress–related terms safety and health professionals should be familiar with:

Heat exhaustion. This is a physical state in which the worker's skin becomes clammy and moist and his or her body temperature is still normal or slightly higher than normal. **Heat exhaustion** results from loss through sweating off fluid and salt that are not properly replaced during exertion.

Heatstroke. This is a physical state in which the worker's skin becomes hot and dry, there is mental confusion, and there may be seizures or convulsions.

Heat cramps. Heat cramps are muscle cramps that can occur when workers exert themselves sufficiently to lose fluids and salt through sweating, but replace only the fluids by drinking large amounts of water containing no salt.

Heat syncope or fainting. Workers who exert themselves in a hot environment will sometimes faint. This is especially the case with workers who are not accustomed to working in such an environment.

Heat rash. Workers who exert themselves in a hot environment in which sweat does not evaporate can develop a prickly rash known as **heat rash**. Before air-conditioning was widely used in the hot and humid summer months in the southeastern United States, children often developed heat rash. Periodic rest breaks in a cool environment that allows sweat to evaporate will prevent heat rash.

Work tolerance time (WTT). WTT is a formula safety and health professionals can use to determine what steps can be taken to allow a worker to safely perform his or her required tasks in the environment in question for the time required. The formula takes into account such factors as temperature, humidity, level of energy that will be expended in performing the task, rest periods, and personal protective equipment (PPE).

Moisture vapor transfer rate (MVTR). The MVTR is a measure of the ability of the fabric used in making PPE to dissipate heat. The best MVTR occurs on an unclothed body. Even the lightest cotton fabric is less capable of dissipating heat when the unclothed body is used for baseline comparisons. The MVTR of impermeable fabric is zero (because the fabric does not allow the skin to "breathe"). The higher the MVTR, the better in hot environments.

Heat Strain Defined

Heat strain is the overall physiological response resulting from heat stress. The physiological adjustments are dedicated to dissipating excess heat from the body. **Acclimatization** is a gradual physiological adaptation that improves an individual's ability to tolerate heat stress.

Recognizing Heat Strain

Safety and health professionals, supervisors, and coworkers should know how to recognize heat strain. The following factors are signs of excessive heat strain. Exposure to heat stress should be stopped immediately for any employee experiencing any of these symptoms:

- A sustained rapid heart rate (180 beats per minutes minus the employee's age in years). For example, a 40-year-old employee has a sustained heart rate of 150 beats per minutes. This is a problem because the heart rate exceeds 140 (180 minus 40) beats per minute.
- Core body temperature is greater than 38.5°C.
- Recovery rate minute after a peak work effort is greater than 110 beats per minute.
- Sudden and severe fatigue, nausea, dizziness, or light-headedness.

These symptoms can be assessed on the spot in real time. In addition, other symptoms can be monitored only over time. Employees are at greater risk of excessive heat strain if they experience any of the following:

- Profuse sweating that continues for hours.
- Weight loss of more than 1.5 percent of body weight during one work shift.
- Urinary sodium excretion of less than 50 moles (24-hour period).

Clothing

Heat is best removed from the body when there is free movement of cool dry air over the skin's surface. This promotes the evaporation of sweat from the skin, which is the body's principal cooling mechanism. Clothing impedes this process, some types more than others. Encapsulating suits and clothing that is impermeable or highly resistant to the flow of air and water vapor multiply the potential for heat strain.

When assessing heat stress hazards in the workplace, safety and health professionals should consider the added effect of clothing. For example, the *wet bulb globe temperature* (WBGT) of working conditions should be increased by 3.5°C for employees wearing cloth overalls. This factor increases to 5°C with double cloth overalls.

Because the WBGT is influenced by air temperature, radiant heat, and humidity, it can be helpful in establishing a threshold for making judgments about working conditions. WBGT values can be calculated using the following formula:

EXPOSED TO DIRECT SUNLIGHT

$$WBGT = 0.7\,T_{nwb} \text{ to } 0.2\,T_g + 0.1\,T_{db}$$

T_{nwb} = Natural wet bulb temperature

T_g = Globe temperature

T_{db} = Dry bulb (air) temperature

NOT EXPOSED TO DIRECT SUNLIGHT

$$WBGT = 0.7\,T_{nwb} + 0.3\,T_g$$

FIGURE 12–1 Criteria for determining the allowable work periods for acclimatized employees.

	Screening Criteria (°C) Acclimatized Employees			
Work Demands	**Light Work**	**Moderate Work**	**Heavy Work**	**Very Heavy Work**
100% Work	29.5	27.5	26.0	—
75% Work 25% Rest	30.5	28.5	27.5	—
50% Work 50% Rest	31.5	29.5	28.5	27.5
25% Work 75% Rest	32.5	31.0	30.0	29.5

Source: www.osha.gov

These formulas for WBGT give safety and health professionals a beginning point for making judgments. The WBGT must be adjusted for clothing, work demands, and the employee's acclimatization state. The key is to ensure that employees never experience a core body temperature of 38°C or higher. Figures 12–1 and 12–2 provide screening criteria for heat stress exposure. Once the WBGT has been calculated and adjusted

FIGURE 12–2 Criteria for determining the allowable work periods for employees who are not acclimatized.

	Screening Criteria (°C) Not-Acclimatized Employees			
Work Demands	**Light Work**	**Moderate Work**	**Heavy Work**	**Very Heavy Work**
100% Work	27.5	25.0	22.5	—
75% Work 25% Rest	29.0	26.5	24.5	—
50% Work 50% Rest	30.0	28.0	26.5	25.0
25% Work 75% Rest	31.0	29.0	28.0	26.5

Source: www.osha.gov

for clothing, these figures may be used for factoring in work demands and acclimatization. To use Figures 12–1 and 12–2, apply the following example:

> Several acclimatized employees have a job to do that has a work demand of 75 percent work and 25 percent rest. The WBGT has been computed as 26. The work is considered "heavy." Because the employees will wear long-sleeved shirts and long trousers made of woven material an additional 3.5° must be added: $26 + 3.5 = 29.5°C$. Using the proper column and row of Figure 12–1, a WBGT of 26 can be determined. Because the calculated and adjusted WBGT is 29.5, there is a problem. In order to work in these conditions, the employees should adjust the work demand to 25 percent work and 75 percent rest.

CLO as a Measurement for Clothing and PPR Selection

In both hot and cold environments, it is important to know the insulation value of clothing, particularly PPE. The insulation of PPE can be expressed in clo units. A clo rating is similar to the R value rating of insulation (1 clo = 0.88 R). One clo is the amount of insulation from clothing or PPE a resting person can wear and be indefinitely comfortable at ambient conditions in a temperature of 21° C (70°F), less than 50 percent humidity, wind velocity of 0.9 kilometers per hour (one half mile per hour). The lowest **clo value** is zero. A zero clo would mean an individual was wearing no clothing, thus had no insulation from clothing. A high clo value is four which would require clothing or PPE with fur lining, a hood, and gloves.

The clo can be used as a measurement to guide safety professionals in selecting the right PPE for conditions in which employees will work, just as the R value is used in selecting insulation. The colder the environment the higher the clo and vice versa. For the purpose of illustration, the following cold-weather garments and their associated clo values are provided:

- Insulated coveralls worn over top and bottom thermal underwear (1.37 clo)
- Sweat pants and a sweat shirt (0.74 clo)
- Light polar PPE (3 clo)
- Heavy polar PPE (4 clo)

It may also be important to know the clo value of clothing that is worn is a hot environment. For the purpose of illustration, the following hot-weather garments and their associated clo values are provided:

- Short sleeve knit sport shirt (0.17 clo)
- Trousers (0.16 clo)
- Sleeveless scoop-neck blouse (0.12)
- Short sleeve shirt dress (0.29)

Heat Stress Management

Safety and health professionals should continually emphasize the importance of paying attention to recognizable symptoms of heat stress. In addition, they should ensure that a comprehensive heat stress management program is in place. Such a program should consist of both general and specific controls.

General Controls

The ACGIH recommends the following general controls:

- Provide accurate verbal and written instructions, training programs, and other information about heat stress and strain.

- Encourage drinking small volumes (approximately one cup) of cool water about every 20 minutes.
- Permit self-limitation of exposure. Encourage coworker observation to detect signs and symptoms of heat strain in others.
- Counsel and monitor those employees who take medications that may compromise normal cardiovascular, blood pressure, body temperature regulation, renal, or sweat gland functions, and those who abuse or who are recovering from the abuse of alcohol and other intoxicants.
- Encourage healthy lifestyles, ideal body weight, and electrolyte balance.
- Adjust expectations of those returning to work after absence from heat-stress situations and encourage consumption of salty foods (with approval of the employee's physician if on a salt-restricted diet).
- Consider replacement medical screening to identify those susceptible to systemic heat injury.[4]

Specific Controls

The ACGIH recommends the following specific controls:

- Establish engineering controls that reduce the metabolic rate, provide general air movement, reduce process heat and water-vapor release, and shield radiant heat sources, among others.
- Consider administrative controls that set acceptable exposure times, allow sufficient recovery, and limit physiological strain.
- Consider personal protection that has been demonstrated to be effective for the specific work practices and conditions at the location.[5]

COLD STRESS

Excessive exposure to cold can lead to **hypothermia**, which can be fatal. The goal of safety and health professionals in protecting employees from acute cold stress is to prevent the deep body temperature from falling below 36°C (96.8°F) and to prevent cold injuries to body extremities, especially the hands, feet, and head. A fatal exposure to cold typically results from failure to remove the employee from a cold air environment or immersion in cold water.[6]

Excessive exposure to cold stress, even when not fatal, can result in impaired judgment, reduced alertness, and poor decision making. Acute cold stress can cause reduced muscular function, decreased tactile sensitivity, reduced blood flow, and thickening of the synovial fluid. Chronic cold stress can lead to reduced functioning of the peripheral nervous system. All these factors increase the likelihood of accidents and injuries. Figure 12–3 shows the effects of allowing the core body temperature to fall to selected levels.

Whether employees are exposed to cold air or are immersed in cold water, wind can magnify the level of cold stress. This phenomenon is often referred to as *windchill*. Figure 12–4 shows the effect of wind on selected temperatures. To read this chart, locate the actual temperature (50, 40, 30...0). Then, find the applicable wind speed. Reading across that row to the right, find the equivalent temperature. For example, if employees are working in an environment that is 30°F and has a wind speed of 15 miles per hour (mph), the equivalent temperature is 9°F.

FIGURE 12–3 The body's response to reducing its core temperature.

Effects of Reducing the Core Body Temperature

Core Temperature		Body's Response
°C	°F	
37.6	99.6	Normal rectal temperature
36.0	96.8	Metabolic rate increases
35.0	95.0	Pronounced shivering
33.0	91.4	Severe hypothermia
30.0	86.0	Progressive loss of consciousness begins
24.0	75.2	Pulmonary edema
20.0	68.0	Cardiac standstill

Preventing Cold Stress

When the equivalent air temperature reaches –32°C (–25.6°F), continuous exposure of exposed skin should not be allowed. At equivalent air temperatures of 2°C (35.6°F), employees who are immersed in water or whose clothing gets wet should be treated for hypothermia immediately. Figure 12–5 shows selected threshold limit values (TLVs) for employees who work in environments with temperature below freezing.

To use Figure 12–5, locate the applicable temperature in the leftmost column. Reading to the right, locate the applicable wind speed. For example, employees working a four-hour shift in an environment with an air temperature of 232°C and a 5-mph

FIGURE 12–4 Effect of wind on the actual temperature.

Cooling Effect of Wind

Wind Speed (in mph)	Actual Temperature (°F) and Equivalent Temperatures (°F)					
	50	40	30	20	10	0
5	48	37	27	16	6	–5
15	36	22	9	–5	–18	–32
25	30	16	0	–15	–29	–44
35	27	11	–4	–20	–35	–51

Source: www.osha.gov

FIGURE 12–5 Partial table for determining TLVs in selected circumstances.

Air Temperature		No Wind		5-mph Wind		10-mph Wind	
		TLVs for a Four-Hour Shift					
C	**F**	**Max. Work Time**	**No. of Breaks**	**Max. Work Time**	**No. of Breaks**	**Max. Work Time**	**No. of Breaks**
−32 to −34	−25 to −29	75 Min.	2	55 Min.	3	40 Min.	4
−35 to −37	−30 to −34	55 Min.	3	40 Min.	4	30 Min.	5
−38 to −39	−35 to −39	40 Min.	4	30 Min.	5	Nonemergency Work Prohibited	

Note: This applies to workers properly dressed in dry clothing.

Source: www.osha.gov

wind should be exposed no longer than 55 minutes at a time and should warm up at least three times during the shift.

When work is to be performed in an environment with an air temperature of 4°C (39.2°F) or less, total body protective clothing is advisable. What follows are several strategies that can be used to decrease the hazards of cold stress:

- When working in a cold environment, provide the right PPE for the conditions that exist. PPE for cold environments might include face protection, head protection, ear protection, wind vests, wind-blocking jackets, over suits, warming vests, hand warmers, foot warmers, gloves, wind-and-water insulated clothing, cold-insulated footwear, ice cleats, slip resistance footwear, and glare protection. In selecting PPE for cold environments, it is advisable to ask employees to participate. This can provide two benefits: (1) employees are more likely to wear PPE they picked out (employees are less likely to wear PPE they do not like) and (2) the PPE is more likely to fit properly if employees participate in its selection (poor fit is a main reason employees do not wear PPE when they should).
- When working in a setting in which wind is a factor, reduce the effect of the wind by (1) erecting a windscreen or (2) wearing wind-breaking clothing.
- When working in a setting in which clothing may get wet, apply one or more of the following strategies: (1) with light work, wear an outer layer of impermeable clothing; (2) with heavier work, wear an outer layer that is water-repellent, but not impermeable (change outerwear as it becomes wet); (3) select outer garments that are ventilated to prevent internal wetting from sweat; (4) if clothing gets wet before going into the cold environment, change first; (5) change socks daily or more often to keep them dry; and (6) use vapor barrier boots to help keep the feet dry.
- When working in a cold setting, use auxiliary heat applied directly to the hands and feet.
- When working in a cold setting, use facial protection to prevent cold stress to the face and lungs.

- If adequate protective clothing that is appropriate for the conditions in question is not available, the work should be modified or suspended until conditions change or the clothing is available.

When work is to be performed in an environment with an air temperature of –12°C (10.4°F), the following additional strategies should be applied:

- Employees should be under continuous observation using either direct supervision or the buddy system.
- The work rate should be paced to avoid sweating. When heavy work is necessary, employees should take frequent warming breaks in heated shelters. If clothing becomes wet—internally or externally—it should be changed during a break.
- Do not allow new employees to work full time in these conditions until they have several days to become accustomed to the conditions and the necessary protective clothing.
- When determining the required work level for employees (light, heavy, or very heavy), consider the weight and bulkiness of protective clothing.
- Organize work in cold environments to minimize long periods of sitting or standing still. Never use unprotected metal chairs or seats.
- Before allowing employees to work in a cold environment, make sure they have been trained in safety and health procedures. Figure 12–6 is a checklist of topics that should be covered as a minimum during employee training.
- When work in a refrigerated room is required, the air velocity should be minimized and maintained at one meter per second (200 feet per minute) or less.
- When work outdoors in snow is required, employees should be provided special safety goggles that protect the eyes from ultraviolet light, glare, and blowing ice crystals.
- Employees who suffer from diseases or take medications that inhibit normal body functions or that reduce normal body tolerances should be prohibited from working in environments where temperatures are at 21°C (69.8°F) or less.
- Employees who are routinely exposed to the following conditions should be medically certified as being suitable for work in such conditions: (1) air temperatures of less than –24°C (–11.2°F) with wind speeds less than 5 mph; and (2) air temperatures of less than –18°C (–0.4°F) with wind speeds greater than 5 mph.

FIGURE 12–6 Checklist for training employees who will work in a cold environment.

Training Checklist for Employees Who Work in a Cold Environment

- ✓ Proper warming procedures
- ✓ Applicable first-aid procedures
- ✓ Protective clothing requirements and proper use of protective clothing
- ✓ Proper eating habits
- ✓ Proper drinking habits (e.g., avoid caffeine and other stimulants)
- ✓ Recognizing of cold stress and strain (e.g., impending frostbite, impending hypothermia, and so on)
- ✓ Safe work processes

BURNS AND THEIR EFFECTS

One of the most common hazards associated with heat in the workplace is the burn, Figure 12–7. Burns can be especially dangerous because they disrupt the normal functioning of the skin, which is the body's largest organ and the most important in terms of protecting other organs. It is necessary first to understand the composition of, and purpose served by, the skin to understand the hazards that burns can represent.

Human Skin

Human skin is the tough, continuous outer covering of the body. It consists of the following two main layers: (1) the outer layer, which is known as the **epidermis** and (2) the inner layer, which is known as the **dermis**, **cutis**, or **corium**. The dermis is connected to the underlying subcutaneous tissue.

The skin serves several important purposes including the following:

- Protection of body tissue
- Sensation
- Secretion
- Excretion
- Heat regulation
- Respiration

Protection from fluid loss, water penetration, ultraviolet radiation, and infestation by microorganisms is a major function of the skin. The sensory functions of touching, sensing cold, feeling pain, and sensing heat involve the skin.

The skin helps regulate body heat through the sweating process. It excretes sweat that takes with it **electrolytes** and certain toxins. This helps keep the body's fluid

FIGURE 12–7 Burns are a common injury in some jobs.

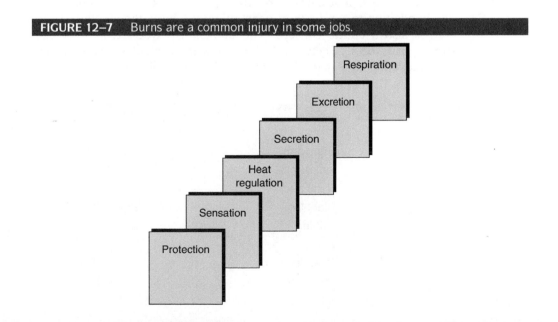

level in balance. By giving off minute amounts of carbon dioxide and absorbing small amounts of oxygen, the skin also aids slightly in respiration.

What makes burns particularly dangerous is that they can disrupt any or all of these functions depending on their severity. The deeper the penetration, the more severe the burn.

Severity of Burns

The severity of a burn depends on several factors. The most important of these is the depth to which the burn penetrates. Other determining factors include location of the burn, age of the victim, and amount of burned area.

The most widely used method of classifying burns is by degree (i.e., first-, second-, or third-degree burns). Modern safety and health professionals should be familiar with these classifications and what they mean.

First-degree burns are minor and result only in a mild inflammation of the skin, known as *erythema*. Sunburn is a common form of first-degree burn. It is easily recognizable as a redness of the skin that makes the skin sensitive and moderately painful to the touch.

Second-degree burns are easily recognizable from the blisters that form on the skin. If a second-degree burn is superficial, the skin will heal with little or no scarring. A deeper second-degree burn will form a thin layer of coagulated, dead cells that feels leathery to the touch. A temperature of approximately 98.9°C can cause a second-degree burn in as little as 15 seconds of contact.

Third-degree burns are very dangerous and can be fatal depending on the amount of body surface affected. A third-degree burn penetrates through both the epidermis and the dermis. A deep third-degree burn will penetrate body tissue. Third-degree burns can be caused by both moist and dry hazards. Moist hazards include steam and hot liquids; these cause burns that appear white. Dry hazards include fire and hot objects or surfaces; these cause burns that appear black and charred.

In addition to the depth of penetration of a burn, the amount of surface area covered is also a critical concern. This amount is expressed as a percentage of **body surface area (BSA)**. Figure 12–8 shows how the percentage of BSA can be estimated. Burns covering more than 75 percent of BSA are usually fatal.

FIGURE 12–8   Estimating percentage of body surface area (BSA) burned.

Right arm	9% of BSA
Left arm	9% of BSA
Head/neck	9% of BSA
Right leg	18% of BSA
Left leg	18% of BSA
Back	18% of BSA
Chest/stomach	18% of BSA
Perineum	1% of BSA

Using the first-, second-, and third-degree burn classifications in conjunction with BSA percentages, burns can be classified further as minor, moderate, or critical. According to Mertz, these classifications can be summarized as described in the following paragraphs.[7]

Minor Burns

All first-degree burns are considered **minor**. Second-degree burns covering less than 15 percent of the body are considered minor. Third-degree burns can be considered minor provided they cover only 2 percent or less of BSA.

Moderate Burns

Second-degree burns that penetrate the epidermis and cover 15 percent or more of BSA are considered **moderate**. Second-degree burns that penetrate the dermis and cover from 15 to 30 percent of BSA are considered moderate. Third-degree burns can be considered moderate provided they cover less than 10 percent of BSA and are not on the hands, face, or feet.

Critical Burns

Second-degree burns covering more than 30 percent of BSA or third-degree burns covering more than 10 percent of BSA are considered **critical**. Even small-area third-degree burns to the hands, face, or feet are considered critical because of the greater potential for infection to these areas by their nature. In addition, burns that are complicated by other injuries (fractures, soft tissue damage, and so on) are considered critical.

CHEMICAL BURNS

Chemical burns are covered in this chapter rather than Chapter 15 (Fire Hazards and Life Safety) because the sources of chemical burns as well as their prevention and their treatment are different from those associated with fire hazards and life safety. Chemicals are widely used in modern industry even by companies that do not produce them as part of their product base. Many of the chemicals produced, handled, stored, transported, or otherwise used in industry can cause burns similar to those caused by heat (i.e., first-, second-, and third-degree burns). The hazards of chemical burns are very similar to those of thermal burns.

Chemical burns, like thermal burns, destroy body tissue; the extent of destruction depends on the severity of the burn. However, chemical burns continue to destroy body tissue until the chemicals are washed away completely.

The severity of the burn produced by a given chemical depends on the following factors:

- Corrosive capability of the chemical
- Concentration of the chemical
- Temperature of the chemical or the solution in which it is dissolved
- Duration of contact with the chemical[8]

Effects of Chemical Burns

Different chemicals have different effects on the human body. The harmful effects of selected widely used chemicals are summarized in Figure 12–9.[9] These are only a few of the many chemicals widely used in industry today. All serve an important purpose; however, all carry the potential for serious injury.

FIGURE 12–9 Harmful effects of selected widely used chemicals.

Chemical	Potential Harmful Effect
Acetic acid	Tissue damage
Liquid bromide	Corrosive effect on the respiratory system and tissue damage
Formaldehyde	Tissue hardening
Lime	Dermatitis and eye burns
Methylbromide	Blisters
Nitric/sulfuric acid mixture	Severe burns and tissue damage
Oxalic acid	Ulceration and tissue damage
White phosphorus	Ignites in air causing thermal burns
Silver nitrate	Corrosive/caustic effect on the skin
Sodium (metal)	Ignites with moisture causing thermal burns
Trichloracetic acid	Tissue damage

The primary hazardous effects of chemical burns are infection, loss of body fluids, and shock, and are summarized in the following paragraphs.[10]

Infection
The risk of **infection** is high with chemical burns—as is it with heat-induced burns—because the body's primary defense against infection-causing microorganisms (the skin) is penetrated. This is why it is so important to keep burns clean. Infection in a burn wound can cause *septicemia* (blood poisoning).

Fluid Loss
Body **fluid loss** in second- and third-degree burns can be serious. With second-degree burns, the blisters that form on the skin often fill with fluid that seeps out of damaged tissue under the blister. With third-degree burns, fluids are lost internally and, as a result, can cause the same complications as a hemorrhage. If these fluids are not replaced properly, the burns can be fatal.

Shock
Shock is a depression of the nervous system. It can be caused by both physical and psychological trauma. In cases of serious burns, it may be caused by the intense pain that can occur when skin is burned away, leaving sensitive nerve endings exposed. Shock from burns can come in the following two forms: (1) primary shock, which is the first stage and results from physical pain or psychological trauma, and (2) secondary shock, which comes later and is caused by a loss of fluids and plasma proteins as a result of the burns.

First Aid for Chemical Burns
There is a definite course of action that should be taken when chemical burns occur, and the need for immediacy cannot be overemphasized. According to the NSC, the proper response in cases of chemical burns is to "wash off the chemical by flooding

the burned areas with copious amounts of water as quickly as possible. This is the only method for limiting the severity of the burn, and the loss of even a few seconds can be vital."[11]

In the case of chemical burns to the eyes, the continuous flooding should continue for at least 15 minutes. The eyelids should be held open to ensure that chemicals are not trapped under them.

Clothing is another consideration when an employee comes in contact with a caustic chemical. If chemicals have saturated the employee's clothes, they must be removed quickly. The best approach is to remove the clothes while flooding the body or the affected area. If necessary for quick removal, clothing should be ripped or cut off.

The critical need to apply water immediately in cases of chemical burns means that water must be readily available. Health and safety professionals should ensure that special eye wash and shower facilities are available wherever employees handle chemicals.

OSHA'S GUIDELINES FOR HAZARDS OF TEMPERATURE EXTREMES

Rather than providing a specific standard relating to hazards of temperature extremes, OSHA provides guidelines safety professionals can use to ensure that workers are properly protected from either heat dangerous levels of heat exposure or cold stress. OSHA produces two documents that are excellent guides for safety professionals, managers, engineers, supervisors, team leaders, and other personnel who are not experts in the prevention of injuries relating to temperature extremes. Both of these documents can be accessed by going to OSHA's website (www.osha.gov) and searching for the following: *Cold Stress Guide* and *Occupational Heat Exposure*.

OSHA's Recommendations for Preventing Cold-Related Injuries and Illnesses

To prevent cold-related injuries and illnesses in workers, OSHA recommends that employers do the following:

- Train workers to recognize cold-stress hazards and how to prevent them including appropriate engineering controls, proper selection and use of PPE, and the application of safe work practices for cold environments. Workers should also be trained in how to provide first aid for cold stress–related injuries and illnesses and how to respond to emergencies.
- Provide appropriate engineering controls such as radiant heaters and wind shields.
- Identify safe work practices relating to cold environments and make sure they are followed by all personnel. Such practices include acclimatizing workers, requiring frequent warming breaks, requiring personnel to work in pairs (buddy system), and ensuring workers remember to stay hydrated.
- Make sure workers use the proper PPE and use it properly. For example, this can mean ensuring that workers wear three layers of clothing, a hat or hood, a knit mask, insulated gloves, and insulated/waterproof boots.

OSHA's Recommendations for Preventing Heat-Related Injuries and Illnesses

To prevent heat-related injuries and illnesses in workers, OSHA recommends that employers do the following:

- Train workers to recognize heat-stress hazards and how to prevent them, including appropriate engineering controls, proper selection of clothing, and the application of safe work practices for environments where heat and humidity are factors. Workers should also be trained in how to provide first aid for heat stress–related injuries and illnesses, and how to respond to emergencies.
- Provide appropriate engineering controls such as air-conditioning, proper ventilation, and shade.
- Identify safe work practices relating to hot/high humidity environments and hot/low humidity environments. Such practices include acclimatizing workers, work/rest cycles, frequent hydration, and working in pairs (buddy system).

Enforcement by OSHA Relating to Hazards of Temperature Extremes

OSHA provides guidelines for preventing injuries and illnesses from hazards of temperature extremes. However, this does not mean OSHA cannot issue citations and assess penalties against employers that do not take appropriate precautions to prevent injuries and illnesses caused by temperature extremes. Of course, OSHA may always use the General Duty Clause of the OSH Act to require employers to provide a safe and healthy workplace. But OSHA can also issue the following standards to ensure employers protect their personnel for hazards of temperature extremes:

- 29 CFR 1910.132(d) or 1926.28 (construction): Personal Protective Equipment
- 29 CFR 1910.141 or 1926.51 (construction): Sanitation
- 29 CFR 1910.151 or 1926.50 (construction): Medical Services and First Aid
- 29 CFR 1926.21 (construction only): Safety Training and Education

Key Terms and Concepts

- Acclimatization
- Body surface area (BSA)
- Chemical burns
- Clo value
- Cold stress
- Conduction
- Convection
- Corium
- Critical
- Cutis
- Dermis
- Electrolytes
- Environmental heat
- Epidermis
- First-degree burns
- Fluid loss
- Heat cramps
- Heat exhaustion
- Heat rash
- Heat strain
- Heat stress
- Heat stress management
- Hypothermia
- Infection
- Metabolic heat
- Minor
- Moderate
- Radiant heat
- Second-degree burns
- Shock
- Third-degree burns

Review Questions

1. Define the following thermal comfort-related terms: *conduction, convection*, and *metabolic heat*.
2. What is heat stress?
3. What is heat strain?
4. Define the following terms: *heat exhaustion, heat cramps, heat syncope, heat rash, work tolerance time*, and *moisture vapor transfer rate*.
5. What are the symptoms of heat exhaustion?
6. How can heat strain be recognized?
7. How does clothing affect the cooling process?
8. What factors influence the WBGT?
9. Describe the various general controls in heat stress management.
10. How can cold stress be prevented?
11. How does wind movement affect the way the body perceives temperature?
12. Describe the symptoms of cold stress and hypothermia.
13. Describe the various components of a cold stress prevention program.
14. Describe the various purposes served by the skin.
15. Describe and differentiate among first-, second-, and third-degree burns.
16. Describe and differentiate among minor, moderate, and critical burns.
17. List the factors that determine the severity of a chemical burn.
18. Explain the hazards of chemical burns besides tissue damage.
19. What should you do if an employee accidentally splashes a caustic chemical on himself or herself?
20. What measures are recommended by OSHA for preventing cold- and heat-related injuries?

Endnotes

1. American Conference of Governmental Industrial Hygienists (ACGIH), 2015, *TLVs and BEIs* (Cincinnati, OH: ACGIH, 2010), 172–181.
2. Ibid., 180–188.
3. Ibid., 181.
4. Ibid., 187.
5. Ibid.
6. Ibid., 171–179.
7. Patricia M. Mertz, *Burn Study*, an unpublished paper, University of Miami School of Medicine, Department of Dermatology and Cutaneous Surgery, Miami, Florida. Retrieved from medscape.com/viewarticle/407569 on July 6, 2017.
8. Ibid.
9. Ibid., 3–4.
10. Ibid., 2.
11. Ibid.

13 PRESSURE HAZARDS

Learning Objectives

- Define the term *pressure hazards*
- List and explain the sources of pressure hazards
- Explain the pressure hazards relating to boilers
- Summarize the hazards of high-temperature water
- Describe the hazards of unified pressure vessels
- Explain the hazards of cracking in pressure vessels
- Describe how to conduct nondestructive testing of pressure vessels
- Summarize pressure hazards to human beings
- Explain decompression procedures
- Explain how to measure pressure hazards
- Summarize the methods for reducing pressure hazards
- Summarize the main elements of OSHA's guidelines for pressure vessel safety

PRESSURE HAZARDS DEFINED

Pressure is defined in physics as the force exerted against an opposing fluid or thrust distributed over a surface. This may be expressed in force or weight per unit of area, such as pounds per square inch (psi). A **hazard** is a condition with the potential of causing injury to personnel, damage to equipment or structures, loss of material, or lessening of the ability to perform a prescribed function. Thus, a **pressure hazard** is a hazard caused by a dangerous condition involving pressure. Critical injury and damage can occur with relatively little pressure. The Occupational Safety and Health Administration (OSHA) defines high-pressure cylinders as those designated with a service pressure of 900 psi or greater.

We perceive pressure in relation to the earth's atmosphere. Approximately 21 percent of the atmosphere is oxygen, with most of the other 79 percent being nitrogen. In addition

317

to oxygen and nitrogen, the atmosphere contains trace amounts of several inert gases: argon, neon, krypton, xenon, and helium.

At sea level, the earth's atmospheric pressure averages 1,013 H (hydrogen) or 10 N/m² or 1.013 millibars or 760 mm Hg (29.92 inches), or 14.7 psi, depending on the measuring scale used.[1] The international system of measurement utilizes newtons per square meter (N/m²). However, in human physiology studies, the typical unit is millimeters of mercury (mm Hg).

Atmospheric pressure is usually measured using a **barometer**. As the altitude above sea level increases, atmospheric pressure decreases in a nonlinear fashion. For example, at 5,486 meters (18,000 feet) above sea level, the barometric pressure is equal to 390 mm Hg. Half of this pressure, around 195 mm Hg, can be found at 2,010 meters (23,000 feet) above sea level.

Boyle's law states that the product of a given pressure and volume is constant at a constant temperature:

$$P_1 V_1 = P_2 V_2, \quad \text{where } T \text{ is constant}$$

Air moves in and out of the lungs because of a pressure gradient or difference in pressure. When atmospheric pressure is greater than pressure within the lungs, air flows down this pressure gradient from the outside into the lungs. This is called **inspiration**, inhalation, or breathing in, and occurs with greater lung volume than at rest. When pressure in the lungs is greater than atmospheric pressure, air moves down a pressure gradient outward from the lungs to the outside. **Expiration** occurs when air leaves the lungs and the lung volume is less than the relaxed volume, increasing pressure within the lungs.

Gas exchange occurs between air in the lung alveoli and gas in solution in blood. The pressure gradients causing this gas exchange are called *partial pressures*. **Dalton's law of partial pressures** states that in a mixture of theoretically ideal gases, the pressure exerted by the mixture is the sum of the pressures exerted by each component gas of the mixture:

$$P_A = P_O + P_N + P_{else}$$

Air entering the lungs immediately becomes saturated with water vapor. Water vapor, although it is a gas, does not conform to Dalton's law. The partial pressure of water vapor in a mixture of gases is not dependent on its fractional concentration in that mixture. Water vapor partial pressure, instead, is dependent on its temperature. From this exception to Dalton's law comes the fact that at the normal body temperature of 37°C (98.6°F), water vapor maintains a partial pressure of 47 mm Hg as long as that temperature is maintained. With this brief explanation of how pressure is involved in human breathing, we now focus on the various sources of pressure hazards.

SOURCES OF PRESSURE HAZARDS

There are many sources of pressure hazards—some natural, most created by humans. Because the human body is made up of approximately 85 percent liquid, which is virtually incompressible, increasing pressure does not create problems by itself. Problems can result from air being trapped or expanded within body cavities.

When sinus passages are blocked so that air cannot pass easily from the sinuses to the nose, expansion of the air in these sinuses can lead to problems. The same complications can occur with air trapped in the middle ear's eustachian tube. As Boyle's law states, gas volume increases as pressure decreases. Expansion of the air in blocked sinus passages or the middle ear occurs with a rapid increase in altitude or rapid ascent underwater. This can cause pain and, if not eventually relieved, disease. Under extreme circumstances of rapid ascent from underwater diving or high-altitude decompression, lungs can rupture.

Nitrogen absorption into the body tissues can become excessive during underwater diving and breathing of nitrogen-enriched air. Nitrogen permeation of tissues occurs in proportion to the partial pressure of nitrogen taken in. If the nitrogen is permeating tissues faster than the person can breathe it out, bubbles of gas may form in the tissues.

Decompression sickness can result from the decompression that accompanies a rapid rise from sea level to at least 5,486 meters (18,000 feet) or a rapid ascent from around 40 to 20 meters (132 to 66 feet) underwater. Several factors influence the onset of decompression sickness:

- **A history** of previous decompression sickness increases the probability of another attack.
- **Age** is a component. Being over 30 increases the chances of an attack.
- **Physical fitness** plays a role. People in better condition have a reduced chance of the sickness. Previously broken bones and joint injuries are often the sites of pain.
- **Exercise** during the exposure to decompression increases the likelihood and brings on an earlier onset of symptoms.
- **Low temperature** increases the probability of the sickness.
- **Speed of decompression** also influences the sickness. A rapid rate of decompression increases the possibility and severity of symptoms.
- **Length of exposure** of the person to the pressure is proportionately related to the intensity of symptoms. The longer the exposure, the greater the chances of decompression sickness.

A reduction in partial pressure can result from reduced available oxygen and cause a problem in breathing known as **hypoxia**. Too much oxygen or oxygen breathed under pressure that is too high is called **hyperoxia**. Another partial pressure hazard, **nitrogen narcosis**, results from a higher-than-normal level of nitrogen pressure.

When breathed under pressure, nitrogen causes a reduction of cerebral and neural activity. Breathing nitrogen at great depths underwater can cause a feeling of euphoria and loss of reality. At depths greater than 30 meters (100 feet), nitrogen narcosis can occur even when breathing normal air. The effects may become pathogenic at depths greater than 60 meters (200 feet), with motor skills threatened at depths greater than 91 meters (300 feet). Cognitive processes deteriorate quickly after reaching a depth of 99 meters (325 feet). Decompression procedures are covered later in this chapter.

BOILERS AND PRESSURE HAZARDS

A boiler is a closed vessel in which water is heated to form steam, hot water, or high-temperature water under pressure.[2] Potential safety hazards associated with boilers and other pressurized vessels include the following:

- Design, construction, or installation errors
- Poor or insufficient training of operators
- Human error
- Mechanical breakdown or failure
- Failure or blockage of control or safety devices
- Insufficient or improper inspections
- Improper application of equipment
- Insufficient preventive maintenance[3]

Through years of experience, a great deal has been learned about how to prevent accidents associated with boilers. OSHA recommends the following daily, weekly, monthly, and yearly accident prevention measures in 29 CFR 1910 Subpart H:

1. **Daily check.** Check the water to make sure that it is at the proper level. Vent the furnace thoroughly before starting the fire. Warm up the boiler using a small fire. When the boiler is operating, check it frequently.
2. **Weekly check.** At least once every week, test the low-water automatic shutdown control and record the results of the test on a tag that is clearly visible.
3. **Monthly check.** At least once every month, test the safety valve and record the results of the test on a tag that is clearly visible.
4. **Yearly check.** The low-level automatic shutdown control mechanism should be either replaced or completely overhauled and rebuilt. Arrange to have the vendor or a third-party expert test all combustion safeguards, including fuel pressure switches, limit switches, motor starter interlocks, and shutoff valves.[4]

HIGH-TEMPERATURE WATER HAZARDS

High-temperature water (HTW) is exactly what its name implies—water that has been heated to a very high temperature, but not high enough to produce steam.[5] In some cases, HTW can be used as an economical substitute for steam (e.g., in industrial heating systems). It has the added advantage of releasing less energy (pressure) than steam does.

In spite of this, there are hazards associated with HTW. Human contact with HTW can result in extremely serious burns and even death. The two most prominent sources of hazards associated with HTW are operator error and improper design. Proper training and careful supervision are the best guards against operator error.

Design of HTW systems is a highly specialized process that should be undertaken only by experienced engineers. Mechanical forces such as **water hammer**, thermal expansion, thermal shock, and faulty materials cause system failures more often than do thermodynamic forces. Therefore, it is important to allow for such causes when designing an HTW system.

The best designs are simple and operator-friendly. Designing too many automatic controls into an HTW system can create more problems than it solves by turning operators into mere attendants who are unable to respond properly to emergencies.

HAZARDS OF UNFIRED PRESSURE VESSELS

Not all pressure vessels are fired. Unfired pressure vessels include compressed air tanks, steam-jacketed kettles, digesters, and vulcanizers, and others that can create heat internally by various means rather than by external fire.[6] The various means of creating internal heat include (1) chemical action within the vessel and (2) application of some heating medium (electricity, steam, hot oil, and so on) to the contents of the vessel. The potential hazards associated with unfired pressure vessels include hazardous interaction between the material of the vessel and the materials that will be processed in it; inability of the filled vessel to carry the weight of its contents and the corresponding internal pressure; inability of the vessel to withstand the pressure introduced into it plus pressure caused by chemical reactions that occur during processing; and inability of the vessel to withstand any vacuum that may be created accidentally or intentionally.

The most effective preventive measure for overcoming these potential hazards is proper design. Specifications for the design and construction of unfired pressure vessels include requirements in the following areas: working pressure range, working temperature range, type of materials to be processed, stress relief, welding or joining measures, and radiography. Designs that meet the specifications set forth for unfired pressure vessels in such codes as the ASME (American Society of Mechanical Engineers) Code (Section VIII) will overcome most predictable hazards.

Beyond proper design, the same types of precautions taken when operating fired pressure vessels can be used when operating unfired pressure vessels. These include continual inspection, proper housekeeping, periodic testing, visual observation (for detecting cracks), and the use of appropriate safety devices.

HAZARDS OF HIGH-PRESSURE SYSTEMS

The hazards most commonly associated with high-pressure systems are leaks, pulsation, vibration, release of high-pressure gases, and whiplash from broken high-pressure pipes, tubing, or hose.[7] Strategies for reducing these hazards include limiting vibration through the use of vibration dampening (use of anchored pipe supports), decreasing the potential for leaks by limiting the number of joints in the system, using pressure gauges, placing shields or barricades around the system, using remote control and monitoring, and restricting access.

CRACKING HAZARDS IN PRESSURE VESSELS

One of the most serious hazards in pressure vessels is the potential for cracking.[8] Cracking can lead to either a complete rupture or leaks. The consequences of a complete rupture include (1) blast effects due to the sudden expansion of the contents of the vessel and (2) possible injuries and damage from fragmentation. The consequences of a leak include (1) suffocation or poisoning of employees depending

FIGURE 13–1 Diagram of a typical pressure vessel showing potential points for leakage or rupture.

on the contents of the vessel, (2) explosion and fire, and (3) chemical and thermal burns from contact with the contents of the vessel.

Pressure vessels are used in many different applications to contain many different types of substances ranging from water to extremely toxic chemicals. Leakage or rupture may occur in welded seams, bolted joints, or at nozzles. Figure 13–1 shows a diagram of a typical pressure vessel showing the potential points of leakage and rupture. The types of vessels that are most susceptible to leakage and rupture, primarily because of the processes they are part of or their contents, are as follows:

Deaerator Vessels
Deaeration is the process of removing noncondensable gases, primarily oxygen, from the water used in steam generation. **Deaerator vessels** are used in such applications as power generation, pulp and paper processing, chemical processing, and petroleum refining. The most common failures associated with deaerator vessels are (1) cracks caused by water hammer at welded joints that were not postweld heat treated and (2) cracks caused by corrosion fatigue.

Amine Vessels
The *amine process* removes hydrogen sulfide from petroleum gases such as propane and butane. It can also be used for removing carbon dioxide in some processes. **Amine vessels** are used in petroleum refineries, gas treatment facilities, and chemical plants. The most common failures associated with amine vessels are cracks in stressed or unrelieved welds.

Wet Hydrogen Sulfide Vessels
Any fluid that contains water and hydrogen sulfide is considered wet hydrogen sulfide. Many of the vessels used to contain wet hydrogen sulfide are made of steel. Hydrogen is generated when steel is exposed to such a mixture. Dissolved hydrogen

can cause cracking, blistering, and embrittlement, particularly in high-strength steels. Consequently, low-strength steels are recommended for **wet hydrogen sulfide vessels**.

Ammonia Vessels

Vessels for the containment of ammonia are widely used in commercial refrigeration systems and chemical processes. Such **ammonia vessels** are typically constructed as spheres of carbon steel. The water and oxygen content in ammonia can cause carbon steel to crack, particularly near welds.

Pulp Digester Vessels

The process used to digest pulp in the manufacture of paper involves the use of a weak water solution of sodium hydroxide and sodium sulfide in a temperature range of 110°C to 140°C (230°F to 284°F). The most common failure in **pulp digester vessels** is cracking along welded seams primarily due to caustic stress corrosion.

NONDESTRUCTIVE TESTING OF PRESSURE VESSELS

To prevent leakage or rupture, it is necessary to examine pressure vessels periodically. There are five widely used nondestructive methods for testing: (1) **visual examination**, (2) **liquid penetration test**, (3) **magnetic particle test**, (4) X-ray radiography, and (5) ultrasonic test. Visual liquid penetration and magnetic particle tests can detect only those defects that are either on the surface or near it. Radiographic and ultrasonic tests can detect problems within the material. Consequently, the visual, liquid penetration, and magnetic particle tests are referred to as *surface tests*. X-ray radiography and ultrasonic tests are called *volumetric tests*.

Visual Examination

A visual examination consists of taking a thorough look at the vessel to detect signs of corrosion, erosion, or hydrogen blistering. In order to conduct a dependable visual examination of a pressure vessel, it is necessary to have a clean surface and good lighting.

Liquid Penetration Test

This test involves placing a specially formulated liquid penetrant over an area and letting it seep in. When the penetrant is removed from the surface, some of it remains entrapped in the area of discontinuity. A developing agent is then applied, which draws out the entrapped penetrant and magnifies the discontinuity. The process can be enhanced by adding fluorescent chemicals to the penetrant to aid in the detection of problems.

Magnetic Particle Test

This test is based on the fact that discontinuities in or near the surface of a pressure vessel disturb magnetic flux lines that are induced in a ferromagnetic material. Disturbances are detected by applying fine particles of ferromagnetic material to the surface of the vessel. The necessary magnetic field is produced most frequently using the "prod" technique in which electric current is run through an area by applying opposing "prods" (contact probes). A drawback of this test is that corners and surface irregularities in the vessel material can produce the same disturbances as defects. Consequently, special care is needed when using this test in a region with corners or welded joints. Because this test works only with ferromagnetic material, its use is limited to vessels made of carbon and low-alloy steels.

X-Ray Radiography

This test amounts to making an X-ray negative of a given portion of the vessel. The process works in the same way as those used by physicians and dentists. Irregularities such as holes, voids, and discontinuities produce a greater exposure (darker area) on the X-ray negative.

Ultrasonic Test

This test is similar to radar and other uses of electromagnetic and acoustic waves for detecting foreign objects. Short signals are induced into the material. Waves that are reflected back from discontinuities are detected by one or more transducers. **Ultrasonic testing** requires an electronic system for generating a signal, a transducer system for converting the electrical into mechanical vibrations and vice versa, and an electronic system for amplifying, processing, and displaying the return signal.

PRESSURE DANGERS TO HUMANS

The term *anoxia* refers to the rare case of a total lack of oxygen. Hypoxia, a condition that occurs when the available oxygen is reduced, can occur while ascending to a high altitude or when oxygen in air has been replaced with another gas, which may happen in some industrial situations.

Altitude sickness is a form of hypoxia associated with high altitudes. Ascent to an altitude of 10,000 feet above sea level can result in a feeling of malaise, shortness of breath, and fatigue. A person ascending to 14,000 to 15,000 feet may experience euphoria, along with a reduction in powers of reason, judgment, and memory. Altitude sickness includes a loss of **useful consciousness** at 20,000 to 25,000 feet. After approximately five minutes at this altitude, a person may lose consciousness. The loss of consciousness comes at approximately one minute or less at 30,000 feet. Over 38,000 feet, most people lose consciousness within 30 seconds and may fall into a coma and possibly die.

Hyperoxia, an increased concentration of oxygen in air, is not a common situation. Hyperbaric chambers or improperly calibrated scuba equipment can create conditions that may lead to convulsions if pure oxygen is breathed for greater than three hours. Breathing air at a depth of around 300 feet can be toxic and is equivalent to breathing pure oxygen at a depth of 66 feet.

At high pressures of oxygen, around 2,000 to 5,000 mm Hg, dangerous cerebral problems such as dizziness, twitching, vision deterioration, and nausea may occur. Continued exposure to these high pressures will result in confusion, convulsion, and eventual death.

Changes in total pressure can induce **trapped gas effects**. With a decrease in pressure, trapped gases will increase in volume (according to Boyle's law). Trapped gases in the body include air pockets in the ears, sinuses, and chest. Divers refer to the trapped gas phenomenon as the *squeeze*. Jet travel causes the most commonly occurring instance of trapped gas effects. Takeoff and landing may cause relatively sudden shifts in pressure, which may lead to discomfort and pain. Very rapid ascent or descent can lead to injury.

Lung rupture can be caused by a swift return to the surface from diving or decompression during high-altitude flight. This event is rare and happens only if the person is holding his or her breath during the decompression.

Evolved gas effects are associated with the absorption of nitrogen into body tissues. When breathed, nitrogen can be absorbed into all body tissues in concentrations proportional to the partial pressure of nitrogen in air. When a person is ascending in altitude, on the ground, in flight, or underwater, nitrogen must be exhaled at a rate equal to or exceeding the absorption rate to avoid evolved gas effects.

If the nitrogen in body tissues such as blood is being absorbed faster than it is being exhaled, bubbles of gas may form in the blood and other tissues. Gas bubbles in the tissues may cause decompression sickness, which can be painful and occasionally fatal. Early symptoms of this disorder occur in body bends or joints such as elbows, knees, and shoulders. The common name for decompression sickness is the **bends**.

When the formation of gas bubbles is due to rapid ambient pressure reduction, it is called **dysbarism**.[9] The major causes of dysbarism are (1) the release of gas from the blood and (2) the attempted expansion of trapped gas in body tissues. The sickness may occur with the decompression associated with rapidly moving from sea level (considered zero) to approximately 20,000 feet above sea level. Dysbarism is most often associated with underwater diving or working in pressurized containers (such as airplanes). Obese and older people seem to be more susceptible to dysbarism and decompression sickness.

Dysbarism manifests itself in a variety of symptoms. The **creeps** are caused by bubble formation in the skin, which causes an itchy, crawling, feeling in the skin. Coughing and choking, resulting from bubbles in the respiratory system, are called the **chokes**. Bubbles occurring in the brain, although rare, may cause tingling and numbness, severe headaches, spasticity of muscles, and in some cases, blindness and paralysis. Dysbarism of the brain is rare. Rapid pressure change may also cause pain in the teeth and sinuses.[10]

Aseptic necrosis of bone is a delayed effect of decompression sickness. Blood in the capillaries supplying the bone marrow may become blocked with gas bubbles, which can cause a collection of platelets and blood cells to build up in a bone cavity. The marrow generation of blood cells can be damaged as well as the maintenance of healthy bone cells. Some bone areas may become calcified with severe complications when the bone is involved in a joint.

DECOMPRESSION PROCEDURES

Employees who work in an environment that is under pressure must undergo decompression procedures before returning to a normal atmosphere.[11] Such procedures are planned based on the amount of pressure to which the employee is subjected and for how long. In 29 CFR 1926 (Subpart S, Appendix A), OSHA provides tables that can be used for planning appropriate decompression procedures for employees. Figure 13–2 is an example of a part of such a table.

FIGURE 13–2 Portion of a table for planning a two-stage decompression.

Partial Table (Two-Stage Decompression)

Working Chamber Pressure (psig)	Working Period (hours)	Stage No.	Pressure Reduction (psig) From	Pressure Reduction (psig) To	Time in Stage (minutes)	Pressure Reduction Rate (min/pound)	Total Time Decompress (minutes)
20	3	1	20	4	3	0.20	—
		2	4	0	12	3.00	15
	4	1	20	4	3	0.20	—
		2	4	0	40	10.00	43
	5	1	20	4	3	0.20	—
		2	4	0	60	15.00	63

Note: Do not interpolate. Always use the next higher value for conditions that fall between numbers in the table.

In most cases, decompression will need to occur in two stages. Figure 13–2 shows a part of a table to be used for planning two-stage decompressions. The following example demonstrates how to use such a table:

An employee will be working for four hours in an environment with a working chamber pressure of 20 pounds per square inch gauge (psig). Locate 20 psig and 4 working period hours in the table in Figure 13–2. Stage 1 of the decompression will require a reduction in pressure from 20 psig to 4 psig over a period of 3 minutes at the uniform rate of 0.20. Stage 2 of the decompression will require a reduction in pressure from 4 psig to 0 psig over a period of 40 minutes at the uniform rate of 10 minutes per pound. The total time for the decompression procedure is 43 minutes.

Decompression procedures are designed to prevent the various effects of decompression sickness that were explained in the previous section. For a complete set of decompression tables, refer to the following Web address: www.osha.gov

MEASUREMENT OF PRESSURE HAZARDS

Confirming the point of pressurized gas leakage can be difficult. After a gas has leaked out to a level of equilibrium with its surrounding air, the symptoms of the leak may disappear. There are several methods of detecting pressure hazards:

- **Sounds** can be used to signal a pressurized gas leak. Gas discharge may be indicated by a whistling noise, particularly with highly pressurized gases escaping through small openings. Workers should not use their fingers to probe for gas leaks as highly pressurized gases may cut through tissue, including bone.
- **Cloth streamers** may be tied to the gas vessel to help indicate leaks. Soap solutions may be smeared over the vessel surface so that bubbles are formed when gas escapes. A stream of bubbles indicates gas release.

- **Scents** may be added to gases that do not naturally have an odor. The odor sometimes smelled in homes that cook or heat with natural gas is not the gas but a scent added to it.
- **Leak detectors** that measure pressure, current flow, or radioactivity may be useful for some types of gases.
- **Corrosion** may be the long-term effect of escaping gases. Metal cracking, surface roughening, and general weakening of materials may result from corrosion.

There are many potential causes of gas leaks. The most common of these are as follows:

- **Contamination** by dirt can prevent the proper closing of gas valves, threads, gaskets, and other closures used to control gas flow.
- **Overpressurization** can overstress the gas vessel, permitting gas release. The container closure may distort and separate from gaskets, leading to cracking.
- **Excessive temperatures** applied to dissimilar metals that are joined may cause unequal thermal expansion, loosening the metal-to-metal joint and allowing gas to escape. Materials may crack because of excessive cold, which may also result in gas escape. Thermometers are often used to indicate the possibility of gas release.
- **Operator errors** may lead to hazardous gas release from improper closure of valves, inappropriate opening of valves, or overfilling of vessels. Proper training and supervision can reduce operator errors.

Destructive and nondestructive methods may be used to detect pressure leaks and incorrect pressure levels. **Nondestructive testing** methods do not harm the material being tested. Nondestructive methods may include mixing dye penetrants and magnetic or radioactive particles with the gas and then measuring the flow of the gas. Ultrasonic and X-ray waves are another form of nondestructive testing methods and are often used to characterize materials and detect cracks or other leakage points.

Destructive testing methods destroy the material being checked. Proof pressures generate stresses to the gas container, typically 1.5 to 1.667 times the maximum expected operating pressure for that container. Strain measurements may also be collected to indicate permanent weakening changes to the container material that remain after the pressure is released. **Proof pressure tests** often call for the pressure to be applied for a specified time and then to be released. Stress and strain tests are then applied to the material. Proof pressure tests may or may not result in the destruction of the container being tested.

REDUCTION OF PRESSURE HAZARDS

The reduction of pressure hazards often requires better maintenance and inspection of equipment that measures or uses high-pressure gases. Proper storage of pressurized containers reduces many pressure hazards. Pressurized vessels should be stored in locations away from cold or heat sources, including the sun. Cryogenic compounds (those that have been cooled to unusually low temperatures) may boil and burst the container when not kept at the proper temperatures. The whipping action of pressurized flexible hoses can also be dangerous. Hoses should be firmly clamped at the ends when pressurized.

Gas compression can occur in sealed containers exposed to heat. For this reason, aerosol cans must never be thrown into or exposed to a fire. Aerosol cans may explode violently when exposed to heat, although most commercially available aerosols are contained in low-melting point metals that melt before pressure can build up.

Pressure should be released before working on equipment. Gauges can be checked before any work on the pressurized system begins. When steam equipment is shut down, liquid may condense within the system. This liquid or dirt in the system may become a propellant, which may strike bends in the system, causing loud noises and possible damage.

Water hammer is a shock effect caused by liquid flow suddenly stopping.[12] The shock effect can produce loud noises. The momentum of the liquid is conducted back upstream in a shock wave. Pipe fittings and valves may be damaged by the shock wave. Reduction of this hazard involves using air chambers in the system and avoiding the use of quick-closing valves.

Negative pressures or **vacuums** are caused by pressures below atmospheric level. Negative pressures may result from hurricanes and tornadoes. Vacuums may cause collapse of closed containers. Building code specifications usually allow for a pressure differential. Vessel wall thickness must be designed to sustain the load imposed by the differential in pressure caused by negative pressure. Figure 13–3 describes several methods to reduce the hazards associated with pressurized containers.

FIGURE 13–3   Reduction of pressure hazards.

- Install valves so that failure of a valve does not result in a hazard.
- Do not store pressurized containers near heat or sources of ignition.
- Train and test personnel dealing with pressurized vessels. Only tested personnel should be permitted to install, operate, maintain, calibrate, or repair pressurized systems. Personnel working on pressure systems should wear safety face shields or goggles.
- Examine valves periodically to ensure that they are capable of withstanding working pressures.
- Operate pressure systems only under the conditions for which they were designed.
- Relieve all pressure from the system before performing any work.
- Label pressure system components to indicate inspection status as well as acceptable pressures and flow direction.
- Connect pressure relief devices to pressure lines.
- Do not use pressure systems and hoses at pressure exceeding the manufacturer's recommendations.
- Keep pressure systems clean.
- Keep pressurized hoses as short as possible.
- Avoid banging, dropping, or striking pressurized containers.
- Secure pressurized cylinders by a chain to prevent toppling.
- Store acetylene containers upright.
- Examine labels before using pressurized systems to ensure correct matching of gases and uses.
- Use dead man's switches on high-pressure hose wands.

OSHA'S STANDARD AND GUIDELINES FOR PRESSURE HAZARDS

OSHA does not publish a standard specifically for pressure hazards. Rather, pressure hazards are covered in 29 CFR 1910 Subpart H: *Hazardous Materials*. Not all hazardous materials are found in pressurized containers, but many are. Consequently, safety and health professionals must be familiar with Subpart H, especially 1910.101 through 1910.105 (compressed gases, acetylene, hydrogen, oxygen, and nitrous oxide) as well as 1910.110 (liquefied petroleum gases) and 1910.111 (anhydrous ammonia). A helpful aid for safety and health professionals concerned with pressure hazards is *OSHA's Technical Manual*, Section IV, Chapter 3, titled "Pressure Vessel Guidelines." This document is the source of much of the information presented in this chapter. It is available at: www.osha.gov/dts/osta/otm/otm_iv/otm_iv_3.html.

Another helpful guide from OSHA specifically for pressure vessel safety assessment is OSHA Instruction Publication 8-1.5: *Guidelines for Pressure Vessel Safety Assessment*. Developed as a guide for Federal and state OSHA compliance officers, consultants, and other safety personnel, the *Guidelines* provide specific information of use to safety and health professionals in the following areas:

- Pressure vessel design
- Deterioration and failure modes
- Inspection methods
- Cracking in pressure vessels
- Periodic inspection requirements and recommendations
- Damage and crack severity evaluation and repair
- Information for conducting safety assessments

The entire publication is available from OSHA and may be downloaded free of charge at OSHA's website: www.osha.gov/OshDoc/Directive_pdf/PUB_8-1_5.pdf

CONFINED SPACES AND PRESSURE VESSELS

A confined space is any workspace that has only limited means of entry and exit, is large enough for a person to fit into, but is not designed for occupancy by human beings. Installing, maintaining, and repairing pressure vessels can involve working in confined spaces. However, pressure vessels represent only one kind of confined space. There are many others such as vaults, vats, silos, ship compartments, train compartments, sewers, storm water run-off pipes, and tunnels. Since there are numerous kinds of confined spaces in addition to pressure vessels, the subject of confined spaces is covered in Chapter 16.

Key Terms and Concepts

- Altitude sickness
- Amine vessels
- Ammonia vessels
- Aseptic necrosis
- Barometer
- Bends
- Boyle's law
- Chokes
- Creeps
- Dalton's law of partial pressures
- Deaerator vessels
- Decompression sickness
- Destructive testing
- Dysbarism
- Evolved gas effects
- Expiration
- Hazard

- Hyperoxia
- Hypoxia
- Inspiration
- Liquid penetration test
- Magnetic particle test
- Negative pressures
- Nitrogen narcosis

- Nondestructive testing
- Pressure
- Pressure hazard
- Proof pressure tests
- Pulp digester vessels
- Trapped gas effects

- Ultrasonic testing
- Useful consciousness
- Vacuums
- Visual examination
- Water hammer
- Wet hydrogen sulfide vessels

Review Questions

1. Against which references is pressure measured? How are these references measured?
2. Define *inspiration* and *expiration*.
3. Explain Dalton's law of partial pressures.
4. Does water vapor conform to Dalton's law?
5. Briefly discuss decompression sickness.
6. What do length of exposure, the bends, the chokes, and aseptic necrosis of bone have in common?
7. Define *hypoxia* and *hyperoxia*.
8. Explain nitrogen narcosis.
9. Discuss altitude sickness.
10. What is the relationship between trapped gas effects and dysbarism?
11. What is the difference between destructive and nondestructive testing?
12. Briefly explain proof pressures.
13. What causes vacuums?
14. Explain three ways to conduct nondestructive testing of pressure vessels.
15. What is the total decompression time for an employee who works for four hours under pressure of 20 psig?

Endnotes

1. Occupational Safety and Health Administration, "Pressure Vessel Guidelines," *OSHA Technical Manual* (Washington, DC: Occupational Safety and Health Administration). Retrieved from www.osha.gov/dts/osta/otm/otm_iv/otm_iv_3.html on July 12, 2017.
2. Ibid., 31.
3. Ibid.
4. Ibid.
5. Ibid.
6. Ibid.
7. Ibid., 38.
8. Ibid., 39.
9. Ibid., 39.
10. Ibid.
11. OSHA Regulations Title 29, *Code of Federal Regulations*, Part 1926 (Subpart S, Appendix A), Decompression Tables. Retrieved from www.osha.gov on July 13, 2017.
12. Ibid.

CHAPTER
14

ELECTRICAL HAZARDS

Learning Objectives

- Define the concept of electrical hazards
- List the most common sources of electrical hazards
- List the most common electrical hazards to human beings
- Explain how to detect electrical hazards
- Summarize methods for reducing electrical hazards
- Explain how to establish an electrical safety program
- Explain how to conduct an electrical hazards self-assessment
- Explain how to prevent arc-flash injuries
- Summarize the training requirements relating to electrical safety
- List and explain the most common electrical safety devices
- List and explain electrical hazards relating to power strips and daisy chains

Consider the following scenario: A textile mill in Massachusetts was fined $66,375 when an employee contacted the Occupational Safety and Health Administration (OSHA) and complained about unsafe conditions at the mill. The Region 1 Office of OSHA conducted an investigation in response to the complaint that uncovered the following willful violation: allowing employees to perform live electrical work without safe work procedures or appropriate personal protective equipment (PPE).[1] In addition, the investigation uncovered several serious violations including (1) storage of flammable materials near emergency exits, (2) improper storage of oxygen and acetylene cylinders, (3) failure to post load ratings, and (4) an exposed live electrical source and unsuitable electrical outlets for wet or damp locations.[2]

ELECTRICAL HAZARDS DEFINED

Electricity is the flow of negatively charged particles called *electrons* through an electrically conductive material. Electrons orbit the nucleus of an atom, which is located approximately in the atom's center. The **negative charge** of the electrons is neutralized by particles called **neutrons**, which act as temporary energy repositories for the interactions between positively charged particles called **protons** and **electrons**.

Figure 14–1 shows the basic structure of an atom, with the positively charged nucleus in the center. The electrons are shown as energy bands of orbiting negatively charged particles. Each ring of electrons contains a particular quantity of negative charges. The basic characteristics of a material are determined by the number of electron rings and the number of electrons in the outer rings of its atoms. A **positive charge** is present when an atom (or group of atoms) in a material has too many electrons in its outer shell. In all other cases, the atom or material carries a *negative charge*.

Electrons that are freed from an atom and are directed by external forces to travel in a specific direction produce **electrical current**, also called *electricity*. **Conductors** are substances that have many free electrons at room temperature and can pass electricity. **Insulators** do not have a large number of free electrons at room temperature and do not conduct electricity. Substances that are neither conductors nor insulators can be called **semiconductors**.

Electrical current passing through the human body causes a **shock**. The quantity and path of this current determine the level of damage to the body. The path of this flow of electrons is from a negative source to a positive point, because opposite charges attract one another.

When a surplus or deficiency of electrons on the surface of a material exists, **static electricity** is produced. This type of electricity is called "static" because there is

---

FIGURE 14–1   Basic structure of an atom.

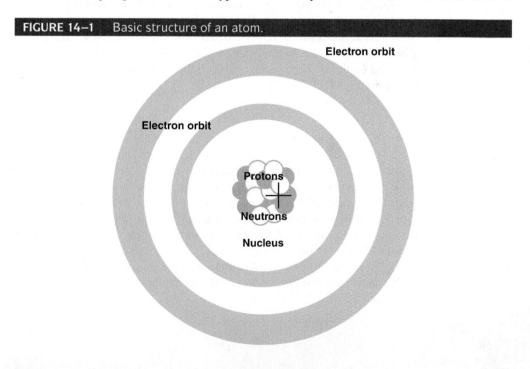

no positive material nearby to attract the electrons and cause them to move. Friction is not required to produce static electricity, although it can increase the charge of existing static electricity. When two surfaces of opposite static electricity charges are brought into close range, a discharge, or spark, will occur. The spark from static electricity is often the first clue that such static exists. A common example is the sparks that come from rustling woolen blankets in dry heated indoor air.

The **potential difference** between two points in a circuit is measured by **voltage**. The higher the voltage, the more likely it is that electricity will flow between the negative and positive points.

Pure conductors offer little **resistance** to the flow of electrons. Insulators, on the other hand, have very high resistance to electricity. Semiconductors have a medium-range resistance to electricity. The higher the resistance, the lower the flow of electrons. Resistance is measured in **ohms**.

Electrical current is produced by the flow of electrons. The unit of measurement for current is **amperes** (or amps). One amp is a current flow of 6.28×10^{18} electrons per second. Current is usually designated by I. **Ohm's law** describes the relationship among volts, ohms, and amps. One ohm is the resistance of a conductor that has a current of one amp under the potential of one volt. Ohm's law is stated as

$$V = IR$$

where

V = potential difference in volts
I = current flow in amps
R = resistance to current flow in ohms

Power is measured in wattage (or **watts**) and can be determined from Ohm's law:

$$W = VI \quad \text{or} \quad W = I^2R$$

where

W = power in watts

Most industrial and domestic use of electricity is supplied by **alternating current** (AC current). In the United States, standard AC circuits cycle 60 times per second. The number of cycles per second is known as **frequency** and is measured in **hertz**. Because voltage cycles in AC current, an **effective current** for AC circuits is computed, which is slightly less than the peak current during a cycle.

A **direct current** (DC current) has been found to generate as much heat as an AC current that has a peak current 41.4 percent higher than the DC. The ratio of effective current to peak current can be determined by

$$(\text{Effective current})/(\text{Peak current}) = (100\%)/(100\% + 41.4\%) = 0.707\% \text{ or } 70.7\%$$

Effective voltages are computed using the same ratios as effective current. A domestic 110-volt circuit has an effective voltage of 110 volts, with peaks of voltage over 150 volts.

The path of electrical current must make a complete loop for the current to flow. This loop includes the source of electrical power, a conductor to act as the path, a device to use the current (called a **load**), and a path to the ground. The earth maintains a relatively stable electrical charge and is a good conductor. The earth is considered to have **zero potential** because of its massive size. Any electrical conductor pushed into the earth is said to have zero potential. The earth is used as a giant common conductor back to the source of power.

Electrocution occurs when a person makes contact with a conductor carrying a current and simultaneously contacts the ground or another object that includes a conductive path to the ground. This person completes the circuit loop by providing a load for the circuit and thereby enables the current to pass through his or her body. People can be protected from this danger by insulating the conductors, insulating the people, or isolating the danger from the people.

The **National Electrical Code (NEC)** is published by the National Fire Protection Association (NFPA). This code specifies industrial and domestic electrical safety precautions. The NEC categorizes industrial locations and gases relative to their degree of fire hazard and describes in detail the safety requirements for industrial and home wiring. The NEC has been adopted by many jurisdictions as the local electrical code. The **National Board of Fire Underwriters** sponsors **Underwriters Laboratories (UL)**. The UL determines whether equipment and materials for electrical systems are safe in the various NEC location categories. The UL provides labels for equipment that it approves as safe within the tested constraints.

Typical 110-volt circuit wiring has a **hot wire** carrying current, a **neutral wire**, and a **ground wire**. The neutral wire may be called a **grounded conductor**, with the ground wire being called a **grounding conductor**. Neutral wires usually have white insulation, hot wires have red or black insulation, and ground wires have green insulation or are bare. Figure 14–2 shows a typical three-wire circuit.

The hot wire carries an effective voltage of 110 volts with respect to the ground, whereas the neutral wire carries nearly zero voltage. If the hot wire makes contact with an unintended conductor, such as a metal equipment case, the current can bypass the load and go directly to the ground. With the load skipped, the ground wire is a low-resistance path to the earth and carries the highest current possible for that circuit.

A **short circuit** is a circuit in which the load has been removed or bypassed. The ground wire in a standard three-wire circuit provides a direct path to the ground, bypassing the load. Short circuits can be another source of electrical hazard if a human is the conductor to the ground, thereby bypassing the load.

FIGURE 14–2 Typical three-wire circuit.

SOURCES OF ELECTRICAL HAZARDS

Short circuits are one of many potential **electrical hazards** that can cause electrical shock. Another hazard is water, which considerably decreases the resistance of materials, including humans. The resistance of wet skin can be as low as 450 ohms, whereas dry skin may have an average resistance of 600,000 ohms. According to Ohm's law, the higher the resistance, the lower the current flow. When the current flow is reduced, the probability of electrical shock is also reduced.

The major causes of electrical shock are as follows:

- Contact with a bare wire carrying current. The bare wire may have deteriorated insulation or be normally bare.
- Working with electrical equipment that lacks the UL label for safety inspection.
- Electrical equipment that has not been properly grounded. Failure of the equipment can lead to short circuits.
- Working with electrical equipment on damp floors or other sources of wetness.
- Static electricity discharge.
- Using metal ladders to work on electrical equipment. These ladders can provide a direct line from the power source to the ground, again causing a shock.
- Working on electrical equipment without ensuring that the power has been shut off.
- Lightning strikes.

Figure 14–3 depicts some of these electrical shock hazards.

Electrostatic Hazards

Electrostatic hazards may cause minor shocks. Shocks from static electricity may result from a single discharge or multiple discharges of static. Sources of electrostatic discharge include the following:

- Briskly rubbing a nonconductive material over a stationary surface. One common example of this is scuffing shoes across a wool or nylon carpet. Multilayered clothing may also cause static sparks.[3]
- Moving large sheets of plastic, which may discharge sparks.
- The explosion of organic and metallic dusts, which have occurred from static buildup in farm grain silos and mine shafts.
- Conveyor belts may cause static sparks. Depending on their constituent material, they can rub the materials being transported and cause static sparks.
- Vehicle tires rolling across a road surface.
- Friction between a flowing liquid and a solid surface.[4]

The rate of discharge of electrical charges increases with lower humidity. Electrostatic sparks are often greater during cold, dry winter days. Adding humidity to the air is not commonly used to combat static discharge, however, because higher humidity may result in an uncomfortable working environment and adversely affect equipment.[5]

Arcs and Sparks Hazards

With close proximity of conductors or contact of conductors to complete a circuit, an electric **arc** can jump the air gap between the conductors and ignite combustible gases or dusts. When the electric arc is a discharge of static electricity, it may be called a **spark**.

FIGURE 14–3 Electrical shock hazards.

Unsafe	Safer
Bare wires or wet floor	Insulated wires and dry floor
Power on; metal ladder	Power off; wooden ladder
Circuit without ground	Grounded circuit

A spark or arc may involve relatively little or a great deal of power and is usually discharged into a small space.

Combustible and Explosive Materials

High currents through contaminated liquids may cause the contaminants to expand rapidly and explode. This situation is particularly dangerous with contaminated oil-filled circuit breakers or transformers. A poor match between current or polarity and capacitors can cause an explosion. In each of these cases, the conductor is not capable of carrying a current of such high magnitude. Overheating from high currents can also lead to short circuits, which in turn may generate fires or explosions.

Lightning Hazards

Lightning is static charges from clouds following the path of least resistance to the earth, involving very high voltage and current. If this path to the earth involves humans, serious disability may result, including electrocution. Lightning may also damage airplanes from intracloud and cloud-to-cloud flashes. Electrical equipment and building structures are commonly subject to lightning hazards. Lightning tends to strike the tallest object on the earth below the clouds. A tree is a common natural path for lightning.

Improper Wiring

Improper wiring permits equipment to operate normally but can result in hazardous conditions. The section of this chapter on detection of electrical hazards discusses tests to identify unsafe wiring practices. One common mistake is to "jump" the ground wire to the neutral wire. In this case, the ground wire is actually connected to the neutral wire. Equipment usually operates in a customary way, but the hazard occurs when low voltages are generated on exposed parts of the equipment, such as the housing. If the neutral circuit becomes corroded or loose, the voltage on the ground wire increases to a dangerous level.

Improper wiring (or miswiring) can cause other hazards. When the ground is connected improperly, the situation is referred to as **open ground**. Usually the equipment with this miswiring will operate normally. If a short occurs in the equipment circuitry without proper grounding, anyone touching that equipment may be severely shocked.

With **reversed polarity**, the hot and neutral wires have been reversed. A worker who is not aware that the black lead (hot) and white lead (neutral) have been reversed could be injured or cause further confusion by connecting the circuit to another apparatus. If a short between the on/off switch and the load occurred, the equipment may run indefinitely, regardless of the switch position. In a reversed polarity light bulb socket, the screw threads become conductors.[6]

Temporary wiring installations sometimes remain in place for years until an accident occurs. Flexible wiring should rarely be substituted for fixed wiring in permanent buildings. A loose knot should be tied in a flexible cord when the plug is installed or replaced. The knot can prevent a pull on the cord from being transmitted to electrical connections such as the plug.

Insulation Failure

The degradation of insulation can cause a bare wire and resulting shock to anyone coming in contact with that wire. Most **insulation failure** is caused by environments toxic to insulation. These environments include the following:

- Direct sunlight or other sources of ultraviolet light, which can induce gradual breakdown of plastic insulation material.
- Sparks or arcs from discharging static electricity, which can result in burned-through holes in insulation.
- Repeated exposure to elevated temperatures, which can produce slow but progressive degradation of insulation material.
- Abrasive surfaces, which can result in erosion of the material strength of the insulation.
- Substance incompatibility with the atmosphere around the insulation and the insulation material, which can induce chemical reactions. Such reactions may include oxidation or dehydration of the insulation and eventual breakdown.
- Animals such as rodents or insects chewing or eating the insulation material, leading to exposure of the circuit. Insects can also pack an enclosed area with their bodies so tightly that a short circuit occurs. This is a common occurrence with electrical systems near water, such as pump housings and television satellite dishes.
- Moisture and humidity being absorbed by the insulation material, which may result in the moisture on the insulation carrying a current.

Equipment Failure

There are several ways in which **equipment failure** can cause electrical shocks. Electrical equipment designers attempt to create devices that are explosion-proof, dust ignition-proof, and spark-proof. Following are some of the more common types of equipment failure:

- Wet insulation can become a conductor and cause an electrical shock.
- Portable tool defects can result in the device's housing carrying an electric current. Workers do not expect tool housings to be charged and may be shocked when they touch a charged tool housing.
- Broken power lines carry great amperage and voltage and can cause severe disability.
- When equipment is not properly grounded or insulated, an unshielded worker may receive a substantial electrical shock.

Hazardous Locations for Electrical Equipment

The NEC classifies **hazardous locations** for electrical equipment. There are three basic classes: Class I for flammable vapors and gases, Class II for combustible dusts, and Class III for ignitable fibers. There are also two divisions of hazard categories. Division I has more stringent requirements for electrical installation than Division II does. Figure 14–4 provides examples for each location category.[7]

FIGURE 14–4 Hazardous electrical equipment location categories.

Class	Group	Division I	Division II
I. Flammable vapors and gases	A. Acetylene B. Hydrogen C. Ether D. Hydrocarbon fuels and solvents	Normally explosive; flammable paint spray areas	Not normally in explosive concentration; adjacent to paint spray area
II. Combustible dusts	E. Metal dusts F. Carbon dusts G. Flour, starch, grain, plastic, or chemical dusts	Conductive or ignitable dusts may be present; grain mills or processors	Not normally in ignitable concentration; grain storage areas
III. Ignitable fibers	Textiles; woodworking	Handled or used in manufacturing; cotton gins	Stored or handled in storage, not in manufacturing; excelsior storage

ELECTRICAL HAZARDS TO HUMANS

The greatest danger to humans suffering electrical shock results from current flow. The voltage determines whether a particular person's natural resistance to current flow will be overcome. Skin resistance can vary between 450 ohms and 600,000 ohms, depending on skin moisture.[8] Some levels of current "**freeze**" a person to the conductor; the person cannot voluntarily release his or her grasp. **Let-go current** is the highest current level at which a person in contact with the conductor can release the grasp of the conductor. Figure 14–5 shows the relationship between amperage dosage and danger with a typical domestic 60-cycle AC current.

The severity of injury with electrical shock depends not only on the dosage of current, as shown in Figure 14–5, but also on the path taken through the body by the

FIGURE 14–5 Current effects on the human body (60-cycle AC current).

Dose in Current in Milliamps	Effect on Human Body
Less than 1	No sensation, no perceptible effect.
1	Shock perceptible, reflex action to jump away. No direct danger from shock but sudden motion may cause accident.
More than 3	Painful shock.
6	Let-go current for women.*
9	Let-go current for men.*
10–15	Local muscle contractions. Freezing to the conductor for 2.5 percent of the population.
30–50	Local muscle contractions. Freezing to the conductor for 50 percent of the population.
50–100	Prolonged contact may cause collapse and unconsciousness. Death may occur after three minutes of contact due to paralysis of the respiratory muscles.
100–200	Contact of more than a quarter of a second may cause ventricular fibrillation of the heart and death. AC currents continuing for more than one heart cycle may cause fibrillation.
Over 200	Clamps and stops the heart as long as the current flows. Heart beating and circulation may resume when current ceases. High current can produce respiratory paralysis, which can be reversed with immediate resuscitation. Severe burns to the skin and internal organs. May result in irreparable body damage.

*Difference between men and women is based on the relative body mass of the "average"-sized man and woman (60-cycle AC current).

current. The path is influenced by the resistance of various parts of the body at the time of contact with the conductor. The skin is the major form of resistance to current flow. Current paths through the heart, brain, or trunk are generally much more injurious than paths through extremities.

DETECTION OF ELECTRICAL HAZARDS

Several items of test equipment can be used to verify electrical equipment safety. A **circuit tester** is an inexpensive piece of test equipment with two wire leads capped by probes and connected to a small bulb. Most circuit testers test at least a 110- to 220-volt range. This simple tester can ensure that power has been turned off before electrical maintenance begins. The tester may also be used to determine whether housings and other equipment parts are carrying a current. When one of the leads makes contact with a hot wire and the other lead connects to a grounded conductor, the bulb lights.

A **receptacle wiring tester** is a device with two standard plug probes for insertion into an ordinary 110-volt outlet and a probe for the ground. Indicator lights show an improperly wired receptacle (outlet). However, there are several types of miswiring that are not disclosed by using this tester, including the ground wire to neutral wire mistake. Figure 14–6 shows the meaning of lit indicator lights on the receptacle wiring tester.

A **continuity tester** may be used to determine whether a conductor is properly grounded or has a break in the circuit. Continuity is checked on circuits that are disconnected from a power source. Continuity testers often have an alligator clip on one end of a wire and a bulb and probe on the other end of the same wire. One terminal of the tester can be connected to the equipment housing; the other terminal is connected to a known ground. If the bulb does not light, the equipment is shown to be improperly grounded. With a circuit, the bulb lights when a current is capable of passing through the complete circuit. The unlit bulb of a continuity tester indicates a break in the circuit.

Infrared thermal imaging is another technique that can be used for detecting electrical hazards.

FIGURE 14–6 Receptacle wiring tester indicator lights.

	Lights		
Situation	1	2	3
Correct wiring	On	On	Off
Ground jumped to neutral	On	On	Off
Neutral and ground reversed	On	On	Off
Reversed polarity	On	Off	On
Open ground	On	Off	Off

REDUCTION OF ELECTRICAL HAZARDS

Grounding of electrical equipment is the primary method of reducing electrical hazards. The purpose of grounding is to safeguard people from electrical shocks, reduce the probability of a fire, and protect equipment from damage. Grounding ensures a path to the earth for the flow of excess current. Grounding also eliminates the possibility of a person being shocked by contact with a charged capacitor. The actual mechanism of grounding was discussed at the beginning of this chapter.

Electrical system grounding is achieved when one conductor of the circuit is connected to the earth. Power surges and voltage changes are attenuated and usually eliminated with proper system grounding. **Bonding** is used to connect two pieces of equipment by a conductor. Bonding can reduce potential differences between the equipment and thus reduce the possibility of sparking. Grounding, in contrast, provides a conducting path between the equipment and the earth. Bonding and grounding together are used for entire electrical systems.

Separate equipment grounding involves connecting all metal frames of the equipment in a permanent and continuous manner. If an insulation failure occurs, the current should return to the system ground at the power supply for the circuit. The equipment ground wiring will be the path for the circuit current, enabling circuit breakers and fuses to operate properly. The exposed metal parts of the equipment shown in Figure 14–7 must be grounded or provided with double insulation.[9]

A **ground fault circuit interrupter (GFCI)**, also called a **ground fault interrupter (GFI)**, can detect the flow of current to the ground and open the circuit, thereby interrupting the flow of current. When the current flow in the hot wire is greater than the current in the neutral wire, a **ground fault** has occurred. The GFI provides a safety measure for a person who becomes part of the ground fault circuit. The GFI cannot interrupt current passing between two circuits or between the hot and neutral wires of a three-wire circuit. To ensure safety, equipment must be grounded and protected by a GFI. A GFI should be replaced periodically based on the manufacturer's recommendations.

There are several options for reducing the hazards associated with static electricity. The primary hazard of static electricity is the transfer of charges to surfaces with lower potential. Bonding and grounding are two means of controlling static discharge. **Humidification** is another mechanism for reducing electrical static; it was discussed in the section on sources of electrical hazards. Raising the humidity above 65 percent reduces charge accumulation.[10] However, when the relative humidity exceeds 65 percent,

FIGURE 14–7 Equipment requiring grounding or double insulation.

Portable electric tools such as drills and saws

Communication receivers and transmitters

Electrical equipment in damp locations

Television antenna towers

Electrical equipment in flammable liquid storage areas

Electrical equipment operated with over 150 V

biological agents can begin to grow in heating, ventilation, and air-conditioning (HVAC) ducts and unventilated areas.

Antistatic materials have also been used effectively to reduce electrical static hazards. Such materials either increase the surface conductivity of the charged material or absorb moisture, which reduces resistance and the tendency to accumulate charges.

Ionizers and **electrostatic neutralizers** ionize the air surrounding a charged surface to provide a conductive path for the flow of charges. **Radioactive neutralizers** include a radioactive element that emits positive particles to neutralize collected negative electrical charges. Workers need to be safely isolated from the radioactive particle emitter.

Fuses consist of a metal strip or wire that melts if a current above a specific value is conducted through the metal. Melting the metal causes the circuit to open at the fuse, thereby stopping the flow of current. Some fuses are designed to include a time lag before melting to allow higher currents during startup of the system or as an occasional event.

Magnetic circuit breakers use a solenoid (a type of coil) to surround a metal strip that connects to a tripping device. When the allowable current is exceeded, the magnetic force of the solenoid retracts the metal strip, opening the circuit. **Thermal circuit breakers** rely on excess current to produce heat and bending in a sensitive metal strip. Once bent, the metal strip opens the circuit. Circuit breakers differ from fuses in that they are usually easier to reset after tripping and often provide a lower time lag or none at all before being activated.

Double insulation is another means of increasing electrical equipment safety. Most double-insulated tools have plastic nonconductive housings in addition to standard insulation around conductive materials.

There are numerous methods of reducing the risk of electrocution by lightning. Figure 14–8 lists the major precautions to take.[11]

FIGURE 14–8   Lightning hazard control.

- Place lightning rods so that the upper end is higher than nearby structures.
- Avoid standing in high places or near tall objects. Be aware that trees in an open field may be the tallest object nearby.
- Do not work with flammable liquids or gases during electrical storms.
- Ensure proper grounding of all electrical equipment.
- If inside an automobile, remain inside the automobile.
- If in a small boat, lie down in the bottom of the boat.
- If in a metal building, stay in the building and do not touch the walls of the building.
- Wear rubber clothing if outdoors.
- Do not work touching or near conducting materials, especially those in contact with the earth such as fences.
- Avoid using the telephone during an electrical storm.
- Do not use electrical equipment during the storm.
- Avoid standing near open doors or windows where lightning may enter the building directly.

FIGURE 14-9 Summary of safety precautions for electrical hazards.

- Ensure that power has been disconnected from the system before working with it. Test the system for deenergization. Capacitors can store current after power has been shut off.
- Allow only fully authorized and trained people to work on electrical systems.
- Do not wear conductive material such as metal jewelry when working with electricity.
- Screw bulbs securely into their sockets. Ensure that bulbs are matched to the circuit by the correct voltage rating.
- Periodically inspect insulation.
- If working on a hot circuit, use the buddy system and wear protective clothing.
- Do not use a fuse with a greater capacity than was prescribed for the circuit.
- Verify circuit voltages before performing work.
- Do not use water to put out an electrical fire.
- Check the entire length of electrical cord before using it.
- Use only explosion-proof devices and nonsparking switches in flammable liquid storage areas.
- Enclose uninsulated conductors in protective areas.
- Discharge capacitors before working on the equipment.
- Use fuses and circuit breakers for protection against excessive current.
- Provide lightning protection on all structures.
- Train people working with electrical equipment on a routine basis in first aid and cardiopulmonary resuscitation (CPR).

Another means of protecting workers is **isolating the hazard** from the workers or vice versa. **Interlocks** automatically break the circuit when an unsafe situation is detected. Interlocks may be used around high-voltage areas to keep personnel from entering the area. Elevator doors typically have interlocks to ensure that the elevator does not move when the doors are open. **Warning devices** to alert personnel about detected hazards may include lights, colored indicators, on/off blinkers, audible signals, or labels.

It is better to design safety into the equipment and system than to rely on human behavior such as reading and following labels. Figure 14–9 summarizes the many methods of reducing electrical hazards.

OSHA'S ELECTRICAL STANDARDS

OSHA's standards relating to electricity are found in 29 CFR 1910 (Subpart S). They are extracted from the NEC. This code should be referred to when more detail is needed than appears in OSHA's excerpts. Subpart S is divided into the following two categories of standards: (1) Design of Electrical Systems and (2) Safety-Related Work Practices. The standards in each of these categories are as follows:

DESIGN OF ELECTRICAL SYSTEMS

1910.302	Electric utilization systems
1910.303	General requirements
1910.304	Wiring design and protection
1910.305	Wiring methods, components, and equipment for general use
1910.306	Specific-purpose equipment and installations
1910.307	Hazardous (classified) locations
1910.308	Special systems
1910.309-330	Reserved

SAFETY-RELATED WORK PRACTICES

1910.331	Scope
1910.332	Training
1910.333	Selection and use of work practices
1910.334	Use of equipment
1910.335	Safeguards for personal protection
1910.336-398	Reserved
1910.399	Definitions applicable to this subpart

ELECTRICAL SAFETY PROGRAM

With electrocution accounting for approximately 6 percent of all workplace deaths in the United States every year, it is important that employers have instituted an effective **electrical safety program**. The National Institute for Occupational Safety and Health (NIOSH) recommends the following strategies for establishing such a program:

- Develop and implement a comprehensive safety program and, when necessary, revise existing programs to address thoroughly the area of electrical safety in the workplace.
- Ensure compliance with existing OSHA regulations, Subpart S of 29 CFR 1910.302 through 1910.399 of the General Industry Safety and Health Standards, and Subpart K of 29 CFR 1926.402 through 1926.408 of the OSHA Construction Safety and Health Standards.
- Provide all workers with adequate training in the identification and control of the hazards associated with electrical energy in their workplace.
- Provide additional specialized electrical safety training to those working with or around exposed components of electric circuits. This training should include, but not be limited to, training in basic electrical theory, proper safe work procedures, hazard awareness and identification, proper use of PPE, proper lockout/tagout procedures, first aid including cardiopulmonary resuscitation (CPR), and proper rescue procedures. Provide periodic retraining as necessary.
- Develop and implement procedures to control hazardous electrical energy that include lockout and tagout procedures. Ensure that workers follow these procedures.

- Provide testing or detection equipment for those who work directly with electrical energy that ensure their safety during performance of their assigned tasks.
- Ensure compliance with the NEC and the National Electrical Safety Code.
- Conduct safety meetings regularly.
- Conduct scheduled and unscheduled safety inspections at work sites.
- Actively encourage all workers to participate in workplace safety.
- In a construction setting, conduct a job site survey before starting any work to identify all electrical hazards, implement appropriate control measures, and provide training to employees specific to all identified hazards.
- Ensure that proper PPE is available and worn by workers where required (including fall protection equipment).
- Conduct job hazard analyses of all tasks that may expose workers to the hazards associated with electrical energy and implement control measures that will adequately insulate and isolate workers from electrical energy.
- Identify potential electrical hazards and appropriate safety interventions during the planning phase of construction or maintenance projects. Planning should address the project from start to finish to ensure that workers have the safest possible work environment.[12]

ELECTRICAL HAZARDS SELF-ASSESSMENT

Even the best safety professional cannot be everywhere at once. Consequently, one of the best strategies for safety personnel is to enlist the assistance of supervisors. After all, helping ensure a safe and healthy work environment is part of every supervisor's job description, or at least it should be. To help prevent accidents and injuries from electrical hazards, safety personnel should consider developing checklists supervisors can use to undertake periodic self-assessments in their areas of responsibility. What follows are the types of questions that should be contained in such checklists:

1. Are all electricians in your company up-to-date with the latest requirements of the NEC?
2. Does your company specify compliance with the NEC as part of its contracts for electrical work with outside personnel?
3. Do all electrical installations located in the presence of hazardous dust or vapors meet the NEC requirements for hazardous locations?
4. Are all electrical cords properly strung (i.e., so that they do not hang on pipes, nails, hooks, etc.)?
5. Are all conduits, BX cables, and so on properly attached to supports and tightly connected to junction boxes and outlet boxes?
6. Are all electrical cords free of fraying?
7. Are rubber cords free of grease, oil, chemicals, and other potentially damaging materials?
8. Are all metallic cables and conduit systems properly grounded?
9. Are all portable electric tools and appliances grounded or double insulated?
10. Are all ground connections clean and tightly made?
11. Are all fuses and circuit breakers the proper size and type for the load on each circuit?

12. Are all fuses free of "jumping" (i.e., with pennies or metal strips)?
13. Are all electrical switches free of evidence of overheating?
14. Are all switches properly mounted in clean, tightly closed metal boxes?
15. Are all electrical switches properly marked to show their purpose?
16. Are all electric motors kept clean and free of excessive grease, oil, or potentially damaging materials?
17. Are all electric motors properly maintained and provided with the necessary level of overcurrent protection?
18. Are bearings in all electrical motors in good condition?
19. Are all portable lights equipped with the proper guards?
20. Are all lamps kept free of any and all potentially combustible materials?
21. Is the organization's overall electrical system periodically checked by a person competent in the application of the NEC?[13]
22. Have solidly grounded electrical systems been upgraded to resistance-grounded systems?
23. Are current-limiting devices being used?
24. Are problem-indicating fuses being used?
25. Are touch-safe fuse holders being used?
26. Are arc-flash assessments conducted as required in NFPA 70E?
27. Have arc-flash relays been installed?

PREVENTION OF ARC FLASH INJURIES

Arc flash injuries occur in the workplace every day in this country. Many of these injuries lead to severe burns and even death, which is doubly tragic because the accidents could have been prevented. An **arc flash** is an electrical short circuit that travels through the air rather than flowing through conductors, bus bars, and other types of equipment. The uncontrolled energy released by an arc flash can produce high levels of heat and pressure. It can also cause equipment to explode, sending dangerous shrapnel flying through the air.[14]

Arc flashes are sometimes produced by electrical equipment malfunctions, but a more common cause is accidental human contact with an electrical circuit or conductor. For example, a person working near a piece of energized electrical equipment might accidentally drop a tool that then makes contact with an electrical circuit or conductor. The result is an arc flash that can injure or even kill the worker, not to mention the equipment damage.

Arc flashes become even more hazardous when workers are wearing flammable clothing instead of appropriate PPE. Arc flashes can produce sufficient heat to easily ignite clothing, cause severe burns, and even damage hearing (hearing damage is caused by the high level of pressure that can be released by an arc flash). The best and most obvious way to prevent arc flash injuries is to de-energize the electrical equipment in question and lock or tag it out before beginning maintenance or service work on it.

However, this is not always possible. Some maintenance and service functions such as troubleshooting require that the equipment being worked on be energized. When this is the case, it is important to consult the NFPA's *Handbook for Electrical Safety in the Workplace* (NFPA 70E) as updated in 2009 and proceed as follows:

- Perform a flash hazard analysis in accordance with **NFPA 70E**, Article 130.3, or use Table 130.7(C)(9)(a) (Hazard/Risk Category Classifications) to identify the hazard/risk category of the job tasks that must be performed on the energized equipment.
- Establish a flash protection boundary around the equipment in question in accordance with NFPA 70E, Article 130.3(A).
- Select the PPE that will be worn by the worker(s) who will perform the tasks in question on the energized equipment from Table 130.7(C)(10) (Protective Clothing and PPE Matrix) based on the level of risk identified for these tasks in Table 130.7(C)(9)(a) in the first step above.

For example, if you determine that a worker must perform tasks on a piece of energized electrical equipment and these tasks are rated in "Hazard/Risk Category 3," Table 130.7(C)(10) would require the following PPE:

1. Cotton underwear.
2. Fire-resistant pants and shirt.
3. Fire-resistant coverall.[15]

Maintenance Requirements of NFPA 70E

The 2009 update of NFPA 70E contains several requirements relating to maintenance. These requirements are as follows:

- **205.3 General Maintenance Requirements.** Overcurrent protective devices shall be maintained in accordance with the manufacturer's instructions or with industry consensus standards.
- **210.5 Protective Devices.** Protective devices shall be maintained to adequately withstand or interrupt available fault current.
- **225.1 Fuses.** Fuses shall be maintained free of breaks or cracks in fuse cases, ferrules, and insulators. Fuse clips must be maintained to provide adequate contact with fuses. Fuseholders for current-limiting fuses must not be modified to allow the insertion of fuses that are not current-limiting.
- **225.2 Molded-Case Circuit Breakers.** Molded-case circuit breakers shall be maintained free of cracks in the cases and free of cracked or broken operating handles.
- **225.3 Circuit Breaker Testing.** Circuit breakers that interrupt faults approaching their interrupting ratings shall be inspected and tested in accordance with the manufacturer's instructions.[16]

TRAINING REQUIREMENTS FOR WORKERS

OSHA's training requirements for all workers are contained in 29 CFR 1910. The training requirements for workers who face the risk of electric shock that is not reduced to a safe level are clarified in OSHA CFR 1910.332. The standard's requirements apply to the following classifications of workers:

- Blue-collar supervisors
- Electrical and electronic engineers
- Electrical and electronic equipment assemblers

- Electrical and electronic technicians
- Electricians
- Industrial machine operators
- Material handling equipment operators
- Mechanics and repairers
- Painters
- Riggers and roustabouts
- Stationary engineers
- Welders[17]

The actual training requirements are as follows: (1) skills and techniques necessary to distinguish exposed live parts from other parts of electric equipment, (2) skills and techniques necessary to determine the nominal voltage of exposed live parts, and (3) clearance distances and corresponding voltages to which they will be exposed.[18]

PERMANENT ELECTRICAL SAFETY DEVICES

An emerging technology in electrical safety is the permanent electrical safety device (PESD). PESDs have excellent potential to help workers safely isolate electrical energy, especially when used as part of an organization's Lockout/Tagout Procedures.

> With PESDs incorporated into safety procedures, installed correctly into electrical enclosures, and validated before and after each use, workers can transition the once-risky endeavor of verifying voltage into a less precarious undertaking that never exposes them to voltage…every electrical incident has one required ingredient: voltage. Electrical safety is radically improved by eliminating exposure to voltage while still validating zero energy from outside the panel.[19]

Traditionally electrical energy that might be present is detected using a voltage detector. To ensure safety, NFPA 70E 120.1(5) requires that voltage detectors be checked before and after every use to ensure that they are operating properly. This requirement means that voltage detectors must be checked against an independent source before being used to detect the present of electrical energy in a machine a worker needs to use. Then after checking the machine for electrical energy, the voltage detector must be validated again to an independent source.

This same principle applies to PESDs, but because they are permanently mounted PESDs cannot be moved between two power sources as voltage meters can. With PESDs the validation technique is different. The PESD validation technique is based on the detection of a small amount of current flowing between two sources. A voltage detector is used to relate this small current flow to actual voltage. The worker is then given an appropriate visual or audible signal to indicate whether it is safe to proceed.

PESDs are designed for one purpose and one purpose only: to ensure electrical safety. Consequently, when used properly they can be even more reliable than the traditional method of using voltage meters. Further, PESDs can be used by the workers themselves without the need for having an electrician make the determination of the presence or absence of electrical energy. PESDs improve electrical safety by eliminating exposure to voltage while still ensuring zero voltage from outside the panel. It has been estimated that using PESDs can allow as much as 75 percent of locked out and tagged out equipment to be put back into service.

HAZARDS OF POWER STRIPS AND DAISY CHAINS

A common occurrence is to have more electrical devices than outlets to plug them in to. This is particularly the case when modern organizations operate out of older facilities that were designed before the age of computers and a multitude of other electrically powered devices. In response to an insufficient number of electrical outlets, personnel often use power strips and even interconnected power strips known as **daisy chains** (see Figures 14–10 and 14–11).

Another common response to insufficient electrical outlets is to connect extension cords and power strips. By creating daisy chains, personnel sometimes introduce electrical hazards into the workplace. In fact, hazards relating to power strips, daisy chains, wiring, switches, and outlets account for more than 250 electrical fires every year. But the hazards of daisy chains are not limited to overloads and fires, extension cords and power strips also introduce trip-and-fall hazards.

Connecting and interconnecting power strips or extension cords and power strips can overload which, in turn, can cause a fire. Consequently, creating daisy chains is prohibited by both OSHA regulations and the National Electrical Code. In spite of these

FIGURE 14–10 Power strips are often used when there are not enough electrical outlets.

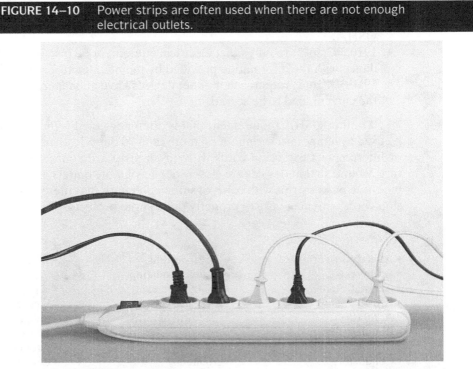

Source: Billion Photos/Shutterstock

FIGURE 14–11   Interconnected power strips are known as daisy chains.

Source: NadyaEugene/Shutterstock

prohibitions, daisy chains represent one of the most common compliance violations found by OSHA personnel. OSHA's 29 CFR 1910 contains the following requirements:

- 1910.303(a) requires that all conductors and related equipment be approved by OSHA.
- 1910.303(b)(2) requires that electrical equipment be installed and used in accordance with the information provided by the manufacturer in the label or listing.
- 1910.304(b)(2) requires that outlet devices have an ampere rating not less than the current load to be served.

Another OSHA requirement that is often neglected is that extension cords are to be used as temporary wiring for no more than 90 days. Rather than create daisy chains or interconnect extension cords with power strips, organizations should move furniture around so that devices are closer to the available outlets, install additional outlets, purchase power strips with cords of sufficient length that there is no need to use extension cords, and make use of properly rated surge protectors.

Key Terms and Concepts

- Alternating current
- Amperes
- Antistatic materials
- Arc
- Arc flash
- Bonding
- Circuit tester
- Conductors
- Continuity tester
- Daisy chain
- Direct current
- Double insulation
- Effective current
- Electrical current
- Electrical hazards
- Electrical safety program
- Electrical system grounding
- Electricity
- Electrons
- Electrostatic hazards
- Electrostatic neutralizers
- Equipment failure
- Freeze
- Frequency
- Fuses
- Ground fault
- Ground fault circuit interrupter (GFCI)

- Ground fault interrupter (GFI)
- Ground wire
- Grounded conductor
- Grounding conductor
- Hazardous locations
- Hertz
- Hot wire
- Humidification
- Improper wiring
- Insulation failure
- Insulators
- Interlocks
- Ionizers
- Isolating the hazard
- Let-go current
- Lightning
- Load

- Magnetic circuit breakers
- National Board of Fire Underwriters
- National Electrical Code (NEC)
- Negative charge
- Neutral wire
- Neutrons
- NFPA 70E
- Ohms
- Ohm's law
- Open ground
- Positive charge
- Potential difference
- Power
- Protons
- Radioactive neutralizers
- Receptacle wiring tester

- Resistance
- Reversed polarity
- Semiconductors
- Separate equipment grounding
- Shock
- Short circuit
- Spark
- Static electricity
- Thermal circuit breakers
- Underwriters Laboratories (UL)
- Voltage
- Warning devices
- Watts
- Zero potential

Review Questions

1. Define *zero potential*. Explain the relationship between zero potential and grounding.
2. Explain what each of the following terms measure: *volt, amp, ohm, hertz*, and *watt*.
3. Briefly discuss the difference between DC and AC using the concept of effective current.
4. What is the relationship between the NEC and UL?
5. Briefly state the relationship among potential difference, lightning, and grounding.
6. Define *open ground*.
7. List at least five lightning hazard control measures.
8. Explain the relationship between circuit load and short circuits.
9. How do ionizers, radioactive neutralizers, and antistatic materials work? Why does humidification work?
10. Discuss how bonding and grounding work together to increase electrical safety.
11. How do continuity testers, circuit testers, and receptacle wiring testers operate?
12. Explain freeze and let-go current.
13. Describe the structure of an atom.
14. Discuss the proper wiring of a three-wire circuit.
15. Why is jumping the ground wire a hazard?
16. Explain reversed polarity.
17. Why are warning devices less effective than designed-in safety precautions?
18. Explain five strategies for establishing an effective electrical safety program.
19. Explain why it is important for safety personnel to help employees and supervisors conduct self-assessments.
20. Explain how to reduce arc flash hazards.
21. What is a "daisy chain" and why is one hazardous?

Endnotes

1. Occupational Safety and Health Administration, "Massachusetts Mill Cited for Safety and Health Violations," *Occupational Health & Safety News* 20, no. 9: 10.
2. Ibid.

3. National Fire Protection Association, *Handbook for Electrical Safety in the Workplace* (Quincy, MA: National Fire Protection Association, 2015), 57.

4. Ibid., 59.

5. Ibid., 60.

6. Ibid., 61.

7. Ibid., 103.

8. Ibid., 104.

9. Ibid., 105.

10. Ibid., 109.

11. Ibid., 110.

12. National Institute for Occupational Safety and Health, *Worker Deaths by Electrocution: A Summary of Surveillance Findings and Investigative Case Reports*, 13–16. Retrieved from cdc.gov/niosh/homepage.html on July 12, 2017.

13. Retrieved from online.migu.nodak.edu/19577/BADM309checklist.htm on July 12, 2017.

14. W. Wallace, "NFPA 70E: Performing the Electrical Flash Hazard Analysis," *Occupational Health & Safety* 74, no. 8: 38–44.

15. National Fire Protection Association, *Handbook for Electrical Safety,* Table 130.7(C)(9)(a).

16. National Fire Protection Association, *Handbook for Electrical Safety,* NFPA 70E updated 2009. Retrieved from nfpa.org on February 17, 2017.

17. OSHA 29 CFR 1910.332. Retrieved from osha.gov on February 17, 2017.

18. Ibid.

19. P. Allen. "Using Permanent Safety Devices," *Occupational Health and Safety*, 81, no. 1: 36.

CHAPTER

15 | FIRE HAZARDS AND LIFE SAFETY

Learning Objectives

- Define the concept of fire hazards
- List the most common sources of fire hazards
- Describe the fire-related dangers to humans
- Explain how to detect fire hazards
- Summarize methods for reducing fire hazards
- Explain how fire safety standards are developed
- List OSHA's fire-related standards
- Summarize OSHA's requirements and recommendation for fire prevention plans
- Summarize OSHA's requirements for exit routes
- Define the concept of life safety
- Summarize the basic requirements of life safety
- Explain the requirements for flame-resistant clothing
- Summarize the basic elements of a fire safety program
- Describe the appropriate response to a fire (including fire drills)
- Explain the types of explosive hazards that might be present in the workplace
- Summarize OSHA's options for fighting fires
- Describe how to conduct a fire protection self-assessment
- Explain the concept and content of a hot work program
- Explain the hazards of chemical burns
- Explain the hazards of heat burns

FIRE HAZARDS DEFINED

Fire hazards are conditions that favor fire development or growth. Three elements are required to start and sustain fire: (1) **oxygen**, (2) **fuel**, and (3) **heat**. Because oxygen is naturally present in most earth environments, fire hazards usually involve the mishandling of fuel or heat.

Fire, or **combustion**, is a chemical reaction between oxygen and a combustible fuel. Combustion is the process by which fire converts fuel and oxygen into energy, usually in the form of heat. By-products of combustion include light and **smoke**. For the reaction to start, a **source of ignition**, such as a spark or open flame, or a sufficiently high temperature is needed. Given a sufficiently high temperature, almost every substance will burn. The **ignition temperature** or **combustion point** is the temperature at which a given fuel can burst into flame.

Fire is a chain reaction. For combustion to continue, there must be a constant source of fuel, oxygen, and heat (see Figure 15–1). The flaming mode is represented by the tetrahedron on the left (heat, oxidizing agent, and reducing agent) that results from a chemical chain reaction. The smoldering mode is represented by the triangle on the right. **Exothermic** chemical reactions create heat. Combustion and fire are exothermic reactions and can often generate large quantities of heat. **Endothermic** reactions consume more heat than they generate. An ongoing fire usually provides its own sources of heat. It is important to remember that cooling is one of the principal ways to control a fire or put it out.

All chemical reactions involve forming and breaking chemical bonds between atoms. In the process of combustion, materials are broken down into basic elements. Loose atoms form bonds with each other to create molecules of substances that were not originally present.

Carbon is found in almost every flammable substance. When a substance burns, carbon is released and then combines with the oxygen that *must* be present to form either **carbon dioxide** or **carbon monoxide**.

Carbon dioxide is produced when there is more oxygen than the fire needs. It is not toxic, but it can be produced in such volumes that it seriously reduces the concentration of oxygen in the air surrounding the fire site. Carbon monoxide—a colorless, odorless, deadly gas—is the result of incomplete combustion of a fuel. It is produced

FIGURE 15–1   Fire tetrahedron (L) and fire triangle (R).

Heat

Oxidizing agent   Reducing agent

Chemical chain reaction

Flaming mode
(TETRAHEDRON)

Fuel   Heat

Fire

Oxygen

Smoldering mode
(TRIANGLE)

when there is insufficient oxygen to burn the fuel present efficiently. In general, most fires have insufficient oxygen and therefore produce large quantities of carbon monoxide. It is important in any intentional industrial fire that the fuel be consumed as completely as possible. This will reduce ash and minimize smoke and gases, including carbon monoxide.

Hydrogen, found in most fuels, combines with oxygen to form water. Synthetic **polymers**, found in plastics and vinyls, often form deadly fumes when they are consumed by fire, or when they melt or disintegrate from being near fire or high heat. Burning, melting, or disintegrating plastic at a fire site should be presumed to be releasing toxic fumes.

Liquids and solids, such as oil and wood, do not burn directly but must first be converted into a flammable **vapor** by heat. Hold a match to a sheet of paper, and the paper will burst into flames. Look closely at the paper, and you will see that the paper is not burning. The flames reside in a vapor area just above the surface of the sheet.

Vapors will burn only at a specific range of mixtures of oxygen and fuel, determined by the composition of the fuel. At the optimum mixture, a fire burns, generates heat and some light, and produces no other by-products. In an unintentional fire, the mixture is constantly changing as more or less oxygen is brought into the flames and more or less heat is generated, producing more or fewer vapors and **flammable gases**.

Remove the fire's access to fuel or remove the oxygen, and the fire dies. Although a spark, flame, or heat may start a fire, the heat that a fire produces is necessary to sustain it. Therefore, a fire may be extinguished by removing the fuel source, starving it of oxygen, or cooling it below the combustion point. Even in an oxygen-rich, combustible environment, such as a hospital oxygen tent, fire can be avoided by controlling heat and eliminating sparks and open flames (see Figure 15–2). The broken lines in the tetrahedron and the triangle indicate that the necessary elements are removed.

An **explosion** is a very rapid, contained fire. When the gases produced exceed the pressure capacity of the vessel, a rupture or explosion must result. The simplest example is a firecracker. The fuse, which usually contains its own source of oxygen, burns into the center of a firecracker. The surrounding powder ignites, and the heat produced vaporizes the balance of the explosive material and ignites it. The tightly wrapped paper of the firecracker cannot contain the expanding gases. The firecracker explodes, in much less time than was required to read about it.

FIGURE 15–2 The broken tetrahedron and broken triangle.

Heat always flows from a higher temperature to a lower temperature, never from a lower temperature to a higher temperature without an outside force being applied. Fires generate heat, which is necessary to sustain the fire. Excess heat is then transferred to surrounding objects, which may ignite, explode, or decompose. **Heat transfer** is accomplished by three means, usually simultaneously: (1) conduction, (2) radiation, and (3) convection.

Conduction is direct **thermal energy** transfer. On a molecular level, materials near a source of heat absorb the heat, raising their kinetic energy. **Kinetic energy** is the energy resulting from a moving object. Energy in the form of heat is transferred from one molecule to the next. Materials conduct heat at varying rates. Metals are very good conductors of heat. Concrete and plastics are poor conductors, hence good insulators. Nevertheless, a heat buildup on one side of a wall will transfer to the other side of the wall by conduction.

Radiation is electromagnetic wave transfer of heat to a solid. Waves travel in all directions from the fire and may be reflected off a surface, as well as absorbed by it. Absorbed heat may raise the temperature beyond a material's combustion point, and then a fire erupts. Heat may also be conducted through a vessel to its contents, which will expand and may explode. An example is the spread of fire through an oil tank field. A fire in one tank can spread to nearby tanks through radiated heat, raising the temperature and pressure of the other tank contents.

Convection is heat transfer through the movement of hot gases. The gases may be the direct products of fire, the results of a chemical reaction, or additional gases brought to the fire by the movement of air and heated at the fire surfaces by conduction. Convection determines the general direction of the spread of a fire. Convection causes fires to rise as heat rises and move in the direction of the prevailing air currents.

All three forms of heat transfer are present at a campfire. A metal poker left in a fire gets red hot at the flame end. Heat is conducted up the handle, which gets progressively hotter until the opposite end of the poker is too hot to touch. People around the fire are warmed principally by radiation, but only on the side facing the fire. People farther away from the fire will be warmer on the side facing the fire than the backs of people closer to the fire. Marshmallows toasted above the flames are heated by convection (see Figure 15–3).

Spontaneous combustion is rare, but it can happen. Organic compounds decompose through natural chemical processes. As they degrade, they release methane gas (natural gas), an excellent fuel. The degradation process—a chemical reaction—produces heat. In a forest, the concentrations of decomposing matter are relatively minimal, and both the gas and the heat vent naturally.

A classic example of spontaneous combustion is a pile of oil-soaked rags. A container of oil seldom ignites spontaneously. A collection of clean fabrics seldom bursts into flames. Rags soaking completely within oil are usually safe. One oil-soaked rag is unlikely to cause a problem. However, in a pile of oil-soaked rags—especially in a closed container—the chemistry is quite different.

The fibers of the rags expose a large surface area of oil to oxidation. The porous nature of rags allows additional oxygen to be absorbed, replacing the oxygen already consumed. When the temperature rises sufficiently, the surfaces of the oil on the rags vaporize.

Hypergolic reactions occur when mixing fuels. Oxidizers produce just such a rapid heat buildup, causing immediate combustion at room temperature with no apparent

FIGURE 15–3   Campfire with convection heat.

Convection

source of ignition. Although the term *hypergolic* originated with rocket propellants, the phenomenon has been around for a long time. **Pyrophor hypergolic fuels** are those that self-ignite in the presence of oxygen found at normal atmospheric concentrations. One example is white phosphorus, which is kept underwater. If it starts to dry out, the phosphorus erupts in flames.

SOURCES OF FIRE HAZARDS

Almost everything in an industrial environment can burn. Metal furniture, machines, plaster, and concrete block walls are usually painted. Most paints and lacquers will easily catch fire. Oxygen is almost always present. Therefore, the principal method of fire suppression is passive—the absence of sufficient heat. Within our environment, various conditions elevate the risk of fire and so are termed *fire hazards*.

For identification, fires are classified according to their properties, which relate to the nature of the fuel. The properties of the fuel directly correspond to the best means of combating a fire (see Figure 15–4).

Without a source of fuel, there is no fire hazard. However, almost everything in our environment can be a fuel. Fuels occur as solids, liquids, vapors, and gases.

Solid fuels include wood, building decorations and furnishings such as fabric curtains and wall coverings, and synthetics used in furniture. What would an office be without paper? What would most factories be without cardboard and packing materials such as Styrofoam molds and panels, shredded or crumpled papers, bubble wrap, and shrink wrap? All these materials easily burn.

Few solid fuels are, or can be made, fireproof. Even fire walls do not stop fires, although they are defined by their ability to slow the spread of fire. Wood and textiles can be treated with fire- or flame-retardant chemicals to reduce their flammability.

Solid fuels are involved in most industrial fires, but mishandling flammable liquids and flammable gases is a major cause of industrial fires. Two often-confused terms applied to flammable liquids are *flash point* and *fire point*. The **flash point** is the lowest temperature for a given fuel at which vapors are produced in sufficient

FIGURE 15–4   Classes of fire.

Class A fires	Solid materials such as wood, plastics, textiles, and their products: paper, housing, clothing.
■ Class B fires	Flammable liquids and gases.
■ Class C fires	Electrical (referring to live electricity situations, not including fires in other materials started by electricity).
■ Class D fires	Combustible, easily oxidized metals such as aluminum, magnesium, titanium, and zirconium.
■ Special categories	Extremely active oxidizers or mixtures, flammables containing oxygen, nitric acid, hydrogen peroxide, and solid missile propellants.
■ Do not use water to extinguish.	

Source: osha.gov

concentrations to flash in the presence of a source of ignition. The **fire point** is the minimum temperature at which the vapors continue to burn, given a source of ignition. The **auto-ignition temperature** is the lowest point at which the vapors of a liquid or solid self-ignite *without* a source of ignition. Other important terms relating to flammable and combustible liquids as defined in OSHA's 29 CFR 1910.106 are (1) *Lower flammable limit*—The percentage of vapor in the air above which a fire cannot occur because there is insufficient fuel (the mixture is too lean); (2) *Upper flammable limit*—The percentage of vapor in the air above which there is insufficient air for a fire (the mixture is too rich); (3) *Vapor density*—The weight of a flammable vapor compared to air in which air = 1; and (4) *Permissible exposure limit (PEL)*—It is expressed in parts of vapor per million parts of contaminated air (important because many vapors present inhalation hazards as well as fire hazards).

Flammable liquids have a flash point below 37.7°C (99.8°F). **Combustible liquids** have a flash point at or higher than that temperature. Both flammable and combustible liquids are further divided into the three classifications shown in Figure 15–5.

As the temperature of any flammable liquid increases, the amount of vapor generated on the surface also increases. Safe handling, therefore, requires both knowledge

FIGURE 15–5   Classes of flammable and combustible liquids.

Flammable Liquids

Class I–A	Flash point below 73°F (22.8°C), boiling point below 100°F (37.8°C).
Class I–B	Flash point below 73°F (22.8°C), boiling point at or above 100°F (37.8°C).
Class I–C	Flash point at or above 73°F (22.8°C), but below 100°F (37.8°C).

Combustible Liquids

Class II	Flash point at or above 100°F (37.8°C), but below 140°F (60°C).
Class III–A	Flash point at or above 140°F (60°C), but below 200°F (93.3°C).
Class III–B	Flash point at or above 200°F (93.3°C).

Source: osha.gov

of the properties of the liquid and an awareness of ambient temperatures in the work or storage place. The **explosive range**, or **flammable range**, defines the concentrations of a vapor or gas in air that can ignite from a source. The auto-ignition temperature is the lowest temperature at which liquids spontaneously ignite.

Most flammable liquids are lighter than water. If the flammable liquid is lighter than water, water cannot be used to put out the fire.[1] The application of water floats the fuel and spreads a gasoline fire. Crude oil fires burn even while floating on fresh or sea water.

Unlike solids (which have a definite shape and location) and unlike liquids (which have a definite volume and are heavier than air), gases have no shape. Gases expand to fill the volume of the container in which they are enclosed, and they are frequently lighter than air. Released into air, gas concentrations are difficult to monitor due to the changing factors of air, current direction, and temperature. Gases may stratify in layers of differing concentrations but often collect near the top of whatever container in which they are enclosed. Concentrations found to be safe when sampled at workbench level may be close to, or exceed, flammability limits if sampled just above head height.

The **products of combustion** are gases, flame (light), heat, and smoke. Smoke is a combination of gases, air, and suspended particles, which are the products of incomplete combustion. Many of the gases present in smoke and at a fire site are toxic to humans. Other, usually nontoxic, gases may replace the oxygen normally present in air. Most fatalities associated with fire are from breathing toxic gases and smoke and from being suffocated because of oxygen deprivation. Gases that may be produced by a fire include acrolein, ammonia, carbon monoxide, carbon dioxide, hydrogen bromide, hydrogen cyanide, hydrogen chloride, hydrogen sulfide, sulfur dioxide, and nitrogen dioxide. Released gases are capable of traveling across a room and randomly finding a spark, flame, or adequate heat source, **flashing back** to the source of the gas.

The National Fire Protection Association (NFPA) has devised the NFPA 704 system for quick identification of hazards presented when substances burn (see Figure 15–6). The NFPA's red, blue, yellow, and white diamond is used on product labels, shipping cartons, and buildings. Ratings within each category are 0 to 4,

FIGURE 15–6 Identification of fire hazards.

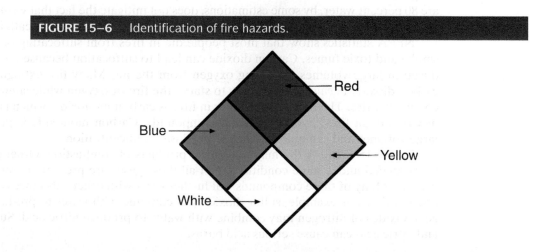

where 0 represents no hazard and 4, the most severe hazard level. The colors refer to a specific category of hazard:

Red   = Flammability
Blue   = Health
Yellow = Reactivity
White   = Special information

Although we do not think of electricity as burning, natural and generated electricity play a large role in causing fires. Lightning strikes cause many fires every year. In the presence of a flammable gas or liquid mixture, one spark can produce a fire.

Electrical lines and equipment can cause fires either by a short circuit that provides an ignition spark, by arcs, or by resistances generating a heat buildup. Electrical switches and relays commonly arc as contact is made or broken.

Another source of ignition is heat in the form of hot surfaces. It is easy to see the flame hazard present when cooking oil is poured on a very hot grill. The wooden broom handle leaning up against the side of a hot oven may not be as obvious a hazard. Irons used in textile manufacturing and dry-cleaning plants also pose a heat hazard.

Space heaters frequently have hot sides, tops, backs, and bottoms, in addition to the heat-generating face. Hotplates, coffee pots, and coffee makers often create heated surfaces. Many types of electric lighting generate heat, which is transferred to the lamp housing.

Engines produce heat, especially in their exhaust systems. Compressors produce heat through friction, which is transferred to their housings. Boilers produce hot surfaces, as do steam lines and equipment using steam as power. Radiators, pipes, flues, and chimneys all have hot surfaces. Metal stock that has been cut by a blade heats up as the blade does. Surfaces exposed to direct sunlight become hot surfaces and transmit their heat by conduction to their other side. Heated surfaces are a potential source of fire.

FIRE DANGERS TO HUMANS

Direct or near direct contact with flame, also known as thermal radiation, is obviously dangerous to humans. Flesh burns, as do muscles and internal organs. The fact that we are 80 percent water, by some estimations, does not mitigate the fact that virtually all the other 20 percent burns. Nevertheless, burns are not the major cause of death in a fire.

NFPA statistics show that most people die in fires from suffocating or breathing smoke and toxic fumes. Carbon dioxide can lead to suffocation because it can be produced in large volumes, depleting oxygen from the air. Many fire extinguishers use carbon dioxide because of its ability to starve the fire of oxygen while simultaneously cooling the fire. The number one killer in fires is carbon monoxide, which is produced in virtually all fires involving organic compounds. Carbon monoxide is produced in large volumes and can quickly reach lethal dosage concentrations.

Figure 15–7 shows the major chemical products of combustion. Other gases may be produced under some conditions. Not all these gases are present at any particular fire site. Many of these compounds will further react with other substances often present at a fire. For example, sulfur dioxide will combine with water to produce sulfuric acid. Oxides of nitrogen may combine with water to produce nitric acid. Sulfuric acid and nitric acid can cause serious acid burns.

FIGURE 15–7 Major chemical products of combustion.

Product	Fuels	Pathology
Acrolein	Cellulose, fatty substances, woods, and paints	Highly toxic irritant to eyes and respiratory system
Ammonia (NH_3)	Wool, silk, nylon, melamine, refrigerants, hydrogen–nitrogen compounds	Somewhat toxic irritant to eyes and respiratory system
Carbon dioxide (CO_2)	All carbon and organic compounds	Not toxic, but depletes available oxygen
Carbon monoxide (CO)	All carbon and organic compounds	Can be deadly
Hydrogen chloride (HCl)	Wool, silk, nylon, paper, polyurethane, rubber, leather, plastic, wood	Quickly lethal asphyxiant
Hydrogen sulfide (H_2S)	Sulfur-containing compounds, rubber, crude oil	Highly toxic gas; strong odor of rotten eggs, but quickly destroys sense of smell
Nitrogen dioxide (NO_2)	Cellulose nitrate, celluloid, textiles, other nitrogen oxides	Lung irritant, causing death or damage
Sulfur dioxide (SO_2)	Sulfur and sulfur-containing compounds	Toxic irritant

DETECTION OF FIRE HAZARDS

Many automatic fire detection systems are used in the industry today. Many systems can warn of the presence of smoke, radiation, elevated temperature, or increased light intensity. **Thermal expansion detectors** use a heat-sensitive metal link that melts at a predetermined temperature to make contact and ultimately sound an alarm. Heat-sensitive insulation can be used, which melts at a predetermined temperature, thereby initiating a short circuit and activating the alarm.

Photoelectric fire sensors detect changes in infrared energy that is radiated by smoke, often by the smoke particles obscuring the photoelectric beam. A relay is open under acceptable conditions and closed to complete the alarm circuit when smoke interferes.

Ionization or **radiation sensors** use the tendency of a radioactive substance to ionize when exposed to smoke. The substance becomes electrically conductive with the smoke exposure and permits the alarm circuit to be completed.

Ultraviolet detectors or **infrared detectors** sound an alarm when the radiation from fire flames is detected. When rapid changes in radiation intensities are detected, a fire alarm signal is given.

The Occupational Safety and Health Administration (OSHA) has mandated the monthly and annual inspection and recording of the condition of fire extinguishers in industrial settings. A hydrostatic test to determine the integrity of the fire extinguisher metal shell is recommended according to the type of fire extinguisher. The hydrostatic test measures the capability of the shell to contain internal pressures and the pressure shifts expected to be encountered during a fire.

REDUCTION OF FIRE HAZARDS

The best way to reduce fires is to prevent them. A major cause of industrial fires is hot, poorly insulated machinery and processes. One means of reducing a fire hazard is the **isolation** of the three triangle elements: fuel, oxygen, and heat. In the case of fluids, closing a valve may stop the fuel element.

Fires may also be prevented by the proper storage of flammable liquids. Liquids should be stored as follows:

- In flame-resistant buildings that are isolated from places where people work. Proper drainage and venting should be provided for such buildings.
- In tanks below ground level.
- On the first floor of multistory buildings.

Substituting less-flammable materials is another effective technique for fire reduction. A catalyst or fire inhibitor can be employed to create an endothermic energy state that eventually smothers the fire. Several ignition sources can be eliminated or isolated from fuels:

- Prohibit smoking near any possible fuels.
- Store fuels away from areas where electrical sparks from equipment, wiring, or lightning may occur.
- Keep fuels separate from areas where there are open flames. These may include welding torches, heating elements, or furnaces.
- Isolate fuels from tools or equipment that may produce mechanical or static sparks. Other strategies for reducing the risk of fires are as follows:
- Clean up spills of flammable liquids as soon as they occur. Properly dispose of the materials used in the cleanup.
- Keep work areas free from extra supplies of flammable materials (e.g., paper, rags, and boxes). Have only what is needed on hand with the remaining inventory properly stored.
- Run electrical cords along walls rather than across aisles or in other trafficked areas. Cords that are walked on can become frayed and dangerous.
- Turn off the power and completely de-energize equipment before conducting maintenance procedures.
- Don't use spark- or friction-prone tools near combustible materials.
- Routinely test fire extinguishers.

Number, Location, and Selection of Portable Fire Extinguishers

Organizations are required by OSHA in 29 CFR 1910.150(c) to have on hand the proper number of portable fire extinguishers of the right type and in the right locations. For example, OSHA has the following requirements concerning the number and locations of portable fire extinguishers: (1) A fire extinguisher rated no less than 2A is required for each 3,000 square feet of building areas to be protected or any major fraction of that area; (2) No fire extinguisher may be located more than 100 feet from any point in the protected area; (3) One or more fire extinguishers rated no less than 2A shall be located on each floor of a multistory building; (4) In multistory buildings at least one fire extinguisher must be located adjacent to stairways; (5) A fire extinguisher rated no less than 10A must be located within 50 feet of any location where five gallons or more of flammable or combustible liquids or five pounds of flammable gas are stored or used in the workplace; (6) Toxic vaporizing liquid fire extinguishers are prohibited; (7) Portable fire extinguishers must be maintained and inspected in accordance with NFPA Number 10A-1970; and (8) Only fire extinguishers that have been approved by a nationally recognized testing laboratory may be used to meet the requirements of 29 CFR 1910.150(c).

Choosing the right fire extinguishers for the setting in question is important. The determination of which type of fire extinguisher is appropriate in a given setting is based on the class of fire that is most likely to occur: Class A, B, C, or D. The following types of fire extinguishers are appropriate for the corresponding class of fire:

- **Class A fires.** These are wood, paper, and trash fires. Appropriate fire extinguishers for Class A fires are water, foam, and multipurpose (ABC) extinguishers.
- **Class B fires.** These are flammable liquid, gasoline, paints, and grease fires. Appropriate fire extinguishers for Class B fires are carbon dioxide, sodium or potassium bicarbonate, or multipurpose (ABC) extinguishers.
- **Class C fires.** These are electrical fires. Appropriate fire extinguishers for Class C fires are carbon dioxide, sodium or potassium bicarbonate, or multipurpose (ABC) extinguishers.
- **Class D fires.** These are combustible metal fires. Class D fires require special extinguishing agents that have been approved by nationally recognized testing organizations.

Fire-Extinguishing Systems

In larger or isolated industrial facilities, an employee **fire brigade** may be created. (See OSHA requirements in next section.) **Standpipe and hose systems** provide the hose and pressurized water for firefighting. Hoses for these systems usually vary from one inch to 2.5 inches in diameter.[2]

Automatic sprinkler systems are an example of a **fixed extinguishing system** because the sprinklers are fixed in position. Water is the most common fluid released from the sprinklers. Sprinkler supply pipes may be kept filled with water in heated buildings; in warmer climates, valves are used to fill the pipes with water when the sprinklers are activated. When a predetermined heat threshold is breached, water flows to the heads and is released from the sprinklers.

Portable fire extinguishers are classified by the types of fire that they can most effectively reduce. Figure 15–8 describes the four major fire extinguisher classifications. Blocking or shielding the spread of fire can include covering the fire with an

FIGURE 15–8 Fire extinguisher characteristics.

Fire Class	Extinguisher Contents	Mechanism	Disadvantages
A	Foam, water, dry chemical	Chain-breaking cooling, smothering, and diluting	Freezes if not kept heated.
B	Dry chemical, bromotrifluoromethane, and other halogenated compounds, foam, CO_2, dry chemical	Chain-breaking smothering, cooling, and shielding	Halogenated compounds are toxic.
C	Bromotrifluoromethane, CO_2, dry chemical	Chain-breaking smothering, cooling, and shielding	Halogenated compounds are toxic; fires may ignite after CO_2 dissipates.
D	Specialized powders such as graphite, sand, limestone, soda ash, sodium chloride	Cooling, smothering	Expensive cover of powder may be broken with resultant reignition.

inert foam, inert powder, nonflammable gas, or water with a thickening agent added. The fire may suffocate under such a covering. Flooding a liquid fuel with nonflammable liquid can dilute this fire element. Figures 15–9, 15–10, and 15–11 are photographs of effective fire prevention equipment.

FIGURE 15–9 Fireproof filing cabinets.

Source: Marynchenko Oleksandr/Shutterstock

FIGURE 15–10 Storing chemicals in fire-resistant barrels.

Source: Isaxar/Shutterstock

OSHA Regulations for Fire Brigades

Fire brigade regulations are covered in 29 CFR 1910.156 (Subpart L, Appendix A). Relevant requirements from the regulations are as follows:

1. **Scope.** Employers are not required to form a fire brigade. However, if an employer does decide to organize a fire brigade, the requirements of this section apply.
2. **Prefire planning.** Have prefire planning conducted by the local fire department or the workplace fire brigade so that they may become familiar with the

FIGURE 15–11 Storing biological hazards in fire-resistant barrels.

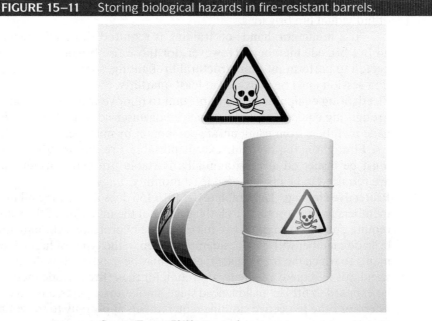

Source: Farmer79/Shutterstock

workplace and process hazards. Involvement with the local fire department or fire prevention bureau is encouraged to facilitate coordination and cooperation between members of the fire brigade and those who may be called upon for assistance during a fire emergency.

3. **Organizational statement.** The organizational statement should contain the following information: a description of the duties that the fire brigade members are expected to perform, the line authority of each fire brigade officer, the number of the fire brigade officers and number of training instructors, and a list and description of the types of awards or recognition that brigade members may be eligible to receive.

4. **Physical capability.** The physical capability requirement applies only to those fire brigade members who perform interior structural firefighting. Employees who cannot meet the physical capability requirement may still be members of the fire brigade as long as such employees do not perform interior structural firefighting. It is suggested that fire brigade members who are unable to perform interior structural firefighting be assigned less stressful and physically demanding fire brigade duties (e.g., certain types of training, record keeping, fire prevention inspection and maintenance, and fire pump operations).

   "Physically capable" can be defined as being able to perform those duties specified in the training requirements of Section 1910.156(c). Physical capability can also be determined by physical performance tests or by a physical examination when the examining physician is aware of the duties that the fire brigade member is expected to perform.

5. **Training and education.** Training and education must be commensurate with those functions that the fire brigade is expected to perform (i.e., those functions specified in the organizational statement). Such a performance requirement provides the necessary flexibility to design a training program that meets the needs of individual fire brigades.

   At a minimum, hands-on training is required to be conducted annually for all fire brigade members. However, for those fire brigade members who are expected to perform interior structural firefighting, some type of training or education session must be provided at least quarterly.

6. **Firefighting equipment.** It is important to remove from service and replace any firefighting equipment that is damaged or unserviceable. This prevents fire brigade members from using unsafe equipment by mistake.

   Firefighting equipment, except portable fire extinguishers and respirators, must be inspected at least annually. Portable fire extinguishers and respirators are required to be inspected at least monthly.

7. **Protective clothing.** Paragraph (e) of 1910.156 does not require all fire brigade members to wear protective clothing. It is not the intention of the standards to require employers to provide a full ensemble of protective clothing for every fire brigade member without consideration given to the types of hazardous environments to which the fire brigade member may be exposed. It is the intention of the standards to require adequate protection for those fire brigade members who may be exposed to fires in an advanced stage, smoke, toxic gases, and high temperatures. Therefore, the protective clothing requirements apply only to those fire brigade members who perform interior structural-firefighting operations.

Additionally, the protective clothing requirements do not apply to the protective clothing worn during outside firefighting operations (brush and forest fires, crash crew operations) or other special firefighting activities. It is important that the protective clothing to be worn during these types of firefighting operations reflect the hazards that may be encountered by fire brigade members.

8. **Respiratory protective devices.** Respiratory protection is required to be worn by fire brigade members while working inside buildings or confined spaces where toxic products of combustion or an oxygen deficiency are likely to be present; respirators are also to be worn during emergency situations involving toxic substances. When fire brigade members respond to emergency situations, they may be exposed to unknown contaminants in unknown concentrations. Therefore, it is imperative that fire brigade members wear proper respiratory protective devices during these situations. Additionally, there are many instances where toxic products of combustion are still present during mop-up and overhaul operations. Therefore, fire brigade members should continue to wear respirators during these types of operations.[3]

Disaster Preparations

Training employees may be the most successful lifesaving preparation for a fire disaster. Company fire brigade members should be trained and tested at least quarterly. Disaster preparation initially requires management commitment and planning and continued response and recovery practice by the fire brigade on a regular basis. Also necessary are regular, but less frequent, fire drills for all personnel.

Disaster preparations also include the integration of company planning with community plans. Community disaster relief agencies such as the police, fire department, Red Cross, and hospitals should be consulted and informed of company disaster preparation plans.

Preventing Office Fires

The shop floor is not the only part of the plant where fire hazards exist. Offices are also susceptible to fires. There are approximately 7,000 office fires a year. The following strategies are helpful in preventing office fires:

- Confine smoking to designated areas that are equipped with nontip ashtrays and fire-resistant furnishings.
- Periodically check electrical circuits and connections. Replace frayed or worn cords immediately.
- Make sure that extension cords and other accessories are Underwriters Laboratories (UL)-approved and used only as recommended.
- Make sure there is plenty of air space left around copying machines and other office machines that can overheat.
- Locate heat-producing appliances away from the wall or anything else that can ignite.
- Frequently inspect personal appliances such as hotplates, coffee pots, and cup warmers. Assign responsibility for turning off such appliances every day to a specific person.
- Keep aisles, stairwells, and exits clear of paper, boxes, and other combustible materials.[4]

DEVELOPMENT OF FIRE SAFETY STANDARDS

The purpose of modern fire safety standards is the protection of life and the prevention of property damage. However, the impetus for developing standards has always been and continues to be the occurrence of major disasters. Typically, standards are developed after a major tragedy occurs in which property is damaged on a large scale and lives are lost. Public shock turns into an outcry for action. A flurry of political activity follows, and agencies and organizations that develop standards are called on to develop new standards.

The trend in fire safety standards is toward performance-based standards and away from the traditional specification-based approach. An example of each type of standard will help illustrate the difference. A specification-based standard may require that brick, concrete, or steel material be used in a given type of building. A performance-based standard may specify that materials used have a one-, two-, or four-hour fire resistance rating.[5] Advances in the testing of engineering materials will help overcome most of the barriers to full development and implementation of performance-based standards.

OSHA FIRE STANDARDS

OSHA standards for fire protection appear in 29 CFR 1910.156 (Subpart L). This subpart contains the standards for fire brigades, fixed fire suppression equipment, and other fire protection systems. Employers are not required to form fire brigades, but those who choose to do so must meet a number of specific requirements. There are other fire-related requirements in other subparts. For example, fire exits, emergency action plans, and means of egress are covered in Subpart E. The standards in Subpart L are as follows:

FIRE PROTECTION
1910.155	Scope, application, and definitions
1910.156	Fire brigades

PORTABLE FIRE SUPPRESSION EQUIPMENT
1910.157	Portable fire extinguishers
1910.158	Standpipe and hose systems

FIXED FIRE SUPPRESSION EQUIPMENT
1910.159	Automatic sprinkler systems
1910.160	Fixed extinguishing systems, general
1910.161	Fixed extinguishing systems, dry chemical
1910.162	Fixed extinguishing systems, gaseous agent
1910.163	Fixed extinguishing systems, water spray and foam

OTHER FIRE PROTECTION SYSTEMS
1910.164	Fire detection systems
1910.165	Employee alarm systems

Figure 15–12 summarizes fire prevention and suppression strategies.

FIGURE 15–12 Fire prevention and suppression summary.

- Use the least-flammable materials whenever possible.
- Analyze the company to determine types of potential fires and provide appropriate sprinklers and/or extinguishers.
- Develop a database of the flammability of materials available in the company.
- Store containers or flammable materials away from sources of heat or sparks and away from humans.
- Do not permit smoking near flammable materials.
- Include a venting mechanism in storage containers and locate them near a drain.
- Minimize fuel storage container size to reduce the size of the fire that may involve those fuels.
- Isolate fuels from sources of heat.
- Include a smoke detection system and portable fire extinguishers in the facility. Extinguishers should be easily available to every workstation.
- Make sure sources of heat have controlling mechanisms and are near fire detection equipment.
- Check fire extinguishing equipment regularly.
- Perform periodic inspections for fire hazards and reappraisal of fire hazards.
- Train plant personnel in basic fire prevention, which should include periodic fire drills.
- Make sure fire brigade personnel are well trained, tested, and regularly practice fire control.
- Stress cleanliness and an organized method of disposal of flammable materials.

OSHA AND FIRE PREVENTION PLANS

OSHA has both requirements and recommendations relating to the development of fire prevention plans. The requirements apply only to organizations that use and/or store ethylene oxide (29 CFR 1910.1047), methylenedianiline (29 CFR 1910.1050), or 1,3 butadiene (29 CFR 1910.1051). However, OSHA recommends that all employers develop fire prevention plans. Those that have 10 or fewer employees may communicate the plan to them verbally. Those with more than 10 employees are to put the plan in writing, kept in the workplace, and made accessible to employees. OSHA's requirements/recommendations concerning fire prevention plans are as follows:

- Make a list of all fire hazards in the workplace.
- Record the proper handling and storage procedures for all hazardous materials in the workplace.
- List potential ignition sources in the workplace and their means of control.
- Record the type(s) of fire protection equipment needed to control all fire hazards in the workplace.
- Develop and record procedures to control flammable and combustible waste materials.

- Develop and record procedures preventing the accidental ignition of combustible material through regular maintenance of safeguards on heat-producing equipment.
- List the names or job titles of all personnel who are responsible for controlling fuel source hazards.
- List the names or titles of all personnel who are responsible for preventing or controlling sources of ignition or fires through regular equipment maintenance.

OSHA REQUIREMENTS FOR EXIT ROUTES

Establishing adequate and proper exit routes is an important part of preventing injuries and deaths from fires. Because having adequate exit routes is so important, OSHA has specific requirements for their provision. However, before considering OSHA's requirements for exit routes, it is important for safety and health professionals and other personnel to understand exit-related terminology. Key concepts relating to exits and exit routes are as follows.

- *Exit routes.* A continuous unobstructed path that leads from any location in the workplace to a place of safety. Exit routes have three components: exit access, exit, and exit discharge.
- *Exit access.* Any portion of an exit route that leads to an exit.
- *Exit.* Any part of the exit route that provides a means of access to the exit discharge.
- *Exit discharge.* Any part of the exit route that leads directly outside of the building or to an open space that has access to the outside. An exit discharge may lead to a stairway, street, sidewalk, public way, or designated refuge area, but it must lead outside the building.

With these terms understood, OSHA's requirements for exit routes may now be summarized under four generic headings: required number of exit routes, design considerations, operational issues, and specific exit requirements.

- *Required number of exit routes.* Under normal conditions, two exit routes are required. However, more are required if the number of employees, size of the building, or arrangement of the workspace will not allow employees to exit safely with just two exit routes. Only one exit route is sufficient by exception if it can be demonstrated that these same factors are such that all employees may safely exit with just one.
- *Examples of design requirements.* Exit routes must be permanent parts of the building and must support the maximum permitted occupancy load for each floor, ceilings of exit routes must be no less than 7' – 6" in height, and doors must be unlocked from the inside.
- *Examples of operational requirements.* Exit routes must be kept free of explosive or highly flammable furnishings and unobstructed by materials, equipment, locked doors, or dead-end corridors. They must also be properly lighted for people of normal vision and equipped with "EXIT" signs in plainly legible letter.
- *Examples of exit requirements.* Exits are to be separated by fire resistant materials and should have only those openings that are necessary to allow access to the exit from occupied areas or to the exit discharge.

What is provided in this section is a representative sample of OSHA's requirements for exit routes. To view all of the requirements, go to OSHA's website (www.osha.gov) and download 29 CFR 1910.33 through 1910.39: *Exit Routes, Emergency Action Plans, and Fire Prevention Plans.*

LIFE SAFETY

Life safety involves protecting the vehicles, vessels, and lives of people in buildings and structures from fire. The primary reference source for life safety is the **Life Safety Code**, published by the NFPA. The code applies to new and existing buildings. It addresses the construction, protection, and occupancy features necessary to minimize the hazards of fire, smoke, fumes, and panic. A major part of the code is devoted to the minimum requirements for design of egress necessary to ensure that occupants can quickly evacuate a building or structure.

BASIC REQUIREMENTS

In this section, the term *structure* refers to a structure or building.

- Every structure, new and existing, that is to be occupied by people must have a **means of egress** and other fire protection safeguards that (1) ensure that occupants can promptly evacuate or be adequately protected without evacuating and (2) provide sufficient backup safeguards to ensure that human life is not endangered if one system fails.
- Every structure must be constructed or renovated, maintained, and operated in such a way that occupants are (1) protected from fire, smoke, or fumes, (2) protected from fire-related panic, (3) protected long enough to allow a reasonable amount of time for evacuation, and (4) protected long enough to defend themselves without evacuating.
- In providing structures with means of egress and other fire protection safeguards, the following factors must be considered: (1) character of the occupancy, (2) capabilities of occupants, (3) number of occupants, (4) available fire protection, (5) height of the structure, (6) type of construction, and (7) any other applicable concerns.
- No lock or other device may be allowed to obstruct egress in any part of a structure at any time that it is occupied. The only exceptions to this requirement are mental health detention and correctional facilities. In these, the following criteria are required: (1) responsible personnel must be available to act in the case of fire or a similar emergency and (2) procedures must be in place to ensure that occupants are evacuated in the event of an emergency.
- All exits in structures must satisfy the following criteria: (1) be clearly visible or marked in such a way that an unimpaired individual can readily discern the route of escape; (2) all routes to a place of safety must be arranged or clearly marked; (3) any doorway and passageway that may be mistaken as a route to safety must be arranged or clearly marked in such a way as to prevent confusion in an emergency; and (4) all appropriate steps must be taken to ensure that occupants do not mistakenly enter a dead-end passageway (see Figure 15–13).

FIGURE 15–13 Fire exits must be clearly marked.

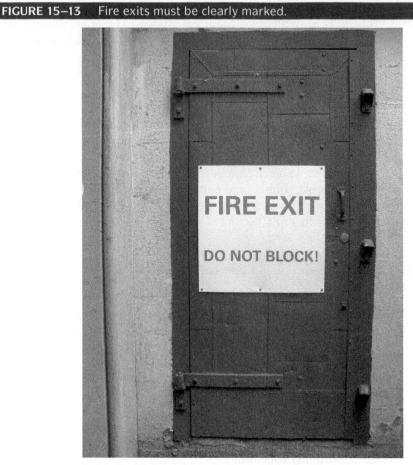

Source: Ira/Shutterstock

- Egress routes and facilities must be included in the lighting design wherever artificial illumination is required in a structure.
- Fire alarm systems must be provided in any facility that is large enough or so arranged that a fire itself may not adequately warn occupants of the danger. Fire alarms should alert occupants to initiate appropriate emergency procedures.
- In any structure or portion of a structure in which a single means of egress may be blocked or overcrowded in an emergency situation, at least two means of egress must be provided. The two means of egress must be arranged in such a way as to minimize the possibility of both becoming impassable in the same emergency situation.
- All stairs, ramps, and other means of moving from floor to floor must be enclosed (or otherwise protected) to afford occupants protection when used as a means of egress in an emergency situation. These means of vertical movement should also serve to inhibit the spread of fire, fumes, and smoke from floor to floor.
- Compliance with the requirements summarized herein does not eliminate or reduce the need to take other precautions to protect occupants from fire

hazards, nor does it permit the acceptance of any condition that could be hazardous under normal occupancy conditions.[6]

The information in this section is a summary of the broad fundamental requirements of the *Life Safety Code* of the NFPA. More specific requirements relating to means of egress and features of fire protection are explained in the following sections.

Means of Egress

This section explains some of the more important issues in the *Life Safety Code* relating to means of egress. Students and practitioners, who need more detailed information are encouraged to refer to the *Life Safety Code*.

1. **Doors.** Doors that serve as exits must be designed, constructed, and maintained in such a way that the means of egress is direct and obvious. Windows that could be mistaken for doors in an emergency situation must be made inaccessible to occupants.
2. **Capacity of means of egress.** The means of egress must have a capacity sufficient to accommodate the occupant load of the structure calculated in accordance with the requirements of the *Life Safety Code*.
3. **Number of means of egress.** Any component of a structure must have a minimum of two means of egress (with exceptions as set forth in the code). The minimum number of means of egress from any story or any part of a story is three for occupancy loads of 500 to 1,000 and four for occupancy loads of more than 1,000.
4. **Arrangement of means of egress.** All exits must be easily accessible at all times in terms of both location and arrangement.
5. **Measurement of travel distance to exits.** The travel distance to at least one exit must be measured on the walking surface along a natural path of travel beginning at the most remote occupied space and ending at the center of the exit. Distances must comply with the code.
6. **Discharge from exits.** All exits from a structure must terminate at a public way or at yards, courts, or open spaces that lead to the exterior of the structure.

SAFETY MYTH

FIRE RATINGS FOR DOORS, WALLS, AND FLOORS

A three-hour fire wall will provide protection for three hours. Right? Not necessarily. In fact, any relationship between a fire rating and the reality of fire resistance may be little more than coincidental. The problem is that the factors used to determine fire ratings are outdated. They are based on materials that were in use many years ago and no longer have any relevance. The materials used for constructing and furnishing buildings today are radically different from those upon which fire ratings are based. Consider fire ratings overvalued, and plan for less time than they allow.

7. **Illumination of means of egress.** All means of egress shall be illuminated continuously during times when the structure is occupied. Artificial lighting must be used as required to maintain the necessary level of illumination. Illumination must be arranged in such a way that no area is left in darkness by a single lighting failure.

8. **Emergency lighting.** Emergency lighting for all means of egress must be provided in accordance with the code. In cases where maintaining the required illumination depends on changing from one source of power to another, there shall be no appreciable interruption of lighting.

9. **Marking of means of egress.** Exits must be marked by readily visible, approved signs in all cases where the means of egress is not obviously apparent to occupants. No point in the exit access corridor shall be more than 100 feet from the nearest sign, Figure 15–14.

10. **Special provisions for high-hazard areas.** If an area contains contents that are classified as highly hazardous, occupants must be able to exit by traveling no more than 75 feet. At least two means of egress must be provided, and there shall be no dead-end corridors.[7]

FIGURE 15–14   Exits must be clearly marked.

Source: igorstevanovic/Shutterstock

11. **Exit Sign Requirements.** All exit signs must contain the word "Exit" in plainly legible letters that are no less than six inches high and 0.75 inches wide (in the main stroke of the letter).

12. **Exit Sign Lighting Requirements.** Light sources for exit signs may be external or internal. Regardless of whether the light source is external or internal, the face of the sign must be continually illuminated while the building is occupied from a reliable source as determined by the Authority Having Jurisdiction (AHJ).

The requirements summarized in this section relate to the fundamental specifications of the *Life Safety Code* relating to means of egress. For more detailed information concerning general requirements, means of egress, and other factors such as fire protection and fire protection equipment, refer to the actual code.

FLAME-RESISTANT CLOTHING

For employees who work in jobs in which flames or electric arcs may occur, wearing **flame-resistant clothing** can be a lifesaver.[8] Electric arcs are the result of electricity passing through ionized air. Although electric arcs last for only a few seconds, during that time they can produce extremely high levels of heat and flash flame.

OSHA's standards relating to flame-resistant clothing are found in CFR 1910.269, paragraph 1. Key elements of paragraph 1 explain the employer's responsibilities regarding personal protective equipment and flame-resistant clothing.

Apparel/PPE

CFR 1910.269, paragraph 1(6) reads as follows:

i. When work is performed within reaching distance of exposed energized parts of equipment, the employer shall ensure that each employee removes or renders nonconductive all exposed conductive articles, such as key or watch chains, rings, or wrist watches or bands, unless such articles do not increase the hazards associated with contact with the energized parts.

ii. The employer shall train each employee who is exposed to the hazards or flames or electric arcs in the hazards involved.

iii. The employer shall ensure that each employee who is exposed to the hazards of flames or electric arcs does not wear clothing that, when exposed to flames or electric arcs, could increase the extent of injury that would be sustained by the employee.

Note: Clothing made from the following types of fabrics, either alone or in blends, is prohibited by this paragraph, unless the employer can demonstrate that the fabric has been treated to withstand the conditions that may be encountered or that the clothing is worn in such a manner as to eliminate the hazard involved: acetate, nylon, polyester, and rayon (see Figure 15–15).

iv. Fuse Handling. When fuses must be installed or removed with one or both terminals energized at more than 300 volts or with exposed parts energized at more than 50 volts, the employer shall ensure that tools or gloves rated for the voltage are used. When expulsion-type fuses are installed with one or both terminals energized at more than 300 volts, the employer shall ensure that each employee wears eye protection meeting the requirements of Subpart I of this Part, uses a tool rated for the voltage, and is clear of the exhaust path of the fuse barrel.[9]

FIGURE 15–15 The proper PPE is critical.

Source: Tyler Olson/Shutterstock

FIRE SAFETY PROGRAMS

Organizations that are interested in protecting their employees from fire hazards should remember the Boy Scouts' motto: Be prepared. The best way to be prepared is to establish a comprehensive fire safety program that encompasses all the functional activities required for being prepared.[10] A comprehensive fire safety program should have at least the following components: assessment, planning, awareness/prevention, and response.

An effective way to develop, implement, and maintain a comprehensive fire safety program is to establish a cross-functional fire safety committee. "Cross-functional" means that it should have members from the organization's various functional units. It should also have at least one executive-level manager to ensure and demonstrate that level of support. This approach has several advantages including the following: (1) it focuses the eyes and ears of a broad cross section of the workforce on fire safety; (2) it ensures a broad base of input; and (3) it ensures executive-level commitment. This committee should be staffed and chaired by the organization's highest ranking safety and health professional.

Assessment

Assessment of the workplace for fire hazards should be continuous and ongoing. Although the organization's safety and health professional will have primary responsibility for this, committee members also need to be involved and involve the departments that they represent. Members of the safety committee should be trained in the fundamentals of fire hazard assessment by the safety and health professional. They should then pass on this knowledge to employees in their departments, units, and teams. In this way, all employees are involved in continually looking for fire hazards and communicating their concerns to the safety committee.

Planning

OSHA requires that an organization's emergency fire safety plan have at least the following components (see Figure 15–16):

- Emergency escape procedures and routes
- Critical "shutdown" procedures
- Employee headcount procedures
- Rescue and medical procedures
- Procedures for reporting fires and emergencies
- Important contact personnel for additional information

Once the plan is in place, it should be reviewed at least annually and updated as necessary.

Awareness and Prevention

After the fire safety committee has completed the emergency plan and upper management has approved it, employees must become acquainted with it. All employees should receive awareness training so that they understand their role in carrying out the emergency plan. The fire safety committee should evaluate the training program periodically, using guidelines such as the following:

- Do all employees know the role they play in implementing the emergency plan?
- How are employees with disabilities provided for?
- Do all employees understand the escape plans? Do they understand the evacuation procedures?
- Do all new and temporary employees receive training?
- Are all employees informed when the plan is revised?
- Is a comprehensive drill undertaken at least once each year?

FIGURE 15–16 Have an emergency evacuation plan in place.

EMERGENCY EVACUATION PLAN

Section 1: General Guidelines

Available Exits:

Meeting Place:

Equipment:

Source: Pixsooz/Shutterstock

- Are all employees familiar with the sound of the alarm system?
- Is the alarm system checked periodically?
- Are sufficient fire-detection devices in place? Are they tested periodically?
- Do all employees know the most likely causes of fires?

RESPONSE AND FIRE DRILLS

Accidents can happen in even the safest organizations. Therefore, it is very important that employees understand the emergency plan and periodically practice responding. Just knowing what the plan says is not sufficient. People do not always think clearly in an emergency situation. They will, however, do what they have learned to do through practice. Consequently, one of the fire safety committee's most important responsibilities is to arrange periodic drills so that employees automatically respond properly.

There is a tendency among people to take fire drills lightly or to simply ignore them. To overcome this tendency, apply the following strategies:

- Have unannounced fire drills periodically and treat them like the real thing in every way. People behave differently when they know it's just a drill.
- Do not allow personnel to simply ignore fire drills.
- Make sure that personnel responsible for shutting down operations in the event of a fire are properly trained to (1) know how to do their shut-down jobs and still get out safely and (2) make a rational decision concerning when to evacuate even if shut-down procedures have not been accomplished.
- Complete procedures for accounting for all personnel. Do not ignore this critical aspect of an evacuation just because it is a drill.
- Periodically use outside third parties to evaluate your evacuation procedures and act immediately on any discrepancies noted.

EXPLOSIVE HAZARDS

Many chemical and toxic substances used in modern organizations are flammable or combustible. Consequently, under certain conditions, they can explode. Working in these conditions involves hazards that require special precautions for handling, storing, transporting, and using such substances.

Explosives-Related Concepts

Safety relating to explosive materials is a highly specialized field. This section discusses terms and concepts used in this field with which modern safety and health professionals should be familiar.

- A **flammable substance** is any substance with a flash point below 37.8°C (100.04°F) and a vapor pressure of less than 40 psi at that temperature. Such liquids are also known as *Class I liquids*. They tend to be compositions of hydrogen and carbon such as crude oil and its numerous by-products.
- A **combustible substance** is any substance with a flash point of 37.8°C (100.04°F) or higher. Such liquids are known as *Class II liquids*. They also tend to be compositions of hydrogen and carbon such as crude oil and its numerous by-products.

- The *flash point* is the lowest temperature at which a substance gives off sufficient vapors to combine with air to form an ignitable mixture. Ignition can be precipitated by a spark.
- The *auto-ignition temperature* is the lowest temperature at which a vapor-producing substance or a flammable gas ignites even without the presence of a spark or a flame. This is sometimes known as *spontaneous ignition*.
- In most cases, a certain amount of oxygen must be present in a vapor-air mixture for an explosion to occur. The amount that must be present for a given substance is the **oxygen limit** for that substance.
- **Volatility** is the evaporation (vaporization) capability of a given substance. The greater the tendency of a substance to vaporize, the more volatile it is.[11]

Common Uses of Flammable and Combustible Substances

Flammable and combustible substances are widely used in modern organizations. Therefore, the hazards associated with them are not limited to the industries producing such materials and substances. The National Safety Council (NSC) lists the following as common uses of flammable and combustible substances in modern industry and specific related precautions that should be taken with each.

- **Dip tanks.** Dipping operations involving flammable or combustible substances should take place in a stand-alone, one-story building constructed of noncombustible materials. The building should be (1) well ventilated, (2) clearly marked as a hazardous area, (3) free of ignition sources, and (4) large. The dip tank itself should be covered and should contain an automatic fire-extinguishing system.
- **Japanning and drying ovens.** Ovens used to evaporate varnish, Japan enamel, and any other combustible substance should be (1) well ventilated, (2) equipped with an automatic fire protection system, and (3) have a shutdown system that activates automatically in the event of a fire or explosion.
- **Oil burners.** Selecting the proper type of fuel for use in an oil burner is the best precaution to prevent the accumulation and potential ignition of soot. The safest fuel to use in an oil burner is one that meets the following criteria: (1) flash point higher than 37.8°C (100.04°F), (2) hydrocarbon-based, and (3) acid- and grit-free. In addition, the supply tank should be located outside the building housing the oil burner and should be underground. The top of the storage tank should be lower than all pipes entering it. Finally, the oil burner should have an automatic system for preventing the discharge of unburned oil into a hot firebox.
- **Cleaning solvents.** Many of the solvents used to clean metal parts are combustible. The primary precaution when cleaning metal parts is selecting substances that are not easily ignited. Additional precautions include ventilation and the selection of a cleaning area that is free of ignition sources.
- **Internal combustion engines.** Internal combustion engines are widely used in modern industry for powering such equipment as forklifts and lift trucks. Because they are typically fueled with gasoline or diesel fuel, there are fire and explosive hazards associated with this operation. Precautions include (1) proper maintenance, (2) good housekeeping, (3) shutdown of engines and cooling of exhaust pipes before filling fuel tanks, and (4) a well-ventilated area for filling fuel tanks.

- **Spray-painting booths.** The hazard associated with spray-painting booths is that an explosive mixture of paint vapor and air can occur. To prevent such occurrences, proper ventilation is critical. Regular cleaning of the booth to remove accumulated spray deposits is also important. Paint booths should be equipped with automatic fire protection systems.[12]

The preceding section lists only some of the many ways in which potentially explosive materials are used in modern organizations.

Other Health Hazards of Explosive Materials

The health hazards associated with explosions and fires are well known. The potential for serious injury or death from the force of a blast or from burns is very high. However, there are other hazards associated with explosive and combustible materials. These include skin irritation, intoxication, and suffocation.

Irritation can occur when the skin comes in contact with hazardous substances. The degree of irritation can range from minor to severe, depending on the type of substance, its concentration, and the duration of contact. Intoxication can occur when an employee breathes the vapors of combustible substances. This can cause impaired judgment, performance, and reaction time, which can, in turn, result in an accident. Finally, the vapors from combustible materials can accumulate in confined spaces. When this happens, the air becomes contaminated and is both toxic and explosive. In such cases, the hazard of suffocation must be added to those associated with explosives.[13]

OSHA'S FIREFIGHTING OPTIONS

Even with the best fire prevention program, and even with the best engineering controls in place, it may still be necessary to manually fight fires at some point. Some companies prefer to have their employees evacuate the premises in the event of a fire. However, for some companies the potential for fires is so much a part of daily operations that they prefer to equip their employees to fight fires. Companies that either allow or expect employees to help fight fires should follow the guidelines set forth by OSHA for manual firefighting.[14]

There are three options available to companies that wish to have their employees participate in firefighting:

1. All employees are involved.
2. Only designated employees are involved.
3. Only employees who are part of an established fire brigade are involved.

Each of these options has its own set of requirements.

Option 1: All Employees Fight Fires

With this option, all employees are allowed to fight fires. However, first they are required to

1. Have and understand an emergency action plan provided by the company.
2. Have and understand a fire prevention plan provided by the company.
3. Complete annual training and refresher training concerning their duties in fighting fires and in the proper use of fire extinguishers.

Option 2: Designated Employees Fight Fires

With this option, only selected employees are allowed to fight fires. First they are required to

1. Have and understand an emergency action plan provided by the company.
2. Have and understand a fire prevention plan provided by the company.
3. Complete annual training and refresher training concerning their duties fighting fires and in how to properly use fire extinguishers.

Option 3: Fire Brigades Fight Fires

With this option, only those employees who are part of an established fire brigade are allowed to fight fires. Fire brigades are divided into two types: incipient and interior structural. An **incipient fire brigade** is used to control only small fires. It requires no special protective clothing or equipment. An **interior structural fire brigade** may fight any type of fire provided it has been issued the appropriate protective clothing and equipment (see Figure 15–17).

The requirements for each type of fire brigade are different. Employees who are part of an incipient fire brigade are required to

1. Have and understand an emergency action plan provided by the company.
2. Have and understand a fire protection plan provided by the company.
3. Have and understand an organizational statement that establishes the scope, organizational structure, training, equipment, and functions of the fire brigade.
4. Have and understand standard operating procedures for the fire brigade to follow during emergencies.
5. Complete annual training and refresher training that is hands-on in nature.

FIGURE 15–17   Fire brigades must be properly trained and equipped.

Source: 108MotionBG/Shutterstock

The requirements of an interior structural fire brigade are the same as those for an incipient fire brigade through the standard operating procedures. However, there are also additional requirements, including

- Satisfactory completion of medical examinations that verify their fitness to participate
- Special protective clothing and equipment of the type used by local fire departments, including self-contained breathing equipment
- Quarterly, as opposed to annual, training and retraining that is hands-on in nature

SELF-ASSESSMENT IN FIRE PROTECTION

Safety and health personnel cannot be everywhere at the same time. Consequently, it is wise to enlist the assistance of supervisors and employees in fire protection. An excellent way to do this is to provide them with a self-assessment checklist that will guide them in scanning their areas of responsibility for fire hazards. Such checklists should contain at least the following questions:

1. Are portable fire extinguishers properly mounted, readily accessible, and available in adequate number and type?
2. Are fire extinguishers inspected monthly for both operability and general condition with appropriate notation made on their respective tags?
3. Are fire extinguishers recharged regularly and are the dates noted on their tags?
4. Are interior standpipes and valves inspected regularly?
5. Is the fire alarm system tested regularly?
6. Are employees trained in the proper use of fire extinguishers?
7. Are employees trained concerning under what conditions they should help fight fires and under what conditions they should evacuate?
8. Are the nearest fire hydrants flushed annually?
9. Are the nearest fire hydrants maintained regularly?
10. Are avenues and ingress and egress clearly marked?
11. Are all avenues of ingress and egress kept free of clutter and other types of obstructions?
12. Are fire doors and shutters in good working condition?
13. Are fusible links in place and readily accessible?
14. Is the local fire department familiar with the facility and any specific hazards?
15. Is the automatic sprinkler system in good working order, maintained on a regular basis, given the proper overhead clearance, and protected from inadvertent contact damage?[15]

HOT WORK PROGRAM

It is important for industrial firms that do hot work to have a comprehensive hot work program. OSHA defines *hot work* as work that involves welding, cutting, chipping, and the use of tools that cause sparks. Even before getting into the specifics of developing a comprehensive hot work program, safety professionals personnel understand the following foundational precautionary measure: "Flammable, combustible, or ignitable materials should be kept a minimum of 20 to 35 feet away from the hot work, or those materials should be covered with a flame-retardant covering for protection."[16]

Components of the Hot Work Program

Safety professionals should take the lead in developing a comprehensive hot work program.[17] Components of the program include safety equipment, work practices, contractor requirements, fire watch, and permits. The details of each of these components should be put in writing and training should be provided so that all stakeholders understand their roles and responsibilities in preventing fires that might be caused by hot work.

Fire Safety Equipment

Does the facility have the proper fire safety equipment in place? The program should require that the following fire safety equipment be in place, operable, and properly maintained over time: (1) fire sprinklers, (2) fire extinguishers of the proper type located throughout the work area, (3) back-up fire extinguishers for extra protection during hot work projects, and (4) fire-retardant tarps or thin sheets of metal for covering combustible and ignitable materials in the hot work area.

Precautionary Work Practices

The program should require at least the following precautionary work practices: (1) a minimum of 35 feet of separation between the hot work and combustible and ignitable material; (2) use of a flammable gas meter to detect gas vapors in the hot work area (hot work should not be allowed if the meter reads anything but zero); (3) do not allow hot work when the sprinkler system is inoperable; (4) keep floors cleanly swept of combustible and ignitable materials in the hot work area; (5) combustible floors should be dampened or covered with an appropriate protective material; (6) when performing hot work on walls, ceilings, or open-rack flooring, openings should be covered and fire-resistive tarpaulins should be suspended beneath the work; (7) use metal sheeting or damp cloths or both when hot work is performed within three feet of a sprinkler; (8) remove all flammable gases and liquids from the hot work area; and (9) use lock-out/tagout procedures and other standard procedures when performing hot work in a confined space.

Contractor Requirements

Contractors that perform hot work in your facility should have their own hot work program that your organization approves or they should be required to follow yours. All of the contractor's employees must know their duties and responsibilities relating to the plan and complete the same type of training required of your employees.

Fire Watch Requirements

During hot work, an individual should be posed in the area with only one responsibility: watching for and responding immediately to fires. This individual should carry with him a properly rated fire extinguisher and a means of communicating with emergency personnel.

Permit Requirements

An in-house permit should be developed so that safety personnel and other stakeholders can assure themselves that all proper steps in the plan have been taken and approved before hot work begins. There are examples of hot work permits on the Internet, but make sure to tailor the permit chosen specifically for your organization. It should ensure that everything required in the plan has been accomplished before hot work begins.

Chemical Burn Injuries

Chemical burn injuries are covered only briefly in this chapter. Readers are referred to Chapter 12 for a more in-depth coverage of this kind of injury. Chemical burn injuries are a special category with which prospective and practicing safety professionals should be familiar. The greatest incidence of chemical burns (approximately one-third) occurs in manufacturing. Other high-incidence industries are services, trade, and construction.

The chemicals that most frequently cause chemical burn injuries include acids and alkalies; soaps, detergents, and cleaning compounds; solvents and degreasers; calcium hydroxide (a chemical used in cement and plaster); potassium hydroxide (an ingredient in drain cleaners and other cleaning solutions); and sulfuric acid (battery acid). Almost 46 percent of all chemical burn injuries occur while workers are cleaning equipment, tools, and vehicles.

What is particularly disturbing about chemical burn injuries is that a high percentage of them occur in spite of the use of personal protective equipment, the provision of safety instruction, and the availability of treatment facilities. In some cases, the personal protective equipment is faulty or inadequate. In others, it is not properly used in spite of instructions.

Preventing chemical burn injuries presents a special challenge to safety and health professionals. The following strategies are recommended:

- Familiarize yourself, the workers, and their supervisors with the chemicals that will be used and their inherent dangers.
- Secure the proper personal protection equipment for each type of chemical that will be used.
- Provide instruction on the proper use of personal protection equipment and then make sure that supervisors confirm that the equipment is used properly every time.
- Monitor that workers are wearing personal protection equipment and replace it when it begins to show wear.

Heat Burn Injuries

Heat burn injuries present a special challenge to safety and health professionals in the modern workplace. Almost 40 percent of all such injuries occur in manufacturing every year. The most frequent causes are flame (this includes smoke inhalation injuries), molten metal, petroleum asphalts, steam, and water. The most common activities associated with heat burn injuries are welding, cutting with a torch, and handling tar or asphalt.

Following are several factors that contribute to heat burn injuries in the workplace. Safety and health professionals who understand these factors will be in a better position to prevent heat burn injuries.

- Employer has no health and safety policy regarding heat hazards.
- Employer fails to enforce safety procedures and practices.
- Employees are not familiar with the employer's safety policy and procedures concerning heat hazards.
- Employees fail to use or improperly use personal protection equipment.
- Employees have inadequate or worn personal protection equipment.
- Employees work in a limited space.
- Employees attempt to work too fast.
- Employees are careless.
- Employees have poorly maintained tools and equipment.

These factors should be considered carefully by safety and health professionals when developing accident prevention programs. Employees should be familiar with the hazards, should know the appropriate safety precautions, and should have and use the proper personal protection equipment. Safety professionals should monitor to ensure that safety rules are being followed, that personal protection equipment is being used correctly, and that it is in good condition.

Key Terms and Concepts

- Auto-ignition temperature
- Automatic sprinkler systems
- Carbon
- Carbon dioxide
- Carbon monoxide
- Chemical burn injuries
- Combustible liquids
- Combustible substance
- Combustion
- Combustion point
- Conduction
- Convection
- Endothermic
- Exothermic
- Explosion
- Explosive range
- Fire
- Fire brigade
- Fire point
- Fixed extinguishing system
- Flame-resistant clothing
- Flammable gases
- Flammable liquids
- Flammable range
- Flammable substance
- Flash point
- Flashing back
- Fuel
- Heat
- Heat burn injuries
- Heat transfer
- Hydrogen
- Hypergolic
- Ignition temperature
- Incipient fire brigade
- Infrared detectors
- Interior structural fire brigade
- Ionization
- Isolation
- Kinetic energy
- *Life Safety Code*
- Means of egress
- Oxygen
- Oxygen limit
- Photoelectric fire sensors
- Polymers
- Products of combustion
- Pyrophor hypergolic fuels
- Radiation
- Radiation sensors
- Smoke
- Source of ignition
- Spontaneous combustion
- Standpipe and hose systems
- Thermal energy
- Thermal expansion detectors
- Ultraviolet detectors
- Vapor
- Volatility

Review Questions

1. What are the three elements of the fire triangle?
2. Fire is a chemical reaction. Explain the reaction.

3. Where is carbon found?
4. Compare and contrast carbon monoxide and carbon dioxide.
5. How is combustion of liquids and solids different from gases?
6. What are the three methods of heat transfer? Describe each method.
7. What can happen to a pile of oil-soaked rags in a closed container? Describe the process.
8. In which direction does a fire normally travel?
9. Name something in your room that will not burn.
10. What are the classes of fires?
11. What property do almost all packing materials share?
12. What are the differences among flash point, fire point, and auto-ignition temperature?
13. Which are more stable: combustible liquids or flammable liquids?
14. Which way do gases usually travel?
15. Describe the NFPA hazards identification system.
16. In what four ways can electricity cause a fire?
17. What are the leading causes of fire-related deaths?
18. What are some of the toxic chemicals often produced by fires?
19. What are some of the systems utilized by smoke detectors?
20. What is the most successful lifesaving preparation for a fire disaster?
21. Explain the best ways to prevent an office fire.
22. What is the trend with regard to future fire safety standards?
23. Define the term *life safety*.
24. What types of fabrics are prohibited in environments that are flame or arc prone?
25. Explain briefly OSHA's regulations for fire brigades.
26. Summarize the key components of a fire safety program.
27. Describe the precautions that should be taken for dip tanks, oil burners, and spray-painting booths.
28. Explain all three of OSHA's manual firefighting options.
29. List five factors that can contribute to heat injuries in the workplace.

Endnotes

1. National Fire Protection Association, *Fire Protection Handbook*, 20th ed. (Quincy, MA: National Fire Protection Association, 2008), 112.
2. Ibid., 132.
3. Title 29, *Code of Federal Regulations,* Part 1910.156 (Subpart L, Appendix A).
4. Ibid., 26–27.
5. Ibid., 42.
6. National Fire Protection Association, *Life Safety Code* (NFPA 101) (Quincy, MA: National Fire Protection Association, 2011), 101–119.
7. Ibid., 101–126 through 101–150.
8. National Fire Protection Association, *Life Safety Code*, 151.
9. *Code of Federal Regulations,* Part 1910.269, Paragraph 1(6).
10. A. Burke, "Before the Fire," *Occupational Health & Safety* 69, no. 2: 50–52.
11. National Fire Protection Association, *Life Safety Code*, 162.
12. Ibid.
13. Ibid.
14. C. Schroll, "Manual Fire Control," *Occupational Health & Safety* 71, no. 2: 27–28.
15. Retrieved from online.misu.nodak.edu/19577/BADM309checklist.htm on July 6, 2017.
16. Ibid.
17. Ibid., 36–37.

CHAPTER

16

INDUSTRIAL HYGIENE AND CONFINED SPACES

Learning Objectives

- Provide a brief overview of industrial hygiene
- List the most pertinent industrial hygiene standards
- Explain how the OSH Act relates to industrial hygiene
- List and explain the four broad categories of environmental hazards in the workplace
- Define the term *toxic substances*
- List and explain the entry points for toxic agents into the human bloodstream
- List and explain the effects of toxic substances
- Describe the relationship between doses of toxic substances and human responses
- List and explain the different types of airborne contaminants
- Define the term *carcinogen* and explain its effects on humans
- Describe safe practices for asbestos containment and removal
- Explain the concept of the *sick-building syndrome*
- Explain the relationship between toxic mold and indoor air quality
- Summary the main tenets of ASTM D7338: Guide for the Assessment of Fungal Growth in Buildings
- Explain the use and calculation of threshold limit values (TLVs)
- Summarize the approach to hazard recognition and evaluation in industrial hygiene
- Summarize prevention and control strategies for industrial hygiene
- Explain the role NIOSH plays in industrial hygiene
- Summarize the NIOSH guidelines for respirators
- List and explain the principle standards and regulations relating to toxic substances

- Summarize the EPA's risk management program
- List and explain the general safety precautions to be observed around explosive and combustible materials
- Explain the concept of nanoscale materials as they relate to industrial hygiene
- Summarize the confined space hazards that might be present in the workplace
- Summarize the requirements of OSHA's Confined Space Standard (29 CFR 1910.146)
- Summarize the requirements of OSHA's Hazard Communication Standard

Industrial hygiene is an area of specialization within the broader field of industrial safety and health. This chapter provides prospective and practicing safety and health professionals with the information they need to know about this area of specialization.

OVERVIEW OF INDUSTRIAL HYGIENE

Industrial hygiene is a safety and health profession that is concerned with predicting, recognizing, assessing, controlling, and preventing environmental stressors in the workplace that can cause sickness or serious discomfort to workers. An environmental stressor is any factor in the workplace that can cause enough discomfort to result in lost time or illness. Common stressors include gases, fumes, vapors, dusts, mists, noise, and radiation.

The Code of Ethics of the American Academy of Industrial Hygiene describes the responsibilities of industrial hygienists:

- To ensure the health of employees
- To maintain an objective approach in recognizing, assessing, controlling, and preventing health hazards regardless of outside pressure and influence
- To help employees understand the precautions that they should take to avoid health problems
- To respect employers' honesty in matters relating to industrial hygiene
- To make the health of employees a higher priority than obligations to the employer[1]

Role of the Safety and Health Professional

The role of modern safety and health professionals vis-à-vis industrial hygiene often depends on the size of the company employing them. Large companies often employ professionals who specialize in industrial hygiene. These specialists have titles such as occupational physician, industrial hygienist, industrial toxicologist, and health physicist. In smaller companies, safety and health professionals often have responsibility for all safety and health matters, including industrial hygiene.

In companies that employ specialists, their recommendations are used by safety and health professionals to develop, implement, monitor, and evaluate the overall safety and health program. If specialists are not employed, safety and health professionals are responsible for seeking the advice and assistance necessary to predict, recognize, assess, control, and overcome environmental stressors that may cause sickness or serious discomfort to employees.

INDUSTRIAL HYGIENE STANDARDS

Industrial hygiene is a broad field encompassing many different areas of specialization. As such, there are a variety of different organizations and agencies that develop standards of practice relating to some aspect of the field. Safety and health professionals need to know how to contact these organizations in order to stay current with their latest standards. Contact information for the most prominent standards-developing organizations in the broad field of industrial hygiene is as follows:

Air-Conditioning and Refrigeration Institute: ahrinet.org
Air Movement and Control Association International, Inc.: amca.org
American Conference of Governmental Industrial Hygienists: acgih.org
American Industrial Hygiene Association: aiha.org
American National Standards Institute: ansi.org
American Society of Heating, Refrigerating and Air-Conditioning Engineers: ashrae.org
American Society of Safety Engineers: asse.org
American Society for Testing Materials: astm.org
International Organization for Standardization: iso.ch
National Air Duct Cleaners Association: nadca.com
National Environmental Balancing Bureau: nebb.org
National Fire Protection Association: nfpa.org
National Institute for Occupational Safety and Health: cdc.gov/niosh
Occupational Safety and Health Administration: osha.gov
Underwriters Laboratories: ul.com
U.S. Department of Energy: energy.gov
U.S. Environmental Protection Agency: epa.gov

OSH ACT AND INDUSTRIAL HYGIENE

The principal federal legislation relating to industrial hygiene is the Occupational Safety and Health Act (OSH Act) of 1970 as amended. The OSH Act sets forth the following requirements relating to industrial hygiene:

- Use of warning labels and other means to make employees aware of potential hazards, symptoms of exposure, precautions, and emergency treatment
- Prescription of appropriate personal protective equipment (PPE) and other technological preventive measures (29 CFR 1910.133 and 1910.134 [Subpart I])
- Provision of medical tests to determine the effect on employees of exposure to environmental stressors
- Maintenance of accurate records of employee exposures to environmental stressors that are required to be measured or monitored
- Accessibility of monitoring tests and measurement activities to employees

- Availability of monitoring tests and measurement activities records to employees on request
- Notification of employees who have been exposed to environmental stressors at a level beyond the recommended threshold and corrective action being taken

OSHA Process Safety Standard and Management Guidelines

The **Occupational Safety and Health Administration (OSHA) Process Safety Standard** is found in 29 CFR 1910.119. Its purpose is to prevent accidents caused by major releases of highly hazardous chemicals (HHCs) and to improve process management. To comply with this standard, companies must have written operating procedures, mechanical integrity programs, and formal incident investigation procedures. Other key elements are as follows:

Application

Process safety management involves identifying and evaluating chemical release hazards and then using what is learned to proactively mitigate or prevent chemical releases that might occur as the result of process, procedural, or equipment failures.

Coverage and Exceptions

Any company that uses the threshold amount of any chemical listed in the standard or 10,000 or more of a flammable material at one site in one location is covered by the standard with the following exceptions: retail facilities, oil or gas drilling or servicing operations, normally unoccupied remote facilities, hydrocarbon fuels used solely for workplace consumption (heating or fuel for vehicles) provided the fuels are not part of a process containing another HHC listed in the standard.

Process Safety Information

Organizations that are covered by the standard must maintain complete, accurate, and up-to-date written information readily on hand explaining the following aspects of process safety:

- Technology of the process
- Equipment in the process
- Employee involvement in the development of process safety programs
- Process hazard analysis
- Operating procedures
- Employee training
- Contractor screening procedures and requirements to ensure that contractors perform their work safely
- Pre-startup safety review
- Mechanical integrity of equipment
- Process defenses
- Written procedures
- Inspection and testing
- Quality assurance
- Non-routine work authorizations
- Managing change
- Accident investigation

- Emergency preparedness
- Compliance audits
- Planning
- Staffing
- Conducting the audit
- Evaluation and corrective action

OSHA Regulation for Chemical Spills

OSHA issues a special regulation dealing with chemical spills. The standard (29 CFR 1910.120) is called the Hazardous Waste Operations and Emergency Response Standard, or *HAZWOPER*. **HAZWOPER** gives organizations two options for responding to a chemical spill. The first is to evacuate all employees in the event of a spill and to call in professional emergency response personnel. Employers who use this option must have an **emergency action plan (EAP)** in place in accordance with 29 CFR 1010.38(a). The second option is to respond internally. Employers using this must have an emergency response plan that is in accordance with 29 CFR 1010.120.

1. *EAPs.* An EAP should have at least the following elements: alarm systems, evacuation plan, a mechanism or procedure for emergency shutdown of the equipment, and a procedure for notifying emergency response personnel.
2. *Emergency response plan.* Companies that opt to respond internally to chemical spills must have an **emergency response plan** that includes the provision of comprehensive training for employees. OSHA Standard 29 CFR 1910.120 specifies the type and amount of training required, ranging from awareness to in-depth technical training for employees who will actually deal with the spill. It is important to note that OSHA forbids the involvement of untrained employees in responding to a spill. The following topics are typical of those covered in up-to-date HAZWOPER courses:

   - Summary of key federal laws
   - Overview of impacting regulations
   - Classification and categorization of hazardous waste
     Definition of hazardous waste
     Characteristics
     Toxic characteristic leaching procedure (TCLP)
     Lists of hazardous wastes
   - Hazardous waste operations
     Definitions
     Levels of response
   - Penalties for noncompliance
     Civil penalty policy
   - Responses to spills
     Groundwater contamination
     Sudden releases
     Clean-up levels
     Risk assessment
     Remedial action

- Emergency response
  Workplan
  Site evaluation and control
  Site-specific safety and health plan
  Information and training program
  Personal protective equipment (PPE)
  Monitoring
  Medical surveillance
  Decontamination procedures
  Emergency response
  Other provisions
- Contingency plans
  Alarm systems
  Action plan
- Personal protective equipment
  Developing a PPE program
  Respiratory equipment
  Protective clothing
  Donning PPE
  Doffing PPE
- **Safety data sheets (SDSs)**
  Introduction
  Preparing SDSs
  SDS information
  Hazardous ingredients
  Physical/chemical characteristics
  Fire and explosion hazard data
  Reactivity data
  Health hazard data
  Precautions for safe handling and use
  Control measures
- Site control
  Site maps
  Site preparation
  Work zones
  Buddy system
  Site security
  Communications
  Safe work practices
- Hazardous waste containers
  Emergency control
  Equipment
  Tools
  Safety
- Decontamination
  Types
  Decontamination plan

Prevention of contamination
Planning
Emergencies
Physical injury
Heat stress
Chemical exposure
Medical treatment area
Decontamination of equipment
Decontamination procedures
Sanitation of PPE
Disposal of contaminated materials

In a given year, industrial facilities in the United States release almost 6 billion pounds of toxic substances into the environment. The Environmental Protection Agency (EPA) began measuring such emissions in 1987. Since that time, the trend in toxic emissions has been downward. Even with the downward trend, however, in a typical year, toxic substances will be released into the environment in the following amounts (according to the EPA):

- 2.3 billion pounds released into the air
- 1.1 billion pounds injected into underground wells
- 910 million pounds placed in treatment and disposal facilities
- 550 million pounds placed in municipal wastewater treatment plants
- 440 million pounds placed in landfills
- 100 million pounds released into the water

People are exposed to a variety of substances every day in the home and at work — paints, paint remover, detergent, cleaning solvents, antifreeze, and motor oil, to name just a few. Most substances with which we interact are not dangerous in small amounts or limited exposure. However, high levels of exposure to certain substances in high concentrations can be dangerous. Levels of exposure and concentration, as well as how we interact with substances, help determine how hazardous substances are. In addition, some substances used frequently in certain industrial settings can explode. Following is some information that modern safety and health professionals need to know about toxic and explosive substances.

HAZARDS IN THE WORKPLACE

The environmental stressors on which industrial hygiene focuses can be divided into the following broad categories: chemical, physical, biological, and ergonomic hazards. Typical **chemical hazards** include mists, vapors, gases, dusts, and fumes. Chemical hazards are either inhaled or absorbed through the skin or both. **Physical hazards** include noise, vibration, extremes of temperature, and excessive radiation (electromagnetic or ionizing). **Biological hazards** come from molds, fungi, bacteria, and insects. Bacteria may be introduced into the workplace through sewage, food waste, water, or insect droppings. **Ergonomic hazards** are related to the design and condition of the workplace. Poorly designed workstations and tools are ergonomic hazards. Conditions that put workers in awkward positions and impair their vision are also hazards.

Safety Data Sheets

What used to be called the *Material Safety Data Sheet* (MSDS) is now called the *Safety Data Sheet* (SDS). The change is part of the move by OSHA to require SDS to comply with the Globally Harmonized System of Chemical Information and Labelling known as **GHS**. An SDS is the principle means of providing employers and employees with the information they need to use chemicals safely and to understand the hazards they represent. Consequently, an SDS contains such information about a chemical as the name of the chemical itself, physical and chemical characteristics, physical hazards, primary routes of entry into the body, permissible exposure limits (PELs), carcinogenic (when applicable), precautions for safe use, control measures, emergency and first aid procedures, date of preparation of the SDS, and contact information (name, address, and telephone number) of the person who prepared the SDS.[2]

Global Harmonization of Hazard Communication

Global Harmonization of Hazard Communication is the new reality for SDS and other forms of hazard communication such as OSHA's Hazard Communication Standard. Imagine if you could travel to foreign countries and everyone spoke the same language. The idea behind global harmonizing of hazard communication is to get people all over the world dealing with the same hazard classification system, labeling procedures, and order of presenting information in SDS. The harmonizing effort is intended to enhance the protection of human health as well as the environment by providing comprehensive internationally used standards for hazard communication.

The purposes of the GHS are as follows: (1) enhance the protection of human health and the environment by providing a comprehensive internationally used standard for hazard communication, (2) provide a recognized framework for countries that do not yet have an established standard, (3) reduce the need for testing of materials and chemicals, and (4) facilitate international trade in materials and chemicals the hazards of which have been identified and assessed.

One of the major steps toward achieving globally harmonized standards for hazard communication is standardizing the format for SDS. Global harmonization requires that the following information be contained in all SDS and in this order:

- Identification
- Hazard identification
- Composition and information on ingredients
- First aid measures
- Firefighting measures
- Accidental release measures
- Handling and storage
- Exposure controls and personal protection
- Physical and chemical properties
- Stability and reactivity
- Toxicology information
- Ecological information
- Disposal procedures
- Transport information
- Regulatory information
- Other applicable information

Hazard communication begins with classification. Classification is achieved in three steps: (1) identification of relevant data relating to the hazards of a given substance, (2) review of the relevant data to determine specific hazards associated with the substance in question, and (3) determination of whether the substance in question will be classified as hazardous and if so to what degree. The physical hazards the GHS is concerned with include the following:

- Explosives
- Flammable bases
- Flammable aerosols
- Oxidizing gases
- Gases under pressure
- Flammable liquids
- Flammable solids
- Self-reactive substances
- Pyrophoric liquids
- Pyroforic solids
- Self-heating substances
- Substances that emit flammable gases when in contact with water
- Oxidizing liquids
- Oxidizing solids
- Organic peroxides
- Substances that corrode metals

Environmental Stressors

Noise is sound that is unwanted or that exceeds safe limits. It can cause problems ranging from annoyance to hearing loss. Acceptable levels of noise have been established by OSHA, the National Institute for Occupational Safety and Health (NIOSH), and the EPA. OSHA mandates that an employee's exposure level be limited to 90 decibels (dB) calculated as an eight-hour, **time-weighted average**. Applying the OSHA recommendation, 14 percent of workers are employed in situations that expose them to excess noise levels.[3] Chapter 18 contains the latest data from NIOSH and OSHA on noise-level standards.

Temperature control is the most basic way to eliminate environmental hazards.

Radiation hazards are increasingly prevalent in the age of high technology. In the category of **ionizing radiation**, safety and health professionals are concerned with five kinds of **radiation** (alpha, beta, X-ray, gamma, and neutron). Of these, alpha radiation is the least penetrating, which makes shielding simple, whereas the others are more difficult to shield against. Meters and other instruments are available to measure radiation levels in the workplace. The greatest risk for **nonionizing radiation** in the modern workplace comes from lasers. Shielding requirements for lasers are described in the construction regulations of the OSH Act.[4] Extremes of pressure also represent a potential hazard in the workplace. Biological hazards from various biological organisms can lead to disease in workers. The now-famous outbreak of what has come to be known as Legionnaire's disease is an example of what can result from biological hazards. This disease first surfaced at a convention where numerous participants became sick and soon died. The cause was eventually traced back to bacteria that grew in the cooling/air-moving systems serving the convention center. That bacterium has since been named *Legionella*.

Ergonomic hazards are conditions that require unnatural postures and unnatural movement. The human body can endure limited amounts of unnatural postures or motions. However, repeated exposure to such conditions can lead to physical stress and injury. Design of tools, workstations, and jobs can lead to or prevent ergonomic hazards.

TOXIC SUBSTANCES DEFINED

A **toxic substance** is one that has a negative effect on the health of a person or animal. Toxic effects are a function of several factors including the following: (1) properties of the substance, (2) amount of the dose, (3) level of exposure, (4) **route of entry**, and (5) resistance of the individual to the substance.

ENTRY POINTS FOR TOXIC AGENTS

The development of preventive measures to protect against the hazards associated with industrial hygiene requires first knowing how toxic agents enter the body. A toxic substance must first enter the bloodstream to cause health problems. The most common routes of entry for toxic agents are inhalation, absorption, injection, and ingestion (see Figure 16–1). These routes are explained in the following paragraphs.

Inhalation

The route of entry about which safety and health professionals should be most concerned is **inhalation**. Airborne toxic substances such as gases, vapors, dust, smoke, fumes, aerosols, and mists can be inhaled and pass through the nose, throat, bronchial tubes, and lungs to enter the bloodstream. The amount of a toxic substance that can be inhaled depends on the following factors: (1) concentration of the substance, (2) duration of exposure, and (3) breathing volume.

Absorption

The second most common route of entry in an industrial setting is **absorption**, or passage through the skin and into the bloodstream. The human skin is a protective barrier

FIGURE 16–1 Common routes of entry of toxic substances.

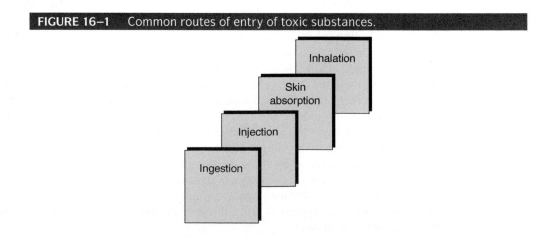

against many hazards. However, certain toxic agents can penetrate the barrier through absorption. Of course, unprotected cuts, sores, and abrasions facilitate the process, but even healthy skin will absorb certain chemicals. Humans are especially susceptible to absorbing such chemicals as organic lead compounds, nitro compounds, organic phosphate pesticides, TNT, cyanides, aromatic amines, amides, and phenols.[5]

With many substances, the rate of absorption and, in turn, the hazard levels increase in a warm environment. The extent to which a substance can be absorbed through the skin depends on the factors shown in Figure 16–2. Another factor is body site. Different parts of the body have different absorption capabilities. For example, the forearms have a lower absorption potential than do the scalp and forehead.

Ingestion

Ingestion, not a major concern in an industrial setting, is entry through the mouth.[6] An ingested substance is swallowed. It moves through the stomach into the intestines and from there into the bloodstream. Toxic agents sometimes enter the body by ingestion when they are accidentally consumed by workers eating lunch or a snack. **Airborne contaminants** can also rest on food or the hands and, as a result, be ingested during a meal or snack. The possibility of ingesting toxic agents makes it critical to confine eating and drinking to sanitary areas away from the work site and to make sure that workers practice good **personal hygiene** such as washing their hands thoroughly before eating or drinking.

As it moves through the gastrointestinal tract, the toxic substance's strength may be diluted. In addition, depending on the amount and toxicity of the substance, the liver may be able to convert it to a nontoxic substance. The liver can, at least, decrease the level of toxicity and pass along the substance to the kidneys, where some of the substance is eliminated in the urine.

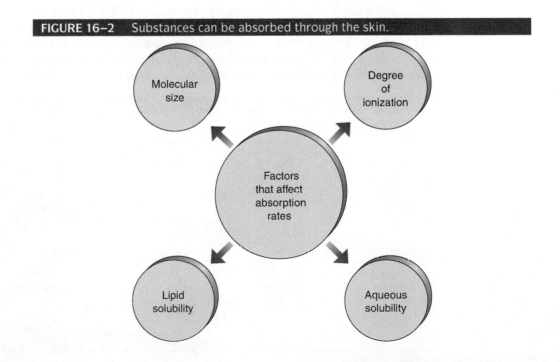

FIGURE 16–2 Substances can be absorbed through the skin.

Injection

Injection involves the introduction of a substance into the body by way of a needle and syringe, compressed air, high-pressure hydraulic leaks, or any other capable medium. Consequently, this is not often a route of entry for a toxic substance in the workplace. Injection is sometimes used for introducing toxic substances in experiments involving animals. However, this approach can produce misleading research results because the needle bypasses some of the body's natural protective mechanisms.

EFFECTS OF TOXIC SUBSTANCES

The effects of toxic substances vary widely, as do the substances themselves. However, all the various effects and exposure times can be categorized as being either acute or chronic.

Acute effects and exposures involve a sudden dose of a highly concentrated substance. They are usually the result of an accident (a spill or damage to a pipe) that results in an immediate health problem ranging from irritation to death. Acute effects and exposures are (1) sudden, (2) severe, (3) typically involve just one incident, and (4) cause immediate health problems. Acute effects and exposures are not the result of an accumulation over time.

Chronic effects and exposures involve limited continual exposure over time. Consequently, the associated health problems develop slowly. The characteristics of chronic effects and exposures are (1) continual exposure over time, (2) limited concentrations of toxic substances, (3) progressive accumulation of toxic substances in the body and progressive worsening of associated health problems, and (4) little or no awareness of exposures on the part of affected workers.

When a toxic substance enters the body, it eventually affects one or more body organs. Part of the liver's function is to collect such substances, convert them to nontoxics, and send them to the kidneys for elimination in the urine. However, when the dose is more than the liver can handle, toxics move on to other organs, producing a variety of effects. The organs that are affected by toxic substances are the blood, kidneys, heart, brain, central nervous system, skin, liver, lungs, and eyes. Figure 16–3 lists some of the more widely used toxic substances and the organs that they endanger most.

FIGURE 16–3 Selected toxic substances and the organs that they endanger most.

Blood	Kidneys	Heart	Brain
Benzene	Mercury	Aniline	Lead
Carbon monoxide	Chloroform		Mercury
Arsenic			Benzene
Aniline			Manganese
Toluene			Acetaldehyde

Eyes	Skin	Lungs	Liver
Cresol	Nickel	Asbestos	Chloroform
Acrolein	Phenol	Chromium	Carbon tetrachloride
Benzyl chloride	Trichloroethylene	Hydrogen sulfide	Toluene
Butyl alcohol		Mica	
		Nitrogen dioxide	

RELATIONSHIP OF DOSES AND RESPONSES

Safety and health professionals are interested in predictability when it comes to toxic substances. How much of a given substance is too much? What effect will a given dose of a given substance produce? These types of questions concern dose–response relationships. A **dose** of a toxic substance can be expressed in a number of different ways depending on the characteristics of the substance, for example, amount per unit of body weight, amount per body surface area, or amount per unit of volume of air breathed. The dose–response relationship may be expressed mathematically as follows:[7]

$$(C) \times (T) = K$$

where

C = concentration
T = duration (time) of exposure
K = constant

Note that in this relationship, C times T is *approximately* equal to K. The relationship is not exact.

Three important concepts to understand relating to doses are dose threshold, lethal dose, and lethal concentration. These concepts are explained in the following paragraphs.

Dose Threshold

The **dose threshold** is the minimum dose required to produce a measurable effect. Of course, the threshold is different for different substances. In animal tests, thresholds are established using such methods as (1) observing pathological changes in body tissues, (2) observing growth rates (are they normal or retarded?), (3) measuring the level of food intake (has there been a loss of appetite?), and (4) weighing organs to establish body weight to organ weight ratios.

Lethal Dose

A **lethal dose** of a given substance is a dose that is highly likely to cause death. Such doses are established through experiments on animals. When lethal doses of a given substance are established, they are typically accompanied by information that is of value to medical professionals and industrial hygienists. Such information includes the type of animal used in establishing the lethal dose, how the dose was administered to the animal, and the duration of the administered dose. Lethal doses do not apply to inhaled substances. With these substances, the concept of lethal concentration is applied.

Lethal Concentration

A **lethal concentration** of an inhaled substance is the concentration that is highly likely to result in death. With inhaled substances, the duration of exposure is critical because the amount inhaled increases with every unprotected breath.

AIRBORNE CONTAMINANTS

It is important to understand the different types of airborne contaminants that may be present in the workplace.[8] Each type of contaminant has a specific definition that must be understood in order to develop effective safety and health measures to protect against it. The most common types of airborne contaminants are dusts, fumes, smoke, aerosols, mists, gases, and vapors (see Figure 16–4).

- **Dusts.** **Dusts** are various types of solid particles that are produced when a given type of organic or inorganic material is scraped, sawed, ground, drilled, handled, heated, crushed, or otherwise deformed. The degree of hazard represented by dust depends on the toxicity of the parent material and the size and level of concentration of the particles.
- **Fumes.** The most common causes of fumes in the workplace are such manufacturing processes as welding, heat treating, and metalizing, all of which involve the interaction of intense heat with a parent material. The heat volatilizes portions of the parent material, which then condenses as it comes in contact with cool air. The result of this reaction is the formation of tiny particles that can be inhaled.
- **Smoke.** Smoke is the result of the incomplete combustion of carbonaceous materials. Because combustion is incomplete, tiny soot or carbon particles remain in the air and can be inhaled.
- **Aerosols.** **Aerosols** are liquid or solid particles so small that they can remain suspended in air long enough to be transported over a distance. These particles can be inhaled.
- **Mists.** **Mists** are tiny liquid droplets suspended in air. Mists are formed in two ways: (1) when vapors return to a liquid state through condensation and (2) when the application of sudden force or pressure turns a liquid into particles.
- **Gases.** Unlike other airborne contaminants that take the form of either tiny particles or droplets, gases are actually formless fluids. Gases become particularly hazardous when they fill a confined, unventilated space. The most common

FIGURE 16–4   Common airborne contaminant.

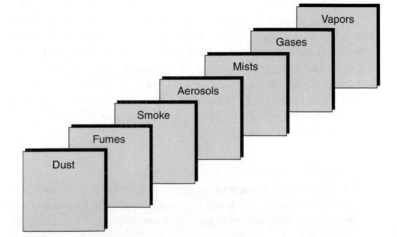

sources of gases in an industrial setting are from welding and the exhaust from internal combustion engines.

- *Vapors.* Certain materials that are solid or liquid at room temperature and at normal levels of pressure turn to **vapors** when heated or exposed to abnormal pressure. Evaporation is the most common process by which a liquid is transformed into a vapor.

In protecting workers from the hazards of airborne contaminants, it is important to know the permissible levels of exposure for a given contaminant and to monitor continually the level of contaminants using accepted measurement practices and technologies. The topic of **exposure thresholds** is covered later in this chapter.

EFFECTS OF AIRBORNE TOXICS

Airborne toxic substances are also classified according to the type of effect they have on the body. The primary classifications are shown in Figure 16–5 and explained in the paragraphs that follow. With all airborne contaminants, concentration and duration of exposure are critical concerns.

Irritants

Irritants are substances that cause irritation to the skin, eyes, and the inner lining of the nose, mouth, throat, and upper respiratory tract. However, they produce no irreversible damages.

Asphyxiants

Asphyxiants are substances that can disrupt breathing so severely that suffocation results. Asphyxiants may be simple or chemical in nature. A simple asphyxiant is an inert gas that dilutes oxygen in the air to the point that the body cannot take in enough air to satisfy its needs for oxygen. Common simple asphyxiants include carbon dioxide, ethane, helium, hydrogen, methane, and nitrogen. Chemical asphyxiants, by chemical action, interfere with the passage of oxygen into the blood or the movement of oxygen from the lungs to body tissues. Either way, the end result is suffocation due to insufficient or no oxygenation. Common chemical asphyxiants include carbon monoxide, hydrogen cyanide, and hydrogen sulfide.

FIGURE 16–5   Airborne toxic substances.

Narcotics and Anesthetics

Narcotics and **anesthetics** are similar in that carefully controlled dosages can inhibit the normal operation of the central nervous system without causing serious or irreversible effects. This makes them particularly valuable in a medical setting. Dentists and physicians use narcotics and anesthetics to control pain before and after surgery. However, if the concentration of the dose is too high, narcotics and anesthetics can cause unconsciousness and even death. When this happens, death is the result of asphyxiation. Widely used narcotics and anesthetics include acetone, methyl-ethyl-ketone, acetylene hydrocarbons, ether, and chloroform.

EFFECTS OF CARCINOGENS

A **carcinogen** is any substance that can cause a malignant tumor or a **neoplastic growth**. A *neoplasm* is cancerous tissue or tissue that may become cancerous. Other terms used synonymously for carcinogen are *tumorigen*, *oncogen*, and *blastomogen*.

Medical researchers are not sure exactly how certain chemicals cause cancer. However, there are a number of toxic substances that are either known or strongly suspected to be carcinogens. These include coal tar, pitch, creosote oil, anthracene oil, soot, lamp black, lignite, asphalt, bitumen waxes, paraffin oils, arsenic, chromium, nickel compounds, beryllium, cobalt, benzene, and various paints, dyes, tints, pesticides, and enamels.[9]

ASBESTOS HAZARDS

The EPA estimates that approximately 75 percent of the commercial buildings in use today contain asbestos in some form.[10] Asbestos was once thought to be a miracle material because of its many useful characteristics, including fire resistance, heat resistance, mechanical strength, and flexibility. As a result, asbestos was widely used in commercial and industrial construction between 1900 and the mid-1970s (see Figure 16–6).[11]

In the mid-1970s, medical research clearly tied asbestos to respiratory cancer, scarring of the lungs (now known as *asbestosis*), and cancer of the chest or abdominal lining (*mesothelioma*).[12] Its use has been severely restricted in the United States since 1989 (see Figure 16–7).

The following quote on **friable asbestos** shows why asbestos is still a concern even though its further use has been banned:

> When asbestos becomes friable (crumbly), it can release fibers into the air that are dangerous when inhaled. As **asbestos-containing material (ACM)** ages, it becomes less viable and more friable. Asbestos can be released into the air if it is disturbed during renovation or as a result of vandalism.[13]

OSHA has established an exposure threshold known as the PEL for asbestos. The PEL for asbestos is 0.1 fiber per cubic centimeter of air for an eight-hour time-weighted average (TWA).[14] Sources of further information about asbestos include:

- OSHA (www.osha.gov)
- EPA (www.epa.gov)
- NIOSH (www.cdc.gov/NIOSH)

FIGURE 16–6 Asbestos use, 1900 to present.

Asbestos is first used as a construction material

Asbestos becomes a popular roofing material in construction

EPA proposes a ban on the use of asbestos and all products containing it

1930s 1960s 1989

1900 1940s 1979

Asbestos is used for insulation in ship building

Medical research first suggests that cancers may be linked to asbestos

EPA bans the use of asbestos in the United States

Source: Centers for Disease Control and Prevention (CDC)

Asbestos Removal and Containment

When an industrial facility is found to contain asbestos, safety and health professionals are faced with the question of whether to remove it or contain it. Before making this decision, the following factors should be considered:

- Is there evidence that the *asbestos-containing material (ACM)* is deteriorating? What is the potential for future deterioration?

FIGURE 16–7 Asbestos use is now severely restricted in the United States.

Source: Krzysztof Slusarczyk/Shutterstock

- Is there evidence of physical damage to the ACM? What is the potential for future damage?
- Is there evidence of water damage to the ACM or spoilage? What is the potential for future damage or spoilage?[15]

Several approaches can be used for dealing with asbestos in the workplace. The most widely used are removal, enclosure, and encapsulation. These methods are explained in the following paragraphs.

Removal

Asbestos removal is also known as *asbestos abatement.* The following procedures are recommended for removal of asbestos: (1) The area in question must be completely enclosed in walls of tough plastic; (2) The enclosed area must be ventilated by **high-efficiency particle absolute (HEPA)** filtered negative air machines (these machines work somewhat like a vacuum cleaner in eliminating asbestos particles from the enclosed area); (3) The ACM must be covered with a special liquid solution to cut down on the release of asbestos fibers; and (4) The ACM must be placed in leakproof containers for disposal.[16]

Enclosure

Enclosure of an area containing ACMs involves completely encapsulating the area in airtight walls.[17] The following procedures are recommended for **enclosing** asbestos: (1) Use HEPA-filtered negative air machines in conjunction with drills or any other tools that may penetrate or otherwise disturb ACMs; (2) Construct the enclosing walls of impact-resistant and airtight materials; (3) Post signs indicating the presence of ACMs within the enclosed area; and (4) Note the enclosed area on the plans of the building.

Encapsulation

Encapsulation of asbestos involves spraying the ACMs with a special sealant that binds them together, thereby preventing the release of fibers.[18] The sealant should harden into a tough, impact-resistant skin. This approach is generally used only on acoustical plaster and similar materials.

Personal Protective Equipment for Asbestos Removal

It is important to use the proper types of personal protective clothing and respiratory devices. Clothing should be disposable and should cover all parts of the body.[19] Respirators used when handling asbestos should be "high-efficiency cartridge filter type (half-and full-face types); any powered-air purifying respirator; any type C continuous-flow, supplied-air, pressure-demand respirator, equipped with an auxiliary positive pressure self-contained breathing apparatus."[20]

Medical Records and Examinations

It is important that employees who handle ACMs undergo periodic medical monitoring. Medical records on such employees should be kept current and maintained for at least 20 years.[21] They should contain a complete medical history of the employees. These records must be made available on request to employees, past employees, healthcare professionals, employee representatives, and OSHA personnel.

Medical examinations, conducted at least annually, should also be required for employees who handle ACMs. These examinations should include front and back chest X-rays that are at least 7 inches by 14 inches. The examination should also test pulmonary function, including forced vital capacity and forced expiratory volume at one second.

INDOOR AIR QUALITY AND "SICK-BUILDING" SYNDROME

A key concept relating to **indoor air quality (IAQ)** is what has come to be called **"sick-building syndrome**." In reality, a "sick" building is one that makes people sick because it has become infested with mold, mildew, spores, and other airborne microorganisms. Although much is still unknown about sick-building syndrome, the EPA estimates that as many as 30 percent of the buildings in the United States have air quality problems. IAQ has been an issue of focused concern for safety and health professionals since the 1990s. Several sets of guidelines have been adopted to help organizations investigate and analyze IAQ problems in the workplace. The most prominent of these guidelines are as follows:

- Supplement to NIOSH Manual of Analytical Methods (1994)
- OSHA Technical Manual, Section III, Chapter 2, "Indoor Air Quality Investigations" (1994)
- ISIAQ Task Force Report on General Principles for Investigation of IAQ Complaints (1998)
- ACGIH Bioaerosols Manual (1999)
- ASTM E2418-06 Standard Guide for Readily Observable Mold and Conditions Conducive to Mold in Commercial Buildings (2006)
- ASTM D 7338-10 (2010) Standard for Mold Assessment (see next section)

Poor IAQ can cause a variety of health problems ranging from the temporary to the long term. Health problems commonly associated with poor IAQ include allergic reactions, respiratory problems, eye irritation, sinusitis, bronchitis, and pneumonia. Often, the cause of poor IAQ can be slipshod maintenance such as failure to keep fans, ductwork, and filters clean. Other contributors are the particles and gases that can be released by office equipment, carpets, paints, cleaning solvents, and office supplies.

One of the keys to preventing sick-building syndrome is air exchange. Important factors in a building's ability to eliminate contaminated air and bring in fresh air are as follows:

- **Ventilation**
- Air infiltration rates
- Airflow rates in ducts
- Airflow patterns
- Fume exhaust

The most accurate methods available for measuring these factors fall under the broad heading of **tracer gas techniques**. A tracer gas is any gas or vapor not normally found in a building. The best have the following characteristics:

- Nontoxic
- Nonallergic at the levels used

- Chemically inert
- Odorless and tasteless
- Nonflammable and nonexplosive
- Easily transported
- Easily dispersed as an atmospheric gas
- Easily and economically measured with a high degree of reliability

The most widely used tracer gases are sulfur hexafluoride, halogenated refrigerants, and perfluorocarbons. To perform a tracer gas test, the following materials and equipment are needed:

- A suitable tracer gas
- A device for measuring tracer gas concentrations
- An air-sampling system
- A tracer gas injection system
- A data acquisition and control system

There are several different types of tracer gas tests, including tracer decay, constant concentration, buildup/decay, CO_2 measurement, and reentrainment/recirculation. Regardless of the type of test used, the testing process involves the following steps:

1. Inject the tracer gas into the building.
2. Measure the concentration of the tracer gas in different parts of the building at different times over a certain period.

The data collected during a tracer gas test can give safety and health professionals the following types of information:

- Total air exchange rate for the building
- Air change rate due to the operation of the building's HVAC system
- Air change rate due to air infiltration and leakage
- Percentage of outside air supplied by the building's HVAC system
- Effectiveness of the ventilation system in removing contaminants
- The distribution of the ventilation air throughout the building

With this type of information available, safety and health personnel can determine whether there are pockets where contaminated air is trapped; whether the ventilation and air infiltration rates are sufficient; whether airflow rates through ducts are sufficient; whether airflow patterns are what they should be; and whether fume hoods are performing as they should. This type of information is needed to detect and prevent IAQ problems.

ANSI's Indoor Air Quality/HVAC Standard

The American National Standards Institute (ANSI) developed a standard specifically addressing IAQ. The standard ANSI Z9.8 carries the following title: "Fundamentals Governing the Management, Operation, Testing, and Maintenance of HVAC Systems for Maintaining Acceptable Indoor Air Quality in Employee Occupancies through Dilution Ventilation." Key components of the standard are summarized in the following paragraphs:

General coverage. ANSI Z9.8 is specifically written to apply primarily to office space. It applies to employee occupancies in nonindustrial spaces, including general office spaces, commercial operations, and office spaces within industrial facilities.

Application flexibility. The requirements set forth in the standard are minimums. Employers may use demonstrably equal or better approaches. Where the provisions of ANSI Z9.8 conflict with other standards, the more stringent standard is to take precedence. If employers deviate from the standard, they must justify the deviation(s) in writing.

Acceptable air quality. The standard does not prescribe or define what is or is not "acceptable" air quality. Employers are required to determine and define "acceptable." Employers may use threshold limit values (TLVs) or PELs as guidelines or they may establish **ceiling** limits (e.g., if more than 2 percent of employees complain about air quality problems the ceiling limit has been reached).

Tobacco smoke. The standard requires employers to evaluate smoking in the workplace and apply whatever management controls are appropriate. This gives employers a great deal of latitude. The attitude of the committee that developed the standard (ANSI Z9.8) seems to be that most employers already have smoking controls established.

TOXIC MOLD AND INDOOR AIR QUALITY

Toxic mold has surfaced as an issue relating to IAQ. The issue is complicated by at least two factors. First, there are thousands of types of molds, but only a few are toxic. Second, different people have different levels of sensitivity to mold. On OSHA's list of the 13 most critical IAQ hazards, mold is the last entry. On the other hand, in those limited instances in which molds are toxic, they can cause coughing, atypical asthma, nasal congestion, sinusitis, rhinitis, skin rashes, and fatigue. In severe cases, toxic molds can be deadly.

OSHA recommends a three-step process for investigating the possibility of the presence of toxic mold in the workplace. Step one involves interviewing employees about symptoms. Step two involves conducting an on-site review of the workplace in question. Step three, which is undertaken only if the symptoms and physical evidence warrant it, involves an on-site environmental evaluation in which a specialist takes samples.

The principal causal factor in most cases of toxic mold is inadequate ventilation. Consequently, an effective approach for preventing the accumulation of toxic mold in the workplace is to apply the following steps: (1) Check outdoor intakes and make sure they are not near trash storage areas, standing water, exhausts, or anything else that might contribute to the growth of mold; (2) Make sure the drip pans are sloped sufficiently to prevent the accumulation of standing water; and (3) Check ductwork regularly to ensure that the lining is dry and clean.

Toxic Mold Assessment and Remediation

Modern safety and health professionals must be prepared to deal with moisture and mold issues. It is important to investigate periodically to identify sources of moisture and mold. The following procedures may be used to guide investigations:

* *Look* for mold in likely locations such as around pipes, drains, windows, and dark, poorly ventilated areas.

- *Listen* to the feedback and comments of employees who might complain about allergies that could be affected by mold or about any aspect of IAQ.
- *Smell* the air in the work environment. You can tell if the air is damp, stale, or musty. If it is, there is a moisture problem, even if it is hidden under floors or behind walls.
- *Train* employees how to be "mold investigators" by showing them how to look, listen, and smell in their work environment.
- *Inventory* the principal areas of moisture and mold risk in your facility and monitor these high-risk areas continually.[22]

When mold is found, it is important to act. Mold remediation, in general terms, proceeds as follows: (1) stop the moisture intrusion, (2) contain and isolate the moisture that is already present, (3) dry and filter the affected area, (4) remove anything in the affected area that cannot be dried, (5) kill existing bacteria with disinfectants and sanitizing agents, (6) clean and then reclean the area, and (7) take whatever steps are necessary to prevent any further moisture intrusion.

According to Alfred Draper III, an industrial hygienist who specializes in mold-related restoration and remediation, mold remediation projects can be divided into four classes ranging from low impact to high. Class I remediation is just good housekeeping (for example, minimizing dust, using drop cloths, and cleaning up with HEPA-filtered vacuums). Class II requires the use of EPA-registered disinfectants, containing construction waste, and limiting access to work areas.

Class III projects are where remediation begins to be more challenging. For Class III projects, Draper recommends the following procedures:

- Remove or isolate the affected area's HVAC system to prevent ductwork contamination.
- Seal off the affected area using hard critical barriers (e.g., plywood and sheetrock).
- Require all personnel who must enter the affected area to wear the proper PPE, including disposable full-body coverings, gloves, and half-mask HEPA filter respirators.
- Contain all remediated waste in tightly covered containers before it is shipped.
- Clean and decontaminate all equipment prior to making the final wipe-down and cleanup of the affected area.

Class IV projects are the most challenging and demanding of moisture or mold remediations. For Class IV, Draper recommends all the Class III procedures plus the following:

- Cover all equipment, structures, and surfaces not being cleaned.
- Require all remediation personnel to wear the PPE, which should include disposable full-body coverings, gloves, rubber boots, and full-face respirators.
- Construct a decontamination facility and require all personnel to pass through it when leaving the affected area.
- Monitor exterior air throughout the duration of the remediation project.
- Evaluate high-risk personnel who work near the affected area, at least during the more hazardous aspects of the remediation.
- Thoroughly "air wash" any materials in the affected room that cannot be disposed of.

- Use negative pressure throughout the cleaning process for **local exhaust** at in the affected area.
- Use air scrubbing throughout the remediation project to remove any mold spores that might be introduced into the air.
- Have all air ducts professionally cleaned.
- Collect cultured and noncultured samples that can be tested to ensure that the remediation process was effective.

ASTM D7338: GUIDE FOR THE ASSESSMENT OF FUNGAL GROWTH IN BUILDINGS

The American Society for Testing Materials (ASTM) developed D7338 as its standard for mold assessment. The standard is specific to indoor mold and contains minimum criteria for making assessments. Much of what the standard does is refer organizations to other ASTN standards for such activities as testing methods for direct moisture control, testing methods for water penetration, and practices for the periodic inspection of buildings. The standard contains information on fungal biology, fungal growth, and indoor building moisture, but the information on these topics is limited.

ASTM D7338 allows observations to be accepted as inspections. The standard also: (1) calls for the development of a hypothesis for evaluation and determination of the exposure of building occupants to mold growth or bioaerosols, (2) suggests that additional investigations be made for materials containing asbestos, and (3) suggests the use of the scientific method in making investigations. As originally published, ASTM D7338 is not yet as comprehensive a standard as it needs to be. However, this is often the case with the initial development of a standard. Over time as the standard is revised it will most likely also be strengthened.

THRESHOLD LIMIT VALUES

How much exposure to a toxic substance is too much? How much is acceptable? Guidelines that answer these questions for safety and health professionals are developed and issued annually by the American Conference of Governmental Industrial Hygienists (ACGIH). The guidelines are known as **threshold limit values (TLVs)**.

ACGIH's Classifications of TLVs and BEIs

The ACGIH develops TLVs and **biological exposure indices (BEIs)** to help safety and health professionals control certain chemical, biological, and physical health hazards in the workplace.[23] TLVs and BEIs are not legal standards and are not intended to be; rather, they are guidelines. However, their impact is increasingly felt, and in a positive way. As government organizations and agencies continue to find that political considerations make it difficult to promulgate legally authorized standards in a timely manner, the ability of the ACGIH to produce TLV guidelines that are updated annually makes its guidelines more and more valuable. Key concepts about TLVs and BEIs that should be understood by safety and health professionals are as follows:

Threshold limit value–time-weighted average (TLV-TWA). The TWA for a conventional eight-hour workday and a 40-hour workweek for a given substance to which it is believed that nearly all workers may be repeatedly exposed on a daily basis without suffering ill effects. For example, the TLV-TWA for liquefied petroleum gas is 1,000 parts per million (ppm).

Threshold limit value–short-term exposure limit (TLV-STEL). The concentration of a given substance to which it is believed that workers may be exposed continuously for short periods without suffering ill effects. A STEL is defined as a 15-minute TWA exposure that should not be exceeded at any time during the workday period. Also, exposures above the TLV-TWA up to the STEL should not exceed 15 minutes and should not occur more than four times in a day (with at least 60 minutes between exposures). For example, the TLV-STEL for isopropyl ether is 310 ppm.

Threshold limit value–ceiling (TLV-C). The concentration of a given substance that should not be exceeded at any point during an exposure period.

Biological exposure indices. The levels of determinants that are expected to be present in specimens taken from healthy workers who have been exposed to selected substances to the same extent as other workers with inhalation exposure to the substance at the TLV. For example, the BEI for acetone in the urine of a worker is 50 milligrams per liter (mg/L).

Physical agents. Substances or factors that can introduce added stress on the human body so that the effects of a given substance at the TLV might be magnified. Physical factors include acoustics (noise), ergonomic conditions, ionizing radiation, lasers, nonionizing radiation, subfrequency and static electric fields, and thermal stress (cold and heat). For example, the TLV for noise of 94 dB (average) is one hour per day.

Calculating a TWA

TWAs can be calculated for exposures to given substances.[24] The following formula can be used for calculating the TWA for an eight-hour day:

$$TWA = \frac{CaTa + CbTb + \cdots CnTn}{8}$$

where

Ta = time of the first exposure period during the eight-hour shift

Ca = concentration of the substance in question in period a

Tb = another time period during the same shift

Cb = concentration of the substance in question in period b

Tn = nth or final time period in the eight-hour shift

Cn = concentration during period n

For the sake of illustration, the following sample case is presented. Assume that three samples were taken over the course of the work day yielding the following results:

Ca = 6.2 ppm and Ta = 220 minutes

Cb = 1.3 ppm and Tb = 96 minutes

Cn = 4.4 ppm and Tn = 50 minutes*

With this information known, it can now be plugged into the TWA equation and the TWA calculated as follows:

$$TWA = \frac{(6.2\,ppm \times 220\,min.) + (1.3\,ppm \times 96\,min.) + (4.4\,ppm \times 50\,min.)}{480\,min. (8\,hours)}$$

$$TWA = 3.56$$

HAZARD RECOGNITION AND EVALUATION

The degree and nature of the hazard must be understood before effective hazard control procedures can be developed. This involves recognizing that a hazard exists and then making judgments about its magnitude with regard to chemical, physical, biological, and ergonomic stresses.

Questions that can be used for recognizing hazards in the workplace are as follows:

- What is produced?
- What raw materials are used in the process?
- What additional materials are used in the process?
- What equipment is used?
- What operational procedures are involved?
- What dust control procedures are involved?
- How are accidental spills cleaned up?
- How are waste by-products disposed?
- Is there adequate ventilation?
- Are processes equipped with exhaust devices?
- How does the facility layout contribute to employee exposure?
- Are properly working personal protective devices available?
- Are safe operating procedures recorded, made available, monitored, and enforced?[25]

All processes should be subjected to the following **hazard recognition** procedures:

- Determine the exposure threshold for each hazardous substance identified when applying the questions just listed, including airborne contaminants.
- Determine the level of exposure to each hazardous substance.
- Determine which employees are exposed to each hazardous material, how frequently, and for how long.
- Calculate the TWAs to the exposure thresholds identified earlier.[26]

---

*Note that Cn and Tn would have been labeled as Cc and Tc had there been more than three samples taken over the course of the day. The letter "n" is used to indicate the final sample of the day regardless of how many samples are taken.

For hazard evaluation, the following considerations are important: the nature of the material or substance involved, the intensity of the exposure, and the duration of the exposure. Key factors to consider are how much exposure is required to produce injury or illness; the likelihood that enough exposure to produce injury or illness will take place; the rate of generation of airborne contaminants; the total duration of exposure; and the prevention and control measures used.[27]

For example, the textile industry was once ranked fifth on a list of 43 industries in terms of preventing workplace injuries and fatalities. The industry improved this ranking to first, a feat attributed to hazard evaluation and recognition efforts within the industry adopted to comply with the OSHA cotton-dust standard (1910.1043).[28]

The textile industry has become a leader in the area of industrial safety and health. Even so, there are problems. According to *Safety & Health*,

> Although the textile industry has stopped the use of cancer-associated dyes, as many as 50,000 current and former industry workers have had significant major exposure during their work lives and are at high risk for developing bladder cancer.[29]

One of the ways in which textile companies are dealing with their industrial hygiene problems is by adding more safety and health professionals to their staffs.

Honeywell (GE) includes industrial hygiene as a major component in its overall safety and health program. A corporate-wide environmental auditing program for recognizing and evaluating hazards is the mainstay of Honeywell's industrial hygiene effort.

As with all effective programs of this nature, Honeywell's program has the involvement and support of the board of directors and higher management. The program came into being in response to what came to be known as the "Kepone tragedy," an incident in which employees at a company in Hopewell, Virginia, suffered Kepone poisoning. The tragic results of this incident motivated Honeywell to establish a rigorous environmental auditing component to its overall industrial hygiene program. Honeywell now conducts approximately 50 environmental audits annually in such areas as occupational health, medical programs, solid waste disposal, hazardous waste disposal, safety/loss prevention, and product safety. The results of the audits are used to develop prevention and control strategies.

PREVENTION AND CONTROL

Most prevention and control strategies can be placed in one of the following four categories: (1) engineering controls, (2) ventilation, (3) PPE, and (4) administrative controls.[30] Examples of strategies in each category are given in the following paragraphs.

Engineering Controls

The category of **engineering controls** includes such strategies as replacing a toxic material with one that is less hazardous or redesigning a process to make it less stressful or to reduce exposure to hazardous materials or conditions. Other engineering controls are **isolating** a hazardous process to reduce the number of people exposed to it and introducing moisture to reduce dust.[31]

For example, exhaust ventilation, which involves trapping and removing contaminated air, is an engineering control. This type of ventilation is typically used with such processes as abrasive blasting, grinding, polishing, buffing, and spray painting or finishing. It is also used in conjunction with open-surface tanks. Dilution ventilation involves simultaneously removing and adding air to dilute a contaminant to acceptable levels.[32]

Personal Protection from Hazards (PPE and Eyewash Stations)

When the work environment cannot be made safe by any other method, **personal protective equipment (PPE)** is used as a last resort. PPE imposes a barrier between the worker and the hazard but does nothing to reduce or eliminate the hazard. Typical equipment includes safety goggles, face shields, gloves, boots, earmuffs, earplugs, full-body clothing, barrier creams, and respirators.[33]

Occasionally, in spite of an employee's best efforts in wearing PPE, his or her eyes or skin will be accidentally exposed to a contaminant. When this happens, it is critical to wash away or dilute the contaminant as quickly as possible. Specially designed eyewash and emergency wash stations should be readily available and accessible in any work setting where contaminants may be present (see Figure 16–8).

For ensuring compliance when installing eyewash stations in the workplace, the relevant standard is ANSI Z358.1. Organizations that comply with this ANSI standard also comply with the Americans with Disabilities Act or ADA. ANSI Z358.a includes a requirement for the clear access to eyewash stations. Station designed that best meet the requirements of ANSI and the ADA include swing-down, sink-mounted eyewash stations; wall-mounted, recessed eyewash stations and drench showers; and floor-mounted combination shower and eyewash stations.

Administrative Controls

Administrative controls involve limiting the exposure of employees to hazardous conditions using such strategies as the following: rotating schedules, required breaks, work shifts, and other schedule-oriented strategies.[34]

FIGURE 16–8   Emergency eyewash station.

Courtesy of David Goetsch

Additional Strategies

The type of prevention and control strategies used will depend on the evaluation of the specific hazards present in the workplace. The following list of generic strategies applies regardless of the setting:

- Practicing **good housekeeping**, including workplace cleanliness, waste disposal, adequate washing and eating facilities, healthful drinking water, and control of insects and rodents
- Using special control methods for specific hazards, such as reduction of exposure time, film badges and similar monitoring devices, and continuous sampling with preset alarms
- Setting up medical programs to detect intake of toxic materials
- Providing training and education to supplement engineering controls[35]

Self-Protection Strategies

One of the best ways to protect employees from workplace hazards is to teach them to protect themselves. Modern safety and health professionals should ensure that all employees are familiar with the following rules of self-protection:

1. ***Know the hazards in your workplace.*** Take the time to identify all hazardous materials and conditions in your workplace and know the safe exposure levels for each.
2. ***Know the possible effects of hazards in your workplace.*** Typical effects of workplace hazards include respiratory damage, skin disease and irritation, injury to the reproductive system, and damage to the blood, lungs, central nervous system, eyesight, and hearing.
3. ***Use personal protective equipment properly.*** Proper use of PPE means choosing the right equipment, getting a proper fit, correctly cleaning and storing equipment, and inspecting equipment regularly for wear and damage.
4. ***Understand and obey safety rules.*** Read warning labels before using any contained substance, handle materials properly, read and obey signs, and do only authorized work.
5. ***Practice good personal hygiene.*** Wash thoroughly after exposure to a hazardous substance, shower after work, wash before eating, and separate potentially contaminated work clothes from others before washing them.[36]

NIOSH AND INDUSTRIAL HYGIENE

NIOSH is part of the Department of Health and Human Services (DHHS). This agency is important to industrial hygiene professionals. The main focus of the agency's research is on toxicity levels and human tolerance levels of hazardous substances. NIOSH prepares recommendations for OSHA standards dealing with hazardous substances, and NIOSH studies are made available to employers.

The areas of research of NIOSH's four major divisions—Biomedical and Behavioral Science; Respiratory Disease Studies; Surveillance, Hazard Evaluations, and Field Studies; and Training and Manpower Development—were discussed in Chapter 5. The results of these divisions' studies and their continually updated lists of toxic materials and recommended tolerance levels are extremely helpful to industrial hygienists concerned with keeping the workplace safe.

NIOSH GUIDELINES FOR RESPIRATORS

The **respirator** is one of the most important types of PPE available to individuals who work in hazardous environments (see Figures 16–9 and 16–10). Because the performance of a respirator can mean the difference between life and death, NIOSH publishes strict guidelines regulating the manufacture of respirators. The standard with which manufacturers must comply is 42 CFR 84. In addition, safety and health professionals must ensure that employees are provided respirators that meet all of the specifications set forth in 42 CFR 84.

There are two types of respirators: air filtering and air supplying. **Air-filtering respirators** filter toxic particulates out of the air. To comply with 42 CFR 84, an air-filtering respirator must protect its wearer from the most penetrating aerosol size of particle, which is 0.3 micron aerodynamic mass in median diameter. The particulate

FIGURE 16–9   Breathing protection device.

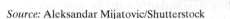

Source: Aleksandar Mijatovic/Shutterstock

FIGURE 16–10   Breathing protection device.

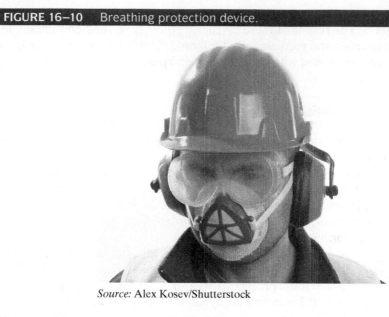

Source: Alex Kosev/Shutterstock

filters used in respirators are divided into three classes, each class having three levels of efficiency as follows:

Class N respirators may be used only in environments that contain no oil-based particulates. They may be used in atmospheres that contain solid or nonoil contaminants.

	Class N (Not Oil Resistant)	Class R (Oil Resistant)	Class P (Oil Proof)
Efficiency	95%	95%	95%
Efficiency	99%	99%	99%
Efficiency	99.97%	99.97%	99.97%

Class R respirators may be used in atmospheres containing any contaminant. However, the filters in Class R respirators must be changed after each shift if oil-based contaminants are present. Class P respirators may be used in any atmosphere containing any particulate contaminant.

If there is any question about the viability of an air-filtering respirator in a given setting, employees should use **air-supplying respirators**. This type of respirator works in much the same way as an air tank does for a scuba diver. Air from the atmosphere is completely blocked out, and fresh air is provided via a self-contained breathing apparatus.

In addition to the NIOSH regulations regarding respirators (42 CFR 84), there are regulations published by OSHA in 29 CFR 1010.134. Key provisions of the OSHA regulations are as follows:

- Respirators, when they are required, must be provided by the employer.
- Medical evaluations must be provided for respirator users.
- Fit testing according to standards must be conducted.
- Respirators must be used in reasonably foreseeable emergency situations.
- Respirators must be properly cleaned and maintained.

- Adequate air quality, quantity, and flow for atmosphere-supplied respirators must be ensured.
- Training and evaluation programs must be provided to ensure effectiveness.

Air Safety Program Elements

Companies with facilities in which fumes, dust, gases, vapors, or other potentially harmful particulates are present should have an **air safety program** as part of their overall safety and health program. The program should have at least the following elements:

- Accurate hazard identification and analysis procedures to determine what types of particulates are present and in what concentration
- Standard operating procedures (in writing) for all elements of the air safety program
- Respirators that are appropriate in terms of the types of hazards present and that are included in 42 CFR 84
- Training, including fit testing, limitations, use, and maintenance of respirators
- Standard procedures for routine maintenance and storage of respirators

STANDARDS AND REGULATIONS

Standards and regulations relating to toxic substances come from a number of different sources. Prominent among these are OSHA and the EPA. Important standards and regulations in this area include the OSHA **Chemical Process Standard**, the EPA Chemical Risk Management Standards, the Hazardous Materials Transportation Act (HMTA), and the Toxic Substances Control Act. These and other standards and regulations are explained in this section.

OSHA Chemical Process Standard

OSHA issued the *Process Safety Management of Highly Hazardous Chemicals* standard (29 CFR 1910.119) to help ensure the safe use of processes using HHCs. A process is any activity or combination of activities involving the use, storage, manufacturing, handling, or off-site movement of HHCs. The goal of OSHA's Chemical Process Standard is to prevent the accidental release of toxic, reactive, or flammable

---

SAFETY MYTH

CONTACT LENSES AND RESPIRATORS

Firefighters and other personnel who wear respirators should not wear contact lenses, Right? Not necessarily. OSHA Standard 29 CFR 1910.134(e)(5)(ii) forbids the practice of wearing contact lenses with a respirator, but the standard is based on old information relating to hard contact lenses. The problem with hard contacts is their tendency to pop out. However, most contact lens wearers now use the soft variety that stay in place better. Employees who wear soft contact lenses can use respirators.

liquids and gases in processes involving HHCs. The standard applies to any facility that uses one or more of the HHCs listed by OSHA or the EPA. The standard requires that organizations do the following:

- Compiling process safety information
- Maintaining safe operating procedures
- Training and educating employees
- Maintaining equipment
- Conducting accident investigations
- Developing emergency response plans
- Conducting safety compliance audits

ENVIRONMENTAL PROTECTION AGENCY RISK MANAGEMENT PROGRAM

The EPA developed its own risk management program for chemical processes that is similar to OSHA's Chemical Process Standard. The EPA's risk management program is contained in 40 CFR Part 68. It requires that organizations that use, store, handle, or transport HHCs have a comprehensive risk management program with the following components:

- Off-site consequence analysis
- Five-year accident history
- Prevention program 2 covering safety information, hazard review, operating procedures, training, maintenance, compliance audits, and incident investigation
- Prevention program 3 covering process safety information, process hazard analysis, operating procedures, training, mechanical integrity, management of change, pre-startup review, compliance audits, incident investigation, employee participation, hot work permit, and provisions for contractors
- Emergency response program
- Certifications
- Updates

Hazardous Materials Transportation Act

The basic provisions of the HTMA are shown in Figure 16–11. In addition to these provisions, the act makes companies that transport hazardous materials partially liable

FIGURE 16–11   Basic provisions of the Hazardous Materials Transportation Act.

- Training relating directly to specific functions for private manufacturers, shippers, and carriers.
- Training for employees of public agencies.
- National registration with registration fees used to help pay for the costs of emergency response training.
- A national permit system for motor carriers.

for damages when an accident occurs and the carrier does not have a satisfactory rating from the Department of Transportation.

The following steps can be used for minimizing the risks associated with transporting hazardous materials:

- Hire personnel who are certified and experienced in dealing with hazardous materials. Then make sure to brief them properly on the hazardous materials that they will be handling and keep them up-to-date on the latest federal regulations.
- Establish a training program that covers at least the following topics: safety equipment, identification of hazardous shipments, routine handling procedures, and emergency procedures.
- At each location where hazardous materials will be loaded or unloaded, name one person who is responsible for making periodic inspections, instructing new employees, meeting with representatives of companies that ship major hazardous materials, maintaining comprehensive records, filing all necessary reports, and contacting the shipper and cosigner when there is a problem.
- Identify all hazardous materials and post a diamond-shaped placard on the outside of all vehicles used to transport hazardous materials.[37]

GENERAL SAFETY PRECAUTIONS

Following are a number of general safety precautions that apply in any settings where explosive and combustible materials are present.

1. *Prohibit smoking.* Smoking should be prohibited in any areas of a plant where explosive and combustible materials are present. Eliminating potential sources of ignition is a standard safety precaution in settings where explosions and fire are possible. In the past, such areas were marked off as restricted, and No Smoking signs were posted. It is becoming common practice to prohibit smoking on the premises altogether or to restrict smoking to designated areas that are well removed from hazard areas.

2. *Eliminate static electricity.* **Static electricity** occurs when dissimilar materials come into contact and then separate. If these materials are combustible or are near other materials that are combustible, an explosion can occur. Therefore, it is important to eliminate static electricity. The potential for the occurrence of static electricity can be reduced substantially by the processes of **grounding** and bonding. **Bonding** involves eliminating the difference in static charge potential between materials. *Grounding* involves eliminating the difference in static charge potential between a material and the ground. Figures 16–12, 16–13, and 16–14 illustrate these concepts.

3. *Use spark-resistant tools.* Sparks from tools used in settings where explosive and combustible materials are present represent a threat that must be addressed. In such settings, spark-resistant tools should be used to the maximum extent possible. Wooden, leather-faced, and rubber-covered tools can help prevent sparks that might ignite volatile materials.[38]

FIGURE 16–12   How static electricity occurs.

► Electrically charged   ► Not electrically charged

► Not grounded   ► Not grounded

► Conductive material   ► Conductive material

FIGURE 16–13   Bonding prevents static spark.

► A and B are connected by a
  bonding wire

► Electrical charge is shared by A
  and B via the bonding wire
  eliminating charge differences
  between them

FIGURE 16–14   Grounding prevents static spark.

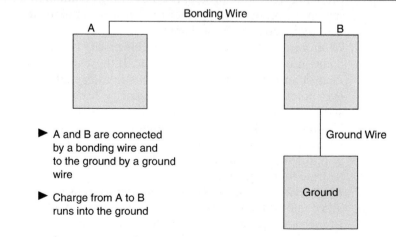

► A and B are connected
  by a bonding wire and
  to the ground by a ground
  wire

► Charge from A to B
  runs into the ground

NANOSCALE MATERIALS AND INDUSTRIAL HYGIENE

A development in the field of industrial hygiene that will confront safety profession-als more and more frequently in the years to come is how to prevent hazards in the workplace relating to toxic nanoscale materials. There is not yet an OSHA standard for nanoscale materials, but the EPA is developing a stewardship program to encour-age responsible development, handling, and use of them. In order to understand this developing subfield of industrial hygiene, it is necessary to understand the language of nanoscale materials. The following definitions are based on ASTM E 2456-06 *Terminology for Nanotechnology* (astm.org):

- *Fine particle.* A particle smaller than about 2.5 micrometers and larger than about 0.1 micrometers.
- *Nanoparticle.* A sub-classification of ultrafine particle, with lengths in two di-mensions greater than 0.001 micrometer (100 nanometers).
- *Nanostructured.* Containing physically or chemically distinguishable compo-nents, at least one of which is nanoscale in one or more dimensions.
- *Nanotechnology.* Technologies that measure, manipulate, or incorporate mate-rials and/or features with at least one dimension between approximately 1 and 100 nanometers (nm).
- *Ultrafine particle.* A particle ranging in size from approximately 0.1 microm-eter (100 nanometers) to 0.001 micrometers (1 nanometer).

Regulatory agencies are still grappling with two key issues relating to nanoscale materials: (1) defining exposure limits and (2) developing methods for accurately measuring exposure. The EPA has taken the lead with its *stewardship program* as developed in accordance with the Toxic Substances Control Act (TSCA) (15 USC 2601). The program is described in a document developed by the EPA entitled, "TSCA Inventory Status of Nanoscale Substances—General Approach, and a pro-posed Information Collection Request." This document is available at epa.gov/oppt/nano/nmspfr.htm. This document states that the purpose of the EPA's Stewardship program is as follows:

- Assemble existing data and information from manufacturers of existing na-noscale materials as well as processors of the materials.
- Identify risk management practices in developing and commercializing na-noscale materials and encourage the use of these practices.
- Encourage the development of the type of test data needed to provide a better scientific foundation for future practices and regulatory/policy decisions.
- Encourage responsible development of nanomaterials.[39]

This subfield of industrial hygiene will become increasingly important to safety and health professionals and should be monitored carefully for developments, espe-cially in the area of exposure measurement methods.

CONFINED SPACE HAZARDS

A **confined space** is any area with limited means of entry and exit that is large enough for a person to fit into but is not designed for occupancy. Examples of confined spaces include vaults, vats, silos, ship compartments, train compartments, sewers, and tunnels. What makes confined spaces hazardous, beyond those factors that define the concept, is their potential to trap toxic and explosive vapors and gases.

OSHA's Confined Space Standard (29 CFR 1910.146, paragraph (b)) defines a hazardous atmosphere as one "that may expose employees to the risk of death, incapacitation, impairment of ability to self-rescue, injury, or acute illness."[40] The definition goes on to describe a hazardous atmosphere as a "flammable gas, vapor, or mist in excess of 10 percent of its lower flammable limit."[41]

Safety and health professionals should be cautious in assuming that 10 percent of a lower flammable limit (LFL) for a given substance constitutes an acceptable atmosphere. Even 10 percent of the LFL may exceed the TLV for gases and vapors. Consequently, safety and health professionals should encourage the use of the most sensitive instruments available to detect airborne contaminants.

Certain substances have upper and lower flammable limits (sometimes referred to as "explosive limits," a less-accurate label because some will flame without exploding). The LFL is the lowest concentration of a gas or a vapor that can generate a flame when in the presence of a sufficient ignition source. The upper flammable limit (UFL) is the highest concentration that can propagate a flame. The range between these two extremes is the explosive/flammable range (see Figure 16–15).

Using a concentration of 10 percent of the LFL can be dangerous for two reasons: (1) the gas sensor used may be inaccurate and (2) the TLV for some substances is exceeded at the LFL. For example, the TLV for acetone is 500 ppm. The 10 percent LFL for acetone is 2,500 ppm, or five times the TLV.

Safety and health professionals are well advised to take the following precautions when dealing with confined spaces that may have a toxic environment. First, use the most sensitive detection instrument available. Detector tubes, portable gas chromatographs, and infrared analyzers are all more accurate than combustible gas sensors are.

FIGURE 16–15 Flammable/explosive range for vapors and gases.

Second, compare the 10 percent LFL for any substance in question with its TLV, and let the TLV take precedence.

To ensure that a confined space is safe, the following questions should be asked and answered in the affirmative before allowing entry. A negative response to even one of these questions means that entry into the confined space in question is not yet safe.

- Are access and exit equipment such as ladders and steps in good working condition?
- Has the confined space been properly purged of the toxic vapors and other toxic substances?
- Are all lines that transport potentially hazardous substances into or through the confined space turned off and properly capped?
- Are all moving equipment and moving parts of equipment in the confined space shut down and locked out?
- Has proper ventilation (natural or mechanical) been provided?
- Has the atmosphere inside the confined space been checked by appropriately sensitive detection devices?
- Have provisions been made to monitor continually the atmosphere inside the confined space during work?

In addition to the toxic and explosive hazards associated with confined spaces, there are often physical hazards. For example, tunnels often contain pipes that can trip an employee or that can leak and cause a fall. Empty liquid or gas storage vessels may contain mechanical equipment or pipes that must be carefully maneuvered around, often in the dark.

OSHA CONFINED SPACE STANDARD

OSHA Confined Space Standard (29 CFR 1910.146)

Confined spaces are those that by their configuration hinder the activities of employees who must enter them, work in them, and exit from them. Employees who work in confined spaces face hazards from the confinement itself, but often from the potential for entrapment, engulfment, and exposure to hazardous atmospheric conditions. The term "permit-required confined space" or the shorter term "permit space" represents concepts that are important to safety and health professionals, managers, supervisors, and employees. A permit-required confined space—hereafter referred to as a *permit space*—is one that meets OSHA's definition of a *confined space* and contains health or safety hazards. Such spaces are off limits to personnel unless they have first been issued an appropriate permit to enter.

Confined Space Defined

Typical confined spaces include vaults, tanks, storage bins, pits, diked areas, vessels, silos, and other areas where employee egress and movement is restricted by space limitations. OSHA's definition of a confined space is specific and detailed. To be considered a confined space, a space must meet the following requirements: (1) Be large enough for an employee to enter fully and perform the work assigned in the space, (2) Not designed for continuous occupancy by employees, and (3) Has limited means of egress (entry and exit).

Permit Space Defined

Not all confined spaces are necessarily permit spaces. To be considered a permit space, a confined space must have one or more of the following characteristics: (1) Contain a hazardous atmosphere or have the potential to contain such an atmosphere; (2) Contain a material with the potential to engulf an employee who enters the space (for example, a silo full of wheat or corn); (3) Has an internal configuration that could trap an employee and/or lead to asphyxiation; or (4) Contain any other recognized serious safety or health hazard.

Exemptions to Entry by Permit Only

Under certain conditions, an organization might be granted an exemption to the requirement that a permit be issued before an employee can enter a certain area. In these cases, the organization must be able to show that it will use alternate procedures to keep employees who enter the space in question safe. For example, an organization might claim that the atmosphere in a confined space can be made safe through an alternative procedure such as continuous forced-air ventilation. If the organization can back this claim with hard data based on monitoring, testing, and inspection, an exemption to the entry-by-permit-only requirement can be made. Even if an exemption is approved, the organization must still test the atmosphere for hazardous conditions before allowing anyone to enter the space in question.

Written-Program Requirements

Employers that plan to allow personnel to enter permit spaces must first develop a written program. The written program must clearly and comprehensively describe how the organization will do the following: (1) Implement the measures needed to prevent unauthorized entry; (2) Identify and evaluate permit space hazards before allowing any personnel to enter the spaces in question; (3) Test atmospheric conditions in the permit space before granting access to it and then monitor conditions, the space, and the employee(s) during the entire course of confinement; (4) Perform appropriate testing for the following atmospheric hazards (in the order listed here): oxygen, combustible gases or vapors, and toxic gases or vapors. If hazardous conditions are detected during an entry, the employee is required to exit the space immediately; (5) If hazards are identified, the employer must then evaluate the space to identify the cause of the hazardous condition and revise the written program to eliminate future instances of exposure to the hazard; (6) Establish and implement effective procedures and practices to eliminate or control hazards in the permit space so that it is safe to enter and work within the space including specifying acceptable entry conditions, isolating the permit space, providing barriers, verifying acceptable entry conditions, and purging, making inert, flushing, or ventilating the permit space; (7) Identify specifically the duties of employees who will enter permit spaces; (8) Provide, maintain, and require the use of—at no cost to the employee—all equipment including PPE necessary to enter and operate safely within the permit space (In addition to appropriate PPE, equipment may include testing, monitoring, ventilating, communication, and lighting equipment; barriers and shields; ladders; and retrieval devices); (9) Ensure that at least one person is stationed as an attendant outside the permit space for the duration of the employee's time inside the permit space; (10) Coordinate entry operations when more than one employee will enter the permit space; (11) Implement effective procedures for summoning rescue personnel and emergency services (Also, implement effective procedures for preventing unauthorized personnel from attempting a rescue); (12) Establish and implement a written system for preparing, issuing, using, and canceling entry permits; (13) Review established entry procedures annually and revise them to as necessary to ensure maximum safety; and (14) Develop and implement the procedures to be followed by an attendant who must monitor more than one permit space at a time in the case of an emergency in one of the spaces.

Informing Contract Employees

If contract employees are allowed to enter permit spaces or are hired for that very purpose, organizations must inform them about (1) the hazards associated with the permit space, (2) entry requirements for the permit space, (3) the organization's experience with the permit space (e.g., knowledge of hazardous conditions), and (4) precautions or procedures to be observed when in or near permit spaces.

Entry Permits

No person should be allowed to enter a permit space without a permit signed by the designated entry supervisor. The permit must verify that pre-entry preparations prescribed in OSHA's Confined Space Standard have been completed. The duration of access to a permit space should be no more than the time required to complete assigned tasks. Entry permits must include the following information: Name of permit space to be entered; name(s) of authorized entrant(s), eligible attendants, and people authorized to be entry supervisors; test results and tester's initials or signature; name and signature of the supervisor who authorizes entry into the permit space; purpose of entry; known hazards in the permit space; measures to be taken to isolate the permit space(s) in question; measures to be taken to eliminate or control hazards in the permit space(s); name and telephone number of rescue and emergency services and the means that will be used to contact them if an emergency arises; date and duration authorized for entry; acceptable entry conditions; communications procedures and equipment for maintaining contact with personnel in the permit space(s); additional permits (if required) for work to be completed in the permit space (e.g., hot work permit); special equipment and procedures required, alarm systems, and PPE; and any other information that might be necessary to ensure the safety of employees who enter a permit space.

Cancellation of Entry Permits

When an assignment is completed in a permit space, the entry supervisor must cancel the entry permit. Permits must also be cancelled if conditions in the permit space change in any way from those documented in the permit. New conditions in the permit space must be recorded on the entry permit and used to revise permit space program accordingly. All cancelled permits must be kept for at least one year.

Training

Any and all personnel who will be required to work in a permit space must first receive the proper training. Further, organizations must ensure that all personnel who complete the prescribed training understand what was taught and have acquired the knowledge and skills to work safely in a permit space. Additional training is required in the following circumstances: (1) Job duties change; (2) A change occurs in the permit space program; (3) A new hazard is found to be present in permit space operations; and (4) An employee's job performance is deficient in any manner.

In addition to employee training, designated rescue personnel must be trained in CPR and first aid, and organizations must certify that the training has been provided. Organizations must maintain comprehensive records of training and make it available to employees and their authorized representatives. Training records must include the employee's name, the trainer's signature, and dates of the training.

Assigned Duties: Authorized Entrant

Personnel authorized to enter permit spaces are responsible for knowing: (1) Space hazards, including the means of exposure such as inhalation, signs of symptoms, and consequences of exposure; (2) How to use appropriate PPE properly; (3) How to maintain communication with attendants so they can monitor the entrant's status and alert the entrant to evacuate if necessary; (4) How to exit from the permit space when ordered by the authorized person, symptoms of exposure are noticed, a prohibited condition exists, or an automatic alarm is activated; and (5) How to alert the attendant when a prohibited condition exists or when warning signs or symptoms of exposure are noticed.

Assigned Duties: Attendant

Personnel authorized to serve as attendants are required to do the following: (1) Remain on duty outside the permit space during entry operations unless relieved by another authorized attendant; (2) Perform non-entry rescues if specified by the organization's rescue procedure; (3) Know all existing and potential hazards for the permit space in question and know the mode of exposure, signs or symptoms, consequences, and physiological effects of the hazards; (4) Maintain communication with personnel in the permit space and account for them at all times; (5) Order evacuation of the permit space when a prohibited condition exists or when a worker shows signs of physiological effects from exposure to a hazard, an emergency outside of the permit space occurs, or when the attendant cannot properly carry out the assigned duties; (6) Summon rescue and other services during an emergency; (7) Ensure that unauthorized personnel are not allowed near the permit space and are ejected immediately if they enter the space; (8) Inform authorized entrants and the entry supervisor if an unauthorized individual enters the permit space; and (9) Perform no other duties that interfere with the attendant's primary duties.

Assigned Duties: Entry Supervisor

Authorized entry supervisors are required to (1) know all hazards in the permit space including mode of exposure, signs and symptoms, and consequences of exposure; (2) verify emergency plans and specified entry conditions such as permits, tests, procedures, and equipment before allowing entry to the permit space; (3) terminate entry and cancel entry permits when operations are completed or when a new condition exists; (4) verify that rescue services are available and that the means of summoning them are operable; (5) take necessary measures to remove unauthorized entrants; and (6) ensure that entry operations remain in accord with the entry permit specifications and that acceptable entry conditions are maintained.

Emergencies: Rescue Service Personnel

Rescue personnel must be able to respond to emergencies promptly. Rescue personnel must be provided with the appropriate PPE and rescue equipment and be trained in how to properly use it. They must also be trained to make authorized entries and to perform assigned rescue duties including performing CPR and administering first aid. At least one member of the rescue squad must be currently certified in CPR and first aid. Organizations must practice that rescue exercises are undertaken at least once per year and that rescue personnel know the hazards associated with the permit space.

Emergencies: Harnesses and Retrieval Lines

Authorized entrants to a permit space must where a chest or full-body harness with a retrieval line attached. The other end of the retrieval line must be attached to a mechanical device or a fixed point outside of the permit space. Wristlets can be used if the organization can demonstrate that the use of a chest or full-body harness is not feasible or that it creates a greater hazard. A mechanical retrieval device must be used with vertical permit spaces more than five feet (1.524 meters) deep.

Safety Data Sheets

If an entrant into a permit space is injured by a substance for which an SDS is required, the SDS must be provided to medical personnel who treat the victim.

OSHA'S HAZARD COMMUNICATION STANDARD

Any organization that uses hazardous materials in the workplace is required to fully inform employees and on-site contractors of the hazards and to provide training concerning the safe handling, storage, and use of the materials. **OSHA's Hazard Communication Standard** is 29 CFR 1910.1200.[42] Organizations that use any of the following types of substances should be especially attentive to complying with this standard: acids, cleaning solvents and compounds, flammable gases and liquids, paints, lacquers, and enamels, insecticides, lubricants and oils, fumigants, fungicides and herbicides, and adhesives. This list is not exhaustive, but it does provide a good start concerning what substances to watch for.

Written Hazard Communication Program

Organizations that use hazardous materials are required by 29 CFR 1910.1200 to develop a written program and to communicate applicable elements of it to employees and on-site contractors. An organization's written hazard communication program should contain at least the following elements:

1. Purpose and objectives
2. Inventory (list) of hazardous chemicals and substances
3. Labels and other forms for warning (prevention procedures)
4. Safety data sheets (procedures and sheets)
5. Hazards of nonroutine tasks (methods of informing)
6. Names of on-site contractors (methods of informing)
7. Training (procedures and elements)

Employee Right-to-Know

The purpose of 29 CFR 1910.1200 is to prevent accidents and injuries by making sure that employees and on-site contractors know about any hazardous substances they might come in contact with in the course of their work, and how to use those substances safely. Those portions of an organization's written hazard communication program that must be shared with employees and on-site contractors are as follows:

1. *Hazard classification.* Chemical manufacturers and importers must determine the hazards of the chemicals they produce or import and properly classify them.
2. *Hazardous product inventory.* A comprehensive list of all hazardous substances found in the workplace in question.
3. *Safety data sheets file.* A master file of data sheets for all hazardous substances in the inventory must be maintained both at the main offices of the organization and any branch sites.
4. *Proper labeling.* All containers — original, immediate-use, and storage — must be properly labeled. The label must contain the name of the hazardous substance, the manufacturer, emergency procedures, and instructions for the safe use of the substance in question. "Immediate-use" containers (those filled by an employee for use during a shift) must have a label that contains the name of the substance and the nature of the hazard (e.g., caustic, corrosive, and flammable).
5. *Emergency plan.* Such plans must contain all the necessary actions to be taken in case of an incident (e.g., fire, spill, accidental release, and reaction). The plan must also list all equipment and materials required to properly respond to an incident, an indication of their locations, and a description of the procedures for using the equipment and materials.
6. *Employee training program.* Training provided to employees and on-site contractors must cover at least the following: (a) discussion of material safety data sheets for all substances in the inventory; (b) right-to-know requirements so that employees and on-site contractors understand what they have a right to know

FIGURE 16–16   Employees must be notified of their right to know.

ATTENTION ALL EMPLOYEES!

As an employee, or on-site contractor, you have the right to be informed of all hazardous chemical substances to which you may be exposed in the workplace.

29 CFR 1910.1200

(see Figure 16–16); (c) explanation of labeling requirements; (d) proper handling and use procedures; (e) emergency procedures; and (f) completion of a standardized form verifying completion of the training. It must also cover the new SDS and labeling requirements.

GHS and OSHA's Hazard Communication Standard

The concept of the Globally Harmonized System (GHS) of labeling hazards substances was covered earlier in this chapter. This section explains only what is necessary concerning GHS implementation to comply with OSHA's Hazard Communication Standard. Employers must now do the following:

- Update their hazard communications programs (including labeling procedures).
- All affected employees must be trained concerning any hazards by the reclassification process of manufacturers, distributors, or importers of hazardous substances.

Employers that have not done these two things are now out of compliance with OSHA's Hazard Communication Standard.

Key Terms and Concepts

- Absorption
- Acute effects and exposures
- Aerosols
- Airborne contaminants
- Air-filtering respirators
- Air safety program
- Air-supplying respirators
- Anesthetics
- Asbestos-containing material (ACM)
- Asbestos removal
- Asphyxiants
- Biological exposure indices (BEIs)
- Biological hazards
- Bonding
- Carcinogen
- Ceiling
- Chemical hazards
- Chemical Process Standard
- Chronic effects and exposures
- Confined space
- Dose
- Dose threshold
- Dusts
- Emergency action plan (EAP)
- Emergency response plan
- Encapsulation
- Enclosing
- Enclosure
- Engineering controls

- Ergonomic hazards
- Exposure thresholds
- Friable asbestos
- GHS
- Good housekeeping
- Grounding
- Hazard recognition
- HAZWOPER
- High-efficiency particle
  absolute (HEPA)
- Indoor air quality (IAQ)
- Industrial hygiene
- Ingestion
- Inhalation
- Injection
- Ionizing radiation
- Irritants

- Isolating
- Lethal concentration
- Lethal dose
- Local exhaust
- Mists
- Narcotics
- Neoplastic growth
- Noise
- Nonionizing radiation
- OSHA Process Safety
  Standard
- OSHA's Hazard
  Communication
  Standard
- Personal hygiene
- Personal protective
  equipment (PPE)

- Physical hazards
- Radiation
- Respirator
- Route of entry
- Safety data sheets
  (SDSs)
- Sick-building syndrome
- Static electricity
- Temperature control
- Threshold limit values
  (TLVs)
- Time-weighted average
- Toxic substance
- Tracer gas techniques
- Vapors
- Ventilation

Review Questions

1. Define the term *industrial hygiene.*
2. Briefly explain the responsibilities of the modern industrial hygienist.
3. What is the role of the safety and health professional regarding industrial hygiene?
4. List five OSHA requirements relating to industrial hygiene.
5. Briefly explain the typical categories of hazards in the workplace.
6. What are the most common routes of entry for toxic agents?
7. Describe the following types of airborne contaminants: dusts, fumes, smoke, mists, and gases.
8. What factors should be considered in deciding whether to remove or contain asbestos?
9. Summarize the various elements of ANSI Z9.8 (indoor air quality/HVAC).
10. Explain the following ways of dealing with asbestos in the workplace: removal, enclosure, and encapsulation.
11. What types of medical examinations should be required of employees who handle ACMs?
12. Briefly explain the following concepts relating to exposure thresholds: time-weighted average, short-term exposure limit, and exposure ceiling.
13. List the most important considerations when evaluating hazards in the workplace.
14. List five generic prevention and control strategies that can be used in any workplace.
15. Give an example of a prevention/control strategy in each of the following categories: engineering controls, ventilation, and personal protective equipment.
16. Explain five self-protection strategies that employees can use in the workplace.
17. How does NIOSH relate to industrial hygiene?
18. What is a toxic substance?
19. List the factors that determine the effect that a toxic substance will have.
20. Describe the most common routes of entry for toxic substances.
21. Explain the mathematical expression of the dose–response relationship.
22. Define the following terms: *dose threshold, lethal dose,* and *lethal concentration.*
23. Differentiate between acute and chronic effects and exposures.

24. List and describe the various classifications of airborne toxics.
25. What is a carcinogen?
26. Describe the basic provisions of the OSHA Chemical Process Standard.
27. What is a threshold limit value?
28. Define the following terms: *time-weighted average* and *ceiling*.
29. Explain the three NIOSH categories of respirators.
30. What is "sick-building" syndrome?
31. Explain the major tenets of the OSHA Confined Space Standard.
32. Explain the various elements that should be included in an organization's confined space management policy.
33. Summarize the requirements of OSHA's Hazard Communication Standard.
34. What are the purposes of GHS?

Endnotes

1. National Safety Council (NSC), "Overview of Industrial Hygiene," in *Fundamentals of Industrial Hygiene*, 6th ed. (Chicago: National Safety Council, 2012), 5.
2. NSC, "Overview of Industrial Hygiene," 160–161.
3. Ibid., 164–165.
4. Ibid., 12–14.
5. Ibid.
6. Ibid., 14.
7. Ibid., 148.
8. Ibid., 18–19.
9. Ibid., 369.
10. Centers for Disease Control and Prevention, National Institute for Occupational Safety and Health, *Selected Topics: Asbestos,* 4–7. Retrieved from cdc.gov/niosh /asbestos.html on June 12, 2017.
11. Ibid.
12. Ibid.
13. Ibid.
14. Ibid.
15. Ibid., 3–4.
16. Ibid., 48.
17. Ibid.
18. Ibid.
19. Ibid., 49.
20. Ibid.
21. Ibid.
22. A. Draper, "I Think It's Mold (Now What?)," *Occupational Health & Safety* 74, no. 5: 67–69.
23. Ibid., 5–6, 45, 44, 96, and 112.
24. NSC, "Overview of Industrial Hygiene," 21.
25. Ibid.
26. Ibid.
27. Ibid., 22.
28. J. Bone, "Textile Industry Weaves a Safety Future," *Safety & Health* 144, no. 3: 17.
29. Ibid., 52.
30. NSC, "Overview of Industrial Hygiene," 25.
31. Ibid.

32. Ibid., 26.
33. Ibid.
34. Ibid., 27.
35. NSC, "Overview of Industrial Hygiene," 24.
36. National Institute for Occupational Safety and Health, Publication No. 90–109. Retrieved from cdc.gov/NIOSH on June 12, 2017.
37. Hazardous Materials Transportation Act.
38. National Institute for Occupational Safety and Health, Publication 90–109.
39. Environmental Protection Agency, "T5CA Inventory Status of Nanoscale Substances—General Approach, and a Proposed Information Collection Request." Retrieved from epa.gov.oppt/nano/nmspfr.htm on March 9, 2017.
40. 29 CFR 1910.146.
41. Ibid.
42. 29 CFR 1910.1200.

CHAPTER

17 | VIOLENCE IN THE WORKPLACE

Learning Objectives

- Explain the relationship between workplace violence and occupational safety
- Define key terms relating to workplace violence
- Summarize the legal considerations relating to workplace violence
- Summarize the strategies for reducing the risk of workplace violence
- Explain the key elements of OSHA's Voluntary Guidelines on Workplace Violence
- Summarize the types of suspicious behaviors workers should be prepared to report
- Explain the three basic responses in an active-shooter situation in the workplace
- Describe the role of supervisors in preventing workplace violence
- Summarize the key components of a workplace violence policy, plan, and program
- Explain the importance of communicating with employees in the aftermath of a violent incident and how to go about it

Workplace violence has emerged as a critical safety and health issue. Homicide is the second leading cause of death to American workers, "accounting for 16 percent of the 6,588 fatal work injuries in the United States."[1] Although more than 80 percent of workplace homicide victims are men, workplace violence is not just a male problem. In fact, workplace homicide is the leading cause of death on the job for women in the United States. Workplace violence includes the following types of crimes: simple assault, aggravated assault, robbery, rape/sexual assault, and homicide.

Almost one million people are injured or killed in workplace-violence incidents every year in the United States, and the number of incidents is on the rise. In fact, according to the U.S. Department of Justice, the workplace is the most dangerous place to be in the United States.[2] Clearly, workplace violence is an issue of concern to safety and health professionals.

OCCUPATIONAL SAFETY AND WORKPLACE VIOLENCE: THE RELATIONSHIP

The prevention of workplace violence is a natural extension of the responsibilities of safety and health professionals. Hazard analysis, records analysis and tracking, trend monitoring, incident analysis, and prevention strategies based on administrative and engineering controls are all fundamental to both concepts. In addition, emergency response and **employee** training are key elements of both. Consequently, occupational safety and health professionals are well suited to add the prevention of workplace violence to their normal duties.

WORKPLACE VIOLENCE: DEFINITIONS

Safety and health professionals should be familiar with the language that has developed around the issue of workplace violence. This section contains the definitions of several concepts as they relate specifically to workplace violence.

- **Workplace violence.** Violent acts, behavior, or threats that occur in the workplace or are related to it. Such acts are harmful or potentially harmful to people, property, or organizational capabilities.
- **Occupational violent crime.** Intentional battery, rape, or homicide during the course of employment.[3]
- **Employee.** An individual with an employment-related relationship (present or past) with the victim of a workplace-violence incident.
- **Outsider.** An individual with no relationship of any kind with the victim of a workplace-violence incident or with the victim's employer.
- **Employee-related outsider.** An individual with some type of personal relationship (past or present) with an employee, but who has no work-related relationship with the employee.
- **Customer.** An individual who receives products or services from the victim of a workplace-violence incident or from the victim's employer.

Each of these terms has other definitions. Those presented here reflect how the terms are used in the language that has evolved around workplace violence.

LEGAL CONSIDERATIONS

Most issues relating to safety and health have legal ramifications, and workplace violence is no exception. The legal aspects of the issue revolve around the competing rights of violent employees and their coworkers (Figure 17–1). These conflicting rights create potential liabilities for employers.

FIGURE 17–1 Conflicting rights.

Rights of Violent Employees

It may seem odd to be concerned about the rights of employees who commit violent acts on the job. After all, logic suggests that in such situations the only concern would be the protection of other employees. However, even violent employees have rights. Remember, the first thing that law enforcement officers must do after taking criminals into custody is to read them their rights.

This does not mean that an employer cannot take the immediate action necessary to prevent a violent act or the recurrence of such an act. In fact, failure to act prudently in this regard can subject an employer to charges of negligence. However, before taking long-term action that will adversely affect the violent individual's employment, employers should follow applicable laws, contracts, policies, and procedures. Failure to do so can serve to exacerbate an already difficult situation.

In addition to complying with all applicable laws, policies, and procedures, it is also important to apply these consistently when dealing with violent employees. Dealing with one violent employee in a particular way and dealing with another in a different way places the employer at a legal disadvantage. Consequently, it is important for employers to be prepared to deal both promptly and properly with violent employees.

Employer Liability for Workplace Violence

Having to contend with the rights of both violent employees and their coworkers, employers often feel as if they are caught between a rock and a hard place. Fortunately, the situation is less bleak than it may first appear primarily due to the **exclusivity provision** of workers' compensation laws. This provision makes workers' compensation the employee's exclusive remedy for injuries that are work-related. This means that even in cases of workplace violence, as long as the violence occurs within the scope of the victim's employment, the employer is protected from civil lawsuits and the excessive jury verdicts that have become so common.

SAFETY MYTH
WORKPLACE VIOLENCE—AN ISOLATED PROBLEM?

Most working Americans are unconcerned about workplace violence. They see it as an isolated problem that happens in other organizations, but not theirs. Right? Not any more. A Time/CNN poll of the general population in the United States found that 37 percent of respondents viewed workplace violence as a growing problem. Of those responding to the poll, 18 percent had personally witnessed some form of violence at work. The same percentage of respondents (19 percent) said they feared for their own safety while at work.

The key to enjoying the protection of the exclusivity provision of workers' compensation laws lies in determining that violence-related injuries are within the scope of the victim's employment—a more difficult undertaking than one may expect. For example, if the violent act occurred at work but resulted from a non-work-related dispute, does the exclusivity provision apply? What if the dispute was work-related, but the violent act occurred away from the workplace?

Making Work-Related Determinations

The National Institute for Occupational Safety and Health (NIOSH) developed the following guidelines for categorizing an injury as being work-related:

- If the violent act occurred on the employer's premises, it is considered an on-the-job event if one of the following criteria applies:

  The victim was engaged in work activity, apprenticeship, or training.

  The victim was on break, in hallway, restrooms, cafeteria, or storage areas.

  The victim was in the employer's parking lots while working, arriving at, or leaving work.

- If the violent act occurred off the employer's premises, it is still considered an on-the-job event if one of the following criteria applies:

  The victim was working for pay or compensation at the time, including working at home.

  The victim was working as a volunteer, emergency services worker, law enforcement officer, or firefighter.

  The victim was working in a profit-oriented family business, including farming.

  The victim was traveling on business, including to and from customer-business contacts.

  The victim was engaged in work activity in which the vehicle is part of the work environment (taxi driver, truck driver, and so on).[4]

RISK-REDUCTION STRATEGIES

Figure 17–2 is a checklist that can be used by employers to reduce the risk of workplace violence in their facilities. The first four of these risk-reduction strategies grow out of the philosophy of *crime prevention through environmental design (CPTED)*.[5] CPTED has the following four major elements, to which the author has added four additional elements.

- Natural surveillance
- **Control of access**
- **Establishment of territoriality**
- Activity support
- Administrative controls
- Identification of high-risk factors
- Training
- Evaluation

FIGURE 17–2   Checklist for workplace-violence risk reduction.

✓   Identify high-risk areas and make them visible. Secluded areas invite violence.

✓   Install good lighting in parking lots and inside all buildings.

✓   Minimize the handling of cash by employees and the amount of cash available on the premises.

✓   Install silent alarms and surveillance cameras where appropriate.

✓   Control access to all buildings (employee badges, visitor check-in and check-out procedure, visitor passes, and so on).

✓   Discourage working alone, particularly late at night.

✓   Provide training in conflict resolution as part of a mandatory employee orientation.

✓   Conduct background checks before hiring new employees.

✓   Train employees how to handle themselves and respond when a violent act occurs on the job.

✓   Develop policies that establish ground rules for employee behavior and responses in threatening or violent situations.

✓   Nurture a positive, harmonious work environment.

✓   Encourage employees to report suspicious individuals and activities or potentially threatening situations.

✓   Deal with allegations of harassment or threatened violence promptly before the situation escalates.

✓   Take threats seriously and act appropriately.

✓   Adopt a *zero-tolerance* policy toward threatening or violent behavior.

✓   Establish a *violence hot line* so that employees can report potential problems anonymously.

✓   Establish a *threat-management team* with responsibility for preventing and responding to violence.

✓   Establish an *emergency response team* to deal with the immediate trauma of workplace violence.

The following explains how these elements can help avoid workplace violence:

1. **Natural surveillance.** This strategy involves designing, arranging, and operating the workplace in a way that minimizes secluded areas. Making all areas inside and outside the facility easily observable allows for **natural surveillance**.
2. **Control of access.** One of the most common occurrences of workplace violence involves an **outsider** entering the workplace and harming employees. The most effective way of stopping this type of incident is to control access to the workplace. Channeling the flow of outsiders to an access-control station, requiring visitor's passes, issuing access badges to employees, and isolating pickup and delivery points can minimize the risk of violence perpetrated by outsiders.
3. **Establishment of territoriality.** This strategy involves giving employees control over the workplace. With this approach, employees move freely within their established territory but are restricted in other areas. Employees come to know

everyone who works in their territory and can, as a result, immediately recognize anyone who shouldn't be there.

4. **Activity support. Activity support** involves organizing workflow and natural traffic patterns in ways that maximize the number of employees conducting natural surveillance. The more employees observing the activity in the workplace, the better.

5. **Administrative controls.** Administrative controls consist of management practices that can reduce the risk of workplace violence. These practices include establishing policies, conducting background checks, providing training for employees, and developing an emergency response plan and team.

6. **High-risk factors.** An important step in preventing workplace violence is to identify any high-risk factors that are present in the organization and eliminate or mitigate those factors. High-risk factors to look for include: job overload because of insufficient staffing or mandatory overtime, employee frustration due to poorly written, ambiguous job descriptions, downsizing and layoffs, antagonistic relations between labor and management, poor leadership and management, no security or inadequate security including poorly training security personnel, failure to provide counseling for employees, and a high injury rate among workers.

7. **Training.** Training for employees, supervisors, and managers should include at least the following topics: familiarization with the organization's workplace violence policy, risk factors to look for, warning signs of potential problems with other employees, how to deal with aggressive behavior, standard responses to violent situations, locations and operation of alarm systems, methods for protecting oneself and others during a violent event, recordkeeping and reporting requirements, and how to obtain medical care, counseling, legal assistance, and other services following a violent incident.

8. **Evaluation.** It is important to evaluate the workplace-violence prevention program so that it can be continually improved. The types of questions that should be included in such an evaluation include: Is there an easy-to-use reporting system that allows employees to report threats, harassment, bullying, and questionable behavior? Are complaints from employees taken seriously and followed up on? Are violent incidents recorded and analyzed to determine if they are increasing or decreasing? Are the results of workplace violence recorded and measured (e.g., lost time and medical costs)? Is the workplace-violence prevention program kept up to date with the latest information and best practices available?

OSHA'S VOLUNTARY GUIDELINES ON WORKPLACE VIOLENCE

The U.S. Department of Labor, working through the Occupational Safety and Health Administration (OSHA), has established *advisory guidelines* relating to workplace violence.[6] Two key points to understand about these guidelines are as follows:

- The guidelines are *advisory* in nature and *informational* in content. The guidelines *do not* add to or enhance in any way the requirements of the general duty clause of the Occupational Safety and Health Act (OSH Act).
- The guidelines were developed with night retail establishments in mind. Consequently, they have a service-oriented emphasis. However, much of the advice contained in the guidelines can be adapted for use in manufacturing, processing, and other settings.

FIGURE 17–3 Broadly applicable elements of OSHA's advisory guidelines on workplace violence.

✓ Management commitment and employee involvement

✓ Workplace analysis

✓ Hazard prevention and control

✓ Safety and health training

✓ Record keeping and evaluation

Figure 17–3 is a checklist of those elements of the specifications that have broader applications. Any management program relating to safety and health in the workplace should have at least these four elements.

Management Commitment and Employee Involvement

Management commitment and employee involvement are fundamental to developing and implementing any safety program, but they are especially important when trying to prevent workplace violence. The effectiveness of a workplace-violence prevention program may be a life-or-death proposition.

Figure 17–4 is a checklist that explains what management commitment means in practical terms. Figure 17–5 describes the practical application of employee

FIGURE 17–4 Elements of management commitment to workplace-violence prevention.

✓ Hands-on involvement of executive management in developing and implementing prevention strategies.

✓ Sincere, demonstrated concern for the protection of employees.

✓ Balanced commitment to both employees and customers.

✓ Inclusion of safety, health, and workplace-violence prevention in the job descriptions of all executives, managers, and supervisors.

✓ Inclusion of safety, health, and workplace-violence prevention criteria in the performance evaluations of all executives, managers, and supervisors.

✓ Assignment of responsibility for providing coordination and leadership for safety, health, and workplace-violence prevention to a management-level employee.

✓ Provision of the resources needed to prevent workplace violence effectively.

✓ Provision of or guaranteed access to appropriate medical counseling and trauma-related care for employees affected physically or emotionally by workplace violence.

✓ Implementation, as appropriate, of the violence-prevention recommendations of committees, task forces, and safety professionals.

FIGURE 17–5 Elements of employee involvement in workplace-violence prevention.

✓ Staying informed concerning all aspects of the organization's safety, health, and workplace-violence program.

✓ Voluntarily complying—in both letter and spirit—with all applicable workplace-violence prevention strategies adopted by the organization.

✓ Making recommendations—through proper channels—concerning ways to prevent workplace violence and other hazardous conditions.

✓ Prompt reporting of all threatening or potentially threatening situations.

✓ Accurate and immediate reporting of all violent or threatening incidents.

✓ Voluntary participation on committees, task forces, and focus groups concerned with preventing workplace violence.

✓ Voluntary participation in seminars, workshops, and other educational programs relating to the prevention of workplace violence.

involvement. Figure 17–6 is a checklist that can be used to ensure that violence prevention becomes a standard component of organizational plans and operational practices. These three checklists can be used by any type of organization to operationalize the concepts of management commitment and employee involvement.

FIGURE 17–6 Checklist for incorporating workplace-violence prevention in strategic plans and operational practices.

✓ Include the prevention of workplace violence in the safety and health component of the organization's strategic plan.

✓ Adopt, disseminate, and implement a *no-tolerance* policy concerning workplace violence.

✓ Adopt, disseminate, and implement a policy that protects employees from reprisals when they report violent, threatening, or potentially threatening situations.

✓ Establish procedures for reporting violent and threatening incidents.

✓ Establish procedures for making recommendations for preventing workplace violence.

✓ Establish procedures for monitoring reports of workplace violence so that trends can be identified and incidents predicted and prevented.

✓ Develop a comprehensive workplace-violence prevention program that contains operational procedures and standard practices.

✓ Develop a workplace-violence component to the organization's emergency response plan.

✓ Train all employees in the application of standard procedures relating to workplace violence.

✓ Conduct periodic emergency-response drills for employees.

Workplace Analysis

Workplace analysis is the same process used by safety and health professionals to identify potentially hazardous conditions unrelated to workplace violence. Worksite analysis should be ongoing and have at least four components (Figure 17–7). An effective way to conduct an ongoing program of workplace analysis is to establish a threat-assessment team with representatives from all departments and led by the organization's chief safety and health professional.

Records Monitoring and Tracking

The purpose of **records monitoring and tracking** is to identify and chart all incidents of violence and threatening behavior that have occurred within a given time frame. Records to analyze include the following: incident reports, police reports, employee evaluations, and letters of reprimand. Of course, individual employees' records should be analyzed in confidence by the human resources member of the team. The type of information that is pertinent includes the following:

- Where specifically did the incident occur?
- What time of day or night did the incident occur?
- Was the victim an employee? Customer? Outsider?
- Was the incident the result of a work-related grievance? Personal?

Trend Monitoring and Incident Analysis

Trend monitoring and incident analysis may prove to be helpful in determining patterns of violence. If there have been enough incidents to create one or more graphs, the team will want to determine if the graphs suggest a trend or trends. If the organization has experienced only isolated incidents, the team may want to monitor

FIGURE 17–7   Four components of workplace analysis.

national trends. By analyzing both local and national incidents, the team can generate information that will be helpful in predicting and, thereby, preventing workplace violence. The team should look for trends in severity, frequency, and type of incidents.

Employee Surveys and Focus Groups

Employees are one of the best sources of information concerning workplace hazards. This is also true when it comes to identifying vulnerabilities to workplace violence. Employee input should be solicited periodically through written **employee surveys and focus groups** or one of these methods. Where are we vulnerable? What practices put our employees at risk? These are the types of questions that should be asked of employees. An effective strategy for use with focus groups is to give participants case studies of incidents that occurred in other organizations. Then ask such questions as, "Could this happen here? Why? Or why not? How can we prevent such incidents from occurring here?"

Security Analysis

Is the workplace secure or could a disgruntled individual simply walk in and harm employees? It is important to ask this question. The team should periodically perform a **security analysis** of the workplace to identify conditions, situations, procedures, and practices that make employees vulnerable. The types of questions to ask include the following:

- Are there physical factors about the facility that make employees vulnerable (for example, isolated, poorly lighted, infrequently trafficked, or unobservable)?
- Is there a process for handling disgruntled customers? Does it put employees at risk?
- Are the prevention strategies already implemented working?
- Is the training provided to employees having a positive effect? Is more training needed? Who needs the training? What kind of training is needed?
- Are there situations in which employees have substantial amount of money in their possession, on- or off-site?
- Are there situations in which employees are responsible for highly valuable equipment or materials late at night or at isolated locations?

Hazard Prevention and Control

Once hazardous conditions have been identified, the strategies and procedures necessary to eliminate them must be put in place. The two broad categories of prevention strategies are engineering controls and administrative controls, just as they are with other safety and health hazards. In addition to these, organizations should adopt postincident response strategies as a way to prevent future incidents.

Engineering Controls

Engineering controls relating to the prevention of workplace violence serve the same purpose as engineering controls relating to other hazards. They either remove the hazard, or they create a barrier between it and employees. Engineering controls typically involve changes to the workplace. Examples of engineering controls include the following:

- Installing devices and mechanisms that give employees a complete view of their surroundings (mirrors, glass or clear plastic partitions, interior windows, and so on)

- Installing surveillance cameras and television screens that allow for monitoring of the workplace
- Installing adequate lighting, particularly in parking lots
- Pruning shrubbery and undergrowth outside and around the facility
- Installing fencing so that routes of egress and ingress to company property can be channeled and, as a result, better controlled
- Arranging outdoor sheds, storage facilities, recycling bins, and other outside facilities for maximum visibility

Administrative Controls

Whereas engineering controls involve making changes to the workplace, **administrative controls** involve making changes to how work is done. This amounts to changing work procedures and practices. Administrative controls fall into four categories (Figure 17–8).

- **Proper work practices** are those that minimize the vulnerability of employees. For example, if a driver has to make deliveries in a high-crime area, the company may employ a security guard to go along, change delivery schedules to daylight hours only, or both.
- **Monitoring and feedback** ensures that proper work practices are being used and that they are having the desired effect. For example, a company has established a controlled access system in which visitors must check in at a central location and receive a visitor's pass. Is the system being used? Are all employees sticking to specified procedures? Has unauthorized access to the workplace been eliminated?
- **Adjustments and modifications** are made to violence-prevention practices if it becomes clear from monitoring and feedback that they are not working or that improvements are needed.
- **Enforcement** involves applying meaningful sanctions when employees fail to follow the established and proper work practices. An employee who has been fully informed concerning a given administrative control, has received the training needed to practice it properly, but consciously decides not to follow the procedure should be disciplined appropriately.

FIGURE 17–8 Categories of administrative controls.

| FIGURE 17–9 | Symptoms of psychological trauma in cases of workplace violence. |

✓ Fear of returning to work

✓ Problems in relationships with fellow
employees and/or family members

✓ Feelings of incompetence

✓ Guilt feelings

✓ Feelings of powerlessness

✓ Fear of criticism by fellow employees,
supervisors, and managers

Postincident Response

Postincident response relating to workplace violence is the same as postincident response relating to traumatic accidents. The first step is to provide immediate medical treatment for injured employees. The second step involves providing psychological treatment for traumatized employees. This step is even more important in cases of workplace violence than with accidents. Employees who are present when a violent incident occurs in the workplace, even if they don't witness it, can experience the symptoms of psychological trauma shown in Figure 17–9. Employees experiencing such symptoms or any others growing out of psychological trauma should be treated by professionals such as psychologists, psychiatrists, clinical nurse specialists, or certified social workers. In addition to one-on-one counseling, employees may also be enrolled in support groups. The final aspect of postincident response is the investigation, analysis, and report. In this step, safety professionals determine how the violent incident occurred and how future incidents may be prevented, just as postaccident investigations are handled.

Training and Education

Training and education are as fundamental to the prevention of workplace violence as they are to the prevention of workplace accidents and health-threatening incidents. A complete safety and health training program should include a comprehensive component covering all aspects of workplace violence (e.g., workplace analysis, hazard prevention, proper work practices, and emergency response). Such training should be provided on a mandatory basis for supervisors, managers, and employees.

Record Keeping and Evaluation

Maintaining accurate, comprehensive, up-to-date records is just as important when dealing with violent incidents as it is when dealing with accidents and nonviolent incidents. By evaluating records, safety personnel can determine how effective their violence-prevention strategies are, where deficiencies exist, and what changes need to be made. Figure 17–10 shows the types of records that should be kept.

- **OSHA log of injury and illness.** OSHA regulations require inclusion in the injury and illness log of any injury that requires more than first aid, is a lost-time injury, requires modified duty, or causes loss of consciousness. Of course, this

FIGURE 17–10   Types of records that should be kept.

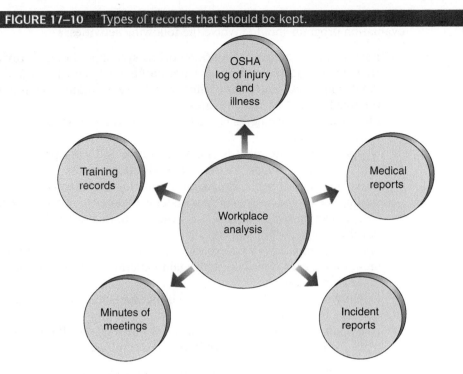

applies only to establishments required to keep OSHA logs. Injuries caused by assaults that are otherwise recordable also must be included in the log. A fatality or catastrophe that results in the hospitalization of three or more employees must be reported to OSHA within eight hours. This includes those resulting from workplace violence and applies to all establishments.

- **Medical reports.** Medical reports of all work injuries should be maintained. These records should describe the type of assault (e.g., unprovoked sudden attack), who was assaulted, and all other circumstances surrounding the incident. The records should include a description of the environment or location, potential or actual cost, lost time, and the nature of injuries sustained.
- **Incidents of abuse.** Incidents of abuse, verbal attacks, and aggressive behavior—which may be threatening to the employee but do not result in injury, such as pushing, shouting, and acts of aggression—should be evaluated routinely by the affected department.
- **Minutes of safety meetings.** Minutes of safety meetings, records of hazard analyses, and corrective actions recommended and taken should be documented.
- **Records of all training programs.** Records of all training programs, attendees, and qualifications of trainers should be maintained.

As part of its overall program, an employer should regularly evaluate its safety and security measures. Top management should review the program regularly, as well as each incident, to determine the program's effectiveness. Responsible parties (managers, supervisors, and employees) should collectively evaluate policies and procedures on a regular basis. Deficiencies should be identified and corrective action taken. An evaluation program should involve the following activities:

- Establishing a uniform violence reporting system and regular review of reports
- Reviewing reports and minutes from staff meetings on safety and security issues
- Analyzing trends and rates of illness or injury or fatalities caused by violence relative to initial or *baseline* rates
- Measuring improvements based on lowering the frequency and severity of workplace violence
- Keeping up-to-date records of administrative and work practice changes to prevent workplace violence to evaluate their effectiveness
- Surveying employees before and after making job or workplace changes or installing security measures or new systems to determine their effectiveness
- Keeping abreast of new strategies available to deal with violence as they develop
- Surveying employees who experience hostile situations about the medical treatment they received initially and, again, several weeks afterward, and then several months later
- Complying with OSHA and state requirements for recording and reporting deaths, injuries, and illnesses
- Requesting periodic law enforcement or outside consultant review of the workplace for recommendations on improving employee safety

Management should share violence-prevention evaluation reports with all employees. Any changes in the program should be discussed at regular meetings of the safety committee, union representatives, or other employee groups. It is also a good idea to develop **conflict guidelines** (see Figure 17–11).

Reporting Suspicious People or Behaviors

Employees should be trained and encouraged to identify and report suspicious looking people or activities. Terrorists are just as likely to be internal to an organization as external to it and they are just as likely to act out their hatred as active shooters as with explosives. Consequently, it is important for employees to be vigilant in watching for the following types of behaviors in others:

- Threats to kill or harm another employee
- Stalking of an employee
- Bullying of employees
- Abusive or aggressive behavior against employees
- Threats to damage property, equipment, or products of the employer
- Defacing of the employer's property
- Unusual behavior (watch for changes in employee behavior)
- Bizarre behavior

In addition to the types of behaviors listed above, employees should be trained and encouraged to identify and report the following types of suspicious activities:

- People drawing maps or making sketches of facilities, particularly entrances, exits, parking garages, etc.

FIGURE 17–11 Sample conflict guidelines.

Conflict Guidelines
Micro Electronics Manufacturing (MEM)

Micro Electronics Manufacturing encourages discussion and debate among employees at all levels concerning better ways to improve continually the quality of our products, processes, people, and work environment. This type of interaction, if properly handled, will result in better ideas, policies, procedures, practices, and decisions. However, human nature is such that conflict can easily get out of hand, take on personal connotations, and become counterproductive. Consequently, in order to promote productive conflict, MEM has adopted the following guidelines. These guidelines are to be followed by all employees at all levels:

- The criteria to be applied when discussing/debating any point of contention is as follows: Which recommendation is most likely to move our company closer to accomplishing its mission?

- Disagree, but don't be disagreeable. If the debate becomes too hot, stop and give all parties an opportunity to cool down before continuing. Apply your conflict resolution skills and anger management skills. Remember, even when we disagree about how to get there, we are all trying to reach the same destination.

- Justify your point of view by tying it to our mission and require others to do the same.

- In any discussion of differing points of view, ask yourself the following question: "Am I just trying to win the debate for the sake of winning (ego), or is my point of view really the most valid?"

- People taking notes about the movements and corresponding times of personnel in an organization
- People taking photographs or videotapes of facilities
- People monitoring facilities that are frequented by large numbers of people such as power plants, schools, shopping centers/malls, sports arenas, religious institutions, airports, and government buildings
- People observing a facility using binoculars and other technological aids
- People attempting to acquire explosives, weapons, detonators, timers, body armor, etc.
- People attempting to purchase large amounts of high nitrate fertilizers or other chemicals
- People putting weapons or explosives in cars or trucks
- People attempting to acquire official gear such as uniforms, badges, and ID cards
- People who display unusual characteristics or behaviors such as nervousness, sweating, clothing that is too heavy for the weather conditions, an unusual focus such as a blank stare or a particularly determined expression, wearing or holding a backpack, and anyone who leaves a bag, box, or backpack unattended and walks off Suspicious people, activities, or behaviors should be reported

immediately to a person in a position of authority. Employees should be taught to err on the side of safety and that a terrorism incident should not be allowed to occur because they were reluctant to report suspicious behavior.

ACTIVE SHOOTER RESPONSE

Although we tend to associate terrorist acts with the use of explosives, it is becoming increasingly common for terrorists to act out their hatred as active shooters. Consequently, it is important for organizations to train their personnel on how to respond in an active-shooter situation. There are three basic responses that employees can take against a terrorist who is an active shooter:

- *Evacuate.* With this response, employees must make a determination if there is a safe route of evacuation that will separate them from the shooter. When this is the case, employees should make use of it as quickly as possible. Delay could be deadly. This means that belongings may have to be left behind. Employees should be trained to know the evacuation routes in their facilities, how to access them quickly, and that going back into the danger zone to retrieve personal belongings could be a fatal mistake. Once safely away, call 911.
- *Hide.* If evacuation is not feasible, employees should be taught to hide. A good hiding place is one that is out of the view of the shooter, provides cover from shots fired, and does not create a trap with no means of exit (e.g., a closet). When this option is taken, it is important that employees silence their cell phones, lock doors if possible, and remain both quiet and still. If the shooter is far enough away, call 911. If not, remain silent.
- *Mount a defense.* When no other option is available, mount a defense. Do not just sit back and allow yourself to be murdered. Attack the shooter with any improvised weapon you can find. If possible, attack as a group with all trapped personnel joining in.

ROLE OF SUPERVISORS

Supervisors can play a pivotal role in the prevention of workplace violence. Following are some rules of thumb that will enhance the effectiveness of supervisors in this regard:

- *Don't* try to diagnose the personal, emotional, or psychological problems of employees.
- *Don't* discuss an employee's drinking unless it occurs on the job. Restrict comments to performance.
- *Don't* preach to employees. Counsel employees about attendance, tardiness, and job performance, not about how they should live their lives.
- *Don't* cover up for employees or make excuses for inappropriate behavior. Misguided kindness may allow problems to escalate and get out of hand.
- *Don't* create jobs to get problem employees out of the way. Stockpiling an employee simply gives him or her more time to brood and to allow resentment to build.

- *Don't* ignore the warning signs explained earlier in this chapter. The problems that they represent will not simply go away. Sooner or later, they will have to be handled. Sooner is better.
- *Do* remember that chemical dependence and emotional problems tend to be progressive. Left untreated, they get worse, not better.
- *Do* refer problem employees to the employee assistance program or to other mental health service providers.
- *Do* make it clear to employees that job performance is the key issue. They are expected to do what is necessary to maintain and improve their performance.
- *Do* make it clear that inappropriate behavior will not be tolerated.

WORKPLACE VIOLENCE: POLICY, PLAN, AND PROGRAMS

In an age when workplace violence has become common, employers are obligated ethically and legally to take responsible steps to protect their personnel. Of course, employees also share in the responsibility to prevent workplace violence. In this regard, workplace violence is like other safety and health issues in the workplace: employers and employees must cooperate in making the workplace as safe an environment as possible.

Employer Responsibilities

Employers play a critical role in preventing workplace violence and in responding properly when, in spite of precautions a violent incident occurs. Key steps employers can take to help prevent workplace violence include the following:

- ***Establish a workplace violence policy statement.*** The organization's policy statement sets the tone for everything else that is done to prevent workplace violence. The statement should make it clear that the organization is committed to providing a safe workplace and acting appropriately on employee complaints concerning harassment, bullying, threats, and any other kind of inappropriate or suspicious behavior. The statement should define the types of behaviors and actions that will not be tolerated (e.g., harassment, intimidation, aggressive behavior, threats, bullying, and bringing weapons on the job). The policy statement should be in writing and all employees should receive a copy of it and an explanation of its meaning and ramifications from a person in a position of authority (e.g., Human Resources officer, manager, or supervisor).
- ***Adopt a set of guiding principles for developing a workplace-violence prevention program.*** It is important to develop a program for implementing the workplace violence policy. In developing the program, the following principles apply: (1) Gain commitment from top management; (2) Tailor the program to the specific type of organization (don't just adopt a one-size-fits-all program); (3) Base the program on proactive actions rather than after-the-fact reactions; (4) Consider the organization's culture (i.e., establishing an effective workplace-violence prevention program might require changing the organization's culture); (5) Use cross-functional teams for planning, threat assessment, emergency response, and evaluation; (6) Establish open, accessible communication channels as part of the program; (7) Practice all elements of the plan

that involve management and employee action; (8) Train all personnel in carrying out the plan; and (9) Include preemployment screening to screen out potentially violent personnel (e.g., history of drug or alcohol abuse, history of violent behavior, past criminal convictions, and record of violent conflicts).

- *Develop a workplace-violence prevention program based on the guiding principles.* The program should have at least the following elements: (1) the organization's workplace violence policy statement; (2) other associated policy statements on such issues as drugs, alcohol use, firearms on the job, etc.; (3) periodic facility surveys and threat assessments/mitigation; (4) written procedures for addressing threats; (5) a cross-functional emergency response team; (6) procedures for seeking professional assistance from outside the organization; (7) periodic training for all levels of personnel; (8) written crisis-response procedures; and (9) written disciplinary measures for inappropriate behavior.

- *Provide training for all personnel.* All personnel at all levels should be trained in violence-prevention measures, recognizing potentially threatening behavior, diffusing potentially violent situations, use of alarm systems, how to report suspicious or threatening behavior, and proper responses during and after a violent event.

- *Support victims of workplace violence.* In workplace-violence incidents, the victims are not the bad guys, but they are sometimes treated this way. Some organizations resent how victims of workplace violence respond in the aftermath of an incident and want them to just snap out of it and get back to work. Organizations in this mold resent having to provide the types of support services that are often required by victims of violence. This kind of treatment of victims is the polar opposite of what is called for by organizations. Victims of workplace violence should be provided the medical care, counseling, and emotional support they need to recover from a violent incident.

- *Establish a culture of safety.* Organizations that make employee safety a top priority and encourage a climate of trust and mutual cooperation between and among personnel at all levels will reduce substantially the risk of workplace-violence incidents.

- *Make effective use of outside professionals and experts.* Experts in preventing workplace violence and dealing with its consequences are available in most communities. These outside experts can be invaluable resources to organizations that are trying to prevent violence or to deal with the effects of one in its aftermath. Law enforcement professionals, psychologists, counselors, and social service agencies can be excellent resources for organizations that are willing to reach out to them for help.

- *Postemergency telephone numbers.* Emergency telephone numbers should be posted conspicuously throughout the facility. Although police and emergency medical personnel can be summoned by dialing 911, that number should still be posted. Even 911 might be forgotten in the heat of the moment during a violent event. Other numbers that should be posted are: (1) Security guards; (2) Internal medical unit (if there is one); (3) Human Resources Office; (4) Union representative; and (5) The organization's Employee Assistance Program (EAP).

- ***Develop an emergency response plan.*** An important part of the violence-prevention program is the *emergency response plan.* The plan should consist of the following elements as a minimum: (1) An emergency response team for carrying out the plan; (2) Training for the emergency response team; (3) Procedures for notifying and handling the media; (4) An emergency evacuation plan (emergency evacuation should be practiced periodically); (5) A backup communication system; (6) Procedures for calming personnel after an incident (see next section); (7) Procedures for debriefing witnesses after an incident; (8) Procedures for helping personnel deal with posttraumatic stress; and (9) Procedures for communicating with personnel about the potential for workplace violence, how to respond when it happens, and how to help prevent it.

Employee Responsibilities

Employees can play a key role in preventing workplace violence. In fact, they can be an employer's most effective asset in this regard. Employee responsibilities in preventing workplace violence include the following:

- ***Know and comply.*** Employees should make a point of learning the organization's workplace violence policy and, then, complying with it in spirit and the letter.
- ***Learn the warning signs of inappropriate behavior in fellow workers and report what is observed.*** Employees should be aware of the types of factors in their fellow workers: (1) Belligerent behavior that seems to be increasing; (2) Threats made publically or quietly between two people; (3) An intolerance for constructive criticism; (4) Recent fascination with weapons; (5) Obsession with a particular grievance or another employee; (6) Focus on violent themes, movies, books, etc.; (7) Obsession with violent events in the news; (8) Angry outbursts or an apparent propensity to lose control; (9) Noticeable changes in behavior; and (10) Suicidal or homicidal threats or just comments. These types of behaviors should be reported to a supervisor or manager.
- ***Follow the procedures in the workplace-violence prevention program.*** The organization's program for preventing workplace violence should be documented in writing. This being the case, employees should follow the procedures set forth in the written plan or document whether reporting suspicious behavior, sounding an alarm, or responding to a violent incident in progress.

Union Responsibilities

In organizations where the workforce is unionized, the union can play an important role in helping prevent workplace violence. Unions can help in this regard by doing the following: (1) Support the employer in developing the workplace-violence prevention policy, (2) Support the employer in establishing and implementing the workplace-violence prevention program, (3) Support the employer in carrying out appropriate disciplinary measures against employees who violate the workplace-violence prevention policy, and (4) Collaborate with the employer on providing violence-prevention training.

COMMUNICATING WITH PERSONNEL IN THE AFTERMATH OF A VIOLENT INCIDENT

Communicating effectively with personnel is always important, but in the aftermath of a violent incident it becomes even more important. In the aftermath of a violent incident, employees are likely to be distracted by feelings of sorrow, anger, uncertainty, and even fear. When employees are distracted, the potential for accidents increases. To overcome the predictable distractions, safety and health professionals and other management personnel should communicate openly and often about the incident to ensure that it does not become the proverbial elephant in the living room that everybody knows is there, but nobody talks about. The following strategies will help distracted workers cope with the emotions, feelings of insecurity, and fear they are likely to feel after a violent incident:

- Make sure employees know that management is concerned and will do whatever is necessary to help them adjust. Give employees plenty of opportunities to discuss their feelings, fears, and uncertainty with management personnel in group meetings and in private, one-on-one meetings. Encourage them to talk about what they are feeling rather than bottling it up inside.
- Let employees know that feelings of anger, fear, grief, shock, and uncertainty are normal. Management personnel should let employees know they experience the same feelings.
- Share all relevant information with employees as soon as it is available. Keep employees fully informed. Do not create a situation in which employees have to speculate or will be distracted wondering what is happening. The motto of managers in these situations should be: *Make sure they hear it from me first.*
- Refer employees who appear to need help to the organization's EAP or to the Human Resources Department. Safety and health professionals and other management personnel should never practice amateur psychology.
- Encourage employees to look out for each other and to support each other.
- Get employee back into the normal routine of work as soon as possible. There is comfort and security in familiar routine.

Key Terms and Concepts

- Activity support
- Administrative controls
- Conflict guidelines
- Control of access
- Employee
- Employee-related outsider
- Employee surveys and focus groups
- Enforcement
- Engineering controls
- Establishment of territoriality
- Exclusivity provision
- Management commitment and employee involvement
- Monitoring and feedback
- Natural surveillance
- Occupational violent crime
- Outsider
- Postincident response
- Proper work practices
- Records monitoring and tracking
- Security analysis
- Trend monitoring and incident analysis
- Workplace analysis
- Workplace violence

Review Questions

1. Define the following terms as they relate to violence in the workplace: *occupational violent crime*, *employee*, and *outsider.*
2. Approximately how many people are direct victims of workplace violence annually?
3. Defend or refute the following statement: Employees who commit violent acts forfeit their rights and can be dealt with accordingly.
4. What is the exclusivity provision of workers' compensation laws? Why is this provision significant?
5. Defend or refute the following statement: A violent act that occurs away from the employer's premises cannot be considered work-related.
6. Explain the concept of crime prevention through environmental design (CPTED).
7. Defend or refute the following statement: A manufacturer must comply with OSHA's guidelines on workplace violence.
8. What elements of OSHA's guidelines on workplace violence can be adapted for future use in most types of business and industrial firms?
9. What are the primary causes of conflict on the job?
10. Explain the four ways in which an angry person may respond in a work setting.
11. Summarize employer responsibilities in preventing workplace violence.
12. Summarize employee responsibilities in preventing workplace violence.
13. Summarize union responsibilities in preventing workplace violence.

Endnotes

1. U.S. Department of Labor, *OSHA Workplace Violence Guidelines*. Retrieved from osha.gov on July 27, 2017.
2. R. Bachman, *Violence and Theft in the Workplace*. U.S. Department of Justice. Retrieved from bjs.ojp.usdoj.gov on July 27, 2017.
3. Lloyd G. Nigro, "Violence in the American Workplace: Challenges to the Public Employees," *Public Administration Review* 56, no. 4: 326.
4. Center for Disease Control and Prevention, *Occupational Violence*. Retrieved from cdc.gov/niosh/topics/violence on July 27, 2017.
5. R. Gardner, *Crime Prevention through Environmental Design*. Retrieved from crimewise.com/library/cpted.html on July 27, 2017.
6. U.S. Department of Labor, Occupational Safety and Health Administration, *Guidelines for Workplace Violence Prevention Programs*. Retrieved from osha.gov /workplaceevidence on July 27, 2017.

CHAPTER

18

NOISE AND VIBRATION HAZARDS

Learning Objectives

- List and define common hearing loss terms
- Explain the characteristics of sound
- Summarize the hazard levels and risks associated with sound
- List the various standards and regulations for noise hazards
- Explain the relationship between noise hazards and workers' compensation
- Demonstrate how to identify and assess hazardous noise conditions
- List and explain common noise control strategies
- Explain the concept of fit testing for HPDs
- List and explain common vibration hazards and how to mitigate them
- Explain the possible detrimental effects of noise other than hearing loss
- Summarize the key elements of a good corporate policy on hearing protection
- Explain how to evaluate a hearing loss prevention program
- Explain how the noise reduction rating of hearing protection devices is used

The modern workplace can be noisy. This poses two safety- and health-related problems. First, there is the problem of distraction. Noise can distract workers and disrupt their concentration, which can lead to accidents. Second, there is the problem of hearing loss. Exposure to noise that exceeds prescribed levels can result in permanent hearing loss.

Modern safety and health professionals need to understand the hazards associated with noise and vibration, how to identify and assess these hazards, and how to prevent injuries related to them. This chapter provides the necessary information for prospective and practicing safety and health professionals to do so.

HEARING LOSS PREVENTION TERMS

There are certain terms common to hearing loss prevention that must be understood by safety and health professionals. You may find the definitions in this section helpful when trying to understand the content of this chapter.

- **Attenuation: Real-world baseline audiogram.** Estimated sound protection provided by hearing protective devices as worn in "real-world" environments.
- **Baseline audiogram.** A valid audiogram against which subsequent audiograms are compared to determine if hearing thresholds have changed. The **baseline audiogram** is preceded by a quiet period to obtain the best estimate of the person's hearing at that time.
- **Continuous noise.** Noise of a constant level measured over at least one second using the "slow" setting on a sound level meter. Note that an intermittent noise (e.g., on for over a second and then off for a period) is both variable and continuous.
- **Decibel (dB).** The unit used to express the intensity of sound. The decibel was named after Alexander Graham Bell. The decibel scale is a logarithmic scale in which 0 dB approximates the threshold of hearing in the midfrequencies for young adults, the threshold of discomfort is between 85 and 95dB, and the threshold of pain is between 120 and 140dB.
- **Dosimeter.** When applied to noise, the instrument that measures sound levels over a specified interval, stores the measures, and calculates the sound as a function of sound level and sound **duration**. It describes the results in terms of dose, **time-weighted average**, and other parameters such as peak level, equivalent sound level, and sound exposure level.
- **Exchange rate.** The relationship between intensity and dose. Occupational Safety and Health Administration (OSHA) uses a 5-dB **exchange rate**. Thus, if the intensity of an exposure increases by 5dB, the dose doubles. This may also be referred to as the *doubling rate*. The U.S. Navy uses a 4-dB exchange rate; the U.S. Army and U.S. Air Force use a 3-dB exchange rate.
- **Fit Testing.** With all PPE fit is an important consideration. A poor fit can undermine the effectiveness of any kind of PPE including **hearing protection devices (HPDs).** Fit testing is a process for ensuring the best possible fit of HPDs. The goal is to ensure that the HPD fits the individual properly as well as the situation.
- **Hazardous noise.** Any sound for which any combination of **frequency**, intensity, or duration is capable of causing permanent hearing loss in a specified population.
- **Conductive and sensorineural loss.** Hearing loss is often characterized by the area of the auditory system responsible for the loss. For example, when injury or a medical condition affects the *outer ear* or *middle ear* (i.e., from the pinna, ear canal, and eardrum to the cavity behind the eardrum, which includes the ossicles), the resulting hearing loss is referred to as a conductive loss. When an injury or medical condition affects the inner ear or the auditory nerve that connects the *inner ear* to the brain (i.e., the cochlea and the VIIIth cranial nerve), the resulting hearing loss is referred to as a sensorineural loss. Thus, a welder's

spark that damages the eardrum causes a conductive hearing loss. Because noise can damage the tiny hair cells located in the cochlea, it causes a sensorineural hearing loss.

- **Hearing threshold level (HTL).** The hearing level, above a reference value, at which a specified sound or tone is heard by an ear in a specified fraction of the trials. HTLs have been established so that dB HTL reflects the best hearing of a group of persons.
- **Hertz (Hz).** The unit measurement for audio frequencies. The frequency range for human hearing lies between 20 Hz and approximately 20,000 Hz. The sensitivity of the human ear drops off sharply below about 500 Hz and above 4,000 Hz.
- **Impulsive noise.** Generally used to characterize impact or impulse noise typified by a sound that rapidly rises to a sharp peak and then quickly fades. The sound may or may not have a "ringing" quality (such as striking a hammer on a metal plate or a gunshot in a reverberant room). Impulsive noise may be repetitive, or may be a single event (as with a sonic boom). Note that if impulses occur in very rapid succession (e.g., some jackhammers), the noise is not described as impulsive.
- **Material hearing impairment.** As defined by the OSHA, a **material hearing impairment** is an average **hearing threshold level** of 25 dB HTL at the frequencies of 1,000, 2,000, and 3,000 Hz.
- **Noise.** Any unwanted sound.
- **Noise dose.** The noise exposure expressed as a percentage of the allowable daily exposure. For OSHA, a 100 percent dose equals an eight-hour exposure to a continuous 90-dBA (A-weighted decibels) noise; a 50 percent dose equals an eight-hour exposure to an 85-dBA noise or a four-hour exposure to a 90-dBA noise. If 85 dBA is the maximum permissible level, an eight-hour exposure to a continuous 85-dBA noise equals a 100 percent dose. If a 3-dB exchange rate is used in conjunction with an 85-dBA maximum permissible level, a 50 percent dose equals a two-hour exposure to 88 dBA or an eight-hour exposure to 82 dBA.
- **Noise-induced hearing loss.** A sensorineural hearing loss that is attributed to noise and for which no other etiology can be determined.
- **Standard threshold shift (STS).** OSHA uses this term to describe a change in hearing threshold relative to the baseline audiogram of an average of 10 dB or more at 2,000, 3,000, and 4,000 Hz in either ear. It is used by OSHA to trigger additional audiometric testing and related **follow-up**.
- **Significant threshold shift.** The National Institute for Occupational Safety and Health (NIOSH) uses this term to describe a change of 15 dB or more at any frequency, 400 through 6,000 Hz, from baseline levels that is present on a retest in the same ear and at the same frequency. NIOSH recommends a confirmation audiogram within 30 days with the confirmation audiogram preceded by a quiet period of at least 14 hours.
- **Time-weighted average (TWA).** A value, expressed in dBA, computed so that the resulting average is equivalent to an exposure resulting from a constant noise level over an eight-hour period.

CHARACTERISTICS OF SOUND

Sound is any change in pressure that can be detected by the ear. Typically, sound is a change in **air pressure**. However, it can also be a change in water pressure or any other pressure-sensitive medium. **Noise** is unwanted sound. Consequently, the difference between noise and sound is in the perception of the person hearing it (e.g., loud rock music may be considered "sound" by a rock fan but "noise" by a shift worker trying to sleep).

What we think of as sound, the eardrum senses as fluctuations in atmospheric pressure. The eardrum responds to these fluctuations in atmospheric pressure by vibrating. These vibrations are carried to the brain in the form of neural sensations and interpreted as sound.

Sound and vibration are very similar. Sound typically relates to a sensation that is perceived by the inner ear as hearing. **Vibration**, on the other hand, is inaudible and is perceived through the sense of touch. Sound can occur in any medium that has both mass and elasticity (air, water, and so on). It occurs as elastic waves that cross over (above and below) a line representing normal atmospheric pressure (see Figure 18–1).

Normal atmospheric pressure is represented in Figure 18–1 by a straight horizontal line. Sound is represented by the wavy line that crosses above and below the line. The more frequently the sound waves cross the normal atmospheric pressure line (the shorter the cycle), the higher the pitch of the sound. The greater the vertical distance above and below the atmospheric pressure line (distance X), the louder or more intense the sound.

The unit of measurement used for discussing the level of sound and, correspondingly, what noise levels are hazardous is the **decibel**, or one-tenth of a bel. One decibel represents the smallest difference in the level of sound that can be perceived by the human ear. Figure 18–2 shows the decibel levels for various common sounds. The weakest sound that can be heard by a healthy human ear in a quiet setting is known as the **threshold of hearing** (1 dBA). The maximum level of sound that can be perceived without experiencing pain is known as the **threshold of pain** (140 dBA).

FIGURE 18–1   Sound waves.

FIGURE 18–2   Selected sound levels.

Source	Decibels (dBA)
Whisper	20
Quiet library	30
Quiet office	50
Normal conversation	60
Vacuum cleaner	70
Noisy office	80
Power saw, lawn mower	90
Chain saw	90
Grinding operations	100
Passing truck	100
Gunshot blast	140
Jet aircraft	150
Rocket launching	180

The three broad types of industrial noise are described as follows: **Wide band noise** is noise that is distributed over a wide range of frequencies. Most noise from manufacturing machines is wide band noise. **Narrow band noise** is noise that is confined to a narrow range of frequencies. The noise produced by power tools is narrow band noise. Finally, **impulse noise** consists of transient pulses that can occur repetitively or nonrepetitively. The noise produced by a jackhammer is repetitive impulse noise.[1]

HAZARD LEVELS AND RISKS

The fundamental hazard associated with excessive noise is hearing loss. Exposure to excessive noise levels for an extended period can damage the inner ear so that the ability to hear high-frequency sound is diminished or lost altogether. Additional exposure can increase the damage until even lower frequency sounds cannot be heard.[2]

A number of different factors affect the risk of hearing loss associated with exposure to excessive noise. The most important of these are

- Intensity of the noise (**sound pressure level**)
- Type of noise (wide band, narrow band, or impulse)
- Duration of daily exposure
- Total duration of exposure (number of years)
- Age of the individual
- Coexisting hearing disease
- Nature of environment in which exposure occurs
- Distance of the individual from the **source** of the noise
- Position of the ears relative to the sound waves[3]

Of these various factors, the most critical are the sound level, frequency, duration, and distribution of noise (see Figure 18–3). The unprotected human ear is at risk when exposed to sound levels exceeding 115 dBA. Exposure to sound levels below 80 dBA

FIGURE 18–3 Critical noise risk factors are sound level, frequency, duration, and distribution.

Source: Aisyaqilumaranas/Shutterstock

is generally considered safe. Prolonged exposure to noise levels higher than 80 dBA should be minimized through the use of appropriate personal protective devices.

To decrease the risk of hearing loss, exposure to noise should be limited to a maximum eight-hour TWA of 85 dBA. The following general rules should be applied for dealing with noise in the workplace:

- Exposures of less than 80 dBA may be considered safe for the purpose of risk assessment.
- A TWA (threshold) of 85 dBA should be considered the maximum limit of continuous exposure over eight-hour days without protection.[4]

OSHA has established the following permissible noise levels for the workplace:

- 90 decibels for 8 hours
- 92 decibels for 6 hours
- 95 decibels for 4 hours
- 97 decibels for 3 hours
- 100 decibels for 2 hours
- 102 decibels for 1.5 hours
- 105 decibels for 1 hour
- 110 decibels for 0.5 hours
- 115 decibels for 0.25 hour or less

Of course, what is more important than these specific permissible noise levels and time duration is TWA over the course of a full workday. To review, a TWA for noise is the worker's average daily exposure over the course of a full work day. Numerous charts are available on the Internet that allow safety and health professionals to simply look up acceptable TWAs. However, TWAs can be calculated. Doing so is a two-step process:

- Step One: Calculate the noise dose
- Step Two: Calculate the TWA

Step One: Calculating the noise dose (ND)

To calculate the noise dose, noise has to be measured at different times across the workday. With this done, the following equation is used to calculate the noise dose:

$$ND = 100 \times \left(C1/T1 + C2/T2 + C3/T3 + \cdots + Cn/Tn \right)$$

where:

C = Time at each noise level measured

T = 8 divided by 2 to the power of (L-90)/5 (T values may also be looked up in tables that are readily available on the Internet)

Assume that a worker is exposed to 90 dB for six hours and 94 dB for two hours. The noise dose could be calculated as follows:

$$ND = 100 \times (6/8 + 2/4.6) = 118 \text{ percent}$$

Step Two: Calculating the TWA

The noise dose percentage is then used to calculate the TWA using the following equation:

$$TWA = 16.61 \, Log_{10} \, (ND/100) + 90$$

where:

TWA is the 8 hour time-weighted average
ND is the noise dose percentage
Log_{10} is the logarithm to base 10

Using the noise dose of 118 percent calculated in the previous step, the TWA can be calculated as follows:

$$TWA = 16.61 \times Log_{10} \, (118/100) + 90$$
$$TWA = 91.19 \, dB$$

Safety and health professionals who are uncomfortable with mathematical calculations or who do not have time to make them can obtain the same information by attaching a noise dosimeter to workers. This device will measure the noise exposure level over the period it is worn by a worker and will calculate both the noise dose and the TWA.

STANDARDS AND REGULATIONS

The primary sources of standards and regulations relating to noise hazards are OSHA, the American National Standards Institute (ANSI), and NIOSH. OSHA regulations require the implementation of **hearing conservation** programs under certain conditions. OSHA's regulations should be considered as minimum standards. ANSI's standard provides a way to determine the effectiveness of hearing conservation programs such as those required by OSHA. The ANSI standard and OSHA regulations are discussed in the following sections. NIOSH bases most of its materials on OSHA regulations, although NIOSH does make recommendations that exceed OSHA regulations in some cases.

ANSI Standard

ANSI published ANSI Standard S12.13, Evaluation of Hearing Conservation Programs. This standard is designed to help safety and health professionals determine if hearing conservation programs work as they are intended to.[5]

Federal regulations require that employees be protected from excessive noise in the workplace. However, they provide no methodology for determining the effectiveness of hearing conservation programs. The primary reason for the development of **ANSI S12.13** was because hearing conservation programs were not actually protecting employees but were only recording their steadily declining hearing ability.

The working group that developed the standard used **audiometric database analysis (ADBA)** to identify procedures for measuring variability in HTLs. The two procedures identified were as follows:

- **Percent worse sequential.** This procedure identifies the percentage of subjects who show a deterioration of 15 dBA or more in their ability to hear at least one test frequency (500 through 6,000 Hz) in either ear between two sequential audiograms.
- **Percent better or worse sequential.** This procedure identifies the percentage of subjects who show either a deterioration or an improvement of 15 dBA or more in thresholds for at least one test frequency (500 through 6,000 Hz) in either ear between two sequential audiograms.[6]

There are important differences between the traditional approaches to measuring the effectiveness of hearing conservation programs and those required by ANSI S12.13. The ANSI approach can be summarized as follows:

- Results of tests are compared in sequence. For example, the results of year 4 are compared with those of year 3. The results of year 3 are compared with those of year 2, and so on. In this way, a current audiogram is compared against an earlier audiogram. The results of the earlier test are used as a baseline for comparison.
- Test results from several employees in a given work unit are examined individually and compared with past results sequentially. If enough employees show hearing loss, the conclusion may be that the work unit's hearing conservation program is ineffective.[7]

OSHA Regulations

OSHA's regulations relating to occupational noise exposure and hearing conservation are found in 29 CFR 1910.95. The basic requirements generated from this standard for hearing conservation programs are as follows:

- Hearing hazards monitoring
- Engineering and administrative controls
- Audiometric evaluation
- Personal hearing protection devices
- Education and motivation
- Record keeping
- Program evaluation

Hearing Hazards Monitoring

As with any health hazard, it is important to determine accurately the nature of the hearing hazard and to identify the affected employees. Those responsible for this aspect of the program must ensure that the exposures of all employees have been properly evaluated and that reevaluations are conducted when changes in equipment or operations significantly alter working conditions. Recent evidence has indicated that aromatic solvents, metals, and petrochemicals may be associated with occupational hearing loss. Although studies are exploring the relationship between hearing loss and chemical exposures, there is insufficient information about this relationship to speculate on potential risk factors. Therefore, this section focuses on monitoring noise exposure, the major factor associated with occupational hearing losses. Hearing hazard exposure monitoring is conducted for various purposes, including

- Determining whether hazards to hearing exist
- Determining whether noise presents a safety hazard by interfering with speech communication or the recognition of audible warning signals
- Identifying employees for inclusion in the hearing loss prevention program
- Classifying employees' noise exposures for prioritizing noise control efforts and defining and establishing hearing protection practices
- Evaluating specific noise sources for noise control purposes
- Evaluating the success of noise control efforts

Various kinds of increment and measurement methods may be used, depending on the type of measurements being conducted. The most common measurements are area surveys, dosimetry, and engineering surveys.

In an area survey, environmental noise levels are measured using a sound level meter to identify work areas where exposures are above or below hazardous levels and where more thorough exposure monitoring may be needed. The result is often plotted in the form of a "noise map," showing noise level measurements for the different areas of the workplace.

Dosimetry involves the use of body-worn instruments (dosimeters) to monitor an employee's noise exposure over the work shift. Monitoring results for one employee can also represent the exposures of other workers in the area with similar noise exposures. It may also be possible to use task-based exposure methods to represent the exposures of other workers in different areas whose exposures result from having performed the same tasks.

Engineering surveys typically employ more sophisticated acoustical equipment in addition to sound-level meters. These may include octave-band analyzers and sound-level recorders that furnish information on the frequency-intensity composition of the noise being emitted by machinery or other sound sources in various modes of operation. These measurements are used to assess options for applying **engineering controls**.

Engineering and Administrative Controls

Engineering and **administrative controls** are essential to achieve an effective hearing loss prevention program. Engineering and administrative controls represent the first two echelons in the hierarchy of controls: (1) remove the hazard and (2) remove the worker. The use of these controls should reduce hazardous exposure to the point where the risk to hearing is eliminated or at least more manageable. Engineering controls are technologically feasible for most noise sources, but their economic feasibility must be determined on a case-by-case basis. In some instances, the application of a relatively simple noise control solution reduces the hazard to the extent that the other elements of the program, such as audiometric testing and the use of HPDs, are no longer necessary. In other cases, the noise reduction process may be more complex and must be accomplished in stages over a period of time. Even so, with each reduction of a few dBs, the hazard to hearing is reduced, communication is improved, and noise-related annoyance is reduced as well.

It is especially important that organizations specify low noise levels when purchasing new equipment. Many types of previously noisy equipment are now available in noise-controlled versions. Consequently, a "buy quiet" purchasing policy should not require new engineering solutions in many cases.

For hearing loss prevention purposes, *engineering controls* are defined as any modification or replacement of equipment or related physical change at the noise source or along the transmission path (with the exception of hearing protectors) that reduces the noise level at the employee's ear. Typical engineering controls involve:

- Reducing noise at the source (e.g., installing a muffler)
- Interrupting the noise path (e.g., erecting acoustical enclosures and barriers)
- Reducing reverberation (e.g., installing sound-absorbing material)
- Reducing structure-borne vibration (e.g., installing vibration mounts and providing proper lubrication)

Assessing the applicability of engineering controls is a sophisticated process. First, the noise problem must be thoroughly defined. This necessitates measuring the noise levels and developing complete information on employee noise exposure and the need for noise reduction. Next, an assessment of the effect of these controls on overall noise levels should be made. Once identified and analyzed, the preceding controls can be considered. Choices are influenced, to some extent, by the cost of purchasing, operating, servicing, and maintaining the control. For this reason, engineering, safety, and industrial hygiene personnel, as well as employees who operate, service, and maintain equipment, must be involved in the noise control plan. Employees who work with the equipment on a daily basis should be asked to provide valuable guidance on such important matters as the positioning of monitoring indicators and panels, lubrication and servicing points, control switches, and the proper location of access doors for operation and maintenance. An acoustical consultant may be hired to assist in the design, implementation, installation, and evaluation of these controls.

Administrative controls, defined as changes in the work schedule or operations that reduce noise exposure, may also be used effectively. Examples include operating a noisy machine on the second or third shift when fewer people are exposed, or shifting an employee to a less noisy job once a hazardous daily noise dose has been reached. Generally, administrative controls have limited use in industry because employee contracts seldom permit shifting from one job to another. Moreover, the practice of rotating employees between quiet and noisy jobs, although it may reduce the risk of substantial hearing loss in a few workers, may actually increase the risk of small hearing losses in many workers. A more practical administrative control is to provide for quiet areas where employees can gain relief from workplace noise. Areas used for work breaks and lunchrooms should be located away from noise.

Audiometric Evaluation

Audiometric evaluation is crucial to the success of the hearing loss prevention program in that it is the only way to determine whether occupational hearing loss is being prevented. When the comparison of audiograms shows temporary threshold shift (a temporary hearing loss after noise exposure), early permanent threshold shift, or progressive occupational hearing loss, it is time to take swift action to halt the loss before additional deterioration occurs. Because occupational hearing loss occurs gradually and is not accompanied by pain, the affected employee may not notice the change until a large threshold shift has accumulated. However, the results of audiometric tests can trigger changes in the hearing loss prevention program more promptly, initiating protective measures, and motivating employees to prevent further hearing loss.

OSHA and NIOSH presently have differing definitions of the amount of change in hearing indicated by repeated audiometry that should trigger additional audiometric testing and related follow-up. OSHA uses the term *standard threshold shift (STS)* to describe an average change in hearing from the baseline levels of 10 dB or more for the frequencies of 2,000, 3,000, and 4,000 Hz. NIOSH uses the term *significant threshold shift* to describe a change of 15 dB or more at any frequency of 500 through 6,000 Hz from baseline levels that is present on an immediate retest in the same ear and at the same frequency. NIOSH recommends a confirmation audiogram within 30 days with the confirmation audiogram preceded by a quiet period of at least 14 hours. The NIOSH STS, called "15 dB twice" (same ear, same frequency), can be tested only if the baseline audiogram is available at the time of the annual audiometric test.

For maximum protection of employees, audiograms should be performed on the following occasions:

- Preemployment
- Prior to initial assignment in a hearing hazardous work area
- Annually as long as the employee is assigned to a noisy job (a TWA exposure level equal to or greater than 85 dBA)
- At the time of reassignment out of a hearing hazardous job
- At the termination of employment

In addition, it is suggested that employees who are not exposed be given periodic audiograms as part of the organization's healthcare program. The audiograms of these employees can be compared to those of the exposed employees whenever the overall effectiveness of the hearing loss prevention program is evaluated. In an optimally effective program, the two employee groups show essentially the same amount of audiometric change.

Personal Hearing Protection Devices

A personal HPD (or "hearing protector") is anything that can be worn to reduce the level of sound entering the ear. Earmuffs, ear canal caps, and earplugs are the three principal types of devices. Each employee reacts individually to the use of these devices, and a successful hearing loss prevention program should be able to respond to the needs of each employee.

There are several different types of earmuffs with which the modern safety and health professional should be familiar:

- **Passive earmuffs.** This kind of device is what most people think of when they think of earmuffs. They consist of ear cups lined with foam and block noise using nothing but the foam-lined cups. The primary weakness of passive ear-muffs is they tend to block out not just unwanted noise, but also certain advantageous sounds such as voices trying to warn of danger.

- **Uniform attenuation earmuffs.** These earmuffs not only block noise, but also attenuate the noise more uniformly within several key octave bands (250 Hz to 4 KHz). This allows employees wearing them to hear certain important sounds such as spoken instructions or warnings, thus reducing one of the main safety risks associated with earmuffs.
- **Electronic earmuffs.** This type of earmuff uses electronic technology to both block and modulate sound. Some of the more popular brands of **electronic earmuffs** can receive AM/FM radio signals or have a wireless connection to a CD or MP3 player.

Regardless of the kind of ear protection device used, it is important to remember the four Cs: *comfort, convenience, communication* (the device should not interfere with the worker's ability to communicate), and *caring* (workers must care enough about protecting their hearing to wear the devices).

Education and Motivation

Training is a critical element of a good hearing loss prevention program. In order to obtain sincere and energetic support by the management and active participation by employees, it is necessary to educate and motivate both groups. A hearing loss prevention program that overlooks the importance of education and motivation is likely to fail because employees will not understand why it is in their best interest to cooperate, and management will fail to make the necessary commitment. Employees and managers who appreciate the precious sense of hearing and understand the reasons for, and the mechanics of, the hearing loss prevention program will be more likely to participate for their mutual benefit, rather than viewing the program as an imposition.

Record Keeping

Records often get the least attention of any of the program's components. However, audiometric comparisons, reports of hearing protector use, and the analysis of hazardous exposure measurements all involve the keeping of records. Unfortunately, records are often kept poorly because there is no organized system in place, and in many cases, those responsible for maintaining the records do not understand their value. People tend to assume that if they merely place records in a file or enter them into a computer, adequate record-keeping procedures are being followed. OSHA's latest version of the Form 300 Log has a column for recording hearing loss.

Many companies have found that their record-keeping system is inadequate only when they discover that they need accurate information. This sometimes occurs during the processing of compensation claims. Problems can be avoided by implementing an effective record-keeping system, in which (1) management encourages that the system be kept active and accessible; (2) hearing loss prevention program implementers make sure that all the information entered is accurate and complete; and (3) employees validate the information.

Hearing loss prevention program records should include all items for each phase of the program: (1) hearing loss prevention audit, (2) monitoring hearing hazards,

(3) engineering and administrative controls, (4) audiometric evaluation, (5) personal hearing protective devices, (6) education and motivation, (7) record keeping, and (8) program evaluation. Each phase must be considered in order to evaluate the effectiveness of the hearing loss prevention program.

Program Evaluation

The primary goal of any hearing loss prevention program must be to reduce, and eventually eliminate, hearing loss due to workplace exposures. Although management may have the best intentions of implementing this goal and a company's hearing loss prevention program may have the appearance of being complete and complying with OSHA's requirements, the program still may not achieve this goal. A thorough evaluation of the effectiveness of all the program's components is necessary to determine the extent to which the hearing loss prevention program is really working.

Management and program implementers should conduct periodic program evaluations to assess compliance with federal and state regulations and to ensure that hearing is being conserved. There are two basic approaches to following program evaluation: (1) assess the completeness and quality of the program's components and (2) evaluate the audiometric data.

WORKERS' COMPENSATION AND NOISE HAZARDS

Hearing loss claims are being covered by state workers' compensation laws. Some states have written hearing loss into their workers' compensation law. Others are covering claims whether hearing loss is in the law or not.

Medical professionals have established a procedure for determining if there is a causal relationship between workplace noise and hearing loss. In making determinations of such relationships, physicians consider the following factors:

- Onset and progress of the employee's history of hearing loss
- The employee's complete work history
- Results of the employee's otological examination
- Results of hearing studies that have been performed
- Determination of whether causes of hearing loss originated outside the workplace

Because, approximately 15 percent of all working people are exposed to noise levels exceeding 90 dBA, hearing loss may be as significant in workers' compensation costs in the future as back injuries, carpal tunnel syndrome, and stress are now significant.

IDENTIFYING AND ASSESSING HAZARDOUS NOISE CONDITIONS

Identifying and assessing **hazardous noise** conditions in the workplace involve (1) conducting periodic noise surveys, (2) conducting periodic audiometric tests, (3) record keeping, and (4) follow-up action. Each of these components is covered in the following sections.

Noise Surveys

Conducting **noise surveys** involves measuring noise levels at different locations in the workplace. The devices that are most widely used to measure noise levels are sound-level meters and dosimeters. A **sound-level meter** produces an immediate reading that represents the noise level at a specific instance in time.[8] A **dosimeter** provides a TWA over a period of time such as one complete work shift. A dosimeter also calculates the sound as a function of sound level and duration and describes the results in terms of dose, peak level, equivalent sound level, sound exposure level, and TWA.[9] The dosimeter is the most widely used device because it measures total exposure, which is what OSHA and ANSI standards specify. Using a dosimeter in various work areas and attaching a personal dosimeter to one or more employees is the recommended approach to ensure dependable, accurate readings.

Audiometric Testing

Audiometric testing measures the hearing threshold of employees. Tests conducted according to ANSI S12.13 can detect changes in the hearing threshold of the employee. A negative change represents hearing loss within a given frequency range.

The initial **audiogram** establishes a baseline hearing threshold. After that, audiometric testing should occur at least annually. Testing should not be done on an employee who has a cold, an ear infection, or who has been exposed to noise levels exceeding 80 dBA within 14 to 16 hours prior to a test. Such conditions can produce invalid results.[10]

When even small changes in an employee's hearing threshold are identified, more frequent tests should be scheduled and conducted as specified in ANSI S12.13. "For those employees found to have standard threshold shift—a loss of 10 dBA or more averaged at 2,000, 3,000, and 4,000 **hertz (Hz)** in either ear—the employer is required to fill out the OSHA 300 Log in which the loss is recorded as a worktime illness."[11]

Record Keeping

Figure 18–4 is an example of an audiometric form that can be used to record test results for individual employees. Such forms should be completed and kept on file to allow for sequential comparisons. It is also important to retain records containing a worker's employment history, including all past positions and the working conditions in those positions.

Follow-Up

Follow-up is critical. Failure to take prompt corrective action at the first sign of hearing loss can lead to permanent debilitating damage.

Hearing loss can occur without producing any evidence of physiological damage. Therefore, it is important to follow up on even the slightest evidence of a change in an employee's hearing threshold.

Follow-up can take a number of different forms. The following would all be appropriate follow-up responses:

- Administering a retest to verify the hearing loss
- Changing or improving the type of **personal protection** used

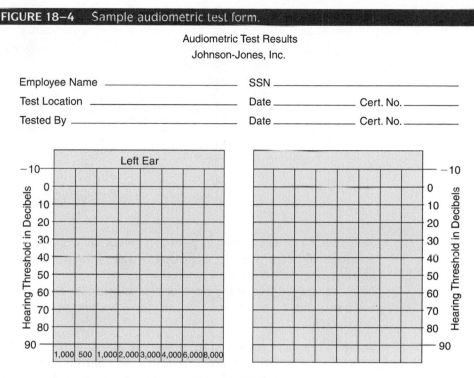

FIGURE 18-4 Sample audiometric test form.

Audiometric Test Results
Johnson-Jones, Inc.

Employee Name _____ SSN _____

Test Location _____ Date _____ Cert. No. _____

Tested By _____ Date _____ Cert. No. _____

Frequency in Hertz

- Conducting a new noise survey in the employee's work area to determine whether engineering controls are sufficient
- Testing other employees to determine if the hearing loss is isolated to the one employee in question or if other employees have been affected

NOISE CONTROL STRATEGIES

Figure 18–5 illustrates the three components of a noise hazard. Noise can be reduced by engineering and administrative controls applied to one or more of these components. The most desirable **noise controls** are those that reduce noise at the source. The second priority is to reduce noise along its path. The last resort is noise reduction at the receiver using personal protective devices. The latter approach should never be substituted for the two former approaches.

The following paragraphs explain widely used strategies for reducing workplace noise at the source, along its path, and at the receiver:

- Noise can be reduced at its source by enclosing the source, altering the acoustical design at the source, substituting equipment that produces less noise, making alterations to existing equipment, or changing the process so that less noisy equipment can be used.

FIGURE 18–5 Three parts of a noise hazard.

- Noise can be reduced along its **path** by moving the source farther away from receivers and improving the acoustical design of the path so that more sound is absorbed as it travels toward receivers.
- Noise can be reduced at the **receiver** by enclosing the worker, using personal protective devices, and changing job schedules so that exposure time is reduced.

Some of the noise reduction strategies explained in the preceding paragraphs are engineering controls; others are administrative controls. For example, enclosing a noise source and substituting less noisy equipment are both examples of engineering controls. Changing job schedules is an example of an administrative control. Safety and health professionals should be familiar with both types of controls.

Engineering Controls

Engineering controls consist of facility and equipment adjustments other than administrative and personal protection strategies made to reduce the noise level either at the source or within the worker's hearing zone. Following are some commonly used engineering controls. All these controls are designed to reduce noise at the source, along its path, or at the receiver. They focus primarily on the noise rather than on the employees who are exposed to it.

MAINTENANCE

- Replacement or adjustment of worn, loose, or unbalanced parts of machines
- Lubrication of machine parts and use of cutting oils
- Use of properly shaped and sharpened cutting tools

SUBSTITUTION OF MACHINES

- Larger, slower machines for smaller, faster ones
- Step dies for single-operation dies
- Presses for hammers
- Rotating shears for square shears
- Hydraulic presses for mechanical presses
- Belt drives for gears

SUBSTITUTION OF PROCESSES

- Compression riveting for impact riveting
- Welding for riveting
- Hot working for cold working
- Pressing for rolling or forging

REDUCE THE DRIVING FORCE OF VIBRATING SURFACES BY

- Reducing the forces
- Minimizing rotational speed
- Isolating

REDUCE THE RESPONSE OF VIBRATING SURFACES BY

- Damping
- Additional support
- Increasing the stiffness of the material
- Increasing the mass of vibrating members
- Changing the size to change resonance frequency

REDUCE THE SOUND RADIATION FROM THE VIBRATING SURFACES BY

- Reducing the radiating area
- Reducing overall size
- Perforating surfaces

REDUCE THE SOUND TRANSMISSION THROUGH SOLIDS BY USING

- Flexible mounting
- Flexible sections in pipe runs
- Flexible-shaft couplings
- Fabric sections in ducts
- Resilient flooring

REDUCE THE SOUND PRODUCED BY GAS FLOW BY

- Using intake and exhaust mufflers
- Using fan blades designed to reduce turbulence
- Using large, low-speed fans instead of smaller, high-speed fans
- Reducing the velocity of fluid flow (air)
- Increasing the cross section of streams
- Reducing the pressure
- Reducing the air turbulence

REDUCE NOISE BY REDUCING ITS TRANSMISSION THROUGH AIR BY

- Using sound-absorptive material on walls and ceiling in work areas
- Using sound barriers and sound absorption along the transmission path
- Completely enclosing individual machines
- Using baffles
- Confining high-noise machines to insulated rooms
- Mufflers
- Wall panels
- Insulation

Administrative Controls

Administrative controls are controls that reduce the exposure of employees to noise rather than reducing the noise itself.

Administrative controls should be considered a second-level approach, with engineering controls given top priority. Smaller companies that cannot afford to reduce noise through engineering measures may use administrative controls instead. However, this approach should be avoided if at all possible.

Hearing Protection Devices

In addition to engineering and administrative controls, employees should be required to use appropriate *hearing protection devices (HPDs)*. It should be noted, however, that such devices are effective only if worn properly. Enforcement of the proper use of HPDs is difficult in some settings. The following four classifications of HPDs are widely used: enclosures, earplugs, superaural caps, and earmuffs.

Enclosures are devices that completely encompass the employee's head, much like the helmets worn by motorcycle riders. **Earplugs** (also known as *aurals*) are devices that fit into the ear canal. Custom-molded earplugs are designed and molded for the individual employee. Premolded earplugs are generic in nature, are usually made of a soft rubber or plastic substance, and can be reused. **Formable earplugs** can be used by anyone. They are designed to be formed individually to a person's ears, used once, and then discarded.

Superaural caps fit over the external edge of the ear canal and are held in place by a headband. Earmuffs, also known as *circumaurals*, cover the entire ear with a cushioned cup that is attached to a headband. Earplugs and earmuffs are able to reduce noise by 20 to 30 dB. By combining earplugs and earmuffs, an additional 3 to 5 dB of blockage can be gained.

Figures 18–6 through 18–9 are examples of various types of HPDs. Figure 18–6 illustrates ear, face, and head protection combined into one comprehensive device. Figure 18–7 contains two types of superaural caps. Figure 18–8 shows an earmuff-style HPD equipped with an FM radio capability. Figure 18–9 displays soft, moldable earplugs.

FIGURE 18–6 Earmuff style HPD with hard hat.

Source: Dmitry Melnikov/Shutterstock

FIGURE 18–7 Canal-cap style HPD.

Source: Kellis/Shutterstock

The effectiveness of HPDs can be enhanced through the use of technologies that reduce noise levels. These **active noise reduction (ANR)** technologies reduce noise by manipulating sound and signal waves. Such waves are manipulated by creating an electronic mirror image of sound waves that tends to cancel out the unwanted noise in the same way that negative numbers cancel out positive numbers in a mathematical equation. Using ANR technologies in conjunction with enclosure devices or earmuffs

FIGURE 18–8 Earmuff style HPD.

Source: Kozini/Shutterstock

FIGURE 18–9 Ear-plug style HPD.

Source: Joe Belanger/Shutterstock

can be an especially effective strategy. Safety and health professionals should know the noise reduction rating (NRR) of all devices and technologies they recommend for employee use.

Traditional, or passive, HPDs can distort or muffle sounds at certain frequencies, particularly high-pitched sounds. *Flat-attenuation HPDs* solve this problem by using electronic devices to block all sound frequencies equally. This eliminates, or at least reduces, the distortion and muffling problems. Flat-attenuation HPDs are especially helpful for employees in settings where high-pitched sound is present that they should be able to hear, as well as for employees who have already begun to lose their ability to hear such sounds. The ability to hear high-pitched sounds is significant because warning signals and human voices can be high-pitched.

A benefit of ANR technologies is **optimization**. The amount of noise protection can be adjusted so that employees can hear as much as they should, but not too much. Too much noise can cause employees to suffer hearing loss. Too little noise can mean that they may not hear warning signals.

Fit Testing of HPDs

Fit testing of HPDs is an important aspect of effective noise reduction that will continue to get more and more attention in the future. The old one-size-fits-all concept does not work any better with HPDs than it does with other personal items. Properly fitted HPDs can reduce noise levels significantly over poorly fitted devices of the same type. When employees know how to properly fit their own HPDs, they are better prepared to be active participants in the protection of their own hearing.

Further, anatomical differences in human ears such as ear canal size and shape can have a marked effect on the effectiveness of a given type of HPD. Fit testing can help employees and employers determine what type of HPD is right for a given individual in a specific environment. There are subjective and objective fit testing methods. Subjective methods typically require the employee to participate in two hearing tests: one with the HPD in place—occluded—and the other without the HPD (unoccluded). Objective tests require a direct measure of sound under the HPD while it is being used as it would be on the job. Regardless of whether the objective or subjective test method is used, fit testing will be an increasingly important aspect of noise reduction.

VIBRATION HAZARDS

Vibration hazards are closely associated with noise hazards because tools that produce vibration typically also produce excessive levels of noise. The strategies for protecting employees against the noise associated with vibrating tools are the same as those presented so far in this chapter. This section focuses on the other safety and health hazards associated with vibration.

The types of injuries associated with vibration depend on its source. For example, workers who operate heavy equipment often experience vibration over the whole body. This can lead to problems ranging from motion sickness to spinal injury. However, the most common vibration-related problem is known as **hand-arm vibration syndrome (HAV)**, a form of Raynaud's syndrome.

Environmental conditions and worker habits can exacerbate the problems associated with vibration. For example, working with vibrating tools in a cold environment is more dangerous than working with the same tools in a warm environment. Gripping a vibrating tool tightly will lead to problems sooner than using a loose grip. Smoking and excessive noise also increase the potential for HAV and other vibration-related injuries. What all these conditions and habits have in common is that they constrict blood vessels, which in turn restricts blood flow to the affected part of the body.[12]

Injury-Prevention Strategies

Modern safety and health professionals should know how to prevent vibration-related injuries. Prevention is especially important with HAV because the disease is thought to be irreversible. This does not mean that HAV cannot be treated. It can, but the treatments developed to date only reduce the symptoms. They do not cure the disease.

Following are prevention strategies that can be used by safety and health professionals in any company regardless of its size.

Purchase Low-Vibration Tools

Interest in producing **low-vibration tools** is relatively new but growing. Since the 1960s, only a limited number of manufacturers produced low-vibration tools. However, a lawsuit filed against three prominent tool manufacturers generated a higher level of interest in producing low-vibration tools.

The suit was filed on behalf of 300 employees of General Dynamics Electric Boat Shipyard in Connecticut, most of whom now suffer from HAV. The case claimed that three predominant tool manufacturers failed to (1) warn users of their tools of the potential for vibration-related injury and (2) produce low-vibration tools even though the technology to do so has been available for many years.[13] As a result of this lawsuit and the potential for others like it, low-vibration tools are becoming more commonplace.

Limit Employee Exposure

Although a correlation between cumulative exposure to vibration and the onset of HAV has not been scientifically quantified, there is strong suspicion in the safety and health community that such a link exists. For example, NIOSH recommends that companies limit the exposure of their employees to no more than four hours per day, two days per week.[14] Until the correlation between cumulative exposure and HAV has been quantified, safety and health professionals are well advised to apply the NIOSH recommendations.

Change Employee Work Habits

Employees can play a key role in protecting themselves if they know how to do so. Safety and health professionals should teach employees who use vibration-producing tools the work habits that will protect them from HAV and other injuries. These work habits include the following: (1) wearing properly fitting thick gloves that can partially absorb vibration; (2) taking periodic breaks (at least 10 minutes every hour); (3) using a loose grip on the tool and holding it away from the body; (4) keeping tools properly maintained (i.e., replacing vibration-absorbing pads regularly); (5) keeping warm; and (6) using vibration-absorbing floor mats and seat covers as appropriate.[15]

Modern safety and health professionals should also encourage higher management to require careful screening of applicants for jobs involving the use of vibration-producing tools and equipment. Applicants who smoke or have other conditions that constrict blood vessels should be guided away from jobs that involve excessive vibration.

Other Injury-Prevention Strategies

In addition to purchasing low-vibration tools, limiting employee exposure, and changing employee work habits, employers can apply the following strategies: work rotation, tool mounting, and using vibration insulation on tool handles. Work rotation involves rotating a worker's job duties between tasks involving vibration and tasks that do not. This is actually another strategy for limiting worker exposure to vibration. Mounting tools wherever appropriate will reduce vibration to workers since the mounting surface will absorb much of the vibration, thereby reducing the amount the work is exposed to. Making sure the handles of tools are coated with an appropriate amount of vibration absorbing material can reduce worker exposure. However, employers should never rely sole on this strategy since it provides only minimal reduction of exposure to vibration.

OTHER EFFECTS OF NOISE HAZARDS

Hearing loss is the principal concern of safety and health professionals relating to noise hazards. However, hearing loss is not the only detrimental effect of excess noise. Noise can also cause communication problems, isolation, and productivity problems.

Noise can also be detrimental to productivity by interfering with an employee's ability to think, reason, and solve problems. Not all employees respond to noise in this way, but many do. Can you concentrate on your studies in a noisy room? Some students can, whereas others cannot. If excessive noise makes it difficult for you to study, you will probably have the same problem on the job.

Isolation is another problem for some employees in a noisy environment. Employees can begin to feel left out and uninformed, the antithesis of the goal of a modern teamwork-oriented organization.

CORPORATE POLICY

Corporate policies relating to hearing loss prevention should be carefully planned and executed to benefit the affected employee and the employer.[16] Organizations with the most successful programs address the following areas of concern:

- The organizational environment should promote a safety culture where employees are empowered to protect their own health and that of coworkers.
- Policies should be based on effective practices rather than on minimum compliance with government regulations.
- The hearing prevention program must be a functional part of the overall company safety and health program. It should not be a stand-alone, separate-budget operation.
- A key individual (or *program implementer*) should have ultimate responsibility for the program. This person may not necessarily perform all the functions of the hearing loss prevention program but is in charge of the overall program. Experience with successful hearing loss prevention programs shows that a single individual often makes the crucial difference between success and failure. This person may be a safety and health officer, a supervisor, or a designated employee. This program implementer acts as the *conscience* and *champion* of the hearing loss prevention program.
- The program implementer should work with management and employees to develop and implement hearing loss prevention plans and policies for an effective program. As a team leader, the program implementer should be given the authority to establish hearing loss prevention provisions that meet or exceed the letter and intent of OSHA's noise control and hearing conservation regulations.
- Employee and administrative compliance with the company's hearing loss prevention program.
- Hearing loss prevention program policies should clearly describe standard operating procedures for each phase of the program. Specific policy statements should be developed for the important elements of the program. For example, company policy should require the participation of all noise-exposed employees in the audiometric program and require the consistent and proper wearing of hearing protectors in posted areas, even if employees or supervisors are only passing through these areas. These requirements should be conditions of employment. Other important policy statements should be written to cover:

  Adoption of a prescribed schedule for monitoring employee noise exposure levels and other risks, including ensuring that equipment and personnel training are appropriate to the task.

  Counseling of employees immediately following each audiometric test, whether it is the initial, annual, retest, threshold-shift confirmation, or termination examination.

  Determining the adequacy and correct use of HPDs by on-site equipment checks.

  Educating, training, and motivating employees to support the company's hearing loss prevention program provisions; assessing employee attitudes and assessing knowledge gained from periodic training.

Establishing a program of quality assurance for the performance of audiometry and management of audiometric records.

Reviewing audiometric data to verify the effectiveness of the hearing loss prevention program.

Encouraging employees to use company-provided hearing protectors for off-the-job exposure.

Purchasing hearing protectors, audiometers, noise measuring equipment, and quieter machinery. This policy should address the reasons why the program implementer responsible for the hearing loss prevention program, not the purchasing department, should have final authority about anticipated purchases.

If services such as noise surveys, employee education, audiometric testing, medical counseling, or the fitting of HPDs cannot be handled by in-house staff, outside vendors or contractors may be required. They should be selected carefully to ensure that their services complement the abilities of the company staff and functional conduct of the in-house program elements. Vendors must understand and agree to abide by the company's hearing loss prevention program policies and standards of operation. On-site personnel must supervise contractors to make sure that they carry out their obligations. Regardless of whether outside vendors or contractors are used, responsibility for the program stays with the program implementer.

Companies that issue clearly defined hearing loss prevention policies and then adhere to these policies consistently will have smoothly running hearing loss prevention programs. Employees will be fully informed, comprehend their functional role, and know what is expected of them. Equipment will be appropriate, hearing protection will be used by the right people in the right places, and the program elements will be implemented in a timely fashion.

EVALUATING HEARING LOSS PREVENTION PROGRAMS

Hearing loss prevention programs should be evaluated periodically to ensure their effectiveness.[17] Such evaluations should have at least the following components: (1) training and education, (2) supervisor involvement, (3) noise measurement, (4) engineering and administrative controls, (5) monitoring and record keeping, (6) referrals, (7) hearing protection devices, and (8) administration. Following are checklists for each of these components.

Training and Education
Failures or deficiencies in hearing conservation programs (hearing loss prevention programs) can often be traced to inadequacies in the training and education of noise-exposed employees and those who conduct elements of the program.

- Has training been conducted at least once a year?
- Was the training provided by a qualified instructor?
- Was the success of each training program evaluated?
- Is the content revised periodically?
- Are managers and supervisors directly involved?

- Are posters, regulations, handouts, and employee newsletters used as supplements?
- Are personal counseling sessions conducted for employees having problems with HPDs or showing hearing threshold shifts?

Supervisor Involvement

Data indicate that employees who refuse to wear hearing protectors or who fail to show up for hearing tests frequently work for supervisors who are not totally committed to the hearing conservation programs.

- Have supervisors been provided with the knowledge required to supervise the use and care of hearing protectors by subordinates?
- Do supervisors wear hearing protectors in appropriate areas?
- Have supervisors been counseled when employees resist wearing protectors or fail to show up for hearing tests?
- Are disciplinary actions enforced when employees repeatedly refuse to wear hearing protectors?

Noise Measurement

Noise measurements, to be useful, should be related to noise exposure risks or the prioritization of noise control efforts, rather than merely filed away. In addition, the results should be communicated to the appropriate personnel, especially when follow-up actions are required.

- Were the essential or critical noise studies performed?
- Was the purpose of each noise study clearly stated? Have noise-exposed employees been notified of their exposures and apprised of auditory risks?
- Are the results routinely transmitted to supervisors and other key individuals?
- Are results entered into health or medical records of noise-exposed employees?
- Are results entered into shop folders?
- If noise maps exist, are they used by the proper staff?
- Are noise measurement results considered when contemplating procurement of new equipment? Modifying the facility? Relocating employees?
- Have there been changes in areas, equipment, or processes that have altered noise exposure? Have follow-up noise measurements been conducted?
- Are appropriate steps taken to include (or exclude) employees in the hearing loss prevention programs whose exposures have changed significantly?

Engineering and Administrative Controls

Controlling noise by engineering and administrative methods is often the most effective means of reducing or eliminating the hazard. In some cases, engineering controls will remove requirements for other components of the program, such as audiometric testing and the use of hearing protectors.

- Have noise control needs been prioritized?
- Has the cost effectiveness of various options been addressed?
- Are employees and supervisors apprised of plans for noise control measures?
- Are they consulted on various approaches?
- Will in-house resources or outside consultants perform the work?

- Have employees and supervisors been counseled on the operation and maintenance of noise control devices?
- Are noise control projects monitored to ensure timely completion?
- Has the full potential for administrative controls been evaluated? Are noisy processes conducted during shifts with fewer employees? Do employees have sound-treated lunch or break areas?

Monitoring Audiometry and Record Keeping

The skills of audiometric technicians, the status of the audiometer, and the quality of audiometric test records are crucial to hearing loss prevention program success. Useful information may be ascertained from the audiometric records and from those who actually administer the tests.

- Has the audiometric technician been adequately trained, certified, and recertified as necessary?
- Do on-the-job observations of the technicians indicate that they perform a thorough and valid audiometric test, instruct and consult the employee effectively, and keep appropriate records?
- Are records complete?
- Are follow-up actions documented?
- Are HTLs reasonably consistent from test to test? If not, are the reasons for inconsistencies investigated promptly?
- Are the annual test results compared to baseline to identify the presence of an OSHA standard threshold shift?
- Is the annual incidence of standard threshold shift greater than a few percent? If so, are problem areas pinpointed and remedial steps taken?
- Are audiometric trends (deteriorations) being identified, both in individuals and in groups of employees? (NIOSH recommends no more than 5 percent of workers showing 15-dB significant threshold shift, same ear, and same frequency.)
- Do records show that appropriate audiometer calibration procedures have been followed?
- Is there documentation showing that the background sound levels in the audiometer room were low enough to permit valid testing?
- Are the results of audiometric tests being communicated to supervisors and managers as well as to employees?
- Has corrective action been taken if the rate of no-shows for audiometric test appointments is more than about 5 percent?
- Are employees incurring STS notified in writing within at least 21 days? (NIOSH recommends immediate notification if retest shows 15-dB significant threshold shift, same ear, same frequency.)

Referrals

Referrals to outside sources for consultation or treatment are sometimes in order, but they can be an expensive element of the hearing loss prevention program and should not be undertaken unnecessarily.

- Are referral procedures clearly specified?
- Have letters of agreement between the company and consulting physicians or audiologists been executed?

- Have mechanisms been established to ensure that employees needing evaluation or treatment actually receive the service (e.g., transportation, scheduling, and reminders)?
- Are records properly transmitted to the physician or audiologist, and back to the company?
- If medical treatment is recommended, does the employee understand the condition requiring treatment, the recommendation, and methods of obtaining such treatment?
- Are employees being referred unnecessarily?

Hearing Protection Devices

When noise control measures are not feasible, or until such time as they are installed, HPDs are the only way to prevent hazardous levels of noise from damaging the inner ear. Making sure that these devices are worn effectively requires continuous attention on the part of supervisors and program implementers as well as noise-exposed employees.

- Have hearing protectors been made available to all employees whose daily average noise exposures are 85 dBA or above? (NIOSH recommends requiring HPD use if noises equal or exceed 85 dBA regardless of exposure time.)
- Are employees given the opportunity to select from a variety of appropriate protectors?
- Are employees fitted carefully with special attention to comfort?
- Are employees thoroughly trained, not only initially but at least once a year?
- Are the protectors checked regularly for wear or defects and replaced immediately if necessary?
- If employees use disposable hearing protectors, are replacements readily available?
- Do employees understand the appropriate hygiene requirements?
- Have any employees developed ear infections or irritations associated with the use of hearing protectors? Are there any employees who are unable to wear these devices because of medical conditions? Have these conditions been treated promptly and successfully?
- Have alternative types of hearing protectors been considered when problems with current devices are experienced?
- Do employees who incur noise-induced hearing loss receive intensive counseling?
- Are those who fit and supervise the wearing of hearing protectors competent to deal with the many problems that can occur?
- Do workers complain that protectors interfere with their ability to do their jobs? Do they interfere with spoken instructions or warning signals? Are these complaints followed promptly with counseling, noise control, or other measures?
- Are employees encouraged to take home their hearing protectors if they engage in noisy nonoccupational activities?
- Are potentially more effective protectors considered as they become available?
- Is the effectiveness of the hearing protector program evaluated regularly?

- Have at-the-ear protection levels been evaluated to ensure that either overprotection or underprotection has been adequately balanced according to the anticipated ambient noise levels?
- Is each hearing protector user required to demonstrate that he or she understands how to use and care for the protector? Are the results documented?

Administration

Keeping organized and current on administrative matters will help the program run smoothly.

- Have there been any changes in federal or state regulations? Have hearing loss prevention programs' policies been modified to reflect these changes?
- Are copies of company policies and guidelines regarding the hearing loss prevention program available in the offices that support the various program elements? Are those who implement the program elements aware of these policies? Do they comply?
- Are necessary materials and supplies being ordered with a minimum of delay?
- Are procurement officers overriding the hearing loss prevention program implementer's requests for specific hearing protectors or other hearing loss prevention equipment? If so, have corrective steps been taken?
- Is the performance of key personnel evaluated periodically? If such performance is found to be less than acceptable, are steps taken to correct the situation?
- Has the failure to hear warning shouts or alarms been tied to any accidents or injuries? If so, have remedial steps been taken?

FUTURE OF HEARING CONSERVATION: NOISE REDUCTION RATING

Noise-induced hearing loss (NIHL) is one of the most common preventable injuries in the workplace. Consequently, noise reduction and hearing protection are high on the list of OSHA's priorities and can be expected to receive ongoing attention in the future. A major change is in how the NRRs of HPDs will be determined and labeled.

To understand NRRs, it is first necessary to understand three important terms: **passive noise reduction**, *active noise reduction*, and *impulsive noise reduction*.

Passive noise reduction is the reduction of sound transmission using nothing but materials and/or structural elements. *Active noise reduction* involves using electronic devices to reduce sound transmission by producing acoustic signals of equal and opposite phase and amplitude. *Impulsive noise reduction* involves reducing peak impulse sound transmission based upon the use of passive or active noise reduction or both.

Historically, HPDs have carried a label established and required by the Environmental Protection Agency (EPA) and certain types of HPDs were not labeled other than with a low NRR because they were not tested in the higher noise ranges. But this is changing and a broader range of devices is being tested at all ranges and labeled in a manner that provides more comprehensive information for employers and users.

FIGURE 18–10 Noise reduction rating label.

Old and New Rating Methods Compared

Historically, the EPA used a single number to rate HPDs. This older method is being replaced by a label that will contain a sliding scale showing the high/low range of estimated protection as shown in Figure 18–10. The new EPA label will identify the HPD's function (active, passive, or impulse) and provide a range that depicts the effectiveness of the HPD in reducing noise. The old rating method estimates the 98th percentile of protection when the HPD is properly fitted. The new method estimates the 80th and 20th of protection. The old method uses a test protocol based on ANSI S.3.19-1974. This protocol requires the use of 10 test subjects for ear plugs and earmuffs that are fit by the experimenter. The new method uses the protocol from ANSI S12.6-2008 which requires 20 subjects for ear plugs and 10 for earmuffs and requires the test subjects to fit their own HPDs after they have been trained in proper fit techniques. The old method does not require periodic re-testing of HPDs. With the new method, HPDs must be re-tested every five years at a minimum.

Using NRR in Making Protected dBA Calculations and in Evaluating HPDs

OSHA uses NRRs in determining compliance. However, this practice is not without controversy. The controversy results from major differences between the noise attenuation of achieved in a laboratory setting and that which is achieved in an actual work setting. These differences are well documented and can be attributed in large measure to the ability of workers to properly use HPDs as well as the differences in conditions in the workplace versus a laboratory.

To compensate, NIOSH recommends applying a concept known as "derating" in which the calculated dBA is reduced by using a multiplier of 75 percent for earmuffs, 50 percent for slow-recovery foam earplugs, and 30 percent for all other earplugs. Regardless of the controversy, safety and health professionals should understand NRR as a concept and how it applies in the workplace since OSHA compliance depends on it. This said, the obvious approach is to ensure that the NRR for HPDs is higher than what is required in the conditions in question based on the NIOSH derating multipliers.

OSHA uses six methods for using NRR to determine is workers are properly protected from noise hazards. However, these six methods can be reduced to just two Mathematical equations: one for use with a C-rated scale and one for use with an A-rated scale as follows:

- *For C-rated measurements: Protected dBA = Unprotected dBC - NRR*
  Assume that an HPD has an NRR of 18 dB and it is used in an environment of 96 dBC, the Protected dBA would be calculated as follows:

$$\text{Protected dBA} = 78 \text{ since } (96 - 18 = 78)$$

- *For A-rated measurements: Protected dBA = Unprotected dBA − (NRR − 7)*
  Assume that an HPD has an NRR of 18 dB and it is used in an environment of 96 dBA, the Protected dBA would be calculated as follows:

$$\text{Protected dBA} = 85 \text{ since } [96 - (18 - 7) = 85]$$

Regardless of which of these formulas is used to determine the Protected dBA of HPDs, remember that the figure determined will be reduced significantly by the application of the multipliers recommended by NIOSH.

DISTRACTION HAZARDS INTRODUCED BY HANDHELD DEVICES WITH EARBUDS

Handheld devices of various kinds equipped with earbuds have become ubiquitous in contemporary society. Consequently, it should come as no surprise to safety and health professionals that there use in the workplace is increasing, and possibly to dangerous levels. There are two issues with using handheld devices with earbuds at work that the worker needs to be aware of: (1) the volume can drown out noises in the work environment and (2) mental distraction. Regardless of whether the problem is volume, distraction, or both, earbuds have introduced a new hazard into the workplace.

To understand how significant the hazards of earbuds are, consider that pedestrian deaths have skyrocketed since earbuds became a commonly used device. Distracted pedestrians are simply walking right into the paths of moving vehicles. Over a seven-year period, more than half of the pedestrians wearing earbuds who were killed walked in front of trains. If a pedestrian wearing earbuds cannot hear an oncoming train, how can workers hear the various types of moving vehicles and equipment used in the workplace (i.e., trucks, cars, forklifts, trams, heavy equipment, etc.). To make matters worse, there have been instance of loose earbuds getting caught in machinery and causing injuries.

A distracted worker is an accident waiting to happen. Consequently, safety and health professionals should ensure that their organizations have policies strictly forbidding the use of earbuds (at least when workers are moving or working with machinery or equipment). Further, earbuds should not be allowed for use as HPDs or in conjunction with HPDs.

Key Terms and Concepts

- Active noise reduction (ANR)
- Administrative controls
- Air pressure
- ANSI S12.13
- Audiogram
- Audiometric database analysis (ADBA)
- Audiometric testing
- Baseline audiogram
- Decibel
- Dosimeter
- Duration
- Electronic earmuffs
- Earplugs
- Enclosures
- Engineering controls
- Exchange rate
- Fit testing
- Follow-up
- Formable earplugs
- Frequency
- Hand-arm vibration syndrome (HAV)
- Hazardous noise
- Hearing conservation
- Hearing protection devices (HPDs)
- Hearing threshold level
- Hertz (Hz)
- Impulse noise
- Impulsive noise reduction
- Low-vibration tools
- Material hearing impairment
- Narrow band noise
- Noise
- Noise controls
- Noise surveys
- Optimization
- Passive noise reduction
- Path
- Personal protection
- Receiver
- Sound
- Sound-level meter
- Sound pressure level
- Source
- Standard threshold shift
- Superaural caps
- Threshold of hearing
- Threshold of pain
- Time-weighted average
- Vibration
- Wide band noise

Review Questions

1. Define the term sound.
2. What is the difference between sound and noise?
3. Differentiate between sound and vibration.
4. Describe the relationship between the pitch of sound and the cycle of sound waves.
5. List and briefly explain the three broad types of industrial noise.
6. Describe the various physiological problems associated with excessive noise.
7. List four factors that affect the risk of hearing loss from exposure to excessive noise.
8. At what sound level is it necessary to begin using some type of personal protection?
9. Give a brief description of ANSI S12.13.
10. Give a brief description of OSHA 29 CFR 1910.95.
11. What factors do medical professionals consider in determining causal relationships of hearing loss?
12. Define the following terms: follow-up, noise survey, and audiometric testing.
13. List three appropriate follow-up activities when an audiometric test reveals hearing loss in an employee.
14. Differentiate between engineering and administrative controls.
15. What is HAV? How can it be prevented?
16. Explain the four classifications of HPDs that are widely used.
17. Explain the main components of an evaluation of a hearing loss prevention program.

Endnotes

1. National Institute for Occupational Safety and Hearing (NIOSH), *A Practical Guide to Preventing Hearing Loss*, Publication 96–110 (Washington, DC: NIOSH), 2. Retrieved from cdc.gov/niosh on July 14, 2017.
2. Ibid.

3. NIOSH, *A Practical Guide*, 6.
4. NIOSH, *A Practical Guide*, 4.
5. Ibid., 6.
6. Ibid.
7. ANSI Technical Report, *Evaluating the Effectiveness of Hearing Conservation Programs through Audiometric Data Base Analysis*. Retrieved from scribd.com/doc/92344634/ASA-S12-13-TR-Hearing-Conservation on July 14, 2017.
8. NIOSH, *A Practical Guide*, 37.
9. NIOSH, *A Practical Guide*, 8.
10. Ibid., 12.
11. Ibid.
12. Ibid., 32–33.
13. Ibid., 34.
14. Ibid., 33.
15. Ibid., 35.
16. NIOSH, *A Practical Guide*, 1–2.
17. Ibid., Appendix B, 1–6.

CHAPTER 19

PREPARING FOR EMERGENCIES, DISASTERS, AND TERRORISM

Learning Objectives

- Explain the rationale for emergency preparation
- Summarize the basic tenets of the Emergency Planning and Community Right-to-Know Act
- Explain the concepts of organization and coordination in emergency planning
- List the OSHA standards relating to emergency prevention, planning, and response
- Summarize the key elements of providing first aid in the aftermath of an emergency
- Explain how to plan for emergencies (the emergency action plan)
- Explain how to plan for workers with disabilities
- Summarize the key elements of evacuation planning
- Describe how emergency plans can be customized to meet local needs
- Explain the concept of the emergency response team and emergency response network
- Explain how computers are used in emergency response
- Describe how organizations should deal with psychological trauma in the aftermath of an emergency
- Describe the role of the employer in preparing for, preventing, and responding to terrorism acts in the workplace
- Explain the proper response options to an active shooter
- Summarize the steps in resuming business after a disaster

Despite the best efforts of all involved, emergencies do sometimes occur. The potential for human-caused emergencies has increased significantly with the rise of world-wide terrorism. It is very important to respond to such emergencies in a way that minimizes harm to people and damage to property. To do so requires plans that can be implemented without delay. This chapter provides prospective and practicing safety and health professionals with the information they need to prepare for emergencies in the workplace. Everything in this chapter pertains to all kinds of emergencies, including natural disasters and terrorism. A special section relating specifically to terrorism is included at the end of the chapter.

RATIONALE FOR EMERGENCY PREPARATION

An *emergency* is a potentially life-threatening situation, usually occurring suddenly and unexpectedly. Emergencies may be the result of natural or human causes. Have you ever witnessed the timely, organized, and precise response of a professional emergency medical crew at an automobile accident? While passers-by and spectators may wring their hands and wonder what to do, the emergency response professionals quickly organize, stabilize, and administer first aid. Their ability to respond in this manner is the result of preparation. As shown in Figure 19–1, preparation involves a combination of **planning**, **practicing**, **evaluating**, and **adjusting** to specific circumstances.

When an emergency occurs, immediate reaction is essential. Speed in responding can mean the difference between life and death or between minimal damage and major damage. Ideally, all those involved should be able to respond properly with a minimum of hesitation. This can happen only if all exigencies have been planned for and planned procedures have been practiced, evaluated, and improved.

A quick and proper response—which results because of proper preparation—can prevent panic, decrease the likelihood of injury and damage, and bring the situation under control in a timely manner. Because no workplace is immune to emergencies, preparing for them is critical. An important component of preparation is planning.

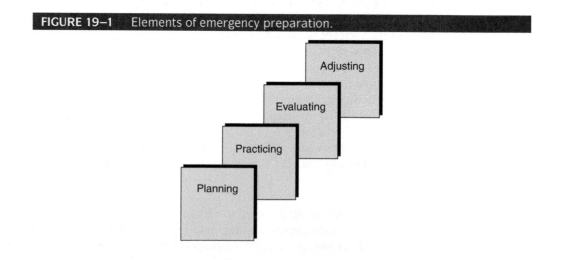

FIGURE 19–1   Elements of emergency preparation.

EMERGENCY PLANNING AND COMMUNITY RIGHT-TO-KNOW ACT

Title III of the Superfund Amendments and Reauthorization Act of 1986 (SARA) is also known as the **Emergency Planning and Community Right-to-Know Act (EPCRA)**. This law is designed to make information about hazardous chemicals available to a community where they are being used so that residents can protect themselves in the case of an emergency. It applies to all companies that use, make, transport, or store chemicals.

Safety and health professionals involved in developing emergency response plans for their companies should be familiar with the act's requirements for **emergency planning**. As shown in Figure 19–2, the EPCRA includes the four major components discussed in the following paragraphs.

Emergency Planning

The emergency planning component requires that communities form **local emergency planning committees (LEPCs)** and that states form **state emergency response commissions (SERCs)**. LEPCs are required to develop emergency response plans for the local communities, host public forums, select a planning coordinator for the community, and work with the coordinator in developing local plans. SERCs are required to oversee LEPCs and review their emergency response plans. Plans for individual companies in a given community should be part of that community's larger plan. Local emergency response professionals should use their community's plan as the basis for simulating emergencies and practicing their responses.

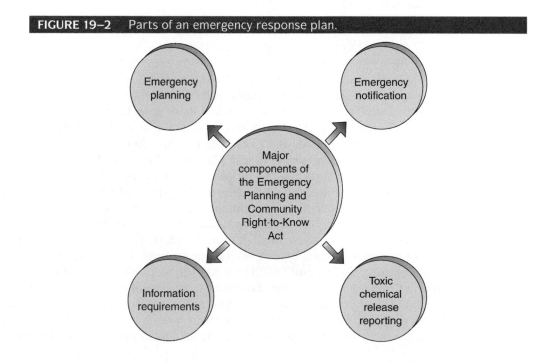

FIGURE 19–2   Parts of an emergency response plan.

Emergency Notification

The **emergency notification** component requires that chemical spills or releases of toxic substances that exceed established allowable limits be reported to appropriate LEPCs and SERCs. Immediate notification may be verbal as long as a written notification is filed promptly. Such reports must contain at least the following information: (1) the names of the substances released, (2) where the release occurred, (3) when the release occurred, (4) the estimated amount of the release, (5) known hazards to people and property, (6) recommended precautions, and (7) the name of a contact person in the company.

Information Requirements

Information requirements mean that local companies must keep their LEPCs and SERCs and, through them, the public informed about the hazardous substances that the companies store, handle, transport, or use. This includes keeping comprehensive records of such substances on file, up-to-date, and readily available; providing copies of safety data sheets for all hazardous substances; giving general storage locations for all hazardous substances; providing estimates of the amount of each hazardous substance on hand on a given day; and estimating the average annual amount of hazardous substances kept on hand.

Toxic Chemical Release Reporting

The **toxic chemical release reporting** component requires that local companies report the total amount of toxic substances released into the environment as either emissions or hazardous waste. Reports go to the Environmental Protection Agency (EPA) and the state-level environmental agency.

ORGANIZATION AND COORDINATION

Responses to emergencies are typically from several people or groups of people, including medical, firefighting, security, and safety personnel as well as specialists from a variety of different fields. People in each of these areas have different but interrelated and often interdependent roles to play in responding to the emergency. Because of their disparate backgrounds and roles, both organization and coordination are critical.

A company's **emergency response plan** should clearly identify the different personnel and groups that respond to various types of emergencies and, in each case, who is in charge. One person should be clearly identified and accepted by all emergency responders as the **emergency coordinator**. This person should be knowledgeable, at least in a general sense, of the responsibilities of each individual emergency responder and how each relates to those of all other responders. This knowledge must include the **order of response** for each type of emergency set forth in the plan.

A company's safety and health professional is the obvious person to organize and coordinate emergency responses. However, regardless of who is designated, it is important that (1) one person is in charge, (2) everyone involved knows who is in charge, and (3) everyone who has a role in responding to an emergency is given ample opportunities to practice in simulated conditions that come as close as possible to real conditions.

OSHA STANDARDS

All Occupational Safety and Health Administration (OSHA) standards are written for the purpose of promoting a safe, healthy, accident-free, hence emergency-free workplace. Therefore, **OSHA standards** play a role in emergency prevention and should be considered when developing emergency plans. For example, exits are important considerations when planning for emergencies. Getting medical personnel in and employees and injured workers out quickly is critical when responding to emergencies. The following sections of OSHA's standards deal with emergency preparedness:

Emergency Action Plan	29 CFR 1910.38
Exit arrangements	29 CFR 1910.37(e)
Exit capacity	29 CFR 1910.37(c), (d)
Exit components	29 CFR 1910.37(a)
Exit workings	29 CFR 1910.37(q)
Exit width	29 CFR 1910.37(c)
Exterior exit access	29 CFR 1910.37(g)
Occupational Health and Environmental Controls	29 CFR 1926.65, Appendix C
Hazardous Materials	29 CFR 1910.120, Appendix C

A first step for companies developing emergency plans is to review these OSHA standards. This can help safety and health personnel identify and correct conditions that may exacerbate emergency situations before they occur.

FIRST AID IN EMERGENCIES

Workplace emergencies often require a medical response. The immediate response is usually first aid. First aid consists of lifesaving measures taken to assist an injured person until medical help arrives.

Because there is no way to predict when first aid may be needed, providing **first-aid training** to employees should be part of preparing for emergencies. In fact, in certain cases, OSHA requires that companies have at least one employee on-site who has been trained in first aid (CFR 1910.151). Figure 19–3 contains a list of the topics that may be covered in a first-aid class for industrial workers.

First-Aid Training Program

First-aid programs are usually available in most communities. The continuing education departments of community colleges and universities typically offer first-aid training. Classes can often be provided on-site and customized to meet the specific needs of individual companies.

The American Red Cross provides training programs in first aid specifically geared toward the workplace. For more information about these programs, safety and health professionals may contact the national office of the American Red Cross at 202-639-3200 (redcross.org).

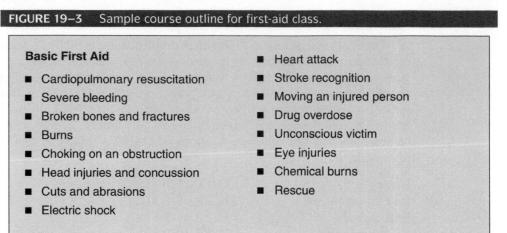

FIGURE 19–3   Sample course outline for first-aid class.

Basic First Aid

- Cardiopulmonary resuscitation
- Severe bleeding
- Broken bones and fractures
- Burns
- Choking on an obstruction
- Head injuries and concussion
- Cuts and abrasions
- Electric shock

- Heart attack
- Stroke recognition
- Moving an injured person
- Drug overdose
- Unconscious victim
- Eye injuries
- Chemical burns
- Rescue

The National Safety Council (NSC) also provides first-aid training materials. The First Aid and Emergency Care Teaching Package contains a slide presentation, overhead transparencies, a test bank, and an instructor's guide. The council also produces a book titled *First Aid Essentials*. For more information about these materials, safety and health professionals may contact the NSC at 800-832-0034 (nsc.org).

Beyond Training

Training employees in first-aid techniques is an important part of preparing for emergencies. However, training is only part of the preparation. In addition, it is important to do the following:

1. **Have well-stocked first-aid kits available.** First-aid kits should be placed throughout the workplace in clearly visible, easily accessible locations. They should be properly and fully stocked and periodically checked to ensure that they stay fully stocked. Figure 19–4 lists the minimum recommended contents for a workplace first-aid kit.
2. **Have appropriate personal protective devices available.** With the concerns about AIDS and hepatitis, administering first aid has become more complicated than in the past. The main concerns are with bleeding and other body fluids. Consequently, a properly stocked first-aid kit should contain rubber surgical gloves and facemasks or mouthpieces for CPR.
3. **Post emergency telephone numbers.** The advent of 911 service has simplified the process of calling for medical care, police, or firefighting assistance. If 911 services are not available, emergency numbers for ambulance, hospital, police, fire department, LEPC, and appropriate internal personnel should be posted at clearly visible locations near all telephones in the workplace.
4. **Keep all employees informed.** Some companies require all employees to undergo first-aid training; others choose to train one or more employees in each department. Regardless of the approach used, it is important that all employees be informed and kept up-to-date concerning basic first-aid information. Figures 19–5 and 19–6 are first-aid fact sheets of the type used to keep employees informed.

FIGURE 19–4 Minimum recommended contents of workplace first-aid kits.

- Sterile gauze dressings (individually wrapped in sizes 4 × 4 and 8 × 10 inches)
- Triangular bandages
- Roll of gauze bandages (at least 2 inches wide)
- Assorted adhesive bandages
- Sealed moistened towelettes
- Adhesive tape
- Absorbent cotton
- Sterile saline solution
- Mild antiseptic for minor wounds
- Ipecac syrup to induce vomiting
- Powdered activated charcoal to absorb swallowed poisons
- Petroleum jelly
- Baking soda (bicarbonate of soda)
- Aromatic spirits of ammonia
- Elastic wraps
- Scissors
- Tweezers
- Needles
- Sharp knife or stiff-backed razor blades
- Medicine dropper (eye dropper)
- Measuring cup
- Oral thermometer
- Rectal thermometer
- Hot water bag
- Wooden safety matches
- Flashlight
- Rubber surgical gloves
- Face masks or mouthpieces
- Blanket
- Splint

HOW TO PLAN FOR EMERGENCIES

Developing an **emergency action plan (EAP)** is a major step in preparing for emergencies. A preliminary step is to conduct a thorough analysis to determine the various types of emergencies that may occur. For example, depending on geography and the types of products and processes involved, a company may anticipate such emergencies as fires, chemical spills, explosions, toxic emissions, train derailments, hurricanes, tornadoes, lightning, floods, earthquakes, or volcanic eruptions.

A company's EAP should be a collection of small plans for each anticipated or potential emergency. These plans should have the following components:

1. **Procedures.** Specific, step-by-step emergency response procedures should be developed for each potential emergency.
2. **Coordination.** All cooperating agencies and organizations and emergency responders should be listed along with their telephone number and primary contact person.
3. **Assignments and responsibilities.** Every person who will be involved in responding to a given emergency should know his or her assignment. Each person's responsibilities should be clearly spelled out and understood. One person may be responsible for conducting an evacuation of the affected area, another for the immediate shutdown of all equipment, and another for telephoning for medical, fire, or other types of emergency assistance. When developing this part of the EAP, it is important to assign a backup person for each area of responsibility. Doing so ensures that

FIGURE 19–5   First-aid fact sheet.

Global Technologies
500 Anchors Street • Industrial Park • Fort Walton Beach, Florida 32748

First-Aid Fact Sheet No. 16
Moving an Injured Person

If a victim has a neck or back injury, do not move him unless it must be done to prevent additional injuries. If it is absolutely essential to move the victim, remember the following rules of thumb:

1. Call for professional medical help.

2. Always pull the body lengthwise, never sideways.

3. If there is time, slip a blanket under the victim and use it to pull him to safety.

4. If the victim must be lifted, support all parts of the body so that it does not bend or jackknife. (Use a spine board and cervical collar.)

FIGURE 19–6   First aid fact sheet.

Global Technologies
500 Anchors Street • Industrial Park • Fort Walton Beach, Florida 32748

First-Aid Fact Sheet No. 12
ABCs of First Aid

If a fellow employee is injured and you are the first person to respond, remember the ABCs of first aid.

A = Airway
Is the airway blocked? If so, clear it quickly.

B = Breathing
Is the victim breathing? If not, begin administering artificial respiration.

C = Circulation
Is the victim bleeding severely? If so, stop the bleeding.
Is there a pulse? If not, begin administering CPR.

the plan will not break down if a person assigned a certain responsibility is one of the victims.

4. **Accident prevention strategies.** The day-to-day strategies to be used for preventing a particular type of emergency should be summarized in this section of the EAP. In this way, the strategies can be reviewed, thereby promoting prevention.

5. **Schedules.** This section should contain the dates and times of regularly scheduled practice drills. It is best to vary the times and dates so that practice drills don't become predictable and boring. Figure 19–7 is a checklist that can be used for developing an EAP.

PLANNING FOR WORKERS WITH DISABILITIES

When developing EAPs it is important to consider the special needs of workers with disabilities. Do not develop a separate EAP specifically for workers with disabilities—they should be included as a normal part of the organization's plan. An excellent set of

FIGURE 19–7 Emergency planning checklist.

Type of Emergency

____Fire	____Explosion
____Chemical spill	____Toxic emission
____Train derailment	____Hurricane
____Tornado	____Lightning
____Flood	____Earthquake
____Volcanic eruption	

Procedures for Emergency Response
1. Controlling and isolating?
2. Communication?
3. Emergency assistance?
4. First aid?
5. Shutdown/evacuation/protection of workers?
6. Protection of equipment/property?
7. Egress, ingress, exits?
8. Emergency equipment (e.g., fire extinguishers)?
9. Alarms?
10. Restoration of normal operations?

Coordination
1. Medical care providers?
2. Fire service providers?
3. LEPC personnel?
4. Environmental protection personnel?
5. Civil defense personnel (in the case of public evacuations)?
6. Police protection providers?
7. Communication personnel?

FIGURE 19–7   *(continued)*

Assignments and Responsibilities
1. Who cares for the injured?
2. Who calls for emergency assistance?
3. Who shuts down power and operations?
4. Who coordinates communication?
5. Who conducts the evacuation?
6. Who meets emergency responders and guides them to the emergency site?
7. Who contacts coordinating agencies and organizations?
8. Who is responsible for ensuring the availability and upkeep of fire extinguishers?
9. Who is responsible for ensuring that alarms are in proper working order?
10. Who is responsible for organizing cleanup activities?

Accident Prevention Strategies
1. Periodic safety inspections?
2. Industrial hygiene strategies?
3. Personal protective equipment?
4. Ergonomic strategies?
5. Machine safeguards?
6. Hand and portable power tool safeguards?
7. Material handling and storage strategies?
8. Electrical safety strategies?
9. Fire safety strategies?
10. Chemical safety strategies?

Schedules
1. Dates of practice drills:_____
2. Times of practice drills:_____
3. Duration of practice drills: _____

guidelines is available to assist organizations in considering the needs of workers with disabilities when developing EAPs.[1]

Titled *Preparing the Workplace for Everyone*, these guidelines were developed for the use of federal agencies but are available to private organizations and other public organizations as well. To obtain a copy of the guidelines, contact the Office of Disability Employment Policy at the following address:

United States Department of Labor
Office of Disability Employment Policy
200 Constitution Ave. NW, Room S-1303
Washington, DC 20210
202-693-7880
dol.gov/odep

The guidelines are presented under numerous subheadings with questions listed relating to each subheading. What follows in the remainder of this section are categories of questions similar to and based on those presented in the actual guidelines. Safety and health professionals are encouraged to obtain a complete copy of the guidelines to ensure that their EAPs account for workers with disabilities appropriately.

Involving Key Personnel

Under this heading, the guidelines pose questions similar to the following:

- Are key personnel familiar with the EAP?
- Are personnel with disabilities involved in all aspects of the development of the EAP?
- Do senior executives support the development and updating of the EAP?
- Has the EAP been reviewed by first responders and facility personnel?
- Does any part of the plan conflict with procedures established by other applicable agencies?

Implementing Shelter-in-Place (SIP) Plans

Shelter-in-place (SIP) means that rather than evacuate in a disaster or emergency, immediate shelter is sought. SIP comes into play when attempting to evacuate might increase the risk of harm or injury. There should be an SIP provision built into the larger EAP containing provisions that speak on the following concerns:

- Are there provisions for shutting down the building's ventilation system quickly?
- Are there provisions for turning off the elevators?
- Are there provisions for closing all exits and entrances and for securing the loading dock and garage areas (as applicable)?
- Are there provisions for notifying all occupants—including visitors—of emergency procedures?
- Are there provisions for asking all occupants—including visitors—to remain in the building until it is safe to leave?

Evaluating Personnel Needs

It is important to determine the needs of workers with disabilities when developing an EAP, but to do so in strict accordance with the Rehabilitation Act. Before evaluating the needs of personnel with disabilities, consult your human resources department to develop guidelines that are consistent with the Rehabilitation Act. Then ask the following types of questions during development of the EAP:

- Has all applicable information about the needs of personnel with disabilities been collected in accordance with the Rehabilitation Act?
- Has the information collected been appropriately protected by sharing it only with those who need to know it as part of emergency planning (e.g., first-aid personnel, safety and health professionals, and those responsible for carrying out the EAP during an emergency)?
- In selecting equipment, have appropriate agencies and individuals with disabilities been consulted as to the ability of personnel with disabilities to use the equipment?

- Have the needs of personnel with service animals been considered in the development of the plan?
- Has the establishment of personal support networks been built into the plan?
- Has the appropriate training been built into the plan for people who need assistance during an emergency as well as for people who will provide the assistance?

Distributing and Communicating the Plan

In developing an EAP, a critical element that should not be overlooked concerns how the completed plan will be distributed to personnel and how those who developed the plan will communicate with personnel about it. The following questions will help ensure proper development of this section of the plan:

- How will the EAP be communicated with the same level of detail to all personnel?
- Is the plan available on the organization's website as well as in hard copy? Are there text explanations of all graphic material contained in the plan?
- Are training sessions contained in the plan offered in accessible locations?
- Are the learning aids necessary to accommodate personnel with disabilities readily available in all training locations (e.g., sign-language interpreters and listening devices)?
- Are hard copies of the EAP placed in various easily accessible locations throughout the facility?
- Are appropriate sections of the plan (evacuation and SIP information) given to frequent visitors to the facility?

Balancing Employer Responsibilities and Employee Right to Self-Determination

On the one hand, employers have specific responsibilities for ensuring the safety of their personnel. On the other hand, employees with disabilities have certain rights of self-determination. To ensure that these two issues do not collide when developing the EAP, ask the following questions:

- Are there employees who impede the evacuation of others during practice drills? If so, has the issue been dealt with privately and have appropriate solutions been developed?
- In determining whether an employee with a disability represents a direct threat in the event of an emergency, have the following factors been considered: nature and severity of the potential harm, likelihood that the potential harm will occur during an emergency, the risk posed by the individual in question, the imminence of the potential harm, and the availability of reasonable accommodations that would mitigate or eliminate the risk?
- Have personnel with disabilities been included in the emergency planning process?
- Has every effort been made to ensure that personnel with disabilities have not been segregated as part of the emergency planning process?
- Have employees with disabilities made requests for reasonable accommodations in the event of an emergency? Have those accommodations been built into the EAP? If not, can the organization show clearly that the requested accommodation would pose an undue hardship?

Working with First Responders

It is important to coordinate emergency planning and the resulting EPA with first-responder agencies and organizations. These personnel and those who carry out the EAP should coordinate closely on all matters pertaining to implementation of the plan.

- Have first responders been made aware of any special issues relating to personnel with disabilities?
- Does the organization have a policy regarding evacuation built into its plan? Are all stakeholders aware of the policy and its ramifications? Has anyone expressed opposition to the policy? Has the opposition been properly addressed?
- Have first responders been included in all steps of the emergency planning process? If not, how will they be made aware of all elements of the EAP?

Implementing an Elevator Policy

Should the organization use its elevators during an emergency or close them down? This can become a critical issue when the needs of personnel with disabilities are considered.

- Does the organization have a policy concerning elevator use during emergencies? Has the policy been built into the EAP?
- Were first responders consulted during the development of the **elevator policy**?
- Under what circumstances may elevators be used during an emergency? Who may operate the elevators under these conditions?
- Who gets priority in the use of elevators during an emergency?
- How are personnel, including those with disabilities, evacuated in the event that elevators are inoperable during an emergency?

Developing Emergency Notification Strategies

Developing the EAP is just one step in an ongoing process. Once the plan is developed, all stakeholders must be notified of its contents and what they mean. The following questions will help ensure appropriate notification:

- What steps have been built into the plan for notifying stakeholders in the event of an emergency and for ensuring that personnel with disabilities have access to the same information that any other stakeholder has?
- Have a variety of notification or communication methods been built into the EAP?
- Do notification or communication methods account for personnel who are away from their desks or the office?
- Do notification strategies take into account the presence of visitors?

Practicing and Maintaining the EAP

The EAP should be viewed as a living document that is updated and maintained continually. In addition, all aspects of the plan should be practiced periodically to ensure that all parties know how to carry out their responsibilities. The following questions will help ensure that the plan is properly maintained and practiced:

- Does the organization have a policy for the regular practice and maintenance of the EAP?
- Does the organization comply with the policy and even exceed compliance requirements?

- Have first responders been involved in all practice drills and have they been consulted to ensure that all applicable equipment is appropriate and in proper working order?
- Have drills been varied in terms of time of day and type of drill?
- Have various unforeseen problems, roadblocks, and inhibitors been built into the drills so that emergency personnel have opportunities to practice improvising in a realistic way?
- Are personnel with disabilities included in all drills?
- Does the organization have a policy built into its EAP for dealing with employees or visitors who might need to leave the facility in the middle of an emergency drill?

EVACUATION PLANNING

The OSHA standard for evacuation planning is 29 CFR 1910.38. This standard requires a written plan for evacuating the facility in the event of an emergency. Critical elements of the plan are marking of exit routes, communications, outside assembly, and training.

Marking of Exit Routes

Clearly identified and marked routes of egress are critical during a time of crisis (fire, natural disaster, terrorist attack, etc.). To ensure that routes of egress and all related evacuation response items are clearly marked, safety professionals should answer the following questions about the facilities for which they are responsible:

1. Are all exit, emergency exit, and nonexit doors clearly identified and marked?
2. Are there up-to-date evacuation route maps mounted at strategic locations throughout the facility?
3. Are all egress route aisles, hallways, and stairs marked clearly and can the markings be seen in the event of darkness?
4. Are there low-level markings posted strategically throughout the facility that can be viewed in the event that smoke fills the facility?
5. Are all items of firefighting equipment clearly marked with directional signs so they can be easily located in the event of an emergency?
6. Is all emergency first-aid equipment clearly marked with directional signs so that it can be easily located in the event of an emergency?
7. Are all electrical, chemical, and physical hazards identified and clearly marked?
8. Are all physical obstructions clearly outlined?
9. Are all critical shutdown procedures and equipment identified and clearly marked?
10. Are the handrails, treads, and risers on all stairs clearly marked?
11. In the case of multicultural workforce settings, are all signs and markings provided in a bilingual or pictogram format?[2]

Ensuring that all signs, pictograms, and other markings relating to facility evacuation are visible during power outages is critical. Battery backup systems are one approach that is widely used. However, there are environments in which even the smallest spark from a battery might set off an explosion. For this reason, some facilities find the use of photoluminescent signs a better alternative.

Photoluminescent signs and markings absorb normal light energy from their surroundings and then release this energy in the form of light during periods of darkness. Some of the better photoluminescent materials will give off light for as long as 24 hours and have a maintenance-free life expectancy of up to 25 years.

Communication and Alarm Procedures

People are so accustomed to false alarms in their lives that when the *real thing* occurs, it can be difficult to convince them that it's not just another drill. In addition, people tend to trust what they can see, smell, and hear. Consequently, if they cannot physically sense an emergency, they tend to ignore the warning. The communication component of a facility's evacuation plan should include procedures for early detection of a problem, procedures for reporting an emergency, procedures for initiating an evacuation, and procedures for providing the necessary information to employees who are being evacuated.[3]

Notifying employees of the emergency is the function of the facility's alarm system. However, just pulling the alarm switch is not sufficient communication. Once the alarm has been given, verbal instructions should be broadcast so that people know that the alarm is real (not just another drill) as well as specific actions they should take immediately. There must also be procedures for informing evacuated employees that the emergency is over and they can return to their work.

External communication procedures are also important. All employees should know how to notify outside authorities of the emergency. With the advent of 911 service, this problem has been simplified. However, do not assume that all employees will remember the number when under the stress of an emergency, or that they will remember to dial "9" or some other code to gain access to an outside line. This problem can be solved by placing clearly marked signs above or near telephones containing such messages as "In an emergency dial 911" or "In an emergency dial 9911" (if it is necessary to first access an outside line).

Outside Assembly

The company's evacuation plan should include an assembly area to which employees go once evacuated.[4] This area should be well known by all employees, and employees should understand that it is critical to assemble there so that a headcount can be taken. In addition, there should be a backup assembly area known to all employees so they know where to go if the primary assembly area has been rendered inaccessible or hazardous.

Part of the evacuation plan relating to assembly areas must be devoted to transient personnel—nonemployees such as vendors, visitors, and contractors. How will they know where to assemble? Who will check to see that they have been notified of the emergency? These issues should be addressed in the evacuation plan.

Training

Training for evacuations is a critical element of the evacuation plan.[5] Developing a plan and then letting it just sit on the shelf gathering dust is a formula for disaster. Everything contained in the plan that requires action or knowledge on the part of employees should be part of the required evacuation training. Training should be provided when employees are first hired, and retraining should be provided periodically as various elements of the plan are updated.

Drills should be a major part of the training provided for employees. The old sports adage that says "What you do in practice you will do in the game" applies here. Make sure employees have practiced their emergency duties before an emergency occurs.

CUSTOMIZING PLANS TO MEET LOCAL NEEDS

Emergency plans must be **location-specific**. General plans developed centrally and used at all plant locations have limited effectiveness. The following rules of thumb can be used to ensure that EAPs are location-specific:

1. **A map of the plant.** A map of the specific plant helps localize an EAP. The map should include the locations of exits, access points, evacuation routes, alarms, emergency equipment, a central control or command center, first-aid kits, emergency shutdown buttons, and any other important elements of the EAP.
2. **Chain of command.** An organization chart illustrating the **chain of command**— who is responsible for what and who reports to whom—also helps localize an EAP. The chart should contain the names and telephone numbers (internal and external) of everyone involved in responding to an emergency. It is critical to keep the organization chart up-to-date as personnel changes occur. It is also important to have a designated backup person shown for every position on the chart.
3. **Coordination information.** All telephone numbers and contact names of people in agencies with which the company coordinates emergency activities should be listed. Periodic contact should be maintained with all these people so that the EAP can be updated as personnel changes occur.
4. **Local training.** All training should be geared toward the types of emergencies that may occur in the plant. In addition, practice drills should take place on-site and in the specific locations where emergencies are most likely to happen.

EMERGENCY RESPONSE

An **emergency response team (ERT)** is a team of personnel designated to ensure and facilitate a proper response to an emergency. The ERT is typically composed of representatives from several different departments such as maintenance, security, safety and health, production and processing, and medical. The actual composition of the team depends on the size and type of company. The ERT should be contained in the assignments and responsibilities section of the EAP.

Not all ERTs are company-based. Communities also have ERTs for responding to emergencies that occur outside of a company environment. Such teams should be included in a company's EAP in the coordinating organizations section. This is especially important for companies that use hazardous materials.

Most dangerous spills occur on the road, and the greatest number of hazardous materials (hazmat) incidents occurs on the manufacturer's loading dock. More than 80 percent of chemical releases are caused by errors in loading and unloading procedures.[6] Community-based teams typically include members of the police and fire departments who have had special training in handling hazmat emergencies.

Another approach to ERTs is the **emergency response network (ERN)**. An ERN is a network of ERTs that covers a designated geographical area and is typically responsible for a specific type of emergency.

Whether the ERT is a local company team or a network of teams covering a geographical region, it should be included in the EAP. In-house teams are included in the assignments and responsibilities section. Community-based teams and networks are included in the coordinating organizations section.

COMPUTERS AND EMERGENCY RESPONSE

Advances in chemical technology have made responding to certain types of emergencies particularly complicated.

Fortunately, the complications brought by technology can be simplified by technology. Expert computer systems especially programmed for use in emergency situations can help meet the challenge of responding to a mixed-chemical emergency or any other type of emergency involving multiple hazards.

An **expert system** is a computer programmed to solve problems. Such systems rely on a database of knowledge about a very particular area, an understanding of the problems addressed in that area, and expertise in solving problems in that area. Expert systems use suppositional rules called **heuristics** to make decisions. Human thought processes work in a similar manner. For example, if our senses provide the brain with input suggesting the stovetop is hot, the decision is made not to touch it. In an if-then format, this may read as follows:

IF hot, THEN do not touch.

This similarity to human thought processes is why the science on which expert systems are based is known as *artificial intelligence*.

A modern expert system used for responding to chemical emergencies provides the types of information needed to respond properly:

- Personal protective equipment
- Methods to be used in cleaning up the toxic release
- Decontamination procedures
- Estimation of the likelihood that employees or the community will be exposed to hazard
- Reactions that may result from interaction of chemicals
- Combustibility of chemicals and other materials on hand
- Evacuation information
- Impact of different weather conditions on the situation
- Recommended first-aid procedures
- Community notification procedures
- Follow-up procedures

Expert systems can be user-friendly so that computer novices have no difficulty interacting with them. The advantages of expert systems are that they do not panic or get confused, they consider every possible solution in milliseconds, they are not biased, they do not become fatigued, and they are detailed.

DEALING WITH THE PSYCHOLOGICAL TRAUMA OF EMERGENCIES

In addition to the physical injuries and property damage that can occur in emergencies, modern safety and health professionals must also be prepared to deal with potential psychological damage. Psychological trauma among employees involved in workplace disasters is as common as it is among combat veterans.

Trauma is psychological stress. It occurs as the result of an event, typically a disaster or some kind of emergency, so shocking that it impairs a person's sense of security or well-being. **Traumatic events** are typically unexpected and shocking, and they involve the reality or threat of death.

Dealing with Emergency-Related Trauma
The typical approach to an emergency can be described as follows: control it, take care of the injured, clean up the mess, and get back to work. Often, the psychological aspect is ignored. This leaves witnesses and other coworkers to deal on their own with the trauma they've experienced. It is important to respond to trauma quickly, within 24 hours if possible and within 72 hours in all cases. The purpose of the response is to help employees get back to normal by enabling them to handle what they have experienced. This is best accomplished by a team of people who have had special training. Such a team is typically called the **trauma response team (TRT)**.

Trauma Response Team
A company's TRT may consist of safety and health personnel who have undergone special training or fully credentialed counseling personnel, depending on the size of the company. In any case, the TRT should be included in the assignments and responsibilities section of the EAP.

The job of the TRT is to intervene as early as possible, help employees acknowledge what they have experienced, and give them opportunities to express how they feel about it to people who are qualified to help. The *qualified to help* aspect is very important. TRT members who are not counselors or mental health professionals should never attempt to provide care that they are not qualified to offer. Part of the trauma training that safety and health professionals receive involves recognizing the symptoms of employees who need professional care and referring them to qualified care providers.

In working with employees who need to deal with what they have experienced, but are not so traumatized as to require referral for outside professional care, a group approach is best.

The group approach shows employees they are not alone and provides comfort in numbers.

Convincing Companies to Respond
Modern safety and health professionals may find themselves having to convince higher management of the need to have a TRT. Some corporate officials may not believe that trauma even exists. Others may acknowledge its presence but view trauma as a personal problem that employees should handle on their own.

In reality, psychological trauma that is left untreated can manifest itself as **post-traumatic stress disorder**, the same syndrome experienced by some combat veterans.

In today's competitive marketplace, companies need all their employees operating at peak performance levels. Employees experiencing trauma-related disorders will not be at their best. Safety and health professionals should use this rationale when it is necessary to convince higher management of the need to provide a company-sponsored trauma response team.

RECOVERING FROM DISASTERS

Many organizations put a great deal of effort into planning for disaster response, including emergency evacuation. But what about after the disaster?

A comprehensive disaster recovery plan is a must. It should have at least the following components: recovery coordinator, recovery team, recovery analysis and planning, damage assessment and salvage operations, recovery communications, and employee support and assistance. The overall goal of a disaster recovery plan is to get an organization fully operational again as quickly as possible.

Recovery Coordinator

There must be one person who has ultimate responsibility and authority for disaster recovery. This person must have both the ability and the authority to take command of the situation, assess the recovery needs, delegate specific responsibilities, approve the necessary resources, interact with outside agencies, and activate the organization's overall response.

Recovery Team

The recovery team consists of key personnel to whom the disaster coordinator can delegate specific responsibilities. These responsibilities include facility management, security, human resources, environmental protection (if applicable), communications, and the various personnel needed to restart operations.

Recovery Analysis and Planning

This phase involves assessing the impact of the disaster on the organization and establishing both short- and long-term recovery goals. The more recovery analysis and planning that can be done, the better. One of the ways to do this is to consider various predictable scenarios and plan for them. This is the business equivalent of the war-gaming activities that take place in the military.

Damage Assessment and Salvage Operations

This component of the plan has two elements: preparedness and recovery. The preparedness element should include the following information: (1) a comprehensive inventory of all property at the facility in question; (2) a checklist of the items on the inventory that are essential for maintaining the facility; (3) a list of all personnel who will aid in the recovery (make sure to have fully trained and qualified backup personnel in case a primary player is not available or is injured during the emergency); (4) a list of all vendors, contractors, and so on whose assistance will be needed during the damage assessment and salvage phase of recovery; (5) a worksheet that can be used to document all actions taken during recovery operations; and (6) procedures for quickly establishing a remote operational site.

The recovery element should include procedures for securing workspace for the recovery team and coordinator; identifying areas of the facility that must be accessible; maintaining security at the facility against looting and vandalism; analyzing and inspecting damage to the facility and reporting it to the recovery coordinator; assessing the extent of damage to goods, supplies, and equipment; photographing and videotaping damage to the facility; taking appropriate action to prevent additional damage to the facility; repairing, restoring, and resetting fire detection and suppression equipment; and investigating accidents.

Recovery Communications

Communication is one of the most important considerations in disaster recovery. This component of the plan should deal with both who is to be notified and how that is to take place. The "how" aspects concern backup procedures for telephone service, e-mail, and so forth. Will cell phones be used? Will radio stations be part of the mix? Will "walkie-talkie" type radios be used for communicating on-site? The following list contains the types of entities who might have to be contacted as part of the disaster recovery operation:

- Customers
- Vendors and suppliers
- Insurance representatives
- Employees' families
- Appropriate authorities
- Media outlets (radio and television stations and newspapers)

Employee Support and Assistance

After a disaster, employees are likely to need various types of assistance, including financial, medical, and psychological. The following steps for developing this component of the disaster recovery plan are recommended:

1. Determine postdisaster work schedules and provide them to employees. Include overtime work if it will be necessary, and make sure employees know that flexibility in scheduling work hours will be important until the recovery is complete.
2. Plan for the whole range of employee-assistance services that might be needed, including medical, transportation, financial, shelter, food, water, clothing, and psychological services (trauma, shock, and stress counseling).
3. Plan for the provision of grief counseling. The best way to handle this is to assign grief counseling to the company's employee assistance program provider.
4. Plan for the possible need to relocate the facility as part of disaster recovery.
5. Plan to give employees opportunities to participate in personal actions taken on behalf of fatally injured employees and their families. Work for employee consensus before deciding what to do for these families.
6. Plan to fully inform all employees about what happened, why, how the company is responding in the short term, and how it will respond in the long term. Be sure to build in ways to let employees know the company cares about them and will do everything possible to protect their safety.[7]

TERRORISM IN THE WORKPLACE

This section describes the roles that employers and their safety and health personnel can play in preparing for, preventing, and responding to terrorist attacks in the workplace.

Role of the Employer

There is no question that the threat of terrorist attacks has become an ever-present reality in today's workplace. Because this is the case, employers clearly have a role to play in preparing for terrorist attacks, taking all prudent precautions to prevent them, and in responding properly should an attack occur. Because terrorism threatens the safety and health of employees, it is more than just a security issue; it is also an occupational safety and health issue.

The roles of employers and safety and health professionals relating to terrorism in the workplace are summarized below:

Run a safe and caring operation. Employees watch the nightly news and read their morning newspapers. They know what is going on in their world. As a result, many are discomfited by the possibility that they or their workplace might become the target of a terrorist attack. Consequently, the first responsibility of the employer is to run a safe operation in which employees know their safety is a high priority. Many of the engineering, administrative, training, and enforcement actions taken to make the workplace safe from occupational hazards will also help mitigate the threat of terrorism.

Listen to employees. Employees are concerned about the threat of terrorism, and they have a right to be. Employers should take the concerns of employees seriously, and deal with them. Answer questions, communicate openly and frequently, and refer employees who need professional help to the employee assistance program.

Train employees. Security and safety procedures do little good unless employees know what they are and how to use them. In addition, personnel in certain positions need to have specialized knowledge relating to terrorism. For example, mailroom personnel need to be trained in how to screen incoming mail for biohazards and explosives.

Communicate. Talk with employees openly and frequently. Let them know what the company knows. It is better for employees to hear news from the company than to receive it in the form of third-party gossip and rumors. Before giving out information, however, it is a good idea to verify it. Check with local authorities or go online to one of the following websites: snopes .com or urbanlegends.com.

Know your personnel. Institute background checks as part of the hiring process. Make sure that supervisors get to know their direct reports well enough to sense when something is wrong with one of them. If inconsistencies in normal behavioral patterns occur, address them right away.

Empower personnel. Empower employees to back away from a situation that does not feel right or that makes them uncomfortable. Be flexible in allowing them to have time off for family activities. This can be especially important following particularly busy times for employees who have worked longer than normal hours.

Harden the site against external threats and restrict access. Insulate the workplace from negative outside influences, control who has access to the workplace, and take all necessary steps to reduce the exposure of employees to potential threats. Call in security experts if necessary to help develop and implement the necessary controls.

Remove any barriers to clear visibility around the facility. The better employees can see around them, the less likely it will be that a terrorist will be able to pull off a surprise attack. The trees, shrubs, and bushes that make the perimeter of the facility and parking lot so attractive can be used by unauthorized personnel to hide while attempting to gain access.

Have and enforce parking and delivery regulations. Arrange parking spaces so that no car is closer to a building than 100 feet. This will lessen the likelihood of damage from a car bomb. In addition, have strict delivery procedures so that terrorists posing as delivery personnel do not gain access through this means.

Make sure that visitors can be screened from a distance. Arrange the facility so that visitors are channeled into a specific area, an area in which they can be viewed by company personnel from a distance. This will lessen the likelihood of a terrorist gaining access by overpowering or killing access-control personnel.

Keep all unstaffed entrance doors locked and alarmed. Employees need to be able to get out of the building through any exit door, but access into the building should be channeled through doors staffed by access-control personnel.

Make air intakes and other utilities inaccessible to all but designated maintenance personnel. Releasing toxic material into the air intake is one way terrorists could harm the highest possible number of people. The likelihood of this happening can be decreased by locating air intakes in inaccessible locations.

Prevent access to roofs and upper stories. Terrorists who cannot gain access on the ground floor might simply gain entry through doors and other openings on the roof. Consequently, it is important to keep roof doors locked from the outside and alarmed. They should open from the inside, but doing so should trigger an alarm. It is also important to establish control procedures for emergency escape routes from the roof and upper stories so that these avenues are not used by terrorists trying to gain access into the building.

Secure trash containers. The grounds should be kept free of debris and clutter. Further, trash containers should be secured either by keeping them inside the building's wall or, if kept outside, at a distance from the building.

Ensure that employees, contractors, and visitors wear badges. Establish a system in which all employees, contractors, and visitors must wear badges in order to gain access to the facility. Require identification of all visitors and an internal sponsor before providing them with badges. Have visitors sign in and out on every visit.

Have an emergency response plan and practice it periodically. Plan for all predictable exigencies and practice the various components of the plan on a regular basis.

Be cautious of information placed on your company's website. Make sure that your company's website does not contain information that can be used by terrorists such as detailed maps, floor plans, or descriptions of hazardous materials stored on-site.

Keep up-to-date with the latest safety and security strategies. Crime Prevention through Environmental Design (CPTED) should become part of the knowledge base of safety and health professionals. For the latest information concerning CPTED, visit the following website periodically: cops.usdoj.gov and search on Crime Prevention through Environmental Design.

Protect the integrity of your facility's key system. Terrorists can use any key from your facility to make their own master key if they know how, and many do. To make matters worse, they share information about how to make master keys via the Internet. The following procedures will help protect the integrity of your facility's key system: (a) restrict access to keys (including restroom keys); (b) consider not using master keys; and (c) switch from a key system to an electronic keycard or fingerprint recognition system.[8]

Securing Hazardous Materials

One of the tactics of terrorists is to convert hazardous materials used in the workplace into weapons of mass destruction. Consequently, it is important for facilities that produce, use, or store any type of hazardous materials to develop, implement, and enforce a security program that denies terrorists access to these materials.

A hazmat security plan should have two broad components: *personnel security* and *physical security.* The fundamental elements of the personnel component of the plan are to (1) determine who should be granted access to the materials; (2) conduct comprehensive background checks on all individuals who require access; (3) submit employees who require access to psychological screening to ensure stability; and (4) require identification badges with photographs or fingerprints of those authorized access.

The physical security component consists of measures taken to prevent or control access to the hazardous materials in question. Security practices in this component should be integrated and layered by using a combination of measures, including fences, lights, electronic alarm systems, guards, reaction forces, and the *two-person rule* that requires two employees to be present in order to gain access to the hazardous materials. A good source of help concerning how to develop a plan for keeping hazardous materials out of the hands of terrorists is as follows: cdc.gov.

REPORTING SUSPICIOUS PEOPLE AND BEHAVIORS

Employees should be trained and encouraged to identify and report suspicious looking people or activities. Workplace violence can be perpetrated by disgruntled employees, suppliers, customers, or even terrorists. Consequently, it is important for employees to be vigilant in watching for the following types of behaviors in others:

- Threats to kill or harm another employee
- Stalking of an employee
- Bullying of employees
- Abusive or aggressive behavior against employees
- Threats to damage property, equipment, or products of the employer
- Defacing of the employer's property

- Unusual behavior (watch for changes in employee behavior)
- Bizarre behavior

In addition to the types of behaviors listed above, employees should be trained and encouraged to identify and report the following types of suspicious activities:

- People drawing maps or making sketches of facilities, particularly entrances, exits, parking garages, etc.
- People taking notes about the movements and corresponding times of personnel in an organization
- People taking photographs or videotapes of facilities
- People monitoring facilities that are frequented by large numbers of people such as power plants, schools, shopping centers/malls, sports arenas, religious institutions, airports, and government buildings
- People observing a facility using binoculars and other technological aids
- People attempting to acquire explosives, weapons, detonators, timers, body armor, and so on
- People attempting to purchase large amounts of high nitrate fertilizers or other chemicals
- People putting weapons or explosives in cars or trucks
- People attempting to acquire official gear such as uniforms, badges, and ID cards
- People who display unusual characteristics or behaviors such as nervousness, sweating, clothing that is too heavy for the weather conditions, an unusual focus such as a blank stare or a particularly determined expression, wearing or holding a backpack, and anyone who leaves a bag, box, or backpack unattended and walks off

Suspicious people, activities, or behaviors should be reported immediately to a person in a position of authority. Employees should be taught to err on the side of safety and that a terrorism incident should not be allowed to occur because they were reluctant to report suspicious behavior.

ACTIVE SHOOTER RESPONSE

Active shooters are becoming more and more common in the workplace. Consequently, it is important for organizations to train their personnel on how to respond in an active-shooter situation. There are three basic responses to a terrorist who is active shooter that employees can take:

- ***Evacuate.*** With this response, employees must make a determination if there is a safe route of evacuation that will separate them from the shooter. When this is the case, employees should make use of it as quickly as possible. Delay could be deadly. This means that belongings may have to be left behind. Employees should be trained to know the evacuation routes in their facilities, how to access them quickly, and that going back into the danger zone to retrieve personal belongings could be a fatal mistake. Once safely away, call 911.
- ***Hide.*** If evacuation is not feasible, employees should be taught to hide. A good hiding place is one that is out of the view of the shooter, provides cover from shots fired, and does not create a trap with no means of exit (e.g., a closet). When this option is taken it is important that employees silence their cell

phones, lock doors if possible, and remain both quiet and still. If the shooter is far enough away, call 911. If not, remain silent.

- *Mount a defense.* When no other option is available, mount a defense. Do not just sit back and allow yourself to be murdered. Attack the shooter with any improvised weapon you can find. If possible, attack as a group with all trapped personnel joining in.

RESUMING BUSINESS AFTER A DISASTER

After a disaster in the workplace—regardless of whether the source of the disaster was natural, accidental, or terrorist related—a comprehensive hazard assessment should be completed before business is resumed. Before resuming business, an organization should consider the following factors:[9]

- *Structural integrity.* Has the structural integrity of the building been checked by competent engineering professionals to ensure that it is safe to enter?
- *Utility checks.* Have all utilities—gas, electricity, water, sewer—been checked by competent professionals to ensure that there are no leaks, cracks, loose wires, and so on? Have the appropriate utility companies given their approval for reopening? Remember that if reopening involves the use of electric generators, these devices should not be used inside the building because they can create a carbon monoxide hazard.
- *Cleanup protection.* Make sure that cleanup crews are properly protected from any hazardous materials or conditions that might have been created by the disaster. Make sure they properly use the appropriate personal protective gear, and comply with all applicable safety and health regulations and procedures as they clean the facility.
- *Health and sanitation.* Kitchens, bathrooms, and any area in which food or potentially hazardous or toxic substances are stored should be checked by competent professionals and thoroughly cleaned to prevent the exposure of employees and customers to hazardous conditions.
- *Air quality.* Make sure the air quality in the facility is tested by competent professionals before allowing personnel to enter the building. Certain types of disasters such as hurricanes, for example, might have caused the proliferation of mold and mildew. The air should be checked for any potentially hazardous biological or chemical agents that could be harmful to humans.
- *Ventilation.* Make sure that all types of ductwork and ventilation have been checked for the presence of potentially harmful biological and chemical agents as well as for dust and debris that might impede airflow. Once it appears that ventilations systems are clean, have the air-conditioning and heating systems started up and all ventilation systems checked again before allowing personnel back into the building. When the air-conditioning and heating systems are restarted, blow cold air through them, even in winter. This will help prevent the growth of mold in the ventilation ducts.
- *Walls, ceilings, and floors.* Check walls and ceilings to ensure that no materials are in danger of falling off—inside for occupants and outside for pedestrians. Check floors for any hazards that might contribute to slipping and falling.

- **Safety equipment.** Check all fire extinguishers, all types of alarms, and any other safety equipment to determine whether it has been damaged. Make sure that all safety equipment is in proper working order before allowing personnel to reenter the facility.
- **Lighting.** Ensure that all illumination devices are in proper working order and that all personnel have the required amount of illumination to do their jobs. Employees should not return to work unless the necessary illumination is available and working.
- **Hazardous waste removal.** Any type of potentially hazardous material that is left lying around after the disaster should be collected and properly disposed of. Broken glass, debris, litter, and sharp-edged material should be removed before employees are allowed to return to work.
- **Machines and equipment.** All machines and equipment should be checked carefully before they are reenergized. All electrical, gas, hydraulic, fill, drain, and plumbing lines should be checked for leaks and proper connections before the machines are energized for use.
- **Furniture.** Check all furniture to ensure its structural integrity. Make sure that fasteners, braces, and supports have not been damaged during the emergency or that furniture has not become unstable due to water damage.

Key Terms and Concepts

- Adjusting
- Chain of command
- Elevator policy
- Emergency action plan (EAP)
- Emergency coordinator
- Emergency notification
- Emergency planning
- Emergency Planning and Community Right-to-Know Act (EPCRA)
- Emergency response network (ERN)
- Emergency response plan
- Emergency response team (ERT)
- Evaluating
- Expert system
- First-aid training
- Heuristics
- Information requirements
- Local emergency planning committees (LEPCs)
- Location-specific
- Order of response
- OSHA standards
- Planning
- Post-traumatic stress disorder
- Practicing
- Shelter-in-place (SIP)
- State emergency response commissions (SERCs)
- Toxic chemical release reporting
- Trauma
- Trauma response team (TRT)
- Traumatic events

Review Questions

1. Define the term *emergency*.
2. Explain the rationale for emergency preparation.
3. List and explain the four main components of the Emergency Planning and Community Right-to-Know Act.
4. Describe how a company's emergency response effort should be coordinated.
5. How do OSHA standards relate to emergency preparation?
6. Explain how you would provide first-aid training if you were responsible for setting up a program at your company.
7. Besides training, what other first-aid preparation should a company take?

8. What are the critical elements of OSHA's standard for evacuation planning?
9. Describe the essential components of an EAP and explain how to build the needs of personnel with disabilities into the plan.
10. How can a company localize its EAP?
11. Define the following emergency response concepts: ERT, ERN, and TRT.
12. What is an expert system? How can one be used in responding to an emergency?
13. What is trauma?
14. Why should a company include trauma response in its EAP?
15. Describe how a company may respond to the trauma resulting from a workplace emergency.
16. What elements should a disaster recovery plan contain?
17. How can employers prepare for the threat of terrorism?
18. Explain how to prepare for and respond to an active-shooter incident.
19. Explain the precautions that should be taken before resuming business after a disaster.

Endnotes

1. United States Department of Labor, Office of Disability Employment Policy, *Preparing the Workplace for Everyone: Accounting for the Needs of People with Disabilities.* Retrieved from dol.gov/odep/pubs/ep/preparing/workplace_final.pdf on July 10, 2017.
2. 29 CFR 1910.38.
3. Ibid.
4. Ibid.
5. Ibid.
6. Ibid.
7. U.S. Office of Personnel Management, *Handling Traumatic Events.* Retrieved from opm.gov/policy-data-oversight/worklife/reference-materials/traumaticevents.pdf on July 12, 2017.
8. Ibid.
9. American Society of Safety Engineers, "ASSE Offers Business Resumption Safety Checklist," *Occupational Health & Safety Online.* Retrieved from ohsonline.com/stevens/ohspub.nsf/d3d5b4f938b22b6e8625670c006bc58/6600 on July 10, 2017.

CHAPTER

20

BLOODBORNE PATHOGENS AND BACTERIAL HAZARDS IN THE WORKPLACE

Learning Objectives

- List the symptoms of AIDS
- Explain the various components of a corporate AIDS policy
- Summarize the legal concerns relating to AIDS
- Explain how to counsel employees who are infected with AIDS
- Summarize strategies for protecting employees from AIDS in the workplace
- Summarize the hazards and prevention strategies relating to Hepatitis B (HBV) and Hepatitis C (HCV) in the workplace
- Summarize OSHA's standard (29 CFR 1910.1030) on occupational exposure to bloodborne pathogens
- Explain how to prevent and respond to needlestick injuries
- Summarize the prevention steps for dealing with MRSA in the workplace

Acquired immunodeficiency syndrome (AIDS) has become one of the most difficult issues that safety and health professionals are likely to face today. It is critical that they know how to deal properly and appropriately with this controversial disease. The major concerns of safety and health professionals with regard to AIDS are knowing the facts about AIDS, knowing the legal concerns associated with AIDS, knowing their role in AIDS education and related employee **counseling**, and knowing how to ease unfounded fears concerning the disease while simultaneously taking the appropriate steps to protect employees from infection. In addition to AIDS, the modern safety and health manager must be concerned with bloodborne pathogens, including **human immunodeficiency virus (HIV)** and **hepatitis B virus (HBV)**.

SYMPTOMS OF AIDS

AIDS and various related conditions are caused when humans become infected with HIV. This virus attacks the human immune system, rendering the body incapable of repelling disease-causing microorganisms. Symptoms of the onset of AIDS are as follows:

- Enlarged lymph nodes that persist
- Persistent fevers
- Involuntary weight loss
- Fatigue
- Diarrhea that does not respond to standard medications
- Purplish spots or blotches on the skin or in the mouth
- White, cheesy coating on the tongue
- Night sweats
- Forgetfulness

How AIDS Is Transmitted

HIV is transmitted in any of the following three ways: (1) **sexual contact**, (2) blood contact, and (3) mother-to-child during pregnancy or childbirth. Any act in which **body fluids** are exchanged can result in infection if either partner is infected. The following groups of people are at the highest level of risk with regard to AIDS: (1) homosexual men who do not take appropriate precautions; (2) **IV drug users** who share needles; (3) people with a history of multiple **blood transfusions** or blood-product transfusions, such as hemophiliacs; and (4) sexually promiscuous people who do not take appropriate precautions.

How AIDS Is *Not* Transmitted

There is a great deal of misunderstanding about how AIDS is transmitted. This can cause inordinate fear among fellow employees of HIV-positive workers. Safety and health professionals should know enough about AIDS transmission so that they can reduce employees' fears about being infected through casual contact with an HIV-positive person.

The following clarifications concerning how AIDS is *not* transmitted are provided:

AIDS is a blood-borne, primarily sexually transmitted disease. It is not spread by casual social contact in schools, workplaces, public washrooms, or restaurants. It is not spread via handshakes, social kissing, coughs, sneezes, drinking fountains, swimming pools, toilet facilities, eating utensils, office equipment, or by being next to an infected person.

No cases of AIDS have been reported from food being either handled or served by an infected person in an eating establishment.

AIDS is not spread by giving blood. New needles and transfusion equipment are used for every donor.

AIDS is not spread by mosquitoes or other insects.

AIDS is not spread by sexual contact between uninfected individuals—whether homosexual or heterosexual—if an exclusive sexual relation has been maintained.[1]

AIDS IN THE WORKPLACE

The first step in dealing with AIDS at the company level is to develop a comprehensive **AIDS policy**. Safety and health professionals should be part of the team that drafts the initial policy and updates an existing policy. If a company has no AIDS policy, the

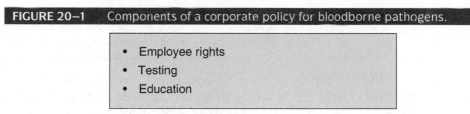

FIGURE 20–1 Components of a corporate policy for bloodborne pathogens.

- Employee rights
- Testing
- Education

safety and health professional should encourage the company to develop one. In all likelihood, most companies won't take much convincing.

AIDS is having a widely felt impact in the workplace, particularly on employers. Employers are feeling the impact of AIDS in increased insurance premiums and healthcare costs, time-on-the-job losses, decreased productivity, AIDS-related lawsuits, increased stress, and related problems that result from misconceptions about AIDS. The starting point for dealing with AIDS in the workplace is the development of a company policy that covers AIDS and other bloodborne pathogens. The policy should cover the following areas at a minimum: employee rights, testing, and education (see Figure 20–1).

Employee Rights

An AIDS policy should begin by spelling out the rights of employees who have tested positive for the disease. The following recommendations can be used for developing the **employee rights** aspects of an AIDS policy:

- Treat HIV-positive employees compassionately, allowing them to work as long as they are able to perform their jobs.
- Develop your company's AIDS policy and accompanying program before learning that an employee is **HIV positive**. This allows the company to act instead of having to react.
- Make reasonable allowances to accommodate the HIV-positive employee. The U.S. Supreme Court has recognized AIDS as a disabling condition. Consequently, reasonable allowances must include modified work schedules and special adaptations to the work environment.
- Ensure that HIV-positive employees have access to private health insurance that covers the effects of AIDS. Also, work with state and federal government insurance providers to gain their support in helping cover the costs of healthcare for HIV-positive employees.
- Include provisions for evaluating the work skills of employees to determine whether there has been any degradation of ability caused by the disease.[2]

Testing

According to the Centers for Disease Control and Prevention (CDC), there is no single test that can reliably diagnose AIDS.[3] However, there is a test that can detect antibodies produced in the blood to fight the virus that causes AIDS. The presence of these antibodies does not necessarily mean that a person has AIDS, as the CDC states:

Presence of HTLV-III antibodies [now called HIV antibodies] means that a person has been infected with that virus... The antibody test is used to screen donated blood and plasma and assist in preventing cases of AIDS resulting from blood transfusions or use of blood products, such as Factor VIII, needed by patients with hemophilia. For people who think they may be infected and want to know their health status, the test is available through private physicians, most state or local health departments and at other sites. Anyone who tests positive should be considered potentially capable of spreading the virus to others.[4]

Whether a company can, or even should, require AIDS tests of employees or potential employees is widely debated. The issue is contentious and controversial. However, there is a growing body of support for mandatory testing. The legal and ethical concerns surrounding this issue are covered in the next section. The testing component of a company's AIDS policy should take these concerns into account.

Education

The general public is becoming more sophisticated about AIDS. People are beginning to learn how AIDS is transmitted. However, research into the causes, diagnosis, treatment, and prevention of this disease is ongoing. The body of knowledge changes continually. Consequently, it is important to have an ongoing education program to keep employees up-to-date and knowledgeable.

Once a comprehensive AIDS policy has been developed and shared with all employees, a company has taken the appropriate and rational approach for dealing with this deadly and controversial disease. If an employer has not yet taken this critical step, safety and health professionals should encourage such action immediately. It is likely that most companies either employ now or will employ in the future HIV-positive employees.

Clearly, one of the major challenges facing safety and health professionals is convincing their employers to develop an AIDS policy.

LEGAL CONCERNS

There are legal considerations relating to AIDS in the workplace with which safety and health professionals should be familiar. These grow out of several pieces of federal legislation, including the Rehabilitation Act, the Occupational Safety and Health Act, and the Employee Retirement Income System Act.

The **Rehabilitation Act** was enacted in 1973 to give protection to people with disabilities (then referred to as handicaps), including workers. Section 504 of the act makes discrimination on the basis of a disability unlawful. Any agency, organization, or company that receives federal funding falls within the purview of the act. Such entities may not discriminate against individuals who have disabilities but are **otherwise qualified**. Through various court actions, this concept has been well defined. A person with a disability is "otherwise qualified" when he or she can perform what the courts have described as the **essential functions** of the job.

When the disability that a worker has is a contagious disease such as AIDS, it must be shown that there is no significant risk of the disease being transmitted in the workplace. If there is a significant risk, the infected worker is not considered "otherwise qualified." Employers and the courts must make these determinations on a case-by-case basis.

Another concept associated with the Rehabilitation Act is the concept of **reasonable accommodation**. In determining whether a worker with a disability can perform the essential functions of a job, employers are required to make reasonable accommodations to help the worker. This concept applies to workers with any type of disabling condition, including a communicable disease such as AIDS. What constitutes "reasonable accommodation," just as what constitutes "otherwise qualified," must be determined on a case-by-case basis.

The concepts growing out of the Rehabilitation Act give the supervisor added importance when dealing with AIDS-infected employees. The supervisor's knowledge of the various jobs in his or her unit is essential in helping company officials make an otherwise qualified decision. The supervisor's knowledge that AIDS is transmitted only by exchange of body fluids coupled with his or her knowledge of the job tasks in question is helpful in determining the likelihood that AIDS may be transmitted to other employees. Finally, the supervisor's knowledge of the job tasks in question is essential in determining what constitutes reasonable accommodation and what the actual accommodations should be. Therefore, it is critical that safety and health professionals work closely with supervisors and educate them in dealing with AIDS in the workplace.

In arriving at what constitutes reasonable accommodation, employers are not required to make fundamental changes that alter the nature of the job or result in undue costs or administrative burdens. Clearly, good judgment and a thorough knowledge of the job are required when attempting to make reasonable accommodations for an AIDS-infected employee. Safety and health professionals should involve supervisors in making such judgments.

The **Occupational Safety and Health Act (OSH Act)** requires that employers provide a safe workplace free of hazards. The act also prohibits employers from retaliating against employees who refuse to work in an environment they believe may be unhealthy (Section 654). This poses a special problem for employers of AIDS-infected employees. Other employees may attempt to use Section 654 of the OSH Act as the basis for refusing to work with such employees. For this reason, it is important that companies educate their employees about AIDS and how it is transmitted. If employees know how AIDS is transmitted, they will be less likely to exhibit an irrational fear of working with an infected colleague.

Even when a comprehensive AIDS education program is provided, employers should not automatically assume that an employee's fear of working with the infected individual is irrational. Employers have an obligation to treat each case individually. Does the complaining employee have a physical condition that puts him or her at greater risk of contracting AIDS than other employees? If so, that employee's fears may not be irrational. However, a fear of working with an AIDS-infected coworker is usually irrational, making it unlikely that Section 654 of the OSH Act could be used successfully as the basis of a refusal to work.

The **Employee Retirement Income Security Act (ERISA)** protects the benefits of employees by prohibiting actions taken against them based on their eligibility for benefits. This means that employers covered by ERISA cannot terminate an employee with AIDS or who is suspected of having AIDS as a way of avoiding expensive medical costs. With ERISA, it is irrelevant whether the employee's condition is considered a disability because the act applies to all employees regardless of condition.

Testing Issue

Perhaps the most contentious legal concern growing out of the AIDS controversy is the issue of testing.

FIGURE 20–2 Disease testing issues.

Health and safety professionals should be familiar with how the following factors might affect the issue of AIDS testing at their companies:

- Applicable state laws
- Applicable federal laws and regulations
- Case law from civil suits
- Company policy

Issues regarding testing for AIDS and other diseases with which safety and health professionals should be familiar are summarized as follows (see Figure 20–2):

1. **State laws.** Control of communicable diseases is typically considered to be the province of the individual state. In response to the AIDS epidemic, several states have passed legislation. Some states prohibit the use of preemployment AIDS tests to deny employment to infected individuals. Because of the differences among states with regard to AIDS-related legislation, safety and health professionals should familiarize themselves with the laws of the state in which their company is located.

2. **Federal laws and regulations.** The laws protecting an individual's right to privacy and due process apply to AIDS testing. These laws fall within the realm of constitutional law. They represent the primary federal contribution to the testing issue.

3. **Civil suits.** **Case law** serves the purpose of establishing precedents that can guide future decisions. One precedent-setting case was taken all the way to the Supreme Court (*School Board of Nassau County v. Arline*, 1987), where it was decided that an employer cannot discriminate against an employee who has a communicable disease.

4. **Company policy.** It was stated earlier that companies should have an AIDS policy that contains a testing component. This component should include at least the following: a strong rationale, procedures to be followed, employee groups to be tested, the use and dissemination of results, and the circumstances under which testing will be done.

On one side of the testing controversy are the issues of fairness, accuracy, and **confidentiality** or, in short, the rights of the individual. On the other side of the controversy are the issues of workplace safety and public health. Individual rights' proponents ask such questions as: What tests will be used? How do test results relate to the maintenance of workplace safety? How will test results be used, and who will see them? Workplace safety proponents ask such questions as: What is the danger of transmitting the disease to other employees? Can the safety of other workers be guaranteed?

The following guidelines for establishing testing programs that satisfy the concerns of both sides of the issue are recommended:

- The purpose of screening must be ethically acceptable.
- The means to be used in the screening program and the intended use of the information must be appropriate for accomplishing the purpose.

- High-quality laboratory services must be used.
- Individuals must be notified that the screening will take place.
- Individuals who are screened must have a right to be informed of the results.
- Sensitive and supportive counseling programs must be available before and after screening to interpret the results, whether they are positive or negative.
- The confidentiality of screened individuals must be protected.[5]

Facts about Testing for AIDS and Other Diseases

In addition to understanding the legal concerns associated with disease testing, safety and health professionals should also be familiar with the latest facts about AIDS tests and testing. A concerned employee may ask for recommendations concerning AIDS testing.

Ensuring the accuracy of an **HIV antibody test** (there is no such thing as an AIDS test) requires two different tests, one for initial screening and one for confirmation. The **screening test** currently used is the **enzyme-linked immunosorbent assay (ELISA) test**. The **confirmation test** is the **immunofluorescent assay (IFA)**, or the **Western Blot test**. The ELISA test is relatively accurate, but it is susceptible to both false positive and false negative results. A **false positive** test is one that shows the presence of HIV antibodies when no such antibodies exist. A **false negative** result is one that shows no HIV antibodies in people who actually are infected. A negative result indicates that no infection exists at the time of the test. A confirmed positive result indicates that HIV antibodies are present in the blood.[6]

The American College Health Association makes the following recommendations concerning the HIV antibody test:

- The test is not a test for AIDS, but a test for antibodies to HIV, the virus that can cause AIDS.
- Talk to a trained, experienced, sensitive counselor before deciding whether to be tested.
- If you decide to be tested, do so *only* at a center that provides both pre- and posttest counseling.
- If possible, use an *anonymous* testing center.
- Be sure that the testing center uses two ELISA tests and a Western Blot or IFA test to confirm a positive result.
- A positive test result is *not* a diagnosis of AIDS. It does mean you have HIV infection and that you should seek medical evaluation and early treatment.
- A positive test result *does* mean that you can infect others and that you should avoid risky or unsafe sexual contact and IV needle sharing.
- It can take six months (and—although rarely—sometimes even longer) after infection to develop antibodies, so the test result may not indicate whether you have been infected during that period.
- A negative test result *does not* mean that you are immune to HIV or AIDS, or that you cannot be infected in the future.[7]

Safety and health professionals need to share this type of information with employees who ask about AIDS tests and testing. Employees who need more detailed information should be referred to a healthcare professional.

AIDS EDUCATION

The public is becoming more knowledgeable about AIDS and how the disease is spread. However, this is a slow process, and AIDS is a complex and controversial disease. Unfortunately, many people still respond to the disease out of ignorance and inaccurate information. For this reason, it is imperative that a company's safety and health program include an **AIDS education program**.

A well-planned AIDS education program can serve several purposes: (1) It can give management the facts needed to develop policy and make informed decisions with regard to AIDS; (2) it can result in changes in behavior that will make employees less likely to contract or spread the disease; (3) it can prepare management and employees to respond appropriately when a worker falls victim to the disease; and (4) it can decrease the likelihood of legal problems resulting from an inappropriate response to an AIDS-related issue. Consequently, safety and health professionals should be prepared to participate in developing AIDS education programs.

Planning an AIDS Education Program

The first step in planning an AIDS education program is to decide its purpose. A statement of purpose for an AIDS education program should be a broad conceptual declaration that captures the company's reason for providing the education program. Following is an example:

> The purpose of this AIDS education program is to deal with the disease in a positive proactive manner that is in the best interests of the company and its employees.

The next step in the planning process involves developing goals that translate the statement of purpose into more specific terms. The goals should tell specifically what the AIDS education program will do. Sample goals are as follows:

- The program will change employee behaviors that may otherwise promote the spread of AIDS.
- The program will help the company's management team develop a rational, appropriate AIDS policy.
- The program will help managers make responsible decisions concerning AIDS issues.
- The program will help employees protect themselves from the transmission of AIDS.
- The program will alleviate the fears of employees concerning working with an AIDS-infected coworker.
- The program will help managers respond appropriately and humanely to the needs of AIDS-infected workers.

Once goals have been set, a program is developed to meet the goals. The various components of the program must be determined. These components may include confidential one-on-one counseling, referral, posters, a newsletter, classroom instruction, self-paced multimedia instruction, group discussion sessions, printed materials, or a number of other approaches. Figure 20–3 shows a suggested outline for a course on AIDS.

FIGURE 20–3   Outline for AIDS education course.

Statement of Purpose

The purpose of this course is to give employees the knowledge they need to deal with AIDS in a positive, proactive manner.

Major Topics

- What is AIDS?
- What causes AIDS?
- How is AIDS transmitted?
- Who is most likely to get AIDS?
- What are the symptoms of AIDS?
- How is AIDS diagnosed?
- Who should be tested for AIDS?
- Where can I get an AIDS test?
- How can I reduce my chances of contracting AIDS?
- How is AIDS treated?
- Can AIDS be prevented?
- What are common myths about AIDS?

COUNSELING INFECTED EMPLOYEES

The employee who learns that he or she has AIDS will be angry, frightened, and confused. Safety and health professionals who are faced with such an employee should proceed as follows:

- Listen.
- Maintain a nonjudgmental attitude.
- Make the employee aware of the company's policy on AIDS.
- Respond in accordance with company policy.

Listen as carefully as you would to an employee who comes to you with any problem. If you must ask a question for clarification, do so, but be objective, professional, and nonjudgmental. Make the employee aware of the company's policy on AIDS and respond in strict accordance with the policy.

The U.S. Public Health Service recommends the following steps for persons who have determined they are HIV positive.

- Seek regular medical evaluation and follow-up.
- Either avoid sexual activity or inform your prospective partner of your antibody test results and protect him or her from contact with your body fluids during sex. (Body fluids include blood, semen, urine, feces, saliva, and women's genital secretions.) Use a condom, and avoid practices that may injure body tissues (e.g., anal intercourse). Avoid oral-genital contact and open-mouthed intimate kissing.

- Inform your present and previous sex partners, and other persons with whom needles may have been shared, of their potential exposure to HTLV-III [HIV] and encourage them to seek counseling and antibody testing from their physician or at appropriate health clinics.
- Don't share toothbrushes, razors, or other items that could become contaminated with blood.
- If you use drugs, enroll in a drug treatment program. Needles and other drug equipment must never be shared.
- Don't donate blood, plasma, body organs, other body tissue, or sperm.
- Clean blood or other body fluid spills on household or other surfaces with freshly diluted household bleach—1 part bleach to 10 parts water. (Don't use bleach on wounds.)
- Inform your doctor, dentist, and eye doctor of your positive HTLV-III status so that proper precautions can be taken to protect you and others.
- Women with a positive antibody test should avoid pregnancy until more is known about the risks of transmitting HTLV-III from mother to infant.[8]

Safety and health professionals can pass along these recommendations to employees who have contracted the virus. Any information that is requested beyond this should be provided by a qualified professional. Safety and health professionals should be prepared to make appropriate referrals. The Public Health Service's AIDS hotline number is 1-800-342-AIDS.

Employee Assistance Programs

Company-sponsored **employee assistance programs (EAPs)** should have an AIDS component so that employees can seek confidential advice and counseling about the disease. EAPs may provide on-site services or contract for them through a private organization or agency. In either case, confidential counseling, referral, the provision of AIDS-related information, seminars, and other forms of assistance should be provided. From an employee assistance perspective, AIDS should be treated like other health problems such as stress, depression, and substance abuse.[9]

EASING EMPLOYEES' FEARS ABOUT AIDS

During his first inaugural address on March 4, 1933, in the depths of the Great Depression, President Franklin D. Roosevelt said, "So first let me assert my firm belief that the only thing we have to fear is fear itself."[10] This is a message that safety and health professionals should be prepared to spread among employees. Although the level of sophistication concerning AIDS is increasing, people tend to react to the disease from an emotional perspective. Typically, their fears are unfounded and are more likely to cause them problems than does AIDS.

Fear, panic, and even hysteria are all common reactions to AIDS. Fellow employees are likely to respond this way when they discover that a coworker has AIDS. Consequently, safety and health professionals need to know how to ease the fears and misconceptions associated with AIDS. The following strategies will help:

- Work with higher management to establish an AIDS education and awareness program that covers the following topics at a minimum: (1) how HIV is transmitted, (2) precautions that workers can take, and (3) concerns about AIDS testing.

- Conduct group round-table discussions that allow employees to express their concerns.
- Correct inaccuracies, rumors, and misinformation about AIDS as soon as they occur.[11]

PROTECTING EMPLOYEES FROM AIDS

Safety and health professionals should be familiar with the precautions that will protect employees from HIV infection on and off the job. The Occupational Safety and Health Administration's (OSHA) guidelines for preventing exposure to HIV infection identify three categories of work-related tasks: Categories I, II, and III. Jobs that fall into Category I involve routine exposure to blood, body fluids, or tissues that may be HIV infected. Category II jobs do not involve routine exposure to blood, body fluids, or tissues, but some aspects of the job may involve occasionally performing Category I tasks. Category III jobs do not normally involve exposure to blood, body fluids, or tissues.

Most industrial occupations fall into Category III, meaning there is very little risk of contracting AIDS on the job. However, regardless of the category of their job, employees should know how to protect themselves, and safety and health professionals should be prepared to tell them how.

The U.S. Public Health Service recommends the following precautions for reducing the chances of contracting AIDS:

- Abstain from sex or have a mutually monogamous marriage, or relationship, with an infection-free partner.
- Refrain from having sex with multiple partners or with a person who has multiple partners. The more partners you have, the greater the risk of infection.
- Avoid sex with a person who has AIDS or whom you think may be infected. However, if you choose not to take this recommendation, the next logical course is to take precautions against contact with the infected person's body fluids (blood, semen, urine, feces, saliva, and female genital secretions).
- Do not use **intravenous drugs** or, if you do, do not share needles.[12]

Safety and health professionals should make sure that all employees are aware of these common precautions. They should be included in the company-sponsored AIDS education program, available through the company's EAP, and posted conspicuously for employee reading.

CPR and AIDS

It is not uncommon for an employee to be injured in a way that requires resuscitation. Consequently, many companies provide employees with CPR training. But what about AIDS? Is CPR training safe?

The HIV virus has been found in human saliva. Because CPR involves using your fingers to clear the airway and placing your mouth over the victim's, there is concern about contracting AIDS while trying to resuscitate someone. Although there is no hard evidence that HIV can be transmitted through saliva, there is some legitimacy to the concern.

Because of the concern, **disposable face masks** and various other types of personal protective devices are now being manufactured. Safety and health professionals should ensure that such devices are used in both training and live situations involving

CPR. These devices should be readily available in many easily accessible locations throughout the company.

Safety Needles

Employees who work with needles for taking blood, giving injections, or inserting intravenous systems should be considered at high risk for becoming infected with HIV. Whether inserting an IV, taking blood, or giving a vaccination, medical personnel live with the scary reality of bloodborne pathogens exposure every minute of the day. Studies consistently have shown that 50 to 85 percent of healthcare workers who have contracted a bloodborne disease cannot identify when or how they were exposed to bloodborne pathogens.

In addition, whereas more than 1 million accidental needlesticks are reported each year, it is estimated that 66 percent of accidental needlesticks go unreported. With the use of injury prevention devices, the CDC estimates that the number of accidental needlesticks could be reduced by 76 percent. Not all devices that are called "safe" are safer. Injury prevention devices on syringes, IVs, catheters, blood-drawing equipment, vaccination instruments, lancets, and scalpels must be one-handed to be considered safer. Some sheathing devices require two-handed operation and often result in an increase in needlestick injuries.

Figure 20–4 contains examples of syringes with safety features.

FIGURE 20–4   Examples of syringes with safety features.

Plastic shield slides over the needle and locks to encase the exposed point

Syringe with protective shield

With an extra push on the plunger...

...the needle retracts into the syringe

Syringe with retractable needle

HEPATITIS B VIRUS (HBV) AND HEPATITIS C VIRUS (HCV) IN THE WORKPLACE

Hepatitis B Virus (HBV)

Although the spread of HIV receives more attention, a greater risk is from the spread of HBV. This bloodborne virus averages approximately 300,000 new cases per year compared with AIDS, which averages approximately 43,000 new cases. HBV is extremely strong compared with HIV. For example, HBV can live on surfaces for up to a week if it is exposed to air. Hepatitis B is also much more concentrated than HIV.

Hepatitis B is caused by a double-shelled virus. It can be transmitted in the workplace in the following ways:

- Contact with blood
- Contact with body fluids, including tears, saliva, and semen

The HBV virus can live in body fluids for years. Carriers of the virus are at risk themselves, and they place others at risk. Persons infected with HBV may contract chronic hepatitis, cirrhosis of the liver, and primary hepatocellular carcinoma. An HBV-infected individual is more than 300 times more likely to develop primary liver cancer than is a noninfected individual from the same environment. Unfortunately, it is possible to be infected and not know it because the symptoms can vary so widely from person to person.

The symptoms of hepatitis B are varied but include

- Jaundice
- Joint pain
- Rash
- Internal bleeding

HBV Vaccination

The OSHA standard covering bloodborne pathogens requires employers to offer the three-injection HBV vaccination series free to all employees who are exposed to blood or other potentially infectious materials as part of their job duties. This includes health-care workers, emergency responders, morticians, first-aid personnel, law enforcement officers, correctional facilities staff, and launderers, as well as others. The vaccination must be offered within 10 days of a person's initial assignment to a job where exposure to blood or other potentially infectious materials can be "reasonably anticipated."

The HBV vaccination is a noninfectious, yeast-based vaccine given in three injections in the arm. It is prepared from recombinant yeast cultures, rather than human blood or plasma. Thus, there is no risk of contamination from other bloodborne pathogens, nor is there any chance of developing HBV from the vaccine.

The second injection should be given one month after the first, and the third injection six months after the initial dose. More than 90 percent of those vaccinated develop immunity to the HBV. To ensure immunity, it is important for individuals to receive all three injections. At this point, it is unclear how long the immunity lasts, so booster shots may be required at some point in the future.

The vaccine causes no harm to those who are already immune or to those who may be HBV carriers. Although employees may opt to have their blood tested for antibodies to determine the need for the vaccine, employers may not make such screening a condition of receiving the vaccination, nor are employers required to provide prescreening.

Each employee should receive counseling from a healthcare professional when the vaccination is offered. This discussion will help an employee determine whether inoculation is necessary.

Hepatitis C Virus (HCV)

Hepatitis C virus (HCV) infection is the most common chronic bloodborne infection in the United States, with approximately 36,000 new cases diagnosed every year.[13] Most of these persons are chronically infected and may be unaware of their infection because they are not clinically ill. Infected persons serve as a source of transmission to others and are at risk for chronic liver disease or other HCV-related chronic diseases during the first two or more decades following the initial infection.

Chronic liver disease is the tenth leading cause of death among adults in the United States, and accounts for approximately 25,000 deaths annually, or approximately 1 percent of all deaths. Population-based studies indicate that 40 percent of chronic liver disease is HCV related, resulting in an estimated 8,000 to 10,000 deaths each year. Because most HCV-infected persons are aged 30 to 49, the number of deaths attributable to HCV-related chronic liver disease may increase substantially during the next 10 to 20 years, as this group of infected persons reaches the age at which complications from chronic liver disease typically occur.

HCV is transmitted primarily through large or repeated direct percutaneous exposures to blood. In the United States, the two most common exposures associated with transmission of HCV are blood transfusion and injecting-drug use. Figure 20–5 summarizes tests that are used to detect HCV infection.

Preventing and controlling HCV requires a comprehensive strategy comprising at least the following activities:

- Primary prevention activities
  Screening and testing blood, plasma, organ, tissue, and semen donors
  Virus inactivation of plasma-derived products
  Risk-reduction counseling and services
  Implementation and maintenance of infection control practices
- Secondary prevention activities
  Identification, counseling, and testing of persons at risk
  Medical management of infected persons
- Education and training
- Monitoring the effectiveness of prevention activities to develop improved prevention methods

The next section explains OSHA's standard on occupational exposure to bloodborne pathogens. This standard applies to all bloodborne pathogens, including HBV and HIV.

FIGURE 20–5 Various tests available for detecting the presence of HCV.

Tests for Hepatitis C Virus (HCV) Infection

Test	Application
Hepatitis C virus antibody (anti-HCV) • EIA (enzyme immunoassay) • Supplemental assay	• Indicates past or present infection, but does not differentiate between acute, chronic, or resolved infection • All positive EIA results should be verified with a supplemental assay
HCA RNA (hepatitis C virus ribonucleic acid) **Qualitative tests** • Reverse transcriptase polymerase chain reaction (RT-PCR) amplification of HCV RNA by in-house or commercial assays	• Detect presence of circulating HCV RNA • Monitor patients on antiviral therapy
Quantitative tests • RT-PCR amplification of HCV RNA by in-house or commercial assays • Branched chain DNA (bDNA) assays	• Determine concentration of HCV RNA • Might be useful for assessing the likelihood of response to antiviral therapy
Genotype • Several methodologies available (e.g., hybridization, sequencing)	• Group isolates of HCV based on generic differences, into 6 genotypes and 90 subtypes • With new therapies, length of treatment might vary based on genotype
Serotype • EIA based on immunoreactivity to synthetic peptides	• No clinical utility

Source: Center for Disease Control (CDC), "Facts about Aids." Retrieved from cdc.gov/hiv/resources/factsheets/ on July 30, 2017.

OSHA'S STANDARD ON OCCUPATIONAL EXPOSURE TO BLOODBORNE PATHOGENS

OSHA's standard on occupational exposure to bloodborne pathogens is contained in 29 CFRPart 1910.1030. The purpose of the standard is to limit the exposure of personnel to blood and to serve as a precaution against bloodborne pathogens that may cause diseases.

Scope of Application

This standard applies to all employees whose job duties may bring them in contact with blood or other potentially infectious material. There is no attempt on OSHA's part to list all occupations to which 1910.1030 applies. The deciding factor is the *reasonably anticipated* theory. If it can be reasonably anticipated that employees may come in contact with blood in the normal course of performing their job duties, the standard applies. It does not apply in instances of *Good Samaritan* acts in which one employee attempts to assist another employee who is bleeding.

The standard covers blood and other infectious materials. These other materials include the following:

- Semen
- Vaginal secretions
- Cerebrospinal fluid
- Synovial fluid
- Pleural fluid
- Peritoneal fluid
- Amniotic fluid
- Saliva
- Miscellaneous body fluids mixed with blood

In addition to these fluids, other potentially infectious materials include the following:

- Unfixed human tissue, or organs other than intact skin
- Cell or tissue cultures
- Organ cultures
- Any medium contaminated by HIV or HBV

Exposure Control Plan

OSHA's 1910.1030 requires employers to have an **exposure control plan** to protect employees from exposure to bloodborne pathogens. It is recommended that such plans have at least the following major components:

Part 1—Administration

Part 2—Methodology

Part 3—Vaccinations

Part 4—Postexposure investigation and follow-up

Part 5—Labels and signs

Part 6—Information and training

Administration

This component of the plan should clearly define the responsibilities of employees, supervisors, and managers regarding exposure control. It should also designate an exposure control officer (usually the organization's safety and health manager or a person who reports to this manager). The organization's exposure control plan must be readily available to all employees, and the administration component of it must contain a list of locations where copies of the plan can be examined by employees. It should also describe the responsibilities of applicable constituent groups and individuals. These responsibilities are summarized as follows:

- **Employees.** All employees are responsible for the following: knowing which of their individual and specific job tasks may expose them to bloodborne pathogens, participating in training provided concerning bloodborne pathogens, carrying out all job duties in accordance with the organization's various control procedures, and practicing good personal hygiene.

- **Supervisors and managers.** Supervisors and managers are responsible for coordinating with the exposure control officer to implement and monitor exposure control procedures in their areas of responsibility.
- **Exposure control officer.** This individual is assigned overall responsibility for carrying out the organization's exposure control plan. In addition to the administrative duties associated with it, this position is also responsible for employee training. In small organizations, exposure control officers may double as the training coordinator. In larger organizations, a separate training coordinator may be assigned exposure control as one more training responsibility.

In either case, the exposure control officer is responsible for the following duties: overall development and implementation of the exposure control plan; working with management to develop other exposure-related policies; monitoring and updating the plan; keeping themselves up-to-date with the latest legal requirements relating to exposure; acting as liaison with OSHA inspectors; maintaining up-to-date training files and documentation showing training; developing the needed training program; working with other managers to establish appropriate control procedures; establishing a HBV vaccination program as appropriate; establishing a postexposure evaluation and follow-up system; displaying labels and signs as appropriate; and maintaining up-to-date, confidential medical records of exposed employees.

Methodology

This section of the plan describes the procedures established to protect employees from exposure. These procedures fall into one of the following five categories:

- General precautions
- Engineering controls
- Work practice controls
- Personal protection equipment
- Housekeeping controls

General precautions include such procedures as assuming that all body fluids are contaminated, and acting accordingly. Engineering controls are design and technological precautions that protect employees from exposure. Examples of engineering controls include self-sheathing needles, readily accessible handwashing stations equipped with antiseptic hand cleaners, leakproof specimen containers, and puncture-proof containers for sharp tools that are reusable. Work practice controls are precautions that individual employees take, such as washing their hands immediately after removing potentially contaminated gloves or refraining from eating or drinking in areas where bloodborne pathogens may be present. Personal protection equipment includes any device designed to protect an employee from exposure. Widely used devices include the following:

- Gloves (disposable and reusable)
- Goggles and face shields

- Respirators
- Aprons, coats, and jackets

Examples of housekeeping controls include the use, disposal, and changing of protective coverings; decontamination of equipment; and regular cleaning of potentially contaminated areas.

Vaccinations

The OSHA standard requires that employers make hepatitis B vaccinations available to all employees for whom the reasonable anticipation rule applies. The vaccination procedure must meet the following criteria: available at no cost to employees; administered at a reasonable time and place within 10 days of assignment to a job with exposure potential; and administered under the supervision of an appropriately licensed healthcare professional, according to the latest recommendations of the U.S. Public Health Service.

Employees may decline the vaccination, but those who do so must sign a *declination form* stating that they understand the risk to which they are subjecting themselves. Employees who decline are allowed to change their minds. Employees who change their minds must be allowed to receive the vaccination.

Part of the exposure control officer's job is to keep accurate, up-to-date records showing vaccinated employees, vaccination dates, employees who declined, and signed declination forms.

Postexposure Investigation and Follow-Up

When an employee is exposed to bloodborne pathogens, it is important to evaluate the circumstances and follow-up appropriately. How did it happen? Why did it happen? What should be done to prevent future occurrences? The postexposure investigation should determine at least the following:

- When did the incident occur?
- Where did the incident occur?
- What types of contaminated substances or materials were involved?
- What is the source of the contaminated materials?
- What types of work tasks were being performed when the incident happened?
- What was the cause of the incident?
- Were the prescribed precautions being observed when the incident occurred?
- What immediate action was taken in response to the incident?

Using the information collected during the investigation, an incident report is written. This report is just like any other accident report. Keeping the exposed employee fully informed is important. The employee should be informed of the likely avenue of exposure and the source of the contaminated material, even if the source is another employee. If the source is another employee, that individual's blood should be tested for HBV or HIV, and the results should be shared with the exposed employee. Once the exposed employee is fully informed, he or she should be referred to an appropriately certified medical professional to discuss the issue. The medical professional should provide a written report to the employer containing all pertinent information and recommendations. The exposed employee should also receive a copy.

Labels and Signs

This section of the plan describes the procedures established for labeling potential biohazards. Organizations may also use warning signs and color-coded containers as appropriate. It is important to label or designate with signs the following:

- Biohazard areas
- Contaminated equipment
- Containers of potentially contaminated waste
- Containers of potentially contaminated material (i.e., a refrigerator containing blood)
- Containers used to transport potentially contaminated material

Information and Training

This section of the plan describes the procedures for keeping employees knowledgeable, fully informed, and up-to-date regarding the hazards of bloodborne pathogens. The key to satisfying this requirement is training. Training provided should cover at least the following:

- OSHA Standard 1910.1030
- The exposure control plan
- Fundamentals of bloodborne pathogens (e.g., epidemiology, symptoms, and transmission)
- Hazard identification
- Hazard prevention methods
- Proper selection and use of personal protection equipment
- Recognition of warning signs and labels
- Emergency response techniques
- Incident investigation and reporting
- Follow-up techniques
- Medical consultation

It is important to document training and keep accurate up-to-date training records on all employees. These records should be available to employees and their designated representatives (for example, family members, attorneys, and physicians) and to OSHA personnel. They must be kept for at least three years and include the following:

- Dates of all training
- Content of all training
- Trainers' names and qualifications
- Names and job titles of all participants

Record Keeping

OSHA Standard 1910.1030 requires that medical records be kept by employers on all employees who are exposed to bloodborne pathogens. These records must be confidential and should contain at least the following information:

- Employee's name and social security number
- Hepatitis B vaccination status
- Results of medical examinations and tests
- Results of incident follow-up procedures

- A copy of the written opinions of healthcare professionals
- A copy of all information provided to healthcare professionals following an exposure incident

PREVENTING AND RESPONDING TO NEEDLESTICK INJURIES

Needlestick injuries are not a major concern in most workplaces outside the healthcare industry, but they are enough of a concern that safety and health professionals should know how to prevent them and how to respond when they occur. An excellent source of help for safety and health professionals concerning needlestick injuries is the National Institute for Occupational Safety and Health (NIOSH). NIOSH maintains a website dedicated specifically to the prevention of needlestick injuries. The Web address is as follows:

cdc.gov/niosh/topics/bbp/safer

This website recommends a five-step model for developing, establishing, and maintaining a needlestick-prevention program:

1. Form a sharps injury prevention team.
2. Identify priorities.
3. Identify and screen safer medical devices.
4. Evaluate safer medical devices.
5. Institute and monitor the use of the safer medical devices selected.[14]

The NIOSH site also provides detailed information that safety and health professionals can use in implementing each step in the development of a needlestick prevention program.

Responding to Needlestick Incidents

When in spite of your best efforts at prevention a needlestick injury does occur, the following steps are recommended:

1. Encourage bleeding where the skin is penetrated.
2. Wash the penetration area thoroughly with copious amounts of warm, soapy water (do not use a scrub brush).
3. If the eyes are somehow involved, wash them immediately with water.
4. If the mouth is somehow involved, rinse it immediately with water, but do not swallow.
5. Get the injured employee to the hospital as soon as possible.
6. Contact a clinical virologist.
7. Make sure that management personnel for the company in question are fully informed of the situation.[15]

METHICILLIN-RESISTANT *STAPHYLOCOCCUS AUREUS* (MRSA) IN THE WORKPLACE

Methicillin-resistant *Staphylococcus aureus* (MRSA) is a potentially life-threatening infection caused by bacteria that mutate and become strongly and more resistant to antibiotics with each mutation. The problems associated with MRSA are now so widespread that it must be considered a workplace hazard. In fact, workers' compensation claims relating to MRSA are on the rise.[16] According to Burris, five events have led to the emergence of MRSA:

- Bacteria are able to mutate
- Antibiotics have been overprescribed for decades (allowing bacteria to become resistant)
- People stop taking their antibiotics as soon as symptoms subside rather than finishing the prescription as instructed
- Antibacterial soaps have proliferated and been overused (killing good bacteria and allowing harmful bacteria to become resistant)
- New tendency of people to skip showers after strenuous exercise or physical activity (dirty, moist, and salty skin is the perfect breeding environment for MRSA)[17]

Prevention Steps

1. **Ensure that employees wash their hands regularly and shower after strenuous physical activity.** MRSA and other bacteria live on the skin making them susceptible to hand washing and showers.
2. **Ensure that employees treat and cover wounds.** Breaks in the skin are the primary way that MRSA gains entry into the body. No matter how minor the cut, scratch, or abrasion may be, make sure it is cleaned, an over-the-counter wound-care treatment is applied, and the wound is covered.
3. **Do not allow employees to share personal items.** If your employees can shower and shave at work, do not allow them to share razors. MRSA can live in razors and on towels, benches, and clothing.
4. **Encourage employees to use a dryer rather than clothesline for drying towels and clothing.** The heat of a dryer can kill bacteria, but items hung out to dry are a potential breeding ground for MRSA.
5. **Ensure that employees wipe down and properly clean tools, work gear, and equipment.** Situations in which workers share tools, equipment, and gear are ripe for the spread of MRSA. Cleaning these items several times during the day and between shift changes can help prevent the infection. This also applies to gym equipment and benches if your employees work out.[18]

Key Terms and Concepts

- Acquired immuno-deficiency syndrome (AIDS)
- AIDS education program
- AIDS policy
- Blood transfusions
- Body fluids
- Case law
- Confidentiality
- Confirmation test
- Counseling
- Disposable face masks
- Employee assistance programs (EAPs)
- Employee Retirement Income Security Act (ERISA)
- Employee rights
- Enzyme-linked immunosorbent assay (ELISA) test
- Essential functions
- Exposure control plan
- False negative
- False positive
- Hepatitis B Virus (HBV)
- Hepatitis C Virus (HCV)
- HIV antibody test
- HIV positive
- Human immunodeficiency virus (HIV)
- Immunofluorescent assay (IFA)
- Intravenous drugs
- IV drug users
- Occupational Safety and Health Act (OSH Act)
- Otherwise qualified
- Reasonable accommodation
- Rehabilitation Act
- Screening test
- Sexual contact
- Western Blot test

Review Questions

1. What does the acronym "AIDS" mean?
2. List five symptoms of AIDS.
3. What are the three known ways that AIDS is transmitted?
4. List the groups of people who are considered high risk with regard to AIDS.
5. List five ways that AIDS is *not* transmitted.
6. Describe the ways in which AIDS is having an impact in the workplace.
7. Briefly explain the minimum components of a corporate AIDS policy.
8. Explain how the following legal concepts relate to AIDS: *otherwise qualified, essential functions,* and *reasonable accommodation.*
9. Briefly explain both sides of the AIDS testing controversy.
10. What are the most widely used HIV antibody tests for screening and confirming AIDS?
11. Name four purposes of a well-planned AIDS education program.
12. How should a safety and health professional proceed when faced with an employee who thinks he or she has AIDS?
13. Briefly explain the steps that a company can take to alleviate the fears of employees about AIDS.
14. Explain how OSHA categorizes work tasks relative to exposure to HIV infection.
15. List four ways to guard against contracting AIDS.
16. How can CPR be administered safely?
17. Explain why HBV poses more of a problem for safety personnel than HIV does.
18. Explain the primary prevention strategies for HCV.
19. What are the recommended steps in a needlestick prevention program?

Endnotes

1. Center for Disease Control and Prevention (CDC), *Facts about AIDS*. Retrieved from cdc.gov/hiv/resources/factsheets/ on July 30, 2017.
2. Ibid.
3. Ibid.
4. Ibid.
5. World Health Organization, *Guidelines on HIV Testing*. Retrieved from who.int/hiv/pub/guidelines/india_art.pdf on July 30, 2017.
6. Ibid.
7. Ibid.
8. CDC, *Facts about AIDS*, 10.
9. Ibid.
10. G. Seldes, *The Great Quotations* (Secaucus, NJ: Castle Books, 1966), 590.
11. CDC, *Facts about AIDS*.
12. Ibid, 8.
13. Centers for Disease Control and Prevention, *Recommendations for Prevention and Control of Hepatitis C Virus (HCV) Infection and HCV-Related Chronic Disease.* Retrieved from cdc.gov on July 29, 2017.
14. Centers for Disease Control and Prevention (CDC), *Bloodborne Infectious Diseases.* Retrieved from cdc.gov/niosh/topics/bbp/emergnedl.html on July 30, 2017.
15. Ibid.
16. G. Burris, *Preventing MSRA at Work*. Retrieved from nsc.org/safetyhealth/pages/3.11 workplacesolutionspreventingMRSA.aspx#.UToKfNbQq9E on July 8, 2017.
17. Ibid.
18. Ibid.

Glossary

abatement period The amount of time given to an employer to correct a hazardous condition that has been cited.

ability to pay Applies when there are a number of defendants in a case but not all have the ability or means to pay financial damages.

absorption Passage through the skin and into the bloodstream.

acceleration Increase in the speed of a falling object before impact.

accident/incident theory Theory of accident causation in which overload, ergonomic traps, or a decision to err lead to human error.

accident prevention The act of preventing a happening that may cause loss or injury to a person.

accident rate A fixed ratio between the number of employees in the workforce and the amount who are injured or killed every year.

accident report Records the findings of an accident investigation, the cause or causes of an accident, and recommendations for corrective action.

accident scene The area where an accident occurred.

accidents Unexpected happenings that may cause loss or injuries to people who are not at fault for causing the injuries.

acclimatization Process by which the body becomes gradually accustomed to heat or cold in a work setting.

accommodation The ability of the eye to become adjusted after viewing the VDT to be able to focus on other objects, particularly objects at a distance.

adjustable guard A device that provides a barrier against a variety of different hazards associated with different production operations.

administrative controls Procedures that are adopted to limit employee exposure to hazardous conditions.

aerosols Liquid or solid particles so small that they can remain suspended in air long enough to be transported over a distance.

aerospace engineering Program of study incorporating a solid foundation of physical and mathematical fundamentals that provides the basis for the development of the engineering principles essential to the understanding of both atmospheric and extra-atmospheric flight.

agreement settlement The injured employee and the employer or its insurance company work out an agreement on how much compensation will be paid and for how long.

air dose Dose measured by an instrument in the air at or near the surface of the body in the area that has received the highest dosage of a toxic substance.

altitude sickness A form of hypoxia associated with high altitudes.

ampere The unit of measurement for current.

analysis and evaluation All potential solutions to a problem are subjected to scientific analysis and careful evaluation.

ancestry A person's line of descent.

anesthetics In carefully controlled dosages, anesthetics can inhibit the normal operation of the central nervous system without causing serious or irreversible effects.

appeals process The process of challenging an OSHA standard as it goes through the adoption process and after formal adoption.

artificial environment One that is fully created to prevent a definite hazardous condition from affecting people or material.

aseptic necrosis A delayed effect of decompression sickness.

asphyxiants Substances that can disrupt breathing so severely as to cause suffocation.

associate degree A two-year college degree.

assumption of risk Based on the theory that people who accept a job assume the risks that go with it. It says employees who work voluntarily should accept the consequences of their actions on the job rather than blaming the employer.

audiogram The results of an audiometric test to determine the noise threshold at which a subject responds to different test frequencies.

audiometric testing Measures the hearing threshold of employees.

auto-ignition temperature The lowest temperature at which a vapor-producing substance or a flammable gas will ignite even without the presence of a spark or a flame.

automatic ejection A system that ejects work pneumatically and mechanically.

automatic feed A system that feeds stock to the machine from rolls.

bachelor's (or baccalaureate) degree A four-year college degree.

barometer Scientific device for measuring atmospheric pressure.

bends Common name for decompression sickness.

537

best-ratio approach   People are basically good and under the right circumstances behave ethically.

biological hazards   Hazards from molds, fungi, bacteria, and insects.

black-and-white approach   Right is right, wrong is wrong, and circumstances are irrelevant.

blackball   To ostracize an employee.

body surface area (BSA)   The amount of surface area that is covered with burns.

bonding   Used to connect two pieces of equipment by a conductor. Also, involves eliminating the difference in static charge potential between materials.

Boyle's law   States that the product of a given pressure and volume is constant with a constant temperature.

cancer   A malignant tumor.

carcinogen   Any substance that can cause a malignant tumor or a neoplastic growth.

carpal tunnel syndrome (CTS)   An injury to the median nerve inside the wrist.

case law   Serves the purpose of establishing precedents that can guide future decisions.

causal relationship   A situation in which an action leads to a certain result.

ceiling   The level of exposure that should not be exceeded at any time for any reason.

central factor   The main issue or factor in a problem or act.

chemical burn injuries   Burn damage to the skin caused by chemicals such as acids and alkalies.

chemical engineer   An engineer concerned with all the physical and chemical changes of matter to produce economically a product or result that is useful to humans.

chemical hazards   Include mists, vapors, gases, dusts, and fumes.

chokes   Coughing and choking, resulting from bubbles in the respiratory system.

circadian rhythm   Biological clock.

circuit tester   An inexpensive piece of test equipment with two wire leads capped by probes and connected to a small bulb.

claim notice   Notice filed to indicate an expectation of workers' compensation benefits owed.

closed environment   One that is completely or almost completely shut off from the natural environment.

closing conference   Involves open discussion between the compliance officer and company and employee representatives.

code   A set of standards, rules, or regulations relating to a specific area.

Code of Hammurabi   Developed around 2000 BC during the time of the Babylonians, the ruler Hammurabi developed this code which encompassed all the laws of the land at that time. The significant aspect is that it contained clauses dealing with injuries, allowable fees for physicians, and monetary damages assessed against those who injured others.

coefficient of friction   A numerical correlation of the resistance of one surface against another surface.

cold stress   Physical or mental stress that results from working in cold conditions.

combination theory   The actual cause of an accident may be explained by combining many models.

combustible substance   Any substance with a flash point of 37.8°C (100.04°F) or higher.

combustion   A chemical reaction between oxygen and a combustible fuel.

combustion point   The temperature at which a given fuel can burst into flame.

commonsense test   Requires person to listen to what instincts and common sense are telling him or her.

competitiveness   The ability to succeed and prosper consistently in the marketplace whether it is local, regional, national, or global.

conduction   The transfer of heat between two bodies that are touching or it is the transfer of heat from one location to another within a body.

conductors   Substances that have many free electrons at room temperature and can pass electricity.

conference method   A problem-solving teaching method in which the trainer serves as a facilitator rather than a teacher.

confined space   An area with limited means of egress that is large enough for a person to fit into, but is not designed for occupancy.

consistency   The rules are enforced in the same manner every time with no regard to any outside factors.

continuity tester   May be used to determine whether a conductor is properly grounded or has a break in the circuit.

contributory negligence   An injured worker's own negligence contributed to the accident. If the actions of employees contributed to their own injuries, the employer is absolved of any liability.

controlled environment   A natural or induced environment that has been changed in some way to reduce or eliminate potential environmental hazards.

convection   The transfer of heat from one location to another by way of a moving medium.

convergence   The coordinated turning of the eyes inward to focus on a nearby point or object.

Cooperative Safety Congress (CSC)   The CSC was a result of the Association of Iron and Steel Electrical Engineers' (AISEE) desire to have a national conference on safety. The first meeting took place in Milwaukee in 1912. The meeting planted the seeds for the eventual establishment of the National Safety Council (NSC).

corium   The inner layer of human skin.

cost   The amount of money needed to produce a product, not to be confused with "price," which is the amount of money needed to purchase a product after the cost has been marked up.

cost allocation   Spread the cost of workers' compensation appropriately and proportionately among industries ranging from the most to the least hazardous.

creeps Caused by bubble formation in the skin, which causes an itchy, crawling, rashy feeling in the skin.

critical burns Second-degree burns covering more than 30 percent of the body and third-degree burns covering over 10 percent are considered critical.

crushing Occurs when a part of the body is caught between two hard surfaces that progressively move together, thereby crushing anything between them.

cutis The inner layer of human skin.

cutting Occurs when a body part comes in contact with a sharp edge.

Dalton's law of partial pressures States that in a mixture of theoretically ideal gases, the pressure exerted by the mixture is the sum of the pressures exerted by each component gas of the mixture.

damages Financial awards assigned to injured parties in a lawsuit.

DBBS See Division of Biology and Biomedical Science.

death rates A fixed ratio between the number of employees in the workforce and the number that are killed each year.

decibel (dB) The unit applied when measuring sound. One-tenth of a bel. One decibel is the smallest difference in the level of sound that can be perceived by the human ear.

decompression sickness Can result from the decompression that accompanies a rapid rise from sea level to at least 18,000 feet or a rapid ascent from around 132 to 66 feet underwater.

demonstration method The instructor shows students how to perform certain skills or tasks.

dermis The inner layer of human skin.

design flaw A defect in a product.

design process A plan of action for reaching a goal.

direct settlement The employer or its insurance company begins making what it thinks are the prescribed payments.

discovery period Period in which evidence is collected, depositions are taken, and products are examined.

Division of Biology and Biomedical Science (DBBS) Conducts research in the areas of toxicology, behavioral science, ergonomics, and the health consequences of various physical agents.

Division of Training and Manpower Development (DTMD) Implements Section 21 of the OSH Act, which sets forth training and education requirements.

document and communicate Engineering drawings, detailed calculations, and written specifications document the design of a product and communicate its various components to interested parties.

domino theory Injuries are caused by the action of preceding factors. Removal of the central factor negates the action of the preceding factors and, in so doing, prevents accidents and injuries.

dose The amount of ionizing radiation absorbed per unit of mass by part of the body or the whole body.

dose threshold The minimum dose required to produce a measurable effect.

dosimeter Provides a time-weighted average over a period such as one complete work shift.

double insulation A means of increasing electrical equipment safety.

drowning The act of suffocating due to submersion in water.

DTMD See Division of Training and Manpower Development.

dusts Various types of solid particles that are produced when a given type of organic or inorganic material is scraped, sawed, ground, drilled, heated, crushed, or otherwise deformed.

dysbarism The formation of gas bubbles due to rapid ambient pressure reduction.

ego strength An employee's ability to undertake self-directed tasks and to cope with tense situations.

electrical engineering A science-oriented branch of engineering primarily concerned with all phases and development of the transmission and utilization of electric power and intelligence.

electrical hazards Potentially dangerous situations related to electricity (for example, a bare wire).

electrical system grounding Achieved when one conductor of the circuit is connected to the earth.

electricity The flow of negatively charged particles called electrons through an electrically conductive material.

electrolytes Minerals that are needed for the body to maintain the proper metabolism and for cells to produce energy.

electromechanical devices Contact bars that allow only a specified amount of movement between the worker and the hazard.

electrons Negatively charged particles.

emergency action plan (EAP) A collection of small plans for every anticipated emergency (for example, fire, hurricane, chemical spill).

emergency coordinator A person who is clearly identified in a company emergency response plan as the responsible party for a specific type of emergency situation.

emergency response management team Composed of shift emergency directors and a full-time fire chief. Unifies all emergency groups and equipment into a single, coordinated effort.

emergency response network (ERN) A network of emergency response teams that covers a designated geographical area and is typically responsible for a specific type of emergency.

emergency response plan A written document that identifies the different personnel or groups that respond to various types of emergencies and, in each case, who is in charge.

emergency response team (ERT) A special team that responds to general and localized emergencies to facilitate personnel evacuation and safety, shut down building services and utilities as needed, work with responding civil authorities, protect and salvage company property, and evaluate areas for safe reentry.

employee   A person who is on the company's payroll, receives benefits, and has a supervisor.

employee responsibilities   Specific obligations of employees relating to safety and health as set forth in OSHA 2056 (revised).

Employee Retirement Income Security Act (ERISA)   Protects the benefits of employees by prohibiting actions taken against them based on their eligibility for benefits.

employee rights   Protection that an employee has against punishment for complaining to an employer union, OSHA, or any governmental agency about hazards on the job.

employer-biased laws   A collection of laws that favored employers over employees in establishing a responsibility for workplace safety.

employer liability   In 1877, the Employer's Liability Law was passed and established the potential for employers to be liable for accidents that occurred in the workplace.

employer responsibilities   Specific responsibilities of employers as specified in OSHA 2056 (revised).

employer rights   The rights of an employer as specified in OSHA 2056 (revised).

environment   The aggregate of social and cultural conditions that influence the life of an individual.

environmental engineer   An individual whose job is to protect and preserve human health and the well-being of the environment.

environmental factors   Characteristics of the environment in which an employee works that can affect his or her state of mind or physical conditions such as noise or distractions.

environmental heat   Heat that is produced by external sources.

environmental management system (EMS)   That component of an organization with primary responsibility for leading, planning, organizing, and controlling as these functions relate specifically to an organization's processes, products, or services and the impact that they have on the environment.

enzyme-linked immunosorbent assay (ELISA) test   Screening test currently used to ensure the accuracy of an HIV antibody test.

epidemiological theory   Holds that the models used for studying and determining epidemiological relationships can also be used to study causal relationships between environmental factors and accidents or diseases.

epidermis   The outer layer of human skin.

ergonomic hazards   Workplace hazards related to the design and condition of the workplace. For example, a workstation that requires constant overhead work is an ergonomic hazard.

ergonomic trap   An unsafe condition unintentionally designed into a workstation.

ergonomics   The science of conforming the workplace and all its elements to the worker.

ERISA   See Employee Retirement Income Security Act.

ethical behavior   That which falls within the limits prescribed by morality.

ethics   The application of morality within a context established by cultural and professional values, social norms, and accepted standards of behavior.

expert system   A computer programmed to solve problems.

expiration   Occurs when air leaves the lungs and the lung volume is less than the relaxed volume, increasing pressure within the lungs.

explosion   A very rapid, contained fire.

explosive range   Defines the concentrations of a vapor or gas in air that can ignite from a source.

exposure threshold   A specified limit on the concentration of selected chemicals. Exposure to these chemicals that exceeds the threshold is considered hazardous.

extraterritorial employees   Those who work in one state but live in another.

failure mode and effects analysis   A formal, step-by-step analytical method that is a spin-off of reliability analysis, a method used to analyze complex engineering systems.

falls   Category of accidents in which an employee unintentionally drops under the force of gravity from a surface that is elevated.

false negative   A result that shows no HIV antibodies in people who, in reality, are infected.

false positive   A result that shows the presence of HIV antibodies when, in reality, no such antibodies exist.

fault tree analysis   An analytical methodology that uses a graphic model to display visually the analysis process.

feedback   Employee's opinions of a project or an employer's opinion about an employee's job performance.

fellow servant rule   Employers are not liable for workplace injuries that result from negligence of other employees.

fire   A chemical reaction between oxygen and a combustible fuel.

fire hazards   Conditions that favor the ignition and spread of fire.

fire point   The minimum temperature at which the vapors or gas in air can ignite from a source of ignition.

fire-related losses   Loss of life caused by fires and instances related to fires such as burns, asphyxiation, and falls.

first-degree burn   Results in a mild inflammation of the skin known as erythema.

fixed guards   Provide a permanent barrier between workers and the point of operation.

flame-resistant clothing   Special clothing made of materials or coated with materials that are able to resist heat and flames.

flammable substance   Any substance with a flash point below 37.8°C (100.04°F) and a vapor pressure of less than 40 pounds per square inch at this temperature.

flash point   The lowest temperature for a given fuel at which vapors are produced in sufficient concentrations to flash in the presence of a source of ignition.

flashing back   When a flame consumes a flammable material such as a gas traveling quickly back to its source.

flexible hours   A work/time scheduling system in which employees are allowed to adopt a work schedule that more closely matches their personal needs.

fluid loss   Depletion of necessary body fluids, primarily through perspiration.

foreign object   Any object that is out of place or in a position to trip someone or to cause a slip.

foreseeability   Concept that a person can be held liable for actions that result in damages or injury only when risks could have been reasonably foreseen.

four-step teaching method   Preparation, presentation, application, and evaluation.

four-to-one ratio   Base one foot away from the wall for every four feet between the base and the support point of a ladder.

free environment   One that does not interfere with the free movement of air.

freeze   The inability to release one's grip voluntarily from a conductor of electricity.

frequency   The number of standard AC circuit cycles per second.

friable asbestos   Asbestos that is in a state of crumbling deterioration. When asbestos is in this state, it is most dangerous.

front-page test   Encourages persons to make a decision that would not embarrass them if it were printed as a story on the front page of their hometown newspaper.

full-potential approach   People are responsible for realizing their full potential within the confines of morality.

fuses   Consist of a metal strip or wire that will melt if a current above a specific value is conducted through the metal.

gases   Formless fluids that are airborne and can be toxic.

gates   Provide a barrier between the danger zone and workers.

global marketplace   The worldwide economic market in which many companies must compete for business.

good housekeeping   Proper cleaning and maintenance of a work area.

ground fault   When the current flow in the hot wire is greater than the current in the neutral wire.

ground fault circuit interrupter (GFCI)   Can detect the flow of current to the ground and open the circuit, thereby interrupting the flow of current.

grounded conductor   Neutral wire.

grounding conductor   The groundwire.

hand-arm vibration (HAV) syndrome   A form of Raynaud's syndrome that afflicts workers who use vibrating power tools frequently over time.

harmful environment   A work environment in which physical or psychological factors exist that are potentially hazardous.

hazard   A condition with the potential of causing injury to personnel, damage to equipment or structures, loss of material, or lessening of the ability to perform a prescribed function.

hazard analysis   A systematic process for identifying hazards and recommending corrective action.

hazard and operability review   An analysis method that was developed for use with new processes in the chemical industry.

hazardous condition   A condition that exposes a person to risks.

hazardous waste reduction   Reducing the amount of hazardous waste generated and, in turn, the amount introduced into the waste stream through the process of source reduction and recycling.

health physicist   Concerned primarily with radiation in the workplace.

hearing conservation   Systematic procedures designed to reduce the potential for hearing loss in the workspace. Employers are required by OSHA to implement hearing conservation procedures in settings where the noise level exceeds a time-weighted average of 85 dBA.

heart disease   An abnormal condition of the heart that impairs its ability to function.

heat burn injuries   Burn damage to the skin caused by high temperatures from some of the following activities: welding, cutting with a torch, and handling tar or asphalt.

heat cramps   A type of heat stress that occurs as a result of salt and potassium depletion.

heat exhaustion   A type of heat stress that occurs as a result of water or salt depletion.

heat rash   A type of heat stress that manifests itself as small raised bumps or blisters that cover a portion of the body and give off a prickly sensation that can cause discomfort.

heat stroke   A type of heat stress that occurs as a result of a rapid rise in the body's core temperature.

heat transfer   The spread of heat from a source to surrounding materials or objects by conduction, radiation, or convection.

high-radiation area   Any accessible area in which radiation hazards exist that could deliver a dose in excess of 100 millirem within one hour.

horizontal work area   One that is designed and positioned so that it does not require the worker to bend forward or to twist the body from side to side.

human error   A mistake that is made by a human, not a machine.

human error analysis   Used to predict human error and not as an after-the-fact process.

human factors theory   Attributes accidents to a chain of events ultimately caused by human error.

humidification   Adding moisture to the air to reduce electrical static.

hyperoxia   Too much oxygen or oxygen breathed under too high a pressure.

hypothermia   The condition that results when the body's core temperature drops to dangerously low levels.

ignition temperature   The temperature at which a given fuel can burst into flame.

impact accidents Involve a worker being struck by or against an object.

impulse noise Consists of transient pulses that can occur repetitively or nonrepetitively.

inanimate power Power that is lacking life or spirit. During the Industrial Revolution, humans and animals were replaced with inanimate power (e.g., steam power).

inappropriate activities Activities undertaken with disregard for established safety procedures.

inappropriate response A response in which a person disregards an established safety procedure.

income replacement Replacement of current and future income (minus taxes) at a ratio of two-thirds (in most states).

independent contractors A person who accepts a service contract to perform a specific task or set of tasks and is not directly supervised by the company.

indirect costs Costs that are not directly identifiable with workplace accidents.

induced environments Environments that have been affected in some way by human action.

industrial hygiene An area of specialization in the field of industrial safety and health that is concerned with predicting, recognizing, assessing, controlling, and preventing environmental stressors in the workplace that can cause sickness or serious discomfort. Concerns environmental factors that can lead to sickness, disease, or other forms of impaired health.

industrial hygiene chemists Often hired by companies that use toxic substances to test the work environment and the people who work in it. Industrial hygiene chemists have the ability to detect hazardous levels of exposure or unsafe conditions so that corrective and preventive measures can be taken at an early stage.

industrial hygienist A person having a college or university degree in engineering, chemistry, physics, medicine, or related physical or biological sciences who, by virtue of special studies and training, has acquired competence in industrial hygiene.

industrial medicine A specialized field that is concerned with work-related safety and health issues.

industrial place accidents Accidents that occur at one's place of work.

industrial safety engineer An individual who is a safety and health professional with specialized education and training. Responsible for developing and carrying out those aspects of a company's overall safety and health program relating to his or her area of expertise.

industrial safety manager An individual who is a safety and health generalist with specialized education and training. Responsible for developing and carrying out a company's overall safety and health program including accident prevention, accident investigation, and education and training.

industrial stress Involves the emotional state resulting from a perceived difference between the level of occupational demand and a person's ability to cope with this demand.

infection The body's response to contamination by a disease-producing microorganism.

ingestion Entry through the mouth.

inhalation Taking gases, vapors, dust, smoke, fumes, aerosols, or mists into the body by breathing in.

inspection tour A facility tour by an OSHA compliance officer to observe, interview, and examine as appropriate.

inspiration When atmospheric pressure is greater than pressure within the lungs, air flows down this pressure gradient from the outside into the lungs.

instructional approach A brief action plan for carrying out the instruction.

interlocked guards Shut down the machine when the guard is not securely in place or is disengaged.

interlocks Automatically break the circuit when an unsafe situation is detected.

internal factors Factors that can add a burden on a person and interfere with his or her work, such as personal problems.

ionizer Ionizes the air surrounding a charged surface to provide a conductive path for the flow of charges.

ionizing radiation Radiation that becomes electrically charged or changed into ions.

irritants Substances that cause irritation to the skin, eyes, and the inner lining of the nose, mouth, throat, and upper respiratory tract.

islands of automation Individual automated systems lacking electronic communication with other related systems.

job autonomy Control over one's job.

job descriptions Written specifications that describe the tasks, duties, reporting requirements, and qualifications for a given job.

job safety analysis A process through which all the various steps in a job are identified and listed in order.

job security A sense—real or imagined—of having the potential for longevity in a job and having a measure of control concerning that longevity.

kinetic energy The energy resulting from a moving object.

lacrimation The process of excreting tears.

learning objectives Specific statements of what the learner should know or be able to do as a result of completing the lesson.

let-go current The highest current level at which a person in contact with a conductor of electricity can release the grasp of the conductor.

liability A duty to compensate as a result of being held responsible for an act or omission.

lifting hazard Any factor that if not properly dealt with may lead to an injury from lifting.

lightning Static charges from clouds following the path of least resistance to the earth, involving very high voltage and current.

line authority The safety and health manager has authority over and supervises certain employees.

litigation The carrying on of legal matters by judicial process.

load A device that uses currents.

lockout/tagout system A system for incapacitating a machine until it can be made safe to operate. "Lockout" means physically locking up the machine so that it cannot be used without removing the lock. "Tagout" means applying a tag that warns employees not to operate the machine in question.

locus of control The perspective of workers concerning who or what controls their behavior.

lost time The amount of time that an employee was unable to work due to an injury.

lost wages The amount that an employee could have earned had he or she not been injured.

Machiavellianism The extent to which an employee will attempt to deceive and confuse others.

malpractice Negligent or improper practice.

meaninglessness The feeling that workers get when their jobs become so specialized and so technology-dependent that they cannot see the meaning in their work as it relates to the finished product or service.

means of egress A route for exiting a building or other structure.

mechanical engineering The professional field that is concerned with motion and processes whereby other energy forms are converted into motion.

mechanical hazards Those associated with power-driven machines, whether automated or manually operated.

mechanical injuries Injuries that have occurred due to misuse of a power-driven machine.

medical expenses Money paid to cover the costs of emergency medical response and follow-up treatment when an employee is injured.

Merit Program Seen as a stepping-stone to recognize companies that have made a good start toward Star Program recognition.

metabolic heat Produced within a body as a result of activity that burns energy.

mindlessness The result of the process of dumbing down the workplace.

minor burns All first-degree burns are considered minor as well as second-degree burns covering less than 15 percent of the body.

mirror test Encourages people to make choices based on how they will feel about their decision when they look in the mirror.

mists Tiny liquid droplets suspended in air.

moderate burns Second-degree burns covering less than 30 percent of the body and third-degree burns covering less than 10 percent are considered moderate.

monetary benefits Actual money owed to injured employees or their relatives under workers' compensation laws.

morning-after test Encourages people to make choices based on how they will feel about their decision the next day.

NACOSH See National Advisory Committee on Occupational Safety and Health.

narrow band noise Noise that is confined to a narrow range of frequencies.

National Advisory Committee on Occupational Safety and Health (NACOSH) Makes recommendations for standards to the secretary of health and human services and to the secretary of labor.

National Council of Industrial Safety (NCIS) Established in 1913, one year after the first meeting of the CSC. In 1915, the NCIS changed its name to the National Safety Council. It is now the premier safety organization in the United States.

National Electrical Code (NEC) Specifies industrial and domestic electrical safety precautions.

National Institute for Occupational Safety and Health (NIOSH) This organization is part of the Centers for Disease Control and Prevention (CDC) of the Department of Health and Human Services (DHHS). It is required to publish annually a comprehensive list of all known toxic substances. It will also provide on-site tests of potentially toxic substances so that companies know what they are handling and what precautions to take.

natural disasters Incidents prompted by nature such as earthquakes, hurricanes, floods, and tornadoes.

natural environment Not human-made. It is the environment that we typically think of as Earth and all its natural components, including the ground, the water, flora and fauna, and the air.

NCIS See National Council of Industrial Safety.

NEC See National Electrical Code.

negative pressures Caused by pressures below atmospheric level.

negligence Failure to take reasonable care or failure to perform duties in ways that prevent harm to humans or damage to property.

negligent manufacture The maker of a product can be held liable for its performance from a safety and health perspective.

neoplastic growth Cancerous tissue or tissue that might become cancerous.

neutrons Particles that neutralize the negative charge of electrons. They act as temporary energy repositories between positively charged particles called protons and electrons.

NIOSH See National Institute for Occupational Safety and Health.

noise Unwanted sound.

nonionizing radiation Radiation on the electromagnetic spectrum that has a frequency of 10^{15} or less and a wavelength in meters of 3×10^{27} or less.

nonskid footwear Shoes that have special nonskid soles.

normlessness The phenomenon in which people working in a highly automated environment can become estranged from society.

notice of contest A written note stating that an employer does not wish to comply with a citation, an abatement period, or a penalty.

notice of proposed rule making Explains the terms of the new rule, delineates the proposed changes to existing rules, or lists rules that are to be revoked.

nuclear engineering Concerned with the release, control, and safe utilization of nuclear energy.

objectivity Rules are enforced equally regardless of who commits an infraction from the newest employee to the chief executive officer.

occupational diseases Pathological conditions brought about by workplace conditions or factors.

occupational health nurse One whose job it is to conserve the health of workers in all occupations.

Occupational Safety and Health Act (OSH Act) Passed by the U.S. Congress in 1970 and updated periodically since that time.

Occupational Safety and Health Administration (OSHA) The government's administrative arm for the Occupational Safety and Health Act. It sets and revokes health and safety standards, conducts inspections, investigates problems, issues citations, assesses penalties, petitions the courts to take appropriate action against unsafe employers, provides safety training, provides injury prevention consultation, and maintains a database of safety and health statistics.

Occupational Safety and Health Review Commission (OSHRC) An independent board whose members are appointed by the president and given quasi-judicial authority to handle contested OSHA citations.

ohms Measure resistance.

opening conference An initial meeting in which an OSHA compliance officer informs company officials and employee representatives of the reason that an on-site inspection is going to occur and what to expect.

organized labor A group of employees who joined together to fight for the rights of all employees (i.e., unions).

OSHA See Occupational Safety and Health Administration.

OSH Act See Occupational Safety and Health Act.

OSHA Form 101 Supplementary Record of Occupational Injuries and Illnesses.

OSHA Form 300 Log of Work-Related Injuries and Illnesses.

OSHA Form 300A Summary of Workplace Injuring and Illnesses.

OSHA Form 301 Injury and Illness Incident Report.

OSHA Poster 2203 Explains employee rights and responsibilities as prescribed in the OSH Act.

OSHRC See Occupational Safety and Health Review Commission.

overexertion The result of employees working beyond their physical limits.

overload Amounts to an imbalance between a person's capacity at any given time and the load that person is carrying in a given state.

oxygen limit The amount of oxygen that must be present in a given substance in order for an explosion to occur.

participatory ergonomics An approach to intervention that combines outside experience and inside experience to design ergonomic interventions that are tailored to a specific workplace.

patent defect One that occurs in all items in a manufactured batch.

permanent partial disability The condition that exists when an injured employee is not expected to recover.

permanent total disability The condition that exists when an injured employee's disability is such that he or she cannot compete in the job market.

permanent variance An exemption from an OSHA standard awarded to an organization that can show it already exceeds the standards.

permissible exposure limit (PEL) OSHA-established exposure threshold.

personal monitoring devices Devices worn or carried by an individual to measure radiation doses received.

personal protective equipment (PPE) Any type of clothing or device that puts a barrier between the worker and the hazard (e.g., safety goggles, gloves, and hard hats).

Petition for Modification of Abatement (PMA) Available to employers who intend to correct the situation for which a citation was issued, but who need more time.

photoelectric devices Optional devices that shut down the machine anytime the light field is broken.

photoelectric fire sensors Detect changes in infrared energy that is radiated by smoke, often by the smoke particles obscuring the photoelectric beam.

physical hazards Include noise, vibration, extremes of temperature, and excessive radiation.

PMA See Petition for Modification of Abatement.

point-of-operation guards Machine guards that provide protection right at the point where the user operates the machine.

poisoning To be injured or killed by a harmful substance.

powerlessness The feeling that workers have when they are not able to control the work environment.

preceding factors Factors that led up to an accident.

predispositional characteristics Human personality characteristics that can have a catalytic effect in causing an accident.

preliminary hazard analysis Conducted to identify potential hazards and prioritize them according to (1) the likelihood of an accident or injury being caused by the hazard and (2) the severity of injury, illness, or property damage that could result if the hazard caused an accident.

pressure The force exerted against an opposing fluid or thrust distributed over a surface.

pressure hazard A hazard caused by a dangerous condition involving pressure.

private insurance Workers' compensation coverage purchased from a private insurance company.

problem identification Engineers must draft a description of a problem before anything else can be done.

produce and deliver Shop or detail drawings are developed, and the design is produced, usually as a prototype. The prototype is analyzed and tested. Design changes are made. The product is then produced and delivered.

product literature Tells users about hazards that cannot be removed by design or controlled by guards and safety devices.

product safety auditor Responsible for evaluating the overall organization and individual departments within it.

product safety committee Consists of representatives within a company having a wide range of expertise. Offers support to the product safety coordinator on safety issues.

product safety coordinator Responsible for coordinating and facilitating a product safety management program in all departments of a company.

product safety program Purpose is to limit as much as possible a company's exposure to product liability litigation and related problems.

productivity The concept of comparing output of goods or services to the input of resources needed to produce or deliver them.

professional societies Typically formed for the purpose of promoting professionalism, adding to the body of knowledge, and forming networks among colleagues in a given field.

proof pressure tests Tests in which containers are "proofed" by subjecting them to specified pressures for specified periods.

property damage Facilities, equipment, or other nonpersonnel items damaged as a result of an accident.

proposed penalty The initial penalty proposed by the OSHA compliance officer after an inspection tour.

protons Positively charged particles.

proximate cause The cause of an injury or damage to property.

psychophysiological techniques Require simultaneous measurement of heart rate and brain waves, which are then interpreted as indexes of mental workload and industrial stress.

psychosocial questionnaires Evaluate workers' emotions about their jobs.

public hearing If an injured worker feels that he or she has been inadequately compensated or unfairly treated, a public hearing can be requested.

pullback devices Pull the operator's hands out of the danger zone when the machine starts to cycle.

puncturing Results when an object penetrates straight into the body and pulls straight out, creating a wound in the shape of the penetrating object.

quality A measure of the extent to which a product or service meets or exceeds customer expectations.

quality management (QM) A way of managing a company that revolves around a total and willing commitment of all personnel at all levels to quality.

quality product One that meets or exceeds customer standards and expectations.

rad A measure of the dose of ionizing radiation absorbed by body tissue stated in terms of the amount of energy absorbed per unit of mass of tissue.

radiant heat The result of electromagnetic nonionizing energy that is transmitted through space without the movement of matter within that space.

radiation Consists of energetic nuclear particles and includes alpha rays, beta rays, gamma rays, X-rays, neutrons, high-speed electrons, and high-speed protons.

radiation area Any accessible area in which radiation hazards exist that could deliver doses (1) within one hour, a major portion of the body could receive more than 5 millirem or (2) within five consecutive days, a major portion of the body could receive more than 100 millirem.

radiation control specialists Monitors the radiation levels to which workers may be exposed, tests workers for levels of exposure, responds to radiation accidents, develops company-wide plans for handling radiation accidents, and implements decontamination procedures when necessary.

radioactive material Material that emits corpuscular or electromagnetic emanations as the result of spontaneous nuclear disintegration.

radio-frequency devices Capacitance devices that brake the machine if the capacitance field is interrupted by a worker's body or another object.

reasonable risk Exists when consumers (1) understand risk, (2) evaluate the level of risk, (3) know how to deal with the risk, and (4) accept the risk based on reasonable risk–benefit considerations.

receptacle wiring tester A device with two standard plug probes for insertion into an ordinary 110-volt outlet and a probe for the ground.

reclamation A process whereby potentially hazardous materials are extracted from the by-products of a process.

rehabilitation Designed to provide the needed medical care at no cost to the injured employee until he or she is pronounced fit to return to work.

Rehabilitation Act Enacted to give protection to people with disabilities.

rem A measure of the dose of ionizing radiation to body tissue stated in terms of its estimated biological effect relative to a dose of one roentgen of X-rays.

repeat violation A violation of any standard, regulation, rule, or order where, upon reinspection, a substantially similar violation is found.

repetitive motion Short-cycle motion that is repeated continually.

repetitive strain injury (RSI) A broad and generic term that encompasses a variety of injuries resulting from cumulative trauma to the soft tissues of the body.

resistance A tendency to block flow of electric current.

response time The amount of time between when an order is placed and the product is delivered.

restraint devices   Hold back the operator from the danger zone.

restricted area   Any area to which access is restricted in an attempt to protect employees from exposure to radiation or radioactive materials.

retrofit   Renovating rather than replacing.

risk analysis   An analytical methodology normally associated with insurance and investments.

role ambiguity   The condition that occurs when an employee is not clear concerning the parameters, reporting requirements, authority, or responsibilities of his or her job.

role-reversal test   Requires a person to trade places with the people affected by the decision that he or she made and to view the decision through their eyes.

safeguarding   Machine safeguarding was designed to minimize the risk of accidents of machine–operator contact.

safety and health movement   Began during World War II when all the various practitioners of occupational safety and health began to see the need for cooperative efforts. This movement is very strong today.

safety and health professional   An individual whose profession (job) is to be concerned with safety and health measures in the workforce.

safety data sheet (SDS)   This sheet contains all the relevant information needed by safety personnel concerning specific hazardous materials.

safety-friendly corporate culture   Exists when the tacit assumptions, beliefs, values, attitudes, expectations, and behaviors that are widely shared and accepted in an organization support the establishment and maintenance of a safe and healthy work environment for all personnel at all levels.

safety policy   A written description of an organization's commitment to maintaining a safe and healthy workplace.

safety trip devices   Include trip wires, trip rods, and body bars that stop the machine when tripped.

SARA   See Superfund Amendments and Reauthorization Act.

SDSs   See safety data sheets.

second-degree burn   Results in blisters forming on the skin.

self-insurance   Workers' compensation coverage in which a company insures itself by building its own fund.

semiautomatic ejection   A system that ejects the work using mechanisms that are activated by the operator.

semiautomatic feed   A system that uses a variety of approaches for feeding stock to the machine.

semiconductors   Substances that are neither conductors nor insulators.

shift work   Employees work at different times of the day instead of during the same hours.

shock   A depression of the nervous system.

short circuit   A circuit in which the load has been removed or bypassed.

short-term exposure limit   The maximum concentration of a given substance to which employees may be safely exposed for up to 15 minutes without suffering irritation, chronic or irreversible tissue change, or narcosis to a degree sufficient to increase the potential for accidental injury, impair the likelihood of self-rescue, or reduce work efficiency.

sick-building syndrome   An internal environment that contains unhealthy levels of biological organisms in the air. A common cause is the introduction of unhealthy outdoor air that is brought in and circulated through the cooling system.

simulation   Involves structuring a training activity that simulates a line situation.

situational characteristics   Factors that can change from setting to setting and can have a catalytic effect in causing an accident.

situational factors   Environmental factors that can affect an employee's safety and that can differ from situation to situation.

smoke   The result of the incomplete combustion of carbonaceous materials.

social environment   The general value system of the society in which an individual lives, works, grows up, and so on.

sound   Any change in pressure that can be detected by the ear.

sound level meter   Produces an immediate reading that represents the noise level at a specific instant in time.

spraining   The result of torn ligaments.

staff authority   The safety and health manager is the staff person responsible for a certain function, but he or she has no line authority over others involved with that function.

standards and testing organizations   Conduct research, run tests, and establish standards that identify the acceptable levels for materials, substances, conditions, and mechanisms to which people may be exposed in the modern workplace.

standpipe and hose systems   Provide the hose and pressurized water for firefighting.

Star Program   Recognizes companies that have incorporated safety and health into their regular management system so successfully that their injury rates are below the national average for their industry.

state funds   Workers' compensation coverage provided by the state.

static electricity   A surplus or deficiency of electrons on the surface of a material.

step and fall   An accident that occurs when a person's foot encounters an unexpected step down.

straining   The result of overstretched or torn muscles.

stress   A pathological, and therefore generally undesirable, human reaction to psychological, social, occupational, or environmental stimuli.

stress claims   Workers' compensation claims that are based on stress-induced disabilities.

stressors Stimuli that cause stress.

stump and fall An accident that occurs when a worker's foot suddenly meets a sticky surface or a defect in the walking surface.

subjective ratings Ratings that are less than objective and can be affected by emotions, human biases, presumptions, and perceptions.

Superfund Amendments and Reauthorization Act (SARA) Act designed to allow individuals to obtain information about hazardous chemicals in their community so that they are able to protect themselves in case of an emergency.

synthesis Second step in the design process wherein engineers combine systematic, scientific procedures with creative techniques to develop initial solutions.

systems theory Views a situation in which an accident may occur as a system comprised of the following components person (host), machine (agency), and environment.

technique of operation review An analysis method that allows supervisors and employees to work together to analyze workplace accidents, failures, and incidents.

technological alienation The frame of mind that results when employees come to resent technology and the impact that it has on their lives.

technology access Access to time and work-saving devices, processes, or equipment that are up to date.

temporary emergency standards OSHA standards that can be adopted on a temporary basis without undergoing the normal adoption procedures.

temporary partial disability The injured worker is incapable of certain work for a period of time but is expected to recover fully.

temporary total disability The injured worker is incapable of any work for a period of time but is expected to recover fully.

temporary variance Employers may ask for this when they are unable to comply with a new standard but may be able to if given time.

thermal expansion detectors Use a heat-sensitive metal link that melts at a predetermined temperature to make contact and ultimately sound an alarm.

third-degree burn Penetrates through both the epidermis and the dermis. It may be fatal.

three E's of safety Engineering, education, and enforcement.

threshold limit values (TLVs) The levels of exposure to which all employees may be repeatedly exposed to specified concentrations of airborne substances without fear of adverse effects. Exposure beyond TLVs is considered hazardous.

threshold of hearing The weakest sound that can be heard by a healthy human ear in a quiet setting.

threshold of pain The maximum level of sound that can be perceived without experiencing pain.

time-weighted average (TWA) The level of exposure to a toxic substance to which a worker can be repeatedly exposed on a daily basis without suffering harmful effects.

tort An action involving a failure to exercise a reasonable care that may, as a result, lead to civil litigation.

total safety management (TSM) The principles of total quality management (TQM) applied to safety management.

toxic substance One that has a negative effect on the health of a person or animal.

trade associations Promote the trade that they represent.

trip and fall An accident that occurs when a worker encounters an unseen foreign object in his or her path.

two-hand controls Require the operator to use both hands concurrently to activate the machine.

ultraviolet detectors Sound an alarm when the radiation from fire flames are detected.

Underwriters Laboratories (UL) Determines whether equipment and materials for electrical systems are safe in the various NEC location categories.

unreasonable risk Exists when (1) consumers are not aware that a risk exists; (2) consumers are not able to judge adequately the degree of risk even when they are aware of it; (3) consumers are not able to deal with the risk; and (4) risk could be eliminated at a cost that would not price the product out of the market.

unrestricted area Any area to which access is not controlled because no radioactivity hazard is present.

unsafe act An act that is not safe for an employee.

unsafe behavior The manner in which people conduct themselves that is unsafe to them or to another.

useful consciousness A state of consciousness in which a person is clear-headed and alert enough to make responsible decisions.

vacuum mentality Workers think that they work in a vacuum and don't realize that their work affects that of other employees and vice versa.

vacuums Caused by pressures below atmospheric level.

value added The difference between what it costs to produce a product and the value the marketplace puts on it.

vapor A mist state into which certain liquids and solids can be converted (e.g., gasoline fumes are vaporized petroleum).

vertical work area One that is designed and positioned so that workers are not required to lift their hands above their shoulders or bend down in order to perform any task.

vocational rehabilitation Involves providing the education and training needed to prepare the worker for a new occupation.

volatility The evaporation capability of a given substance.

voltage Measures the potential difference between two points in a circuit.

wage-loss theory Requires a determination of how much the employee could have earned had an injury not occurred.

water hammer A series of loud noises caused by liquid flow suddenly stopping.

wellness program Any program designed to help and encourage employees to adopt a healthier lifestyle.

whistle-blowing   Act of informing an outside authority or media organ of alleged illegal or unethical acts on the part of an organization or individual.

whole-person theory   What the worker can do after recuperating from the injury is determined and subtracted from what he or she could do before the accident.

wide band noise   Noise that is distributed over a wide range of frequencies.

willful/reckless conduct   Involves intentionally neglecting one's responsibility to exercise reasonable care.

work envelope   The total area within which the moving parts of a robot actually move.

work injuries   Injuries that occur while an employee is at work.

work stress   A complex concept involving physiological, psychological, and social factors.

worker negligence   Condition that exists when an employee fails to take necessary and prudent precautions.

workers' compensation   Developed to allow injured employees to be compensated appropriately without having to take their employer to court.

workplace accidents   Accidents that occur at an employee's place of work.

workplace inspection   An on-site inspection conducted by OSHA personnel.

workplace stress   Human reaction to threatening situations at work or related to the workplace.

Index